Advances in Pest Management and Pest Control

Advances in Pest Management and Pest Control

Edited by **Edwin Tan**

R Callisto
Reference

New York

Published by Callisto Reference,
106 Park Avenue, Suite 200,
New York, NY 10016, USA
www.callistoreference.com

Advances in Pest Management and Pest Control
Edited by Edwin Tan

International Standard Book Number: 978-1-63239-052-3 (Hardback)

Printed in the United States of America.

Contents

Preface

Over the recent decade, advancements and applications have progressed exponentially. This has led to the increased interest in this field and projects are being conducted to enhance knowledge. The main objective of this book is to present some of the critical challenges and provide insights into possible solutions. This book will answer the varied questions that arise in the field and also provide an increased scope for furthering studies.

This book elaborates on the advancements made in pest management and control. Integrated pest management is an efficient and environmentally sensitive approach that depends on an integration of common-sense practices. Its programs employ recent and descriptive information regarding the life cycles of pests and their interplay with the environment. This information, along with the available pest control techniques, is used to manage pest damage by the most economical means and with the least possible hazard to property, people, and the environment. This book discusses approaches related to pest management, current applications and future challenges in pest control.

I hope that this book, with its visionary approach, will be a valuable addition and will promote interest among readers. Each of the authors has provided their extraordinary competence in their specific fields by providing different perspectives as they come from diverse nations and regions. I thank them for their contributions.

Editor

Part 1

Integrated Pest Management – Current Applications

Bark Beetles Control in Forests of Northern Spain

Arturo Goldazarena[1], Pedro Romón[2] and Sergio López[1]
[1]Neiker-Basque Institute of Agricultural Research and Development
[2]FABI- Forestry and Agricultural Biotechnology Institute
[1]Spain
[2]South Africa

1. Introduction

Bark beetles (Coleoptera: Curculionidae: Scolytinae) are an insect group that contains at least 6000 species from 181 genera around all the world (Wood, 1982). It is well known that some species are among the most destructive insects of coniferous forests representing a continuous threat (Ayres & Lombardero, 2000). The knowledge of these insects in Spain is very far away of desirable. Some studies have advanced in the knowledge of the taxonomic and faunistic composition of diverse forest types (Plaza & Gil, 1982; Plaza, 1983; Gil & Pajares, 1986; Pajares, 1987; González, 1990; Lombardero, 1994; Lombardero & Novoa, 1994; Riba, 1994; Fernández, 1996; López et al., 2007a), but the understanding of their dynamics and interrelationship biotic factors is very low. The main objective of the present work is to present a synthesis of the research works that have been performed so far in Spanish mainland, particularly in its northern plateau. First, we will introduce the situation of the conifer species *Pinus radiata* D.Don, the most widely species planted in this geographic zone, and its relationship with both bark beetles and associated phytopathogenic fungi (mainly ophiostomatoid fungi and *Fusarium circinatum*). Secondly, a brief and overall approach to the chemical ecology of bark beetles is given and finally there will be exposed some particular cases of species considered as serious threats for spanish forestry, focusing on studies aimed to the development of integrated pest management strategies.

2. The status of *Pinus radiata* D.Don in Northern Spain and its association with bark beetles (Coleoptera: Scolytinae) and ophiostomatoid fungi (Sordariomycetes: Ophiostomatales) and *Fusarium circinatum* (Sordariomycetes: Hypocreales)

The commonly known as Monterey pine (*Pinus radiata*), arising from Mexico, was introduced in Spain during the first half of the XIX century, initiating its plantation in the Basque Country. The first introduction was carried out by Mr. Carlos Adán de Yarza in his botanical garden of Zubieta (Lekeitio, Vizcaya), planting the first stands in 1840 near Amoroto (Vizcaya). The common name "insignis" became popular, both in Spanish and Basque language, due to be more frequent at that time the scientific synonymy *Pinus insignis*. Monterey pine has been planted in Spain in the North Atlantic orle between the parallels 42 and 44 ° N in altitudes below 800 m (Autonomic Communities of Galicia, Asturias, Cantabria, Basque Country and Navarra), showing up also in the Mediterranean province of Gerona (Cataluña), with similar latitude, and promptly in the botanic grounds of the laurisilva and the fayal-brezal (*Myrica-Erica*) of the Canary islands (28-29 ° N). The worldwide and national areas occupied with commercial plantations of *P. radiata* are indicated in **Table 1**.

Country	Area	Autonomic Community	Area
Chile	1,400,000 ha.	Basque Country	160,000 ha.
New Zealand	1,200,000 ha.	Galicia	60,000 ha.
Australia	650,000 ha.	Asturias	26,000 ha.
Spain	270,000 ha.	Cantabria	15,000 ha.
South Africa	55,000 ha.	Navarra	5,500 ha.
Others	30,000 ha.	Canarias	3,000 ha.
		Gerona	500 ha.
Worldwide total	3,605,000 ha.	National total	270,000 ha.

Table 1. Worldwide and national areas (hectares) intended to the lumber exploitation of Monterey pine (adapted from Michel 2004).

This area of artificial plantation gathers in its whole the characteristics that favour the appearing of pests and diseases. Thus, all of them form (1) pure masses of more or less extension where the nutrient abundance and the facility of dispersion support the population increase of the harmful species; (2) contemporary masses that favour the appearance of pests and diseases related to different ages intervals; and (3) exotic masses which are submitted to intercontinental transports with high risk of introduction of exotic pests (Alonso Zarazaga & Goldarazena, 2005; López et al., 2007b) and diseases (Landeras et al., 2005) of greater potential virulence due to be free of natural enemies or under favourable environmental conditions. Besides, the economic profitability of Monterey pine, at first like cellulose pulp and now also as saw wood (50% for each intention nowadays in Northern Spain), has sponsored its plantation in geographic areas that are far from the optimum conditions. This has caused physiologically weak forestry masses, favouring the attack of pests and diseases due to a higher incidence of abiotic noxious agents such as frosts and droughts.

Bark beetles can cause damage both by the direct boring action and by the inoculation of phytopathogenic fungi. Once the host tree has been localized, pioneer specimens produce aggregation pheromones, which joined with the volatile kairomones of the recently attacked tree, attract other members of the same species. While they feed and build their galleries under the bark, some species cut the vascular flow causing the dead of the tree. At the same time, they inoculate spores and fragments of mycelium into the phloem, playing an important role in the transmission of some phytopathogenic fungi such as the causal agents of Dutch elm disease and blue-staining fungi that alter the structure of the wood and reduce its worth (Pajares, 1987).

Some species of bark beetles are able to cause significant damage to forests. For example, during the 1970's, outbreaks of *Ips typographus* destroyed 2 million m³ of timber in

Scandinavia (Bakke, 1983). Regarding to United States, *Dendroctonus ponderosae* has caused an average annual loss of about 1.5 billion board feet particularly in *Pinus contorta* in western North America since 1895 (Wood, 1982). According to Carter et al. (1991), outbreaks of *Dendroctonus frontalis* in southeastern Texas covered 3,200,000 hectares with heavy economic losses from 1974 to 1980. Furthermore, from 1999 to 2003, *Dendroctonus frontalis* caused losses of about USD $1.5 billion in the southeastern U.S.A. (Nowak, 2005). The current outbreak of *Dendroctonus ponderosae* in Canada has impacted about 12 of the 14.3 million ha of *Pinus contorta* in the British Columbia (Westfall, 2006). In Spain, the most damaging species is *Tomicus piniperda*, which can cause annual losses of up to 72 million € in the Basque Country region (northern Spain) (Amezaga, 1993). Among the scolytid species present in the Iberian Peninsula, *Tomicus piniperda* and *Ips sexdentatus* represent the most relevant directly damaging species, as it will be explained later (see section 4).

Many sapstain fungi, especially ophiostomatoid fungi (Sordariomycetes: Ophiostomatales), are associated with phloeophagous bark beetles and might help to overcome the defences of attacked trees (Kirisits, 2004). They are commonly called "blue-stain fungi" because the discoloration they cause, namely blue, gray, brown or even black on the sapwood of trees, mostly on conifers (Kirisits, 2004). This sap stain is caused by the growth of the hyphae in the ray parenchima cells and resin ducts, disrupting the sap flow, and tracheids are also colonized in later stages of infection. As a consequence of this discoloration, the lumber defect is largely cosmetic. Most ophiostomatoid fungi that cause sapstain are moderately to weakly virulent pathogens, but some species can display relative high virulent levels and cause the death of infected trees when they are inoculated in high doses (Kirisits, 2004). Although some blue-staining fungi can cause strong damage on the strength properties of the wood assigned to furniture industry (Seifert, 1993) and these losses can amount to a 50% price reduction in the Basque Country (Maderas Elorriaga Company, Muxika, Vizcaya, personal communication, from about 180 €/m^3 to 85 €/m^3), the knowledge of conifer bark beetle-associated blue-staining fungi in the Iberian Peninsula is very limited (Fernández et al., 2004; Villareal et al., 2005). Romón et al. (2007a) isolated 16 species of *Ophiostoma* sensu lato or their asexual states from 13 bark beetle species and the root weevil *Brachyderes incanus* (Coleoptera: Curculionidae: Entiminae). Among the isolated taxa, species such as *Ophiostoma ips, O. minus, O. piceae* and *O. pluriannulatum* are important agents of blue-staining (Seifert, 1993), whereas *O. ips, O. minus* and *L. wingfieldii* pathogenicity is well recognized (Raffa & Smalley, 1988; Lieutier et al., 1989; Yamaoka et al., 1990; Fernández et al., 2004). Sixty-nine of the reported associations by Romón et al. (2007a) had not been previously recorded until them. Studies based on fungus-conifer scolytids association have not been only carried out in *P. radiata* inhabiting bark beetle populations, but also in *P. pinaster* (Bueno et al., 2010). They identified twenty-five taxa belonging to the Eumycotina group from isolations of *I. sexdentatus* adults and naturally colonized tissues (sapwood and phloem).

On the other side, pitch canker disease, caused by the fungus *Fusarium circinatum*, is one of the most important pathogens of *Pinus radiata*. This fungus species is endemic to southeastern United States (Dwinell et al., 1985). Pitch canker disease was recently identified and reported in California, predominately in planted urban *P. radiata* and in native Monterey pine forest (Correll et al., 1991). Since being first reported in the United States, it has been also found in different countries around the world: Japan (Muramoto et al., 1989),

Mexico (Rodriquez, 1989), South Africa (Viljoen et al., 1994) and Spain (Landeras et al., 2005). Its accidental introduction in these areas is probably due to softwood lumber, seedling and seed exports. In California, wounding has not resulted. The transmission of the pathogen to cones seems to be not (Correll et al., 1991) or branches (Fox et al., 1991), despite the presence of significant airborne inoculum. In contrast, a complex of insects have been demonstrated to be able to transmit *F. circinatum* (Fox et al., 1991; Hoover et al., 1995; Storer et al., 2004). Within this species-complex there are included some bark beetles species, like *Ips paraconfusus, I. mexicanus, I. plastographus, Pityophthorus carmeli, P. nitidulus, P. setosus* and *Conophthorus radiatae* and the anobid beetle *Ernobius punctulatus* (Coleoptera: Anobiidae). All of them appear to be phoretically associated with the fungus, and are known to visit and infest non-diseased trees. Romón et al. (2007b) isolated *F. circinatum* from adult bark beetles collected from baiting logs in two stands of *P. radiata* located in two plots of Biscay province (Morga and Muxika). Five conifer bark beetles were found to carry the inoculum: *Pityophthorus pubescens* (25.00%), *Hylurgops palliatus* (11.96%), *Ips sexdentatus* (8.57%), *Hypothenemus eruditus* (7.89%), *Hylastes attenuatus* (7.40%) and *Orthotomicus erosus* (2.73%) (Table 2). Frecuency of occurrence of *F. circinatum* is given within parentheses. In addition, *Brachyderes incanus* (14.28%) had the second highest frequency of occurrence of the fungus. Frequencies of occurrence of fungi were computed using the formula of Yamaoka where $F = (NF / NT) 100$ (%) and F represents the frequency of occurrence (%) of the fungus, NT represents the total number of samples from which isolation attempts were made and NF represents the number of samples from which the fungus was isolated.

Insect species	Total no. samples	NF	F (%)*
Pityophthorus pubescens	32	8	25.00
Brachyderes incanus	42	6	14.28
Hylurgops palliatus	117	14	11.96
Ips sexdentatus	35	3	8.57
Hypothenemus eruditus	38	3	7.89
Hylastes attenuatus	54	4	7.40
Orthotomicus erosus	73	2	2.73
Dryocoetes autographus	45	-	-
Hylastes ater	32	-	-
Tomicus piniperda	18	-	-
Xyleborus dryographus	9	-	-
Hylurgus ligniperda	5	-	-

* Frequency of occurrence $F = (NF / NT) 100$ (%), where NT represents the total number of samples from which isolations attempts were made, and NF represents the number of samples from which F. circinatum was isolated.

Table 2. Variation about frecuency of occurrence of *Fusarium circinatum* of several bark beetles (Coleoptera: Curculionidae: Scolytinae) and weevils (Coleoptera: Curculionidae: Entiminae) species in northern Spain.

Table 3 includes the relationship of different fungi species, belonging to the Orders Ophiostomatales, Sphaeropsidales and Hypocreales associated with conifer inhabiting bark beetles (and the weevil *Brachyderes incanus*), with special emphasis to isolations detected in the Basque Country (northern Spain).

Bark beetles species	Order Ophiostomatales	Order Sphaeropsidales	Order Hypocreales
Hylastes attenuatus	*Ophiostoma ips*** *O. olivaceum* *O. piceae*** <u>*O. piliferum-like*</u> *O. quercus* <u>*O. stenoceras*</u> <u>*Leptographium guttulatum*</u>** *L. truncatum-like* *L. wingfieldii*** *Pesotum fragans*	*Diplodia pinea***	<u>*Fusarium circinatum*</u>*** *F. oxysporum*** *F. lateritium**
Hylastes ater	<u>*Ophiostoma ips*</u>** *O. minus* *O. penicillatum* *O. piceae*** *O. piliferum* *O. floccosum** *O. olivaceum** *O. piliferum-like** <u>*O. pluriannulatum*</u>** <u>*O. quercus*</u>* *O. rectangulosporium-like** *O. stenoceras** *Leptographium stenoceras* <u>*L. guttulatum*</u>** *L. lundbergii* *L. serpens* <u>*L. wingfieldii*</u>***		<u>*Fusarium moniliformis*</u>**
Hylurgops palliatus	<u>*Ophiostoma ainoae*</u> *O. bicolor* <u>*O. cucullatum*</u> *O. galeiformis* *O. japonicum* <u>*O.neglectum*</u> <u>*O. penicillatum*</u> <u>*O. piceae*</u>** <u>*O. piceaperdum*</u> *O. simplex*	*Diplodia pinea***	<u>*Fusarium circinatum*</u>*** *F. proliferatum***

Bark beetles species	Order Ophiostomatales	Order Sphaeropsidales	Order Hypocreales
Hylurgops palliatus	*O. stenoceras* *O. ips*** *O. olivaceum** <u>*O. piliferum-like**</u> <u>*O. pluriannulatum***</u> *O. quercus** *O. rectangulosporium-like** *Ceratocystiopsis alba* *C. minuta* <u>*Leptographium guttulatum***</u> <u>*L. lundbergii*</u> *L. procerum* <u>*L. wingfieldii****</u> *L. truncatum-like** *Pesotum fragans**		
Ips sexdentatus	<u>*Ophiostoma ainoae*</u> <u>*O. araucariae*</u> <u>*O. brunneo-ciliatum*</u> *O. clavatum* <u>*O. ips***</u> <u>*O. japonicum*</u> <u>*O. minus**</u> *O. obscura* *O. piceae* *O. piceaperdum* *O. olivaceum** *O. pluriannulatum*** *O. rectangulosporium-like** *O. stenoceras** *Ceratocystiopsis minuta* *Leptographium guttulatum*** *L. truncatum-like** *Pesotum fragans*		<u>*Fusarium circinatum****</u> *F. moniliformis***
Dryocoetes autographus	<u>*Ophiostoma ainoae*</u> *O. araucariae* <u>*O. cucullatum*</u> *O. galeiformis* *O. japonicum* *O. obscura* <u>*O. neglectum*</u> <u>*O. piceae*</u>		

Bark beetles species	Order Ophiostomatales	Order Sphaeropsidales	Order Hypocreales
Dryocoetes autographus	*O. piceaperdum* *O. simplex* *O. ips*** *O. minus** *O. olivaceum** *O. piliferum-like** *O. rectangulosporium-like** *Ceratocystiopsis alba* *C. minuta* *Leptographium guttulatum***		
Orthotomicus erosus	*Ophiostoma ips*** *O. piceae*** *O. pluriannulatum*** *O. canum-like* *O. floccosum* *O. olivaceum* *O. stenoceras* *O. rectangulosporium-like* *Leptographium guttulatum*** *L. wingfieldii**** *Pesotum fragans*		*Fusarium circinatum**** *F. moniliformis*** *F. culmorum** *F. lateritium**
Tomicus piniperda	*Ophiostoma canum* *O. clavatum* *O. floccosum* *O. galeiformis* *O. huntii* *O. ips*** *O. minus* *O. piceae*** *O. piceaperdum* *O. piliferum* *O. pluriannulatum*** *Ceratocystiopsis minuta* *Leptographium euphyes* *L. guttulatum*** *L. lundbergii* *L. procerum* *L. wingfieldii****	*Diplodia pinea***	
Pityophthorus pubescens	*Ophiostoma ips*** *O. piliferum-like** *Leptographium guttulatum***		*Fusarium circinatum**** *F. lateritium**

Bark beetles species	Order Ophiostomatales	Order Sphaeropsidales	Order Hypocreales
Hypothenemus eruditus	*Ophiostoma pluriannulatum*** *O. quercus** *Pesotum fragans**		*Fusarium circinatum**** *F. culmorum** *F. lateritium**
Xyleborus dispar	*Ophiostoma pluriannulatum*	*Diplodia pinea*	
Xyleborus dryographus	*Ophiostoma olivaceum* *O. piceae*** *Sporothrix schenckii-like*		*Fusarium moniliformis***
Hylurgus ligniperda	*Leptographium guttulatum*		
Pityogenes calcaratus	*Ophiostoma ips* *Leptographium guttulatum***		
Brachyderes incanus	*Ophiostoma piceae*** *O. pluriannulatum*** *O. quercus* *Leptographium guttulatum*		*Fusarium circinatum**** *Fusarium moniliformis*

Table 3. Fungal species frequently associated with bark beetles colonizing conifers. Fungal taxa underlined indicate high percentage of isolation with the corresponding insect within a row. Fungal species with an upper symbol are present in the Basque Country, as follows: Saprophytic species (*), facultative pathogens (**) and strictly pathogens (***)(Adapted from Kirisits, 2004; Romón et al., 2007a, 2007b).

3. Chemical ecology of the host tree colonization by bark beetles: Basis for the development of a sustainable strategy for the protection of forestry masses

Each species of bark beetles is adapted to only one or a few host tree species probably due to natural selection driven by trees biochemicals. It is likely that each species of tree has coevolved chemicals to defend against the selection pressures of bark beetles and other insects (Berryman et al., 1985; Byers, 1995). Plant chemicals can be attractive, repellent, toxic or nutritious to bark beetles and have effects on: (1) finding and accepting the host tree (selection and suitability); (2) feeding stimulation and deterrence; (3) host resistance; (4) pheromone biosynthesis and communication; and (5) attraction of predators, parasites and competitors of bark beetles (Byers, 2004).

Bark beetles must locate a suitable host from among a relatively few that are widely scattered in the forest. During dispersal flight, insect must discriminate between potential conifer hosts and avoid any unsuitable host and non-hosts decidious trees, so the ability to detect and recognize different olfactory signals in a complex olfactory landscape represents an important cue for bark beetles colonization processes (Raffa, 2001; Zhang & Schlyter,

2004). The host tree is restricted usually to one or a few species and in most cases the insects seek weakened, less resistant trees. It is expected that insects can detect certain volatile host plant chemicals that indicate its suitability [see Fig. 1 for massive attack of *Tomicus piniperda*, the aggregation of this bark beetle species is considered to be predominantly mediated by host tree kairomonal compounds blend (Vité et al., 1986) although *trans*-verbenol has been recently suggested as its potential pheromone (Poland et al., 2003)]. Pheromones and/or kairomones involved in the semiochemical communication of bark beetle species most commonly colonizing *Pinus radiata* in Spain are indicated in Table 4.

Fig. 1. Massive attack of *Tomicus piniperda*, causing, after one year since the population outbreak, a scattered distribution pattern of nearly dead trees within a *Pinus radiata* plantation near Morga, Biscay province, Basque Country (Spain).

Bark beetle species	Pheromone	Kairomone
Hylurgops palliatus	-	Alpha-pinene (Perttunen, 1957) Beta-pinene (Volz, 1988; Byers, 1992) Terpinolene Myrtenol Ethanol 3-carene *cis*-verbenol *trans*-verbenol
Ips sexdentatus	Ipsdienol (Vité et al., 1972) Ipsenol	Myrcene (Hughes, 1974) 2-methyl-3-buten-2-ol (Serez, 1987) *cis*-verbenol
Tomicus piniperda	*trans*-verbenol (Poland et al., 2003)	Alpha-terpineol (Kangas et al., 1970) *cis*-carveol *trans*-carveol Beta-pinene (Volz, 1988) Myrtenol Ethanol (Byers, 1992) Alpha-terpinolene Delta-3-carene Alpha-pineneoxide (Czokajlo, 1998) Alpha-pinene (Song et al., 2005)

Table 4. Pheromonal and kairomonal components of most common conifer bark beetles distributed among *Pinus radiata* plantations in northern Spain, compendium revision.

Host plant's suitability to bark beetles varies with its nutritional quality and composition of anti-insect toxins (Scriber, 1984). Nonhost trees are probably less nutritional, and the beetle would not be adapted to detoxifying some of the nonhost toxins that have evolved for use against other herbivorous insects. A beetle would save much time and energy if it can discriminate the host from the nonhost and determine the intraspecific suitability of a host by olfactory means, deciding not to land due to the detection of volatile compounds called allomones (Byers, 1995). It is more difficult to isolate repellents and inhibitors used in avoidance behavior than to isolate attractants since tests of avoidance require one to first isolate the attractive host odors and then present these with and without the possibly inhibitory odor compounds. Several studies indicate that at least some species of bark beetles avoid nonhost volatiles during their search for host trees. For example, the attraction of both *T. piniperda* and *Hylurgops palliatus* to ethanol (1-6 g/day) was reduced by odors from cut logs of *Betula pendula* and *Populus tremula* (Schroeder, 1992). In this way, one monoterpenic compound called verbenone has shown relatively promising results as an anti-aggregant pheromone (Lindgren et al., 1989). Verbenone (4,6,6-trimethylbicyclo[3.1.1]-

hept-3-en-2-one) is a simple oxidation product of *trans*-verbenol, which in turn is a biological oxidation product of α-pinene (Birgersson & Leufvén, 1988), one of the most ubiquitous monoterpenes in the Pinaceae. *a*-Pinene is quite toxic to a number of coniferophagous insects, whereas *trans*-verbenol and verbenone appear to be less toxic (Lindgren et al., 1996). Thus, insects inhabiting environments high in *a*-pinene could be expected to have either a high tolerance or an effective detoxification system. Verbenone has been found in relatively large amounts (mg) in hindguts of *Dendroctonus ponderosae* (Pierce et al., 1987), *Dendroctonus frontalis* (Renwick & Vité, 1968), *Dendroctonus brevicomis* (Byers et al., 1984) and *Dendroctonus pseudotsugae* (Rudinsky et al., 1974), and in low amounts (ng) in *T. piniperda* (Lanne et al., 1987), but it appears to be absent in *I. paraconfusus*, *I. typographus* and *Pityogenes chalcographus* (Byers, 1983; Birgersson et al. 1984, 1990). During the last two decades, verbenone has been demonstrated as a good natural repellent for the control of damages caused by several insect species such as *Hylobius pales* (Salom et al., 1994), *Hylobius abietis* (Lindgren et al., 1996), *D. ponderosae* (Lindgren et al., 1989), *Ips pini* and *Ips latidens* (Lindgren & Miller, 2002). It has been hypothesised that bark beetles species which required relatively fresh host tissue would be more affected by the presence of verbenone, whereas species inhabiting aged tissues would have higher tolerance to verbenone and/or a more efficient detoxifying system (Lindgren, 1994; Lindgren & Miller 2002; Romón et al., 2007b).

Bark beetle pheromones used in aggregation and for avoidance of competition consist of many varied structures. Plant compounds, predominantly the monoterpenes α-pinene or myrcene, are used as kairomonal precursors for their pheromonal components (Hendry et al., 1980). Many of the same pheromonal compounds are used by species in throme same genus, such as ipsenol, ipsdienol, and *cis*-verbenol in the genus *Ips* (Byers, 1995) or *exo*-brevicomin, frontalin, *trans*-verbenol and verbenone in the genus *Dendroctonus* (tribe Tomicini) (Borden, 1982). Some compounds such as *cis*- and *trans*-verbenol may be found in *Ips* (Tribe Ipini) as well as *Tomicus* (Tribe Tomicini). However, *cis*-verbenol has so far only been proven as an aggregation pheromone component for species in the tribe Ipini, whereas *trans*-verbenol has semiochemical activity only in the tribe Tomicini. The base structure of ipsenol, ipsdienol, and myrcenol resembles the plant monoterpene myrcene; likewise, *cis*- and *trans*-verbenol resemble α-pinene. These structural similarities support the hypothesis that in many cases bark beetles use plant compounds as precursors for their pheromone components (Byers, 1995).

Aggregation pheromone components were first identified in bark beetles from males of *Ips paraconfusus* as a synergistic blend of (*S*)-(-)-ipsenol, (*S*)-(+)-ipsdienol, and (4*S*)-*cis*-verbenol (Wood et al., 1968). Several other *Ips* species were soon discovered to produce and respond to various blends of these compounds (Vité et al., 1972). The similarity of chemical structure between myrcene and ipsenol and ipsdienol led to propose, and demonstrate by gas chromatography and mass spectrometry (GC-MS), that myrcene is precursor of pheromones in *Ips* spp. (Hughes, 1974; Byers et al., 1979).

More resistant tree genotypes may have evolved through natural selection lower levels of pheromonal precursor terpenes or attractive kairomones and/or higher concentrations of other toxic monoterpenes. Evolution of plant chemicals that increase tree's resistance to colonization by bark beetles requires that (1) the plant chemicals are detrimental to the beetle; (2) the host chemistry is genetically driven; (3) population variation in genotypes of these trees exists; and (4) the bark beetle exerts selection pressure on the tree by killing or reducing fertility. The beetle population should coevolve, if possible, by shifting their genotype frequencies to those that offer more protection against the plant chemicals.

Host tree chemistry affects most aspects of bark beetle biology, moreover, bark beetles probably differentially affect survival of host trees and alter genotypic frequencies and host chemistry (Byers, 1995). Geographic and intraspecific variation in toxicity of host compounds has been little studied. Thus, more studies are needed in stands with ongoing outbreaks of bark beetles to determine if natural selection can slant trees intraspecific variation that will determine their monoterpene properties.

4. An overview to conifer-inhabiting bark beetle species with most forestry importance in spanish mainland, particularly northern Spain: Concrete cases

Alonso-Zarazaga (2002) listed 128 bark beetles species present across iberian and balearic area, including both conifer and decidious-inhabiting species. Two more alien species to Iberian Peninsula should be added to this checklist, i.e. *Gnathotrichus materiarius* and *Xylosandrus germanus* (López et al., 2007b). *Gnathotrichus materiarius* is considered as a nearctic native species, whereas the origin of *Xylosandrus germanus* is asiatic. According to Kirkendall and Faccoli (2010) and references therein, owing to commercial trading, bark beetles mainly travel in wood and in wooden packing materials such as crating, dunnage and pallets. Both alien species were found in sawmills and wood-proccesing companies of Basque Country that use imported lumber from France. Nowdays, their establishment to different *P. radiata* stands in the Basque Country is fully confirmed (Goldarazena et al., personnal observation). Although these two species are not considered as highly dangerous species, the prevention of the entry and early detection of invasive species, through different pathways including treatment of imported commodities, should be a priority task, in order to avoid potential negative environmental impacts that would be generated.

Within all of these species present in spanish mainland, few species are capable of killing healthy trees (Gil & Pajares, 1986). The register of produced damage data are very scarce, because there have not been studies focused on tracing the economical incidence of these insects in contrast with other countries. First reports were produced in 1907, related to some attacks of *Tomicus piniperda* and other decidious bark beetle species (Gil & Pajares, 1986). During mid-1950´s and mid-1970´s several attacks of *Ips acuminatus* (Fig. 2) were registered in *P. sylvestris* stands of Guadalajara province (1954-57) and Cuenca province (1972-1973) (Gil & Pajares, 1986) but there are no economic data available for these events. In addition, less important sporadic attacks of *I. sexdentatus* and *O. erosus* had occurred in north and central Spain in last decades (Gil & Pajares, 1986). According to Grégoire & Evans (2004), three species of conifer-inhabiting bark beetles species, i.e. *Ips sexdentatus*, *Ips acuminatus* and *Tomicus piniperda* should be considered as significant pests for Spanish mainland. Not only these species, but also *Pityophthorus pubescens* is a serious candidate to be considered as potential forest pest, taking into account its importance as vector of *Fusarium circinatum*, fact that has been previously mentioned in text (Romón et al, 2007b). Following sub-sections will deepen about different studies carried out with these species in Spanish mainland so far. Although quantitative data of economical damage produced by bark beetles are difficult to estimate, some data are provided.

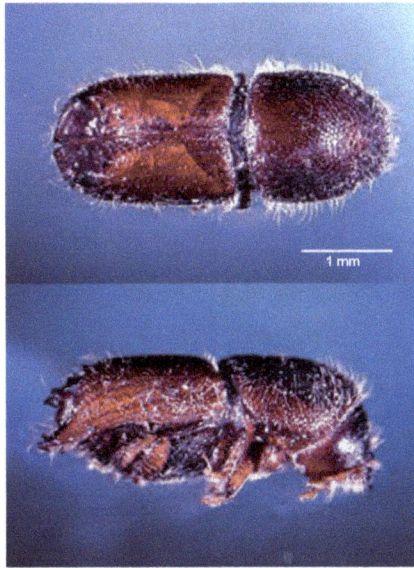

Fig. 2. *Ips acuminatus*, dorsal and lateral view. Photographs taken from López et al., 2007a.

4.1 The six-toothed bark beetle *Ips sexdentatus* (Börner) (Fig. 3)

This species is considered as one of the forest pests that higher damages cause in conifer stands of Iberian Peninsula (Gil & Pajares, 1986). Although endemic populations of *I. sexdentatus* tend to colonize weakened or dead trees, it is well reported that healthy trees can be attacked under epidemic conditions. Frecuently, improper management of logs, for example storaging them for long time, adverse abiotic and climatic conditions (storms, fires, droughts) generate breeding resources for *I. sexdentatus* and favor the generation of these population outbreaks. As a significant example, 11,997 ha of a mixed forest with predominant presence of *P. pinaster* were affected by a fire during 2005 in Gualadajara province (Sánchez et al., 2008). Due to this fact, *Ips sexdentatus* populations significantly increased causing severe damages in some zones. Thus, a massive trapping program was carried out in following years, setting 99 and 237 Theysohn traps in 2006 and 2007 respectively. This trapping methodology led to the capture of 4,928,270 beetles. Moreover, different silvicultural techniques (extraction of affected timber) were applied in parallel. All of these measurements contributed to reduce the negative effects of this population outbreak. On the other hand, 25,000 ca. trees were killed by *I. sexdentatus* in Castilla y León province in 2000 (Consejería de Medio Ambiente, Junta de Castilla y Léon 2001, as cited in Bueno et al., 2010).

Fig. 3. *Ips sexdentatus*, dorsal and lateral view. Photographs taken from López et al., 2007a.

In addition, it is remarkable that the incidence of "Klaus" named windstorm during January 2009 affected 37.9 million m³ of maritime pine (*P. pinaster*) in Aquitanie (southern France) (Inventaire Forestiere Nationale, 2009). As a consequence a great amount of timber was left as suitable breeding material for *I. sexdentatus*. The importation of those *P. pinaster* logs to different sawmills and timber-processing industries located at the Basque Country is a common commercial activity. So, it must be taken into account that the importation and storage of such infested logs for a long time would have consequences for the forest management and put into risk the adjacent *P. radiata* stands (Goldarazena et al., personnal observations).

Some studies focusing on verbenone have been carried out in order to test it as a potential component of IPM strategies for the protection of different pine species stands against this species. Two compounds, verbenone and *trans*-conophthorin, have been mainly considered as the most potential anti-aggregative semiochemicals. Biological implication of verbenone in bark beetles has been previously mentioned (see subsection 3), so it is worthwhile to remark the bioactivity of the second compound. The spyroketal conophthorin [5S,7S-(-)-7-methyl-1,6-dioxaspiro(4.5)decane] is a non-host bark volatile found in angiosperm trees, such as *Betula pubescens* and *B. pendula* (Betulaceae) in Europe (Byers et al., 1998), and *Populus tremuloides*, *P. trichocarpa* (Salicaceae), *B. papyrifera* and *Acer macrophyllum* (Aceraceae) in North America (Huber et al., 1999). On insects, it was first identified from the abdomina of workers of some Hymenoptera: Vespidae, i.e. *Paravespula vulgaris* (Francke et al., 1978), and *P. germanica* and *Dolichovespula saxonica*, together with males of the bark beetle species *Leperisinus fraxini* (Francke et al., 1979). Later, Kohnle et al. (1992) found it in the frass of the fir bark beetle *Cryphalus piceae*, reducing field response of the insect to attractants. The name conophthorin comes from the genus *Conophthorus*, that includes species known to produce it with an inhibitor effect to aggregation pheromones or host kairomones, that is, *Conophthorus coniperda*

(Birgersson et al. 1995; de Groot et al., 1998, Rappaport et al., 2000) and *C. resinosae* (Pierce et al., 1995; de Groot & DeBarr, 2000; Rappaport et al., 2000). In addition, there are several studies about the repellent effect of *(E)-(-)-conophthorin* and racemic conophthorin to pheromone baited traps in bark beetles species that are not known to produce it, as seen in *Xylosandrus germanus* (Kohnle et al. 1992), *Dendroctonus ponderosae* (Huber et al., 1999), *D. pseudotsugae* (Huber et al., 1999, 2000, 2001), *Dryocoetes confusus* (Huber et al., 2000) *Pityophthorus setosus* (Dallara et al., 2000), *C. cornicolens* and *C. teocotum* (Rappaport et al. 2000), *I. pini* (Huber et al., 2000, 2001), *I. duplicatus* (Zhang et al., 2001), *I. sexdentatus* (in France) (Jactel et al., 2001) and *I. typographus* (Zhang & Schlyter, 2003). In spite of this fact, there are reports referring to conophthorin as attractant as well, according to results observed in *I. mexicanus, Lasconotus pertenuis* (Coleoptera: Colydiidae) and *P. carmeli* (Dallara et al. 2000) and in *Epuraea thoracica* (Coleoptera: Nitidulidae) (Kohnle et al., 1992).

Romón et al. (2007b) detected a significant negative dose-dependent relationship between verbenone release rate and catches of *I. sexdentatus* in *P. radiata* stands with traps baited with an specific *Ips sexdentatus* attractant blend (Myrcene 250 mg/day + Ipsdienol 0.20 mg/day + Ipsenol 0.40 mg/day) (Fig. 4). Four verbenone release rate were used, as follows: 0.01, 0.2, 1.8 and 3.1 mg/24h (at 22-24°C). It has been previously mentioned that bark beetles species which requiere relative fresh host tissues would have a less tolerance to verbenone. So, if we asume that *I. sexdentatus* requires relatively fresh phloem and may attack healthy and live trees when endemic populations outbreak, it is feasible to show a negative response to the presence of verbenone, as these results show. In contrast with these results, bark beetle species (*Tomicus piniperda, Orthotomicus erosus, Dryocoetes autographus, Hypothenemus eruditus, Xyleborus dryographus, Hylastes ater, H. attenuatus* and *Hylurgus ligniperda*), trapped accidentally while carrying out that bioassay, have been shown to be not affected significantly by verbenone.

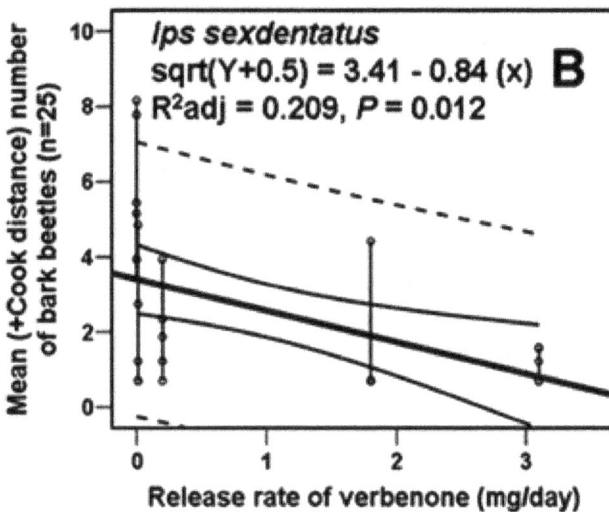

Fig. 4. Effect of four different release rates of verbenone on the attraction of *Ips sexdentatus*. Slope of regression line is significantly different from zero (*t*-test, $P < 0.001$). Confidence limits (95%) (thin solid lines) are associated with the regression line (thick solid line). Dashed lines represent confidence limits (95%) for catches in control trap.

Moreover, Etxebeste and Pajares (2011) tested verbenone and *trans*-conophthorin against *I. sexdentatus* populations present in mixed pine stands (*P. sylvestris*, *P. nigra* and *P. pinaster*). Two verbenone release rate (2, 40 and 60 mg/day) and one *trans*-conophthorin release rate (0.3 mg/day) were tested in two different field trapping bioassays with Ipsdienol 2.35 mg/day as attractant. Both verbenone and conophthorin and their combination significantly elicited a reduction of trap catches. Verbenone at 2 and 40 mg/day reduced the catches in a similar percentage (73% and 82% respectively) whereas *trans*-conophthorin reduced them by 45-49%. The strongest effect was showed by the blend of both compounds (verbenone at 40 mg/day plus *trans*-conophthorin at 0.3 mg/day) with a trap reduction rate of 90%. Another experiment was conducted to determine the potential of verbenone (at 60 mg/day) as a tool for tree protection. All control considered trees, that is, with no verbenone releasing device, were attacked by *I. sexdentatus*, whereas verbenone treated trees were less attacked.

4.2 The pine shoot beetle *Tomicus piniperda* L. (Fig. 5A) and Mediterranean pine shoot beetle *Tomicus destruens* (Wollaston) (Fig. 5B)

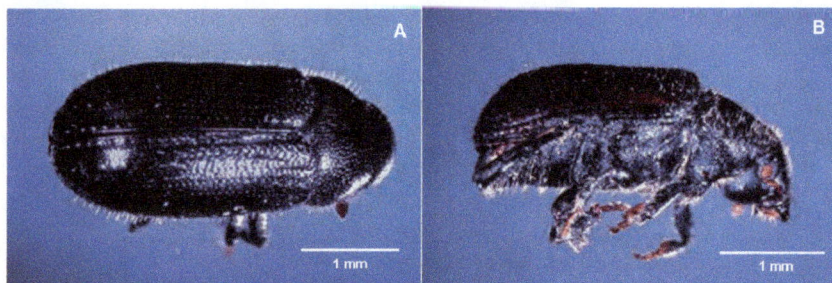

Fig. 5. (A) *Tomicus piniperda*, dorsal view; (B) *Tomicus destruens*, lateral view. Photographs taken from López et al., 2007a.

Genus *Tomicus* is represented by *Tomicus piniperda*, *Tomicus minor* and *Tomicus destruens* (Mediterranean pine shoot beetle) in Spain, since the latter taxa was definitively distinguished from *T. piniperda* and considered as a distinct species by the use of molecular techniques (Gallego & Galián, 2001; Kerdelhué et al., 2002). Morphologically, the colour of the elytra, colour of the antennal club, distribution of the antennal setae and distribution of the punctures along the elytral declivity seem to be the most useful diagnostic characters to differ both species, but only between mature exemplars (Faccoli, 2006). Concerning their distribution, *T. destruens* seems to be the predominant species in Spanish mainland, living in low an hot areas, whereas *T. piniperda* inhabit wet and cold areas of north-central Spain (Northern Plateau, the Pyrenees and perphaps in the Betic Mountains, at the South) (Gallego et al., 2004). However, there would be an overlapping of both species in the Atlantic Coast and the Bay of Biscay Coast, where they apparently coexist in sympatry. On the other hand, these authors consider *T. minor* as a less abundant species, with a fragmented distribution through high and wet areas. These potential distributions were suggested after applying predictive General Additive Models and Ecological Niche Factor Analysis models from 254 specimens of 81 different plots of Spanish mainland.

Although *P. sylvestris* is considered as the natural host of *T. piniperda*, its development in *P. radiata* (including maturation feeding in Monterrey pine shoots) is well reported. Host preference between both pine species has been tested in northern Spain. Amezaga (1996) observed that *T. piniperda* is able to exploit *P. radiata* as well as *P. sylvestris*. Even though brood production (number of progeny adults per gallery) was not significantly affected by tree species, the development was slowlier in *P. radiata*, and callows weighed less. In addition, after sampling two study areas at different altitudes (at 250 and 650 m) no sister generation was detected, so Amezaga (1996) hypothesized with the small chance of ocurring a second generation in northern Spain. She observed that attacks of *T. piniperda* began in March. Thus, according to her results *T. piniperda* might start its swarmming flight approximately in January over the entire altitudinal range in which pine stands are present in Northern Plateau.

Regardig to damage data, in 1989 massive outbreaks of *Tomicus piniperda* caused losses of 72 million € in *Pinus radiata* in Basque Country region (Northern Spain) (Amezaga, 1993). It has been estimated that a total of 200,000 ha have been affected by *T. piniperda* from 1990 to 1999 in Spain (Grégoire & Evans, 2004). The monitoring of *T. piniperda* populations is a common task carried out in different spanish provinces, but to our knowledge there has not been any control program with potential antiaggregant compounds until now. In contrast, researches aimed to *T. destruens* have been conducted in order to test the role of non host volatiles in its behaviour. Guerrero et al. (1997) showed by single-cell electrophysiological technique that *T. destruens* antennae possess specific olfactory cells capable of detecting benzyl alcohol. Furthermore, a ca. 700 mg/day release rate of this compound can significantlly reduce the attraction of *T. destruens* to host logs. Besides, the development of an effective lure for monitoring populations of *T. destruens* has been another objective. Gallego et al. (2008) tested different releases rates of ethanol and *a*-pinene, alone or in combination (with an upper threshold of 1800 and 900 mg/day respectively) in monospecific *P. halepensis* stands of southern Spain. The addition of *trans*-verbenol was also tested, but it did not affect the response. *a*-pinene alone did not showed a strong attraction effect, but a synergistic effect when adding it to ethanol. The most attractive blend appeared to when releasing 300 mg/day of *a*-pinene and 900 mg/day of ethanol.

4.3 Twig beetle *Pityophthorus pubescens* (Marsham) (Fig. 6)

Genus *Pityophthorus* has been typically not considered as a major pest species-complex, due to its life habits and development in branches of mainly dead and decaying trees (Bright, 1981; Wood, 1982).

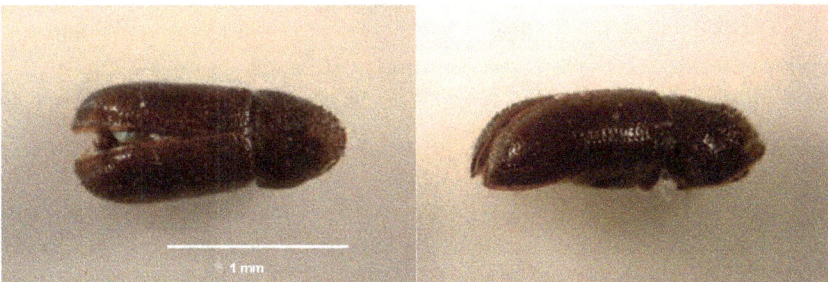

Fig. 6. *Pityophthorus pubescens*, dorsal and lateral view. Photographs taken from López et al., 2007a.

However, *Pityophthorus setosus* and *P. carmeli* has been associated with the causal agent of pitch canker *Fusarium circinatum* (Storer et al., 2004; Sakamoto et al., 2007). In Spanish mainland, it has been previously reported that *Pityophthorus pubescens* is also associated with *F. circinatum* in *P. radiata* stands of the Basque Country (northern Spain) (Romón et al., 2007b). Regarding the chemical ecology of *P. pubescens*, López et al. (2011) showed that both sexes emit (2*R*,5*S*)-2-(1-hydroxy-1-methylethyl)-5-methyltetrahydrofuran, also known as (*E*)-pityol, through different techniques of volatile collection (PORAPAK-Q and Solid Phase Microextraction/SPME). Positive enantiomer of this compound is also a component of the aggregation pheromone of other species of the genus, such as *P. pityographus* (Francke et al., 1987) and *P. carmeli*, *P. nitidulus* and *P. setosus* (Dallara et al., 2000), and the female-produced aggregation pheromone of the cone beetles *Conophthorus resinosae*, *C. coniperda* and *C. ponderosae* (Pierce *et al.* 1995; Birgersson *et al.* 1995; Miller *et al.*, 2000). However, in contrast with *P. pubescens* only one of the sexes of these species seems to emit (*E*)-(+)-pityol, as follows: males of *P. pityographus* and *P. carmeli* and females of *P. nitidulus* and *P. setosus*. Electroantennographic assays has revealed that both males and females of *P. pubescens* are able to detect (*E*)-(+)-pityol (López et al., 2011) (Fig. 7).

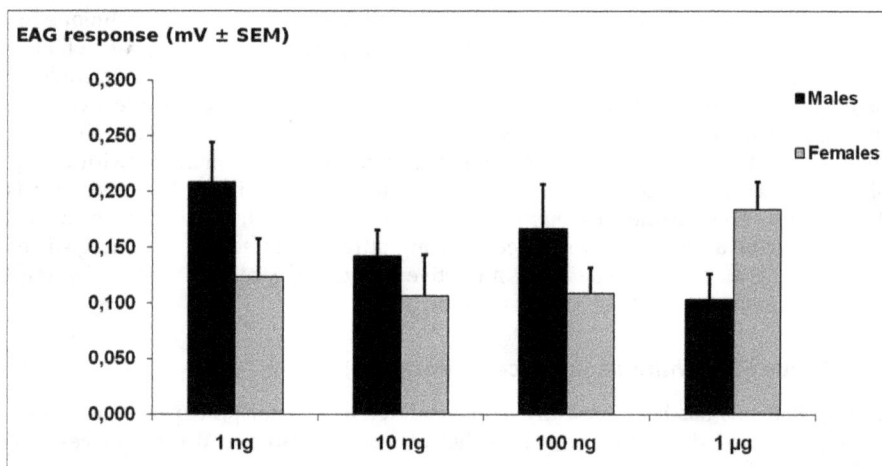

Fig. 7. Absolute EAG (mV± SE) responses of *P. pubescens* males (dark grey) and females (light grey) to serial dilutions containing 1 ng, 10 ng, 100 ng and 1 µg of (*E*)-(+)-pityol. Means followed by different letters were significantly different (Two-way ANOVA followed by Tukey multiple range test ($P \leq 0.05$), n = 8).

Moreover males were more attracted to (*E*)-(+)-pityol and (*E*)-(±)-pityol in olfactometric bioassays when testing three different doses (from 1 to 100 ng in decadeic steps) (López et al., 2011) (Fig. 8). In addition, sex-ratio appears to be male-biased in field trapping performed in different *Pinus* spp. stands of the Basque Country with multiple funnel traps baited with (*E*)-(+)-pityol and racemic pityol (López et al., unpublished data). Thus, the use of (*E*)-(+)-pityol or its cheaper racemate form might be an useful tool for monitoring *P. pubescens* populations and even to trap out male beetles in *P. radiata* stands.

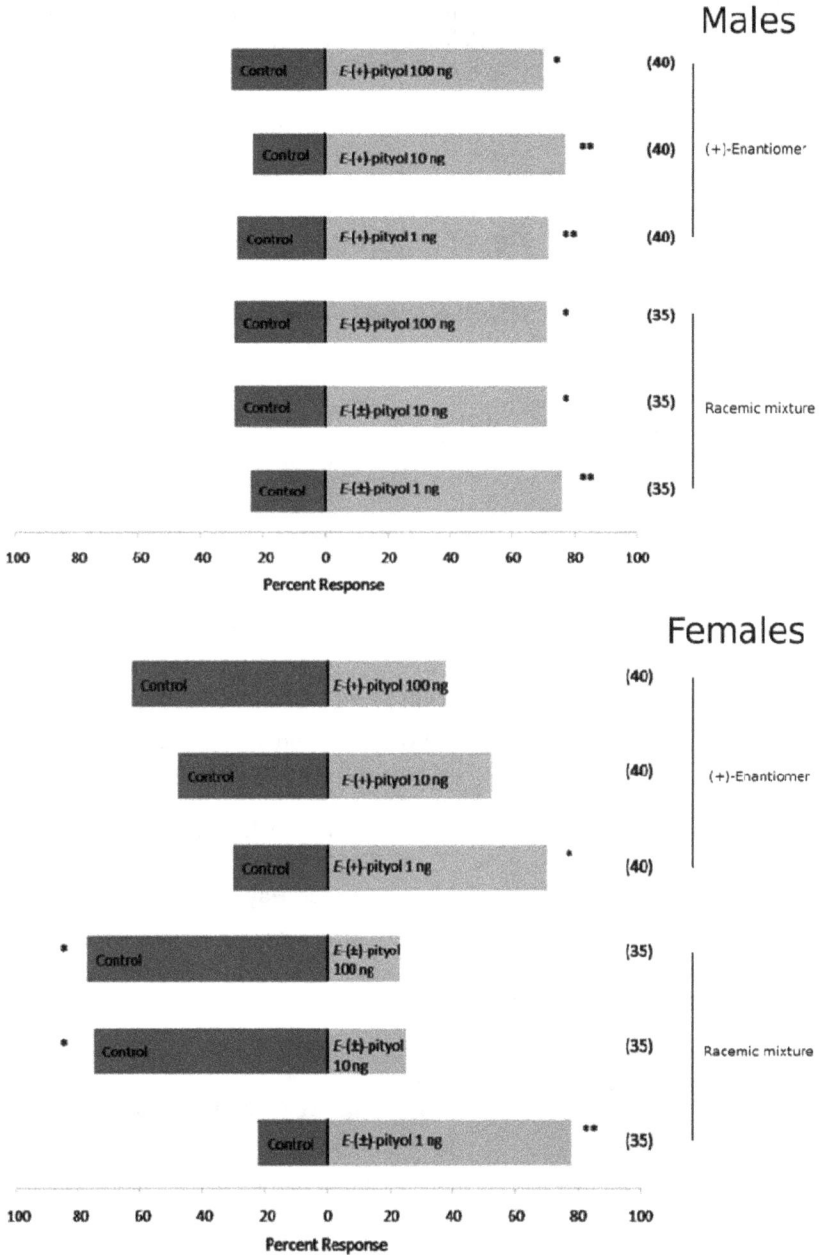

Fig. 8. Response of *P. pubescens* males (upper panel) and females (lower panel) at different doses of (*E*)-(+) and racemic (*E*)-pityol in Y-tube olfactometer trials. One and two asterisks indicate significant differences at $P<0.05$ and $P<0.01$, respectively (Chi-square test, with a significance level of $a = 0.05$). Number in parentheses indicates number of beetles responding.

On the other hand, verbenone shows promise as a disruptant of the aggregation of *P. pubescens*. Field studies has been undertaken in *P. radiata* stands testing four different release rates (0.01, 0.20, 1.80 and 3.10 mg/day) (at 22-24°C) of this compound (Romón et al., 2007b). This work revealed a significant negative dose-dependent relationship in captured insects when comparing with control traps baited with a racemic pityol-releasing device at 0.14 mg/day (Fig. 9).

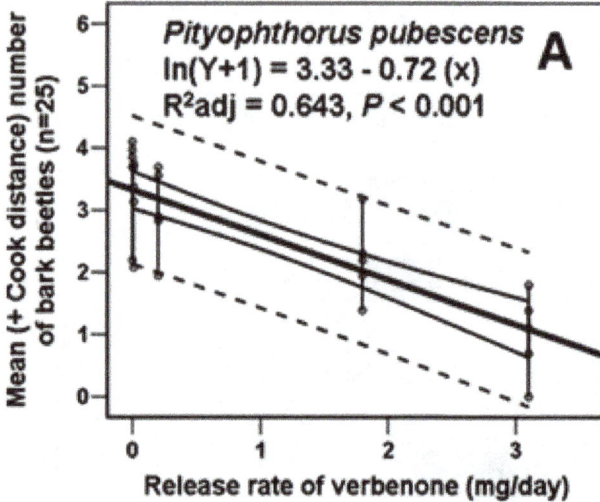

Fig. 9. Effect of four different release rates of verbenone on the attraction of *Pityophthorus pubescens*. Slope of regression line is significantly different from zero (t-test, $P < 0.001$). Confidence limits (95%) (thin solid lines) are associated with the regression line (thick solid line). Dashed lines represent confidence limits (95%) for catches in control trap.

4.4 *Hylurgops palliatus* (Gyllenhal) and *Hylaster ater* (Paykull)

Even though these two species (Fig 10 & 11) are considered as secondary, due to their colonization of dying or decaying trees, it is remarkable to mention their association with different species of pathogenic fungi (see subsection 2, table 3).

Fig. 10. *Hylurgops palliatus*, dorsal and lateral view. Photographs taken from López et al., 2007a.

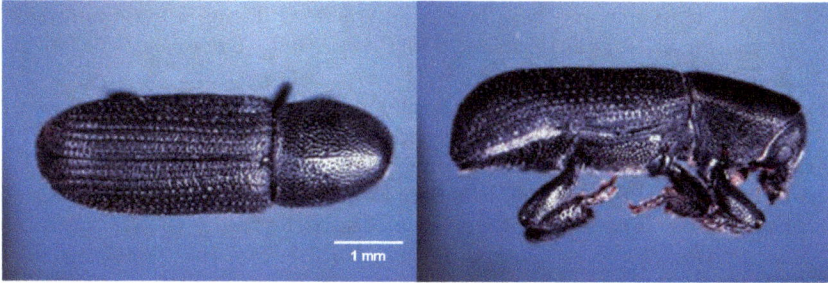

Fig. 11. *Hylastes ater*, dorsal and lateral view. Photographs taken from López et al., 2007a.

5. Conclusions

At the light of what has been exposed, it is apparent that further extensive studies are needed to determine many gaps that still remain unclear. Excluding taxonomic and faunistic composition, which have been widely studied, little is known about the chemical ecology (including insect-host and insect-non host detailed interactions) of the species perceived as pest, although substantial progresses have been made. A deeper study of populations dynamics would be neccesary, in order to characterize better the brood production, flight periods and number of generations per year. This information would aid to a major understanding of which control methods should be applied. In addition, the development of useful IPM strategies, especially in the field of semiochemicals which might act as effective anti-aggregants, represents an important research line with many questions to be responded. Moreover, proper forestry management should be also recommended to be combined with, due to its relative influence on favoring the generation of populations outbreaks.

6. References

Alonso Zarazaga, M.A. (2002). Lista preliminar de los Coleoptera Curculionoidea del área ibero-balear, con descripción de *Melicius* gen. nov. y nuevas citas. *Boletín de la Sociedad Entomológica Aragonesa*, Vol.31, (October 2002), pp. 9-33, ISSN 1134-6094

Alonso Zarazaga, M.A. & Goldarazena, A. (2005). Presencia en el País Vasco de *Rhyephenes humeralis* (Coleoptera, Curculionidae), plaga de *Pinus radiata* procedente de Chile. *Boletín de la Sociedad Entomológica Aragonesa,* Vol.36, (April 2005), pp. 143-152, ISSN 1134-6094

Amezaga, I. (1993). The ecology and pest status of *Tomicus piniperda* L. (Coleoptera: Scolytidae) on pines. Ph.D. dissertation. University of London. United Kingdom.

Amezaga, I. (1996). Monterrey Pine (*Pinus radiata* D.Don) suitability for the pine shoot beetle (*Tomicus piniperda* L.) (Coleoptera: Scolytidae). Vol.86, pp. 73-79, ISSN 0378-1127

Ayres, M.P., Lombardero, M.J. (2000). Assessing the consequences of global change for forest disturbance from herbivores and pathogens. *The Science of Total Environment*, Vol.262, No.3, (November 2000), pp. 263-286, ISSN 0048- 9697

Bakke, A. (1983). Host tree and bark beetles interactions during a mass outbreak of *Ips typographus* in Norway. *Zeitschrift Fuer Angewandte Entomologie*, Vol.96, pp. 118-125, ISSN 0044-2240

Berryman, A.A., Dennis, B., Raffa, K.F. & Stenseth, N.C. (1985). Evolution of optimal group attack with particular reference to bark beetles (Coleoptera: Scolytidae). *Ecology*, Vol.66, No.3 (June 1985), pp.898-903, ISSN 0012-9658

Birgersson, G., Schlyter, F., Löfqvist, J. & Bergström, G. (1984). Quantitative variation of pheromone components in the spruce bark beetle *Ips typographus* from different attack phases. *Journal of Chemical Ecology*, Vol.10, No.7, (July 1984), pp. 1029-1055, ISSN 0098-0331

Birgersson, G. & Leufvén, A. (1988). The influence of host tree response to *Ips typographus* and fungal attack on production of semiochemicals. *Insect Biochemistry*, Vol. 18, pp. 761-770, ISSN 0020-1790

Birgersson, G., Byers, J.A., Bergström, G. & Löfqvist, J. (1990). Production of pheromone components, chalcogran and methyl (E,Z)-2,4-decadienoate, in the spruce engraver *Pityogenes chalcographus*. *Journal of Insect Physiology*, Vol. 36, No.6, pp. 391-395, ISSN 0022-1910

Birgersson, G., DeBarr, G.L., de Groot, P., Dalusky, M.J., Pierce Jr, H.D., Borden, J.H., Meyer H., Francke, W., Espelie, K.E. & Berisford, C.W. (1995) Pheromones in white pine cone beetle, *Conophthorus coniperda* (Schwarz) (Coleoptera: Scolytidae). *Journal of Chemical Ecology*, Vol.21, No.2, (February 1995), pp. 143-167, ISSN 0098-0331

Borden, J.H. (1982). Aggregation pheromones. In: *Bark Beetles in North American Conifers*, B. Mitton and K.B. Sturgeon (Eds.), 74-139, University of Texas Press, ISBN 0292707444, Austin, Texas

Bright, D.E. 1981. Taxonomic monograph of the genus *Pityophthorus* Eichhoff in North and Central America (Coleoptera: Scolytidae). *Memoirs of the Entomological Society of Canada*, Vol.118, pp. 1-378, ISSN 1920-3047

Bueno, A., Díez, J.J. & Fernández, M.M. (2010) Ophiostomatoid fungi transported by *Ips sexdentatus* (Coleoptera: Scolytidae) in *Pinus pinaster* in NW Spain. *Silva Fennica*, Vol.44, No.3, pp. 387-397, ISSN 0037-5330

Byers, J.A. (1983). Influence of sex, maturity and host substances on pheromones in the guts of the bark beetles, *Dendroctonus brevicomis* and *Ips paraconfusus*. *Journal of Insect Physiology*, Vol.29, No.1, (January 1983), pp. 5-13, ISSN 0022-1910

Byers, J.A. (1992). Attraction of bark beetles, *Tomicus piniperda*, *Hylurgops palliatus*, and *Trypodendron domesticus* and other insects to short-chain alcohols and monoterpenes. *Journal of Chemical Ecology*, Vol.18, No.12, (December 1992), pp. 2385-2402, ISSN 0098-0331

Byers, J.A. (1995). Host tree chemistry affecting colonization in bark beetles. In: *Chemical ecology of insects 2*, R.T. Cardé and W.J. Bell (Eds.), 154-213, Chapman and Hall, ISBN 978-0412039614, New York, United States.

Byers, J.A. (2004). Chemical ecology of bark beetles in a complex olfactory landscape In: *Bark and wood boring insects in living trees in Europe, a synthesis*, F. Lieutier, K.R. Day, A. Battisti, J.C. Grégoire and H.F. Evans (Eds.), 89-134, Kluwer Academic Press, ISBN 1-4020-2240-9, Dordrech, The Netherlands.

Byers, J.A., Wood, D.L., Browne, L.E., Fish, R.H., Piatek, B. & Hendry, L.B. (1979). Relationship between a host plant compound, myrcene and pheromone production in the bark beetle, *Ips paraconfusus*. *Journal of Insect Physiology*, Vol.25, pp. 477-482, ISSN 0022-1910

Byers, J.A., Wood, D.L., Craig, J. & Hendry, L.B. (1984). Attractive and inhibitory pheromones produced in the bark beetle, *Dendroctonus brevicomis*, during host colonization: Regulation of inter- and intraspecific competition. *Journal of Chemical Ecology*, Vol.10, No.6, (June 1984), pp. 861-877, ISSN 0098-0331

Byers, J.A., Zhang, Q.H. & Schlyter, F. (1998). Volatiles from non-host birch trees inhibit pheromone response in spruce bark beetles. *Naturwissenschaften*, Vol.85, No.11, (December 1998), pp. 557–561, ISSN 0028-1042

Carter, D.R., O'Laughlin, J. & McKinney, C. (1991). Southern pine beetle impacts and control policy in the 1982-1986 Texas epidemic. *Southern Journal of Applied Forestry*, Vol.15, No.3, (August 1991), pp. 145-153, ISSN 0148-4419

Correll, J.C., Gordon, T.R., McCain, A.H., Fox, J.W., Koehler, C.S., Wood D.L. & Schultz, M.E. (1991). Pitch Canker Disease in California: pathogenicity, distribution, and canker development on Monterrey pine (*Pinus radiata*). *Plant Disease*, Vol.75, No.7, (July 1991), pp. 676-682, ISSN 0191-2917

Czokajlo, D. (1998). Semiochemicals for the larger pine shoot beetles (*Tomicus piniperda* L.) and its clerid predators. Ph.D. Dissertation, State University of New York, Syracuse, NY. United States.

Dallara, P.L., Seybold, S.J., Meyer, H., Tolasch, T., Francke, W. & Wood, D.L. (2000). Semiochemicals from three species of *Pityophthorus* (Coleoptera: Scolytidae): Identification and field response. *The Canadian Entomologist*, Vol.132, No.6, (November 2000), pp. 889-906, ISSN 0008-347X

De Groot, P., DeBarr, G.L. & Birgersson, G. (1998). Field bioassays of synthetic pheromones and host monoterpenes for *Conophthorus coniperda* (Coleoptera: Scolytidae). *Environmental Entomology* Vol.27, No.2, (April 1998), pp. 382-387, ISSN 0046-225X

De Groot, P. & DeBarr, G.L. (2000). Response of cone and twig beetles (Coleoptera: Scolytidae) and a predator (Coleoptera: Cleridae) to pityol, conophthorin, and verbenone. *The Canadian Entomologist*, Vol.132, No.6, (November 2000), pp. 843-851, ISSN 0008-347X

Dwinell, L.D., Barrows-Broaddus, J.B. & Kuhlman, E.G. (1985). Pitch canker: a disease complex of southern pines. *Plant Disease*, Vol.69, No.3, (March 1985), pp. 270-276, ISSN 0191-2917

Etxebeste, I. & Pajares, J.A. (2011). Verbenone protects pine trees from colonization by the six-toothed pine bark beetle, *Ips sexdentatus* Boern. (Col.: Scolytinae). *Journal of Applied Entomology*, Vol.135, No.4, (May 2011), pp. 258-268, ISSN 0931-2048

Faccoli, M. (2006). Morphological separation of *Tomicus piniperda* and *T. destruens* (Coleoptera: Curculionidae: Scolytinae): new and old characters. *European Journal of Entomology*, Vol.103, No.2, pp. 433–442, ISSN 1210-5759

Fernández, M.M.F. (1996). Estructura biológica y ecológica del barrenillo de pino *Tomicus minor* (Coleoptera, Scolytidae). Ph.D. dissertation. University of León, Spain.

Fernández, M.M.F., García A.E. & Lieutier, F. (2004). Effects of various densities of *Ophiostoma ips* inoculations on *Pinus sylvestris* in north-western Spain. *Forest Pathology*, Vol.34, No.4, pp. 213–223, (August 2004), ISSN 1437-4781

Fox, J.W., Wood, D.L., Koehler, C.S. & O'Keefe, S.T. (1991). Engraver beetles (Scolytidae: *Ips* species) as vectors of the pitch canker fungus, *Fusarium subglutinans*. *The Canadian Entomologist*, Vol.123, No.6, (November, 1991), pp. 1355-1367, ISSN 0008-347X

Francke, W., Hindorf, G. & Reith, W. (1978). Methyl-1,6-dioxaspiro[4.5]decanes as odors of *Paravespula vulgaris* (L.). *Angewandte Chemie International Edition*, Vol.17, No.11, (November 1978), pp. 862, ISSN 0044-8249

Francke, W., Hindorf, G. & Reith, W. (1979). Alkyl-1,6-dioxaspiro[4.5]decanes- a new class of pheromones. *Naturwissenschaften*, Vol. 66, No.12, (December 1979), pp. 618-619, ISSN 0028-1042

Francke, W., Pan, M.L., König, W.A., Mori, K., Puapoomchareon, P., Heuer, H. & Vité, J.P. (1987). Identification of 'pityol' and 'grandisol' as pheromone components of the bark beetle, *Pityophthorus pityographus*. *Naturwissenschaften*, Vol.74, pp. 343-345, ISSN 0028-1042

Gallego, D. & Galián, J. (2001). The internal transcribed spacers (ITS1 and ITS2) of the rDNA diferenciate the bark beetle forest pests *Tomicus destruens* and *T. piniperda*. *Insect Molecular Biology*, Vol.10, No.5, (October 2001), pp. 415-420, ISSN 0962-1075

Gallego, D., Cánovas, F, Esteve, M.A. & Galián, J. (2004). Descriptive biogeography of *Tomicus* (Coleoptera: Scolytidae) species in Spain. *Journal of Biogeography*, Vol.31, No.12, (December 2004), pp. 2011-2024, ISSN 0305-0270

Gallego, D., Galián, J., Díez, J.J. & Pajares, J.A. (2008). Kairomonal responses of *Tomicus destruens* (Col., Scolytidae) to host volatiles α-pinene and ethanol. *Journal of Applied Entomology*, Vol.132, No.8, (September 2008), pages 654–662, ISSN 0931-2048

Gil L.A. & Pajares J.A. (1986). *Los Escolítidos de las Coníferas de la Península Ibérica*. Instituto Nacional de Investigaciones Agrarias, Ministerio de Agricultura, Pesca y Alimentación. ISBN 84-7498-250-2, Madrid, Spain.

González, R. (1990). Estudio bio-ecológico de *Phloeotribus scarabaeoides* (Bernard, 1788) en la provincia de Granada. Ph.D. dissertation. University of Granada, Spain.

Grégoire, J.C. & Evans H.F. (2004). Damage and control of BAWBILT organisms an overview. In: *Bark and Wood Boring Insects in Living Trees in Europe, A Synthesis*, F. Lieutier, K.R. Day, A. Battisti, J.C. Grégoire and H.F. Evans (Eds.), 19-37. Kluwer Academic Press, ISBN 1-4020-2240-9, Dordrech, The Netherlands.

Guerrero, A., Feixas, J., Pajares, J.A., Wadhams, L.J., Pickett,. J.A & Woodcock, C.M. (1997). Semiochemically Induced Inhibition of Behaviour of *Tomicus Destruens*. *Naturwissenschaften*, Vol.84, No.4, (April, 1997), pp. 155-157, ISSN 0028-1042

Hendry, L.B., Piatek, B., Browne, L.E., Wood, D.L., Byers, J.A., Fish, R.H. and Hicks, R.A. (1980). In vivo conversion of a labelled host plant chemical to pheromones of the bark beetle *Ips paraconfusus*. *Nature*, Vol.284, pp. 485, 0028-0836

Hoover, K., Wood, D.L., Fox, J.W. & Bros W.E. (1995). Quantitative and seasonal association of the pitch canker fungus, *Fusarium subglutinans* f.sp. *pini* with *Conophthorus radiatae* (Coleoptera: Scolytidae) and *Ernobius punctulatus*

(Coleoptera: Anobiidae) which infest *Pinus radiata*. *The Canadian Entomologist*, Vol.127, No.1, (January 1995), pp. 79-91, ISSN 0008-347X

Huber, D.P.W., Gries, R., Borden, J.H. & Pierce Jr., H.D. (1999). Two pheromones of coniferophagous bark beetles found in the bark of nonhost angiosperms. *Journal of Chemical Ecology*, Vol.25, No.4, (April 1999), pp. 805–816, ISSN 0098-0331

Huber, D.P.W., Borden, J.H., Jean-Willians, N.L. & R. Gries. (2000). Differential bioactivity of conophthorin on four species of North American bark beetles (Coleoptera: Scolytidae). *The Canadian Entomologist*, Vol.132, No.5, (September 2000), pp. 649-653, ISSN 0008-347X

Huber, D.P.W., Borden, J.H. & Stastny, M. (2001). Response of the pine engraver, *Ips pini* (Say) (Coleoptera: Scolytidae), to conophthorin and other angiosperm bark volatiles in the avoidance of non-host. *Agricultural and Forest Entomology*, Vol.3, No.3, (August 2001), pp. 225-232, ISSN 1461-9563

Hughes, P.R. (1974). Myrcene: a precursor of pheromones in *Ips* beetles. *Journal of Insect Physiology*, Vol.20, pp. 1271-1275. ISSN 0022-1910

Inventaire Forestier Nationale (2009) Estimations pour l'ensemble de la zone évaluée. In: *Tempête Klaus du 24 janvier 2009* , 01.08.2011. Available from http://www.ifn.fr/spip/spip.php?article618

Jactel, H., Van Halder, I., Menassieu, P., Zhang, Q.H. & Schlyter, F. (2001). Non-host volatiles disrupt the response of the stenographer bark beetle, *Ips sexdentatus* (Coleoptera: Scolytidae), to pheromone-baited traps and maritime pine logs. *Integrated Pest Management Reviews* Vol.6, No.3-4, (September 2001), pp. 197-207, ISSN 1353-5226

Kangas, E., Oksanen, H. & Perttunen, V. (1970). Responses of *Blastophagus piniperda* L. (Col., Scolytidae) to trans-verbenol, cis-verbenol, and verbenone, known to be population pheromones of some American bark beetles. *Annals of Entomologica Fennica*, Vol.36, pp. 75-83, ISSN 0785-8760

Kerdelhué, C., Roux-Morabito, G., Forinchon, J., Chambon, J.M., Robert, A. & Lieutier F. (2002). Population genetic structure of *Tomicus piniperda* L. (Coleoptera: Curculionidae: Scolytinae) on different pine species and validation of *T. destruens* (Woll.). *Molecular Ecology*, Vol.11, No.3, (March 2002), pp. 483-495, ISSN 0962-1083

Kirisits, T. (2004). Fungal associates of european bark beetles with specian emphasis to ophiostomatoid fungi. In: *Bark and Wood Boring Insects in Living Trees in Europe, A Synthesis*, F. Lieutier, K.R. Day, A. Battisti, J.C. Grégoire and H.F. Evans (Eds.), 181-235. Kluwer Academic Press, ISBN 1-4020-2240-9, Dordrech, The Netherlands.

Kirkendall, L.R. & Faccoli, M. (2010). Bark beetles and pinhole borers (Curculionidae, Scolytinae, Platypodinae) alien to Europe. In: *Sixty years of discovering scolytine and platypodine diversity: A tribute to Stephen L. Wood*. A.I. Cognato and M. Knížek (Eds), pp. 227-251. *ZooKeys*, Vol.56, ISSN 13132989

Kohnle, U., Densborn, S., Kölsch, P., Meyer, H. & Francke W. (1992). E-7-methyl-1,6-dioxaspiro[4.5]decane in the chemical communication of European Scolytidae and Nitidulidae. *Journal of Applied Entomology*, Vol.114, pp. 187-192, ISSN 0931-2048

Landeras, P., García, Y., Fernández, L. & Braña, M. (2005). Outbreak of pitch canker caused by *Fusarium circinatum* on *Pinus* spp. in Northern Spain. *Plant Disease*, Vol.89 No.9, pp. 1015, ISSN 0191-2917

Lanne, B.S., Schlyter, F., Byers, J.A., Löfqvist, J., Leufvén, A., Bergström, G., Van Der Pers, J.N.C., Unelius, R., Baeckström, P. & Norin. T. (1987). Differences in attraction to semiochemicals present in sympatric pine shoot beetles, *Tomicus minor* and *T. piniperda*. *Journal of Chemical Ecology*, Vol.13, No.5, (May 1987), pp. 1045-1067, ISSN 0098-0331

Lieutier, F., Yart, A., Garcia, J., Ham, M.C., Morelet, M. & Levieux, J. (1989). Champignons phytopathogènes associés à deux coléoptères scolytidae du pin sylvestre (*Pinus sylvestris* L.) et étude préliminaire de leur agressivité envers l'hôte. *Annals of Forest Science*, Vol.46, No.3, pp. 201-216, ISSN 1286-4560

Lindgren, B.S. (1994) Research needs for anti-aggregation pheromones of bark beetles. In: *Research, Development, and Commercialization of Semiochemicals in Insect Pest Management in Canada. Canada Pest Management Alternatives Office and Pest Management Alternatives Program* pp. 62-67. Ottawa, Canada.

Lindgren, B.S., Borden, J.H., Cushon, G.H., Chong, L.J. & Higgins C.J. (1989). Reduction of mountain pine beetle (Colcoptera: Scolytidae) attacks by verbenone in lodgepole pine stands in British Columbia. *Canadian Journal of Forest Research*, Vol.19, No.1 (January 1989), pp. 65-68, ISSN 1208-6037

Lindgren, B.S., Nordlander G. & Birgersson, G. (1996). Feeding deterrency and acute toxicity of verbenone to the pine weevil, *Hylobius abietis*. *Journal of Applied Entomology*, Vol.120, No.5, (December 1996), pp. 397-403. ISSN 0931-2048

Lindgren, B.S. & Miller, D.R. (2002). Effect of verbenone on five species of bark beetles (Coleoptera: Scolytidae) in lodgepole pine forests. *Environmental Entomology*, Vol.31, No.5, (October 2002), pp. 759-765, ISSN 0046-225X

Lombardero, M.J. (1994). Estudio de los Scolytidae (O. Coleoptera) de Galicia. Ph.D. dissertation. University of Santiago de Compostela, Spain.

Lombardero, M.J. & Novoa, F. (1994). Datos faunísticos sobre escolítidos ibéricos (Coleoptera: Scolytidae). *Boletín de la Sociedad Entomológica Aragonesa*, Vol.18, No.1-2, pp. 181-186, ISSN 1134-6094

López, S., Romón, P., Iturrondobeitia, J.C. & Goldarazena A. (2007a). *Los escolítidos de las coníferas del País Vasco. Guía práctica para su identificación y control.* Servicio Central de Publicaciones del País Vasco. ISBN 978-84-457-2650-1. Vitoria, Spain.

López S., Iturrondobeitia J.C. and Arturo Goldarazena A. (2007b). Primera cita de la Península Ibérica de *Gnathotrichus materiarius* (Fitch, 1858) y *Xylosandrus germanus* (Blandford, 1894) (Coleoptera: Scolytinae). *Boletín de la Sociedad Entomológica Aragonesa*, Vol.40, (October 2007), pp. 527-532, ISSN 1134-6094

López, S., Quero, C., Iturrondobeitia, J.C., Guerrero, A. & Goldarazena A. (2011). Evidence for (*E*)-Pityol as an aggregation pheromone of *Pityophthorus pubescens* (Coleoptera: Curculionidae: Scolytinae). *The Canadian Entomologist* Vol.143, No.5, pp. 447-454, ISSN 0008-347X

Michel, M. (2004). El pino radiata (*Pinus radiata* D.Don) en la historia de la Comunidad Autónoma de Euskadi. Análisis de un proceso de forestalismo intensivo. Ph.D. dissertation. University of Madrid, Spain

Miller, D.R, Pierce Jr, H.D., de Groot, P., Jean-Williams, N., Bennett, R., & Borden, J.H. (2000). Sex pheromone of *Conophthorus ponderosae* (Coleoptera: Scolytidae) in a coastal stand of western white pine (Pinaceae). *The Canadian Entomologist*, Vol.132, No.2, (March 2000), pp. 243-245, ISSN 0008-347X

Muramoto, M., Minamihashi,. H., Mitsuishi, K. & Kobayashi, T. (1989). Occurrence of resiniferous damage on *Pinus luchuensis* Mayr – Symptom and analysis of the damage. In: *Proceedings, Transnational 99th annual meeting of Japan Forestry Society*

Nowak, J. (2005). Southern pine beetle prevention and restoration program. USDA Forest Service Technical Report, 02.08.2011, Available from http://www.fs.fed.us/foresthealth/publications/spb_success_story.pdf.

Pajares JA, 1987. Contribución al conocimiento de los escolítidos vectores de la grafiosis en la Península Ibérica. Ph.D. dissertation. University of Madrid, Spain.

Perttunen, V. (1957). Reactions of two bark beetle species, *Hylurgops palliatus* Gyll. and *Hylastes ater* Payk. (Col., Scolytidae) to the terpene alpha-pinene. *Annals of Entomologica Fennica*, Vol.23, pp. 101-110, ISSN 0785-8760

Pierce, H.D., Conn, J.E., Oehlschlager, A.C. & Borden, J.H. (1987). Monoterpene metabolism in female mountain pine beetles, *Dendroctonus ponderosae* Hopkins attacking *ponderosa* pine. *Journal of Chemical Ecology* Vol.13, No.6, (June 1987), pp. 1455-1480, ISSN 0098-0331

Pierce, H.D., de Groot, P., Borden, J.H., Ramaswamy, S., & Oehlschlager, A.C. (1995). Pheromones in red pine cone beetle, *Conophthorus resinosae* Hopkins, and its synonym, *C. banksianae* McPherson (Coleoptera: Scolytidae). *Journal of Chemical Ecology*, Vol.21, No.2, (February 1995), pp. 169-185, ISSN 0098-0331

Plaza, E. (1983). Los representantes españoles de las tribus *Crypturgini* y *Pityophthorini* (Col. Scolytidae). *Eos*, Vol.59, pp. 223-241, ISSN 0013-9440

Plaza, E. & Gil. L. (1982). Los *Ipini* (Col. Scolytidae) de la Península Ibérica. *Eos*, Vol.58, pp. 237-269, ISSN 0013-9440

Poland, T.M., de Groot, P., Burke, S., Wakarchuk, D., Haack, R.A., Nott, R. & Scarr, T. (2003). Development of an improved attractive lure for the pine shoot beetle, *Tomicus piniperda* (Coleoptera: Scolytidae). *Agricultural and Forest Entomology*,Vol.5, No.4 (November 2003), pp. 293-300, ISSN 1461-9563

Raffa, K.F. (2001). Mixed messages across multiple trophic levels: the ecology of bark beetle chemical communication systems. *Chemoecology*, Vol.11, No2, (April 2001), pp. 49-65, ISSN 0937-7409

Raffa, K.F. & Smalley, E.B. (1988). Response of red and jack pines to inoculations with microbial associates of the pine engraver, *Ips pini* (Coleoptera: Scolytidae). *Canadian Journal of Forest Research*, Vol.18, No.5, (May 1988), pp. 581-586, ISSN 1208-6037

Rappaport, N.G., Stein, J.D., del Rio Mora, A.A., DeBarr, G., de Groot P. & Mori S. (2000). Responses of *Conophthorus* (Coleoptera: Scolytidae) to behavioral chemicals in field trials: A transcontinental perspective. *The Canadian Entomologist*, Vol.132, No.6, (November 2000), pp. 925-937, ISSN 0008-347X

Renwick, J.A.A. & Vité, J.P. (1968). Isolation of the population aggregating pheromone of the southern pine beetle. *Contributions of the Boyce Thompson Institute*, Vol.24, pp. 65-68, ISSN 0006-8543

Riba, J.M. (1994). Bioecología de los Scolytidae (Coleoptera) que nidifican en los abetales del Valle de Arán (Pirineos Orientales). Ph.D. dissertation. University of Barcelona.

Rodriquez, R.G. (1989). Pitch canker on *Pinus douglasiana*, pines indigenous to San Andres Milpillas, Municipal of Huajicori, Nay. Forest. In: *Proceedings, Parasitology Symposium, City of Juarez, Chihuahua*, Mexico, 4–6 October 1989

Romón, P., Zhou, X.D., Iturrondobeitia J.C., Wingfield, M.J. & Goldarazena, A., (2007a). *Ophiostoma* species (Ascomycetes: Ophiostomatales) associated with bark beetles (Coleoptera: Scolytinae) colonizing *Pinus radiata* in Northern Spain. *Canadian Journal of Microbiology*, Vol.53, No.6, (2007a), pp. 756-767, ISSN 0008-4166

Romón, P., Iturrondobeitia, J.C., Gibson, K., Lindgren, B.S. & Goldarazena A. (2007b). Quantitative association of bark beetles with pitch canker fungus and effects of verbenone on their semiochemical comunication in Monterey pine forests in Northern Spain. *Environmental Entomology*, Vol.36, No.4, (2007b), pp. 743-750, ISSN 0046-225X

Rudinsky, J.A., Morgan, M.E., Libbey, L.M. & Putnam, T.B. (1974). Antiaggregative-rivalry pheromone of the mountain pine beetle, and a new arrestant of the southern pine beetle. *Environmental Entomology*, Vol. 3, No.1, (February 19874), pp. 90-98, ISSN 0046-225X

Sakamoto, J.M., Gordon, T.R., Storer, A.J. & Wood, D.L. 2007. The role of *Pityophthorus* spp. as vectors of pitch canker affecting Monterey pines (*Pinus radiata*). *The Canadian Entomologist*, Vol.139, No.6, (November 2007), pp. 864-871, ISSN 0008-347X

Salom, S.M., Carlson, J.A., Ang, B.A., Grossman, D.M. & Day, E.R. (1994). Laboratory evaluation of biologically based compounds as antifeedants for the Pales weevil, *Hylobius pales* (Herbst) (Coleoptera: Curculionidae). *Journal of Entomological Science*, Vol.29, pp. 407-419. ISSN 0749-8004

Sánchez, G., González E., Vela, A., Ayuso, S., (2008). *Control de la población de Ips sexdentatus en el área afectada por el incendio del Rodenal*. Consejería de Medio Ambiente y Desarrollo Rural de Castilla la Mancha. ISBN 978-84-7788-521-4. Castilla la Mancha, Spain.

Schroeder L.M., 1992. Olfactory recognition of nonhosts aspen and birch by conifer bark beetles *Tomicus piniperda* and *Hylurgops palliatus*. *Journal of Chemical Ecology*, Vol.18, No.9, (September 1992), pp. 1583-1593, ISSN 0098-0331

Scriber, J.M. (1984). Host-plant suitability. In: *Chemical Ecology of Insects*, W.J. Bell and R.T. Cardé (Eds.), 159-202. Chapman and Hall, ISBN 978-087-8930-708, London, United Kingdom.

Seifert, K.A. (1993). Sapstain of commercial lumber by species of *Ophiostoma* and *Ceratocystis*. In: *Ceratocystis and Ophiostoma: Taxonomy, Ecology, and Pathogenicity*. M.J. Wingfield, K.A. Seifert and J.F. Webber (Eds.), 141-151, APS Press, ISBN 978-0-89054-156-2, St. Paul, United States.

Serez, M. (1987). Use of the aggregation pheromone preparation 'Ipslure' against the Mediterranean pine bark-beetle *Ips* (*Orthotomicus*) *erosus* (Woll.) (Col., Scolytidae). *Anzeiger für Schädlingskunde, Pflanzenschutz und Umweltschutz*, Vol.60, pp. 94-95, ISSN 0340-7330

Song, L.W., Ren, B.Z., Sun, S.H., Zhang, X.J., Zhang, K.P. & Gao, C.Q. (2005). Field trapping test on semiochemicals of pine shoot beetle *Tomicus piniperda* L. *Journal of Northeast Forestry University*, Vol.33, No,1, (January 2005), pp. 38-40, ISSN 1000-5382

Storer, A.J., Wood, D.L. & Gordon, T.R. (2004). Twig beetles, *Pityophthorus* spp. (Coleoptera : Scolytidae), as vectors of the pitch canker pathogen in California. *The Canadian Entomologist*, Vol.136, No.5, (September 2004), pp. 685-693, ISSN 0008-347X

Viljoen, A., Wingfield, M.J. & Marasas, W.F.O. (1994). First report of *Fusarium subglutinans* f.sp. *pini* on pine seedlings in South Africa. *Plant Disease*,Vol. 78, No.4, (April 1994), pp. 309-312, ISSN 0191-2917

Villarreal, M., Rubio, V., De Troya, M.T. and Arenall, F. (2005). A new *Ophiostoma* species isolated from *Pinus pinaster* in the Iberian Peninsula. *Mycotaxon*, Vol.92, No., pp. 259-268, ISSN 0093-4666

Vité, J.P., Bakke, A. & Renwick, J.A.A. (1972). Pheromones in *Ips* (Coleoptera: Scolytidae): occurrence and production. *The Canadian Entomologist*, Vol.104, No.12, (December 1972), pp. 1967-1975, ISSN 0008-347X

Vité, J.P., Volz, H.A., Paiva, M.R. & Bakke, A. (1986). Semiochemicals in host selection and colonization of pine trees by the pine shoot beetle, *Tomicus piniperda*. *Naturwissenschaften*, Vol.73, No.1 (January 1986), pp. 39-40, ISSN 0028-1042

Volz, H.A. (1988). Monoterpenes governing host selection in the bark beetles *Hylurgops palliatus* and *Tomicus piniperda*. *Entomologia Experimentalis et Applicata*, Vol. 47, No.1 (January 1988), pp. 31-35, ISSN 0013-8703

Westfall, J. (2006). Summary of forest health conditions in British Columbia. In: *Forest Health Conditions in 2006*, 01.08.2011. Available from http://www.for.gov.bc.ca/ftp/HFP/external/!publish/Aerial_Overview/2006/Aer_OV_final.pdf

Wood, D.L. (1982). The role of pheromones, kairomones, and allomones in the host selection and colonization behavior of bark beetles. *Annual Review of Entomology*, Vol.27, pp. 411-446, ISSN 0066-4170

Wood, D.L., Browne, L.E., Bedard, W.D., Tilden, P.E., Silverstein, R.M. & Rodin, J.O. (1968). Response of *Ips confusus* to synthetic sex pheromones in nature. *Science*, Vol.159, No.3860, pp. 1373-1374, ISSN 0036-8075

Wood, S.L. (1982). The bark and ambrosia beetles of North and Central America, a taxonomic monograph. *Great Basin Naturalist Memoirs*, Vol.6, pp. 1-1359, ISSN 0160-239X

Yamaoka, Y., Swanson, R.H. & Hiratsuka, Y. (1990). Inoculation of lodgepole pine with four blue-stain fungi associated with mountain pine beetle, monitored by a heat pulse velocity (HPV) instrument. *Canadian Journal of Forest Research*, Vol.20, No.1 (January 1990), pp. 31-36, ISSN 1208-6037

Zhang, Q.H., Liu, G.T., Schlyter, F., Birgersson, G., Anderson, P. & Valeur, P. (2001). Olfactory Responses of *Ips duplicatus* from Inner Mongolia, China to Nonhost Leaf and Bark Volatiles. *Journal of Chemical Ecology*, Vol.27, No.5, (May 2001), pp. 995-1009, ISSN 0098-0331

Zhang, Q.H. & Schlyter, F. (2003). Redundancy, synergism, and active inhibitory range of non-host volatiles reducing pheromone attraction in European spruce bark beetle *Ips typographus*. *Oikos*, Vol.101, No.2, (May 2003), pp. 299-310, ISSN 0030-1299

Zhang, Q.H. & Schlyter, F. (2004). Olfactory recognition and behavioural avoidance of angiosperm nonhost volatiles by conifer-inhabiting bark beetles. *Agricultural and Forest Entomology*, Vol.6, No.1, (January 2004), pp. 1-19, ISSN 1461-9563

Integrated Pest Management of Eucalypt Psyllids (Insecta, Hemiptera, Psylloidea)

Dalva Luiz de Queiroz[1], Daniel Burckhardt[2] and Jonathan Majer[3]
[1]Embrapa Florestas, Colombo-PR
[2]Naturhistorisches Museum, Basel
[3]Department of Environment and Agriculture Biology,
Curtin University of Technology, Perth
[1]Brazil
[2]Switzerland
[3]Australia

1. Introduction

Eucalypts, which are native to Australia, grow rapidly and have multiple uses. Because of these properties, they are planted on all continents except for Antarctica. In Brazil, many eucalypt species find suitable weather and environmental conditions to develop and today they are commercially grown on a large scale. In 2010, plantations covered a surface of over 4.75 million hectares (ABRAF, 2011). In Australia, eucalypts host an abundant fauna of phytophagous insects, among which the jumping plant-lice (Hemiptera: Psylloidea) are particularly species rich.

Psyllids are tiny sap-sucking insects resembling minute cicadas and they generally develop on woody dicotyledons (Hodkinson, 1974; Burckhardt, 1994; Hollis, 2004). Most species have very restricted host plant ranges. Larvae can be free-living or develop in open or closed galls, whereas others build waxy coverings, called lerps, under which they develop (Hodkinson, 1984, 2009; Hollis, 2004; Burckhardt, 2005). The subfamily Spondyliaspidinae (Aphalaridae *sensu* Burckhardt & Ouvrard, 2011) is almost exclusively restricted to Australia and to host plants of the Myrtaceae, in particular *Eucalyptus* species (Burckhardt, 1991; Hollis, 2004). Unsurprisingly, several Australian spondyliaspidine species have been introduced into other continents where they have become pests (Burckhardt, 1994; Burckhardt et al., 1999; Hollis, 2004).

In Brazil, the first spondyliaspidine found was *Ctenarytaina spatulata*, when it was observed in 1994 in a *Eucalyptus grandis* plantation in Paraná State (Iede et al., 1997). Another three species infesting eucalypts were introduced into Brazil within a decade: *Blastopsylla occidentalis* in 1997, *Ctenarytaina eucalypti* in 1998, and *Glycapis brimblecombei* in 2003 (Maschio et al., 1997; Burckhardt et al., 1999; Wilcken et al., 2003).

In this chapter, we present information on eucalypt psyllids that are pests outside Australia with particular emphasis on the situation in Brazil. This information constitutes the necessary base to control these pests. Some of the control options are also discussed.

2. Jumping plant-lice

Psyllids, or jumping plant-lice, are small phloem-feeding insects, measuring 1–10 mm body length. Together with the white flies, aphids and scale insects they constitute the suborder Sternorrhyncha within the Hemiptera. Currently, 3850 species have been described worldwide (Li, 2011), which is probably less than half of the existing number of species. Bekker-Migdisova (1973) suggested that psyllids are a very old group already represented in the Permian by the extinct family Protopsyllidiidae. According to Klimaszewski (1964), the major diversification of Psylloidea occurred between the Middle Jurassic and the Middle Cretaceous, which would coincide with the major diversification of angiosperms. More recent studies, however, imply that origin and diversification of Psylloidea may be much younger. Grimaldi (2003) showed that Protopsyllidiidae may constitute the sister-group of all Sternorrhyncha. Host plant and biogeographic evidence presented by White and Hodkinson (1985) further suggests that the modern psyllids evolved in Gondwana from an ancestor associated with the plant order Sapindales. The Mesozoic Liadopsyllidae form a potential sister-group of modern Psylloidea and the latter are represented in the fossil record only from the Eocene and later (Ouvrard et al., 2010).

As with other hemipterans, psyllids have piercing-sucking mouth-parts. When feeding, the mandibular and maxillary stylets are inserted into the host tissue, saliva is injected and then the liquid food is absorbed. Before feeding, the insects probe more or less extensively. The probing also involves injection of saliva, which is particularly relevant in species which transmit bacterial or viral pathogens. As with other Sternorrhyncha and many Fulgoromorpha, psyllids are specialised phloem-feeders and display several adaptations for coping with this unbalanced, nitrogen-poor diet. The anterior and posterior portions of the mid-gut form a loop which permits water to pass directly from the fore- to the hind-gut. The excess water with dissolved sugars is excreted as honeydew. The often copious production of waxy secretions is a result of the hydrocarbon-rich phloem sap. Psyllids possess bacterial endosymbionts which are situated in the mycetome, a specialised organ in the abdomen. Thao et al. (2000, 2001) showed that psyllids and their primary prokaryotic endosymbionts co-speciated.

Unlike in the related aphids, the life cycle of psyllids usually consists of the egg stage, five larval instars and sexually reproducing adults. The ratio of males/females is near to 1:1 (Hodkinson, 2009). Most psyllid species complete their larval development on one or a few closely related plant species. Adult psyllids are always winged and are easily dispersed by wind. They also have large metacoxae, fused to the metathorax, containing strong muscles enabling them to jump, hence their names 'psylla' from Greek 'flea', or 'jumping plant-louse'. Most psyllids are well camouflaged on the substrate they live and remain generally unnoticed. As a result of these properties, they are predisposed to be accidentally transported by humans together with their host plants or accidentally dispersed by wind over large distances. Psyllids are attracted by yellow colours, a fact which is exploited in monitoring pests with yellow water or sticky traps.

Free living psyllid larvae, depending on the species, sit in the leaf or flower buds or along the veins of young leaves. Some species are covered in copious waxy secretions, which provide defence against soiling with honeydew, desiccation, and also protection against predators and parasitoids. An extreme form of this is found in some species of Spondyliaspidinae, Pachypsyllinae or Macrocorsinae which construct lerps, i.e. shields consisting of wax and sugar which often display a very characteristic shape, structure and chemical composition (Moore, 1961; Hollis, 2004). Many psyllids develop in galls (Hodkinson, 1984; Burckhardt, 2005), which are generally induced by the first instar larva. The insertion of the stylets triggers increased cell growth around the larva, which may cover it completely. In *Apsylla cistellata*, the feeding of the first instar larva on leaves induces cell growth in the leaf buds farther away. Later, the second instar migrates into the already formed gall (Raman et al., 2009). Galls can be open, called pit galls, or closed and resemble small nuts, discs or globes (Hodkinson, 1984; Yang & Mitter, 1994). Yang et al. (2000) described a case of inquilinism. Newly hatched first instar larvae of *Pachypsylla cohabitans*, which are unable to induce galls, feed next to larvae of several other gall-inducing *Pachypsylla* species and become incorporated into the growing gall.

Depending on the climatic conditions, psyllid species of temperate regions tend to be univoltine or bivoltine, while those of the tropics are often polyvoltine, with several overlapping generations per year (Burckhardt, 1994; Hollis, 2004; Santana et al., 2010). Hodkinson (2009) analysed the life history characteristics of 342 psyllid species from all over the world and concluded that environmental temperatures and water availability acting on the psyllids, directly or via the host plant, are the major determinants of psyllid life cycles. The phenology of psyllids is, therefore, well synchronised with that of their hosts. Psyllid populations are generally controlled by a whole range of predators and parasitoids (Hodkinson, 1974; Hollis, 2004). Birds and small mammals are known to occasionally eat psyllids. In agricultural systems, anthocorids, syrphids, coccinellids and chrysopids are sometimes capable of controlling psyllid populations, but parasitoids such as encyrtids and eulophids are usually more efficient. Psyllids are also affected by entomopathogenic fungi, with which they may be controlled (Dal Pogetto et al., 2011).

Psyllid host plants are mostly perennial dicotyledonous angiosperms. There are only a few psyllids associated with monocots, such as the holarctic genus *Livia*. Even less species develop on gymnosperms, most notably some triozids in New Zealand and two species of *Ehrendorferiana* from southern Chile (Hodkinson, 2009). Restricted host ranges characterise not only psyllid species but, to a certain extent, also higher taxonomic ranks. Related psyllid species tend to breed on related plant species. Members of the tropical family Homotomidae are all associated with *Ficus* species and other Moraceae (Hollis & Broofield, 1989) and those of its putative sister-group, the Carsidaridae, are associated with members of the Malvales (Hollis, 1987). This pattern suggests that there may be co-speciation between angiosperms and psyllids. Detailed phylogenetic studies, however, show that the observed species richness in psyllids is better explained by geographic vicariance than by co-speciation with the host-plants (Burckhardt & Basset, 2000; Percy, 2003; Burckhardt & Ouvrard, 2007).

Within modern Psylloidea, three probably monophyletic lineages exist (Burckhardt & Ouvrard, 2011), namely: 1. Phacopteronidae; 2. Aphalaridae + (Carsidaridae + Homotomidae); and 3. Liviidae + (Calophyidae + (Psyllidae + Triozidae). Host associations with the plant family Myrtaceae occur in the following taxa: Aphalaridae (Spondyliaspidinae), Liviidae (Diaphorininae) and Triozidae. This suggests that psyllids colonised Myrtaceae at least three times independently. The Spondyliaspidinae comprises 24 genera (Burckhardt, 1991; Hollis, 2004), with species which breed almost exclusively on Myrtaceae and which are restricted to the Australian region. The monotypic Oriental genus *Eurhinocola* is an exception, as it is not represented in the Australian fauna. The genus *Boreioglycaspis* has species in Australia and southeast Asia and is associated with Myrtaceae and the related Lythraceae. *Ctenarytaina* has an even wider distribution, including species native to New Zealand, southeast Asia, China, the Indian subcontinent, and possibly even tropical Africa. It is also wider in its host associations and has some hosts outside the Myrtales.

3. Eucalyptus

Members of the eucalypt group (Myrtaceae) of trees and mallees (shrub forms) represent an old lineage that can be related back to the Late Cretaceous, indicating their Gondwanan origin. There are well in excess of 700 species, with most being confined to Australia, although 15 species naturally occur outside of Australia in southeast Asia, as distant as Sulawesi and the Philippines, and related genera occur in New Caledonia. Australian representatives of the group were originally regarded as two genera, *Eucalyptus* and *Angophora* (Burbidge, 1960). More recently, seven genera within the group have been recognised (Ladiges, 1997), with the main genus being split into *Eucalyptus* L'Hér. *sensu stricto* (+600 species) and the bloodwoods, *Corymbia* Hill & Johnson (+100 species), which are of equal taxonomic rank to *Angophora*. Within the genus *Eucalyptus*, a number of subgenera are recognised (Brooker 2000), namely *Acerosae, Alveolata, Cruciformes, Cuboidea, Eucalyptus, Eudesmia, Idiogenes, Minutifructus, Primitiva*, and *Symphyomyrtus*. The other genera within the eucalypt group which, with one exception, are monotypic, are *Arillastrum* (New Caledonia), *Allosyncarpia* (northern Australia), *Eucalyptopsis* (New Guinea, Moluccan Archipelago, Woodlark Island) and *Stockwellia* (northern Australia).

Considering the antiquity of the 'eucalypts', it is not surprising that invertebrates have co-evolved to feed on or utilize other resources from these trees. Working in the canopy alone, Majer et al. (2000) sampled 641 and 726 invertebrate species from *Eucalyptus moluccana* and *E. crebra* in NSW, eastern Australia and 448 and 444 from *E. marginata* and *Corymbia calophylla* in WA, western Australia. In all instances, the percentage of herbivores in the count ranged from 20.5–25%. The richness of herbivores on 'eucalypts' is accompanied by high degrees of specificity of certain groups to 'eucalypt' species or subgenera. This has been observed in gall-forming eriococcid scale insects (Cook, 2001; Cook & Gullen, 2004), where particular species have been found to be specific to certain species of *Eucalyptus* or *Corymbia*. Another example is the Australian psyllids (Aphalaridae: Spondyliaspidinae), which are renowned for their association with 'eucalypts', with 67% of species utilising 'eucalypts' as host plants (Majer et al., 1997). Working on the 'eucalypt'-feeding genus *Glycaspis*, Moore (1961, 1970) was able to divide the genus into subgenera on the basis of morphology and host plant association. This is but one example of the high degree of host-specificity of Australian psyllids on 'eucalypts'.

Hosts	C. spatulata	C. eucalypti	B. occidentalis	G. brimblecombei
E. benthamii		x		
E. bicostata		x		
E. blakelyi			x	
E. brassiana			x	
E. bridgesiana				x
E. camaldulensis		x		x
E. camphora				x
E. dealbata				x
E. diversicolor				x
E. dunnii		x		
E. forrestiana			x	
E. globulus		x	x	x
E. gomphocephala			x	
E. grandis	x		x	
E. maidenii		x		
E. mannifera				x
E. mannifera maculosa				x
E. microneura			x	
E. microtheca			x	
E. nicholii			x	
E. nicholsii			x	
E. nitens		x		x
E. oleosa			x	
E. pellita	x			
E. pulverulenta		x		
E. robusta	x			
E. rudis			x	
E. saligna			x	
E. sideroxylon				x
E. spathulata			x	
E. tereticornis			x	x
E. urophylla	x		x	

Table 1. Lists the four Australian psyllids that have become established in Brazilian eucalypt plantations, showing their main hosts in Brazil (in bold) and also the other species that they can exploit (sources: Burckhardt et al., 1999; Brennan et al., 2001; Hollis, 2004; Meza & Baldini, 2001). All host species are members of the *Symphyomyrtus* subgenus.

4. Psyllid pests and eucalypts

Due to the close association with their hosts, some psyllids are of economic relevance. While most of these are minor pests (Burckhardt, 1994), a few species are responsible for huge economic losses, such as the species transmitting bacterial phytopathogens (e.g. *Diaphorina citri*, some *Cacopsylla* spp., *Bactericera cockerelli*) (Hodkinson, 2009) or the eucalypt psyllids (Santana & Burckhardt, 2007).

Over 350 species of Psylloidea are reported from Australia (Hollis, 2004), of which 79% are associated with Myrtaceae and 71% with eucalypts. In Australia most populations of eucalypt psyllids are in balance (Collet, 2001a). However, when introduced into other continents, their populations may increase and become a serious problem (Paine & Millar, 2002). This has been the case with *Ctenarytaina eucalypti* (Dahlsten et al., 1998a; Pinzón et al., 2002) and *Glycaspis brimblecombei* (Bouvet et al., 2005). Currently, seven Australian species are known from other continents (*Blastopsylla occidentalis*, *Cardiaspina fiscella*, *Cryptoneossa triangula*, *Ctenarytaina eucalypti*, *C. spatulata*, *Eucalyptolyma maideni*, and *Glycaspis brimblecombei*). An eighth species probably also originates from Australia but has not yet been found there. It is *Ctenarytaina peregrina*, which was described from Eire (Hodkinson, 2007) and has later been found in Germany (K. Schrameyer, pers. comm.).

4.1 *Ctenarytaina spatulata* Taylor, 1997 (rose gum psyllid)

Ctenarytaina spatulata originates from southeast Australia and has been introduced into New Zealand, the USA (California), Uruguay, Brazil, Portugal and Spain (Burckhardt et al., 1999; Santana et al., 1999; Hollis, 2004; Valente et al., 2004). It was observed for the first time in Brazil in Arapoti – Paraná State in 1994, in an *E. grandis* plantation (Burckhardt et al., 1999). It is now commonly found in the states of São Paulo, Paraná, Santa Catarina and Rio Grande do Sul.

The adults of *C. spatulata* (Fig. 1A) are yellowish or orange with dark spots or dark brown stripes. The forewings are transparent and yellowish with slightly darker veins. They remain most of the time on the leaves and new apical shoots, where they feed and reproduce. The females lay their eggs (Fig. 1B) on the newly growing leaf axils (Santana & Zanol, 2006).

All five larval instars have a dorso-ventrally flattened body. They live in colonies and feed on young plants or shoots. They secrete large amounts of honeydew which is secreted as small wax-covered globules together with a large amount of flocculent waxy secretions along the sides of the abdomen, which spread out all over the colony (Santana & Zanol, 2005). Honeydew and waxy secretions harm the development of the young plants, particularly in the first two years of planting (Collet, 2001a). In the first larval instar, the body is entirely yellow, except for the small red eyes. In the final instar larva, the length of the body is 1.40–1.43 mm. The body is brown to yellowish with brown patches (Fig. 1C). The caudal plate bears five lanceolate setae on each side of the anus. The wing buds are well-developed; the forewing buds lack a humeral lobe, and their fore margin lies posterior to the posterior eye margin (Taylor, 1997; Santana & Zanol, 2005).

C. spatulata is associated with several *Eucalyptus* spp., in particular with *E. grandis*, *E. saligna*, *E. robusta*, *E. pellita*, *E. resinifera* and *E. urophylla*, but it has also been observed in lower numbers on *E. deanei, E. saligna, E. tereticornis, E. microcorys, E. viminalis, E. camaldulensis, E. alba* and *E. nitens* (Santana, 2003). Among the species of *Corymbia*, only eggs of *C. spatulata* have been observed on adult leaves of *C. citriodora*. Neither eggs nor larvae of *C. spatulata* have been observed on native Myrtaceae in Brazil (Santana, 2003). Valente et al. (2004) found *C. spatulata* mostly on *E. globulus*, the main eucalypt species planted on the Iberian Peninsula. This eucalypt is also a major host of *C. spatulata* in California (Brennan et al., 2001). These authors also observed large populations of the psyllid on *E. nitens, E. dalrympleana* and *E. maidenii*. Other host species mentioned in the literature are *E. leucoxylon, E. mannifera maculosa, E. pauciflora, E. longifolia, E. rodwayi, E. ovata, E. nitida* (Taylor, 1997), *E. amplifolia, E. dunnii, E. robusta, E. rostrata* and *E. tereticornis* (Burckhardt et al., 1999).

In Brazil, this species has many generations per year, leading to a higher number of individuals during the cold and dry months (Santana & Burckhardt, 2007).

The damage of this pest was observed on *E. grandis* in Paraná state and hybrids of *E. grandis* x *E. urophylla* in São Paulo, with symptoms including sooty mould on the leaves and tips, dieback, loss of apical dominance, super sprouting and decrease of growth (Santana et al., 2005). *C. spatulata* completes its life cycle from egg to adult at temperatures around 20 °C in approximately 45 days (Santana & Zanol, 2006). In São Paulo and Paraná all life stages of the species can be observed during the entire year (Santana, 2003). The first damage by *C. spatulata* to *E. grandis* is caused by oviposition (Santana et al., 2005). A small black spot appears where the egg is inserted, which may increase and lead to withering of the terminal buds. The larvae secrete large amounts of honeydew, which accumulates on leaves and buds, allowing the growth of sooty mould and phytopathogenic fungi. The successive piercing of the substrate and the extraction of plant sap causes deformations and curling of the leaf, thus reducing the leaf surface. High populations weaken the plant further by the extraction of plant sap, causing the death of the terminal buds, the loss of apical dominance and super sprouting of the lateral buds (similar to formation of witches' brooms by excessive growth of lateral buds) (Cadahia, 1980; Zondag, 1982; Meza & Baldini, 2001; Santana et al., 2005).

In addition, the sucking of *C. spatulata* can reduce the increment of stem diameter and of the internodes, resulting in more branches on the stem and rendering the wood of the stem more fragile (Santana et al., 2005). The damage caused by *C. spatulata* in Brazil has been estimated in a green-house (Santana et al., 1999). By studying the nutritional stress, the authors observed that there is an interaction between the insects and Mg deficiency. Both factors together may stop growth and production of biomass, in addition to affecting root development. The research done in the greenhouse and the field shows that *C. spatulata* appeared in all samples, with 100% occurrence of larvae and eggs. Its presence was observed during all months of the year, in every stage, which is typical for a polyvoltine life cycle. It also has a population peak in the colder months with low rainfall (Santana et al., 2005).

Water stress is one of the environmental factors that may improve the development of the psyllid population (White, 1969) due to higher nitrogen concentration in the plants. Santana et al. (2003a) simulated water stress in *E. grandis* plants, with and without the presence of *C. spatulata*, and noticed that the insects may cause a 20% loss of height growth in *E. grandis*. The combination of water stress and presence of the insects may cause considerable damage.

Fig. 1. *Ctenarytaina spatulata* (A - adult, B - eggs, C - larva); *C. eucalypti* (D - adult, E - larva, F – colony of larvae).

4.2 *Ctenarytaina eucalypti* (Maskell, 1890) (blue gum psyllid)

This is a species from southeast Australia that has been introduced into many countries throughout the world. The first mention of *C. eucalypti* from Brazil is by Burckhardt et al. (1999), who reported it from Colombo – Paraná on seedlings of *E. dunnii*. Subsequently, it has also been detected in São Paulo, Santa Catarina and Rio Grande do Sul. Its geographical distribution, apart from its origin in Australia, includes Bolivia, Brazil, Chile, Colombia, Eire, France, Germany, Italy, New Zealand, Papua New Guinea, Portugal, South Africa, Spain, Sri Lanka, Switzerland, UK, Uruguay, and the USA (Burckhardt, 1998; Hodkinson, 1999; Durán & Urrutia, 2001; Burckhardt & Mühlethaler, 2003).

The adults of *C. eucalypti* (Fig. 1D) measure from 1.5–2.0 mm and have dirty whitish forewings with contrasting brown veins and a clear membrane which are normally folded over the body (Burckhardt et al., 1999). The body colour is usually dark brown to black (Burckhardt et al., 1999), with darker transverse stripes on the abdomen, both dorsally and ventrally. The antennae are yellow with dark apices of the individual segments. The compound eyes are dark brown and prominent. The legs are dark yellow. The female proctiger is longer and more slender in comparison with *C. spatulata*, with relatively acute apico-lateral peg setae and strongly curved valvula ventralis (Burckhardt et al., 1999). The female lays 20 to 100 eggs, normally in groups, on the leaf buds and in the axils of young leaves in nurseries or on young trees. Eggs are also laid in small fissures between the bud and the leaf pedicel. Several females may contribute to one egg mass. In summer, the incubation lasts about one week, whereas it is longer in the cold periods (Cadahia, 1980).

The first instar larvae are light yellow with red eyes and thick legs. There are thick setae along the margin of the caudal plate (Zondag 1982). The last instar larvae are yellow with dark patches. The eyes, antennae and wing buds are reddish-brown (Fig. 1E).

C. eucalypti has been observed in Brazil on many *Eucalyptus* spp., such as *E. globulus*, *E. maidenii*, *E. bicostata*, *E. dunnii* and *E. nitens*. Of particular importance are *E. dunnii* and *E. benthamii*, which are planted in temperate region on the south of Brazil where a high psyllid attack has been observed (Santana et al., 1999).

Hodkinson (1999) described the damage of *C. eucalypti*, including direct effects of psyllid feeding such as severe shoot dieback, leaf curl and leaf discoloration. A further factor damaging the plants is sooty mould growing on the honeydew, which is secreted in large amounts.

In Brazil, the development is continuous throughout the year, with several overlapping generations. All developmental stages have been observed to occur together at any particular time in the same population. All five larval instars excrete large amounts of honeydew and white waxy secretions (Fig. 1F).

4.3 *Blastopsylla occidentalis* Taylor, 1985 (eucalypt shoot psyllid)

Blastopsylla occidentalis originates from South and West Australia. Its geographical distribution is Argentina, Brazil, Cameroon, Chile, Hong Kong, Kenya, Mexico, New Zealand, Paraguay, South Africa and USA (Hodkinson, 1991; Burckhardt et al., 1999; Burckhardt & Elgueta, 2000; Hollis, 2004; Bouvet et al., 2005; anonymous, 2007; Tamesse et al., 2011). According to the original description by Li (2011), *Blasopsylla barbara* from China may be a synonym of *B. occidentalis*.

The adults are small insects, measuring 1.5–2.0 mm. They have a yellow head and thorax with dark pattern, the head is as large as the thorax and strongly inclined relative to the longitudinal body axis, with short antennae (Fig. 2A). The forewings have brown veins and a grey membrane. The male terminalia are yellow, while those of the females are dark brown with a yellow base of the subgenital plate. The males are usually yellow, the females are darker coloured. The last instar larvae are yellow with a dark brown antennal tip; lacking humeral lobe on forewing buds or specialized setae; with 9-segmented antennae (Burckhardt & Elgueta, 2000; Meza & Baldini, 2001; Durán & Urrutia, 2001). *Blastopsylla* differs from other spondyliaspidines by the very long posterior lobes on the basal segment of the male proctiger and often in the presence of a single spur on the metabasitarsus (Burckhardt & Elgueta 2000; Hollis, 2004). *B. occidentalis* can be easily separated from *Ctenarytaina* species by the lack of an outer apical comb of bristles on the mesotibia, and the apically curved vein Rs in the forewings (Taylor, 1990; Burckhardt et al., 1999; Burckhardt & Elgueta, 2000). Within *Blastopsylla*, *B. occidentalis* is similar to *B. moorei* from which it differs in the shape of the apical portion of the aedeagus, which is spherical rather than elongate and weakly curved as in *B. moorei*, and the fewer black setae on the apical portion on the inner surface of the paramere and the 4–6 dark sclerotised setae which form a line along hind margin, rather than a group of 5–7 as in *B. moorei* (Taylor, 1985). The females of *B. occidentalis* lay the eggs on the shoots, the leaf axils, on small branches and young leaves.

The larvae secrete large amounts of small wax covered globules containing honeydew and copious white flocculence (Fig. 2B), which enhances growth of sooty mould (Taylor, 1985; Meza & Baldini, 2001). The wax often sticks to the end of the larval abdomen.

B. occidentalis was observed for the first time in Brazil in Goiás in 1997 on hybrids of *Eucalyptus urophylla* and *E. grandis* (Burckhardt et al., 1999). Other hosts are *E. microtheca*, *E. rudis*, *E. gomphocephala*, *E. camaldulensis*, *E. microneura*, *E. nicholsii*, *E. spathulata*, *E. forrestiana*, *E. oleosa*, *E. rudis*, *E. tereticornis*, *E. saligna*, *E. globulus* and *E. nicholii* (Meza & Baldini 2001; Hollis, 2004).

Contrary to the two *Ctenarytaina* spp., *B. occidentalis* occurs in central Brazil, where the dry season is longer than in southern Brazil, with over four months without rain.

4.4 *Glycaspis brimblecombei* Moore, 1964 (red gum lerp psyllid)

Glycaspis brimblecombei (Fig. 2C), the red gum lerp psyllid, has an Australian origin and was introduced into the USA in June 1998 (Brennan et al., 1999). In 2000 it was observed in Mexico, in Chile in 2002 (Dahlsten, 2003), in Brazil in 2003, in Argentina in 2004 (Bouvet et al., 2005), in Ecuador in 2007 (Onore & Gara, 2007), in 2007 in Venezuela (Rosales et al., 2008), in Peru in 2008 (Burckhardt et al., 2008), in 2008 on the Iberian Peninsula (Hurtado & Reina, 2008; Valente & Hodkinson, 2009; Prieto-Lillo et al. in 2009) and in Italy in 2010 (Laudonia & Garonna, 2010; Peris-Felipo et al., 2011). In Brazil, *G. brimblecombei* was first detected in São Paulo and now it is wide-spread in the whole country in areas where its host is planted.

G. brimblecombei develops on various *Eucalyptus* spp., especially *E. camaldulensis* and *E. tereticornis* but also on *E. blakelyi*, *E. brassiana*, *E. bridgesiana*, *E. camphora*, *E. dealbata*, *E. diversicolor*, *E. globulus*, *E. mannifera*, *E. mannifera maculosa*, *E. nitens* and *E. sideroxylon* (Brennan et al., 2001; Hollis, 2004).

The eggs are laid in groups, usually in rows or circles (Fig. 2D). The larvae of *G. brimblecombei* (Fig. 2E) differ from those of the other three species discussed above, as they are not free-living. They live under a shield of wax, called lerp (Fig. 2F). Adults can easily be differentiated from *Blastopsylla* and *Ctenarytaina* spp. by the long genal processes, which are much longer than the vertex along the mid-line (Olivares et al., 2004; Burckhardt et al., 2008).

The damage of this species is similar to that made by the other species discussed above, but *G. brimblecombei* is a more aggressive exploiter of resources. There are reports of outbreaks in many South American countries were the species caused the death of the trees, resulting in serious production losses.

4.5 *Cardiaspina fiscella* Taylor, 1962 (brown lace lerp or brown basket lerp psyllid)

Cardiaspina ficella is native to Australia, where it occurs naturally in the Australian Capital Territory, New South Wales and Victoria (Hollis, 2004). It uses as hosts *Eucalyptus botryoides*, *E. saligna*, *E. robusta*, *E. grandis* (Collet, 2001a), *E. camaldulensis*, *E. blakelyi* (Collet, 2001b), *E. globulus* and *E. tereticornis*. According to Withers (2001), 57 insect species originating from Australia have been introduced in New Zealand, among them, *C. fiscella*. This species was detected in New Zealand in 1996 and it quickly diffused to most of the North Island, with severe infestations, successive defoliation, dieback and general decline of plantings of *Eucalyptus botryoides* and *E. saligna*. According to Berry (2006), *C. fiscella* is the only introduced species of Australia that has become a significant pest on *E. botryoides*, *E. grandis* and *E. saligna* in New Zealand.

There are many species of *Cardiaspina* using eucalyptus as a host, among which *C. fiscella* is one which most often reaches high populations (Campbell, 1992). Adults of this Australian psyllid species are 3.4 mm in length, from head to wing tip. They have a short, more robust body than the other species mentioned previously. Antennae are almost equal to width to the head. Legs are stout with two very small claws. The general colour is light brown; the head straw coloured; the thorax bears several medium brown patches; the abdomen is brownish black, with a yellow or red caudal margin on each segment. The forewings are transparent with uniformly light brown veins (Fig. 3A).

The larvae (Fig. 3B) build very elaborate lerps, which are shell-shaped (Fig. 3C), consisting of a tangled lattice, usually on the dorsal surface of the leaves (Hollis, 2004). Dimensions are approximately 4.1 mm from hinge to apex, and approximately 5.4 mm across. General colour is light brown, darker near the hinge with a darker band where the ribs begin to fan out. It is moderately convex. Larvae can move freely underneath the lerp and can sometimes be seen moving on the undersides of leaves outside of the lerp (Appleton, 2009). The larvae of *C. fiscella* have glands that secrete wax filaments to produce the lerp. Larvae of all *Cardiaspina* spp. form fan-shaped basket lerps, with unique characteristics for each species. Often, necrosis on foliage of mature leaves occurs beneath the lerp (Fig. 3D, E), as can be found on many eucalypt species, especially of *E. camaldulensis* and *E. blakelyi* (Collet, 2001a).

Fig. 2. *Blastopsylla occidentalis* (A - adult, B – colony of larva); *Glycaspis brimblecombei* (C - adult, D - eggs, E - larvae, F - lerps on leaf).

Fig. 3. *Cardiaspina fiscella* (A - adult, B- larva, C - lerp, D, E - damage on leaves); *Eucalyptolyma maideni* (F - adult, G - lerp on leaf).

4.6 *Cryptoneossa triangula* Taylor, 1990 (lemon gum psyllid)

Cryptoneossa triangula was described from *Corymbia citriodora* and *C. maculata* trees collected in the Australian Capital Territory and New South Wales (Taylor, 1990). This species is morphologically similar to *C. minuta and C. leptospermi* and derives its name from the triangular shape of the parameres. Adults have a general ochraceous to brown colour and the vertex has a narrow black margin in some specimens. The forewings are relatively longer and narrower than those of *C. minuta* and the marginal cell cu$_1$ is not as flat as in *C. leptospermi*. Males of *C. triangula* are 1.8 to 2.2 mm length. The head bears spheroid genal processes and anterior margins of vertex expanded on each side of median ocellus; anteriorbital lobes small, conical. The basitarsus has two black spurs. Females are 2.3 to 2.6 mm long with rounded anterorbital lobes (Taylor, 1990).

In August 2000, this psyllid was discovered on California's lemon-scented gum (*C. citriodora*) and spotted gum (*C. maculata*) trees at Disney Resort and surrounding areas of Anaheim, California. The lemon gum psyllid (*C. triangula*) is a free-living psyllid causing leaf damage and inducing leaf drop, which may stress trees and make them susceptible to fatal attack by other insects. The psyllids also produce sticky honeydew, which soils the ground, cars and sidewalks (Dahlsten, 2001; Daane & Paine, 2005).

4.7 *Eucalyptolyma maideni* Froggatt, 1901 (spotted gum lerp psyllid)

Eucalyptolyma maideni was described from *Eucalyptus* spp. and later redescribed by Taylor (1987) from *C. citriodora, C. gummifera* and *C. maculata,* with records in New South Wales, Australian Capital Territory and South Australia. The adults have general bright green colour tinged with yellow (Fig. 3F). The lerp is white, corniculate, with laciniate margins (Taylor, 1987). According to Morgan (1984), this psyllid is widespread in eucalypt forests from Queensland and South Australia to Tasmania. It is trivoltine and prefers fully mature leaves upon which to feed, oviposit and develop.

This psyllid was introduced in USA in 2000 and collected from *C. citriodora* in Los Angeles, California. Since then, it was also discovered near Anaheim (at Disney Resort) and heavy infestations have been reported from many locations within Los Angeles, Orange County and San Diego. Although this insect does not cause the death of the plants, it is considered to be an ornamental pest of lemon scented gum and spotted gum in California. Larvae build a flattened and elongated triangular lerp (Fig. 3G) and produce a copious amount of honeydew which soils the ground beneath trees, similar to that mentioned for *C. triangula,* where a blackish sooty mould grows (Garrison, 2001).

5. Integrated Pest Management (IPM)

The integrated management of the psyllids should start with monitoring, which should be continuous, with more attention being paid in peak periods. The monitoring of the psyllids can be done by installing sticky traps or by manual sampling. These should be run continuously and at regular intervals. Yellow sticky traps are the best way to monitor psyllids infesting eucalypts. Adult psyllids and psyllid parasites are attracted by the yellow colour and become stuck to the surface. The traps should be inspected once a week and the number of adult psyllids and their parasites should be counted and recorded (Paine et al., 2007).

Adult psyllids can also be monitored by shaking or beating plants over a collecting sheet to knock them onto the collecting surface, where they can be easily seen and counted. The sample should be taken about once a week during the season of new plant growth, when adult psyllids are expected (Dreistadt & Dahlsten, 2007).

Through such monitoring, it is possible to determine the moment of population peak, the occurrence of natural enemies and other factors that affect the insect population. In an Integrated Pest Management Programme (IPM), monitoring the pest's development is one of the main components, which uses different techniques to quantify the population and predict outbreaks of the pest. Besides, it can be also used to determine the geographic distribution of the pest, to detect a risk to the area before any damage actually occurs and to determine the effectiveness of the control treatment.

5.1 Biological control

The use of parasitism is a major component of psyllid IPM, with several biological control programme having been successfully carried out for different species of introduced psyllid. A biological control programme of *C. eucalypti* was supported by the University of California IPM project. The parasitoid *Psyllaephagus pilosus* (Hymenoptera: Encyrtidae) was collected in Australia, reared at UC Berkeley and released in California. The parasitoids quickly established in the release sites, and sampling in 1994 revealed parasitism rates of 50–100%. By 1995, the parasitoids had become broadly distributed throughout many parts of the State. An economic analysis of the benefit to the cut-foliage industry alone indicated that the biological control programme generated a benefit-cost ratio ranging from a minimum of 9:1 to a maximum of 24:1, based solely on elimination of pesticide treatments (Dahlsten et al., 1998a). A similar IPM programme was adopted to manage the populations of the lerp psyllid *G. brimblecombei* in California (Paine et al., 2007).

In Brazil, in the end of 2000, beginning of 2001, a sudden population decrease was observed, because of the hymenopterous parasitoids which attacked *C. eucalypti*. The parasitoid was identified as *Psyllaephagus* sp. (Hymenoptera: Encyrtidae), the same genus which has been introduced into many European and American countries for the biological control of *C. eucalypti* (Santana et al., 2002). The parasitoids of *C. eucalypti* (*P. pilosus*) and *G. brimblecombei* (*P. bliteus*) were introduced accidentally together with their hosts in Brazil (Santana et al., 2002; Berti Filho et al., 2003). These parasitoids adapted and dispersed over the whole Brazilian territory. Now they can be found in practically all the areas where the pest is present. The natural control of these psyllids has been successfully facilitated using these specific parasitoids (Santana & Burckhardt, 2007).

The decrease of the *G. brimblecombei* population observed in São Paulo, Brazil in the end of spring, beginning of summer in 2003, was attributed to the rainfall and certainly because of the presence of parasitoids and predators (Fig. 4). The parasitoid *P. bliteus* (Fig. 4 A, B) was detected in the same year of introduction of *G. brimblecombei* in São Paulo State, Paraná and Minas Gerais. With the detection of this parasitoid, it is expected that the population of the psyllid will also stabilize (Santana et al., 2003b). Now, the management of the pest and its natural enemies is established.

Fig. 4. Psyllid IPM. A, B - Parasitoid *P. bliteus*; C-F – predators, G - clone plantation: *G. brimblecombei* resistant (left) susceptible (right); H – detail of infestation on susceptible clone.

Withers (2000) reported the occurrence of *Psyllaephagus gemitus* in *C. fiscella*. According to this author, the health of eucalypt trees in the North Island of New Zealand can be expected to improve as populations of *C. fiscella* continue to be killed by *P. gemitus*, a nymphal parasitoid. This same parasitoid, *P. gemitus*, was tested in 1997 as a possible biological control agent for *C. fiscella* by entomologists funded by the New Zealand Farm Forestry Association. A parasitic microwasp associated with brown lace lerp (*C. fiscella*) has been recorded from New Zealand. This microwasp, *Coccidoctonus psyllae*, is a hyperparasitoid and its presence will probably contribute to an increase in brown lace lerp populations (Berry, 2006).

The eucalypt psyllids are also attacked by many predators, including the ladybird beetles (*Cycloneda sanguinea* , *Olla v-nigrum* (Fig. 4C), *Hippodamia convergens, Eriopsis connexa* and *Harmonia axyridis* (Fig. 4D), the green lacewings (*Chrysoperla* spp.) (Fig. 4E), syrphid flies, pirate bugs (*Anthocoris* sp.) (Fig. 4F) and spiders (Santana, 2003; Santana et al., 2004). Although predators do not provide complete biological control, they can reduce psyllid abundance. Whenever possible, management efforts should be selected that have less adverse effects on these beneficial species (Paine et al., 2007). Santana (2003) observed the following potential natural enemies for *C. spatulata*: *Coccinella ocelligera, Curinus coeruleus, Cycloneda pulchella, Cycloneda sanguinea, Eriopsis connexa, Harmonia axyridis, Hyppodamia convergens, Hyperaspis* sp., *Scymnus (Pullus)* sp. and *Olla v-nigrum* (Coleoptera: Coccinellidae), *Chrysoperla externa* (Neuroptera: Chrysopidae); *Allograpta exotica, Pseudodorus clavatus, Syrphus phaeostigma* (Diptera: Syrphidae), as well as spiders and the fungus *Verticilium lecanii*. The same predators observed for *C. spatulata* also attack *B. occidentalis*, including Syrphidae, Dolichopodidae (both Diptera), Chrysopidae (Neuroptera) and Coccinellidae (Coleoptera) (Santana & Burckhardt; 2007).

Azevedo and Figo (1979) list the following natural enemies of *C. eucalypti* in Portugal: *Syrphocotonus abdominator* (Hymenoptera: Ichneumonidae); *Haematopota ocelligera* (Diptera: Tabanidae); *Sphaerophoria scripta, Meliscaeva cinctellus, Pipizella* sp., *Eumerus* sp. (Diptera: Syrphidae); and *Bradysia* sp. (Diptera: Sciaridae). The predator complex feeding on *C. eucalypti* includes hoverfly larvae, ladybirds, lacewings, anthocorids and spiders (Hodkinson, 1999). According to Zondag (1982), *C. eucalypti* is frequently attacked by a small black wasp in New Zealand. In Tasmania, a small ladybird (*Cleobora mellyi*), used to control a beetle, has been observed feeding on larvae and eggs of psyllids, showing a high potential to control *C. eucalypti*.

5.2 Plant resistance

The search for tolerant or resistant plants has been an economically viable alternative to contain populations of these insects in large plantations. Tolerance varies among eucalypt species and location of plants. Prevention of planting susceptible species is the best way to avoid damage (Dreistadt et al., 2007). According to White (1970), who studied aspects of the life history of *Cardiaspina densitexta*, the physiological and physical characteristics of plants influence the selection of hosts.

Brennan and Weinbaum (2001a) studied the performance of psyllid adults on leaves of *E. globulus* and noted that the epicuticular wax of juvenile leaves plays an important role in resistance to *C. spatulata* and *G. brimblecombei*; these species tend to avoid more waxy leaves. Brennan and Weinbaum (2001b) suggested that the tarsi of *C. eucalypti* are more adapted for adhering to the epicuticular wax-coated surfaces than those of the others two psyllid species. Continuing these studies, Brennan and Weinbaum (2001c) concluded that the epicuticular wax on juvenile *E. globulus* leaves reduces stylet probing by *C. spatulata* and *G. brimblecombei* and this psyllid avoids oil glands on the leaves.

In breeding programmes for improving physical and chemical properties of plants, all aspects related to resistance should be analysed. In Brazil, where most eucalypt plantations are clonal, the forestry companies already have some materials selected in the search for resistance (Fig. 4G, H).

In order to verify the preference of *C. spatulata* for laying eggs and feeding, Santana et al. (2010) carried out a green-house trial on 19 *Eucalyptus* species, one *Eucalyptus* hybrid (Cambiju), three *Corymbia* species and four native Myrtaceae species (*Hexaclamys edulis*, *Plinia edulis*, *Plinia trunciflora* and *Psidium* sp.). As a result of this trial, they found that the largest populations of *C. spatulata* were observed on *E. robusta* and *E. pellita*. *E. grandis* and *E. resinifera*, however, presented the largest number of plants with symptoms of damage. *E. cinerea*, *E. cloeziana*, *E. dunnii*, *E. benthamii*, *E. nitens*, *E. viminalis*, *E. pilularis*, *E. camaldulensis* and *E. dunnii* did not suffer infestations of *C. spatulata*. Among the *Corymbia* species, eggs of *C. spatulata* were only observed on *C. citriodora* with adult leaves. None of the native Myrtaceae had eggs or larvae of *C. spatulata*.

Camargo et al. (2009) evaluated the resistance of different clones of *E. camaldulensis* to attack by *G. brimblecombei* and observed that the largest mean for eggs and larvae were observed on commercial clones 7, 58, 62, 10 and 6. The same ones did not differ statistically, all being considered as highly susceptible to the attack of the red gum lerp psyllid. The clones GG100, 36, 2, I042 and I224 did not differ statistically to each other and they were classified as resistant to psyllid attacks. Clone 19 presented an intermediate average among the two groups, being classified as susceptible. Clone 58 is known as one of the most productive in the Cerrado region of Minas Gerais but, because of its high susceptibility to red gum lerp psyllids, companies are being discouraged from planting it in areas of greatest occurrence of this pest.

5.3 Cultural control

According to White (1986), the decline of some eucalypt forests in Australia is primarily caused by changes in rainfall patterns, which induce physiological stress in plants. In physiologically stressed plants, the amount of nitrogen available as a food source for psyllids is higher. The eucalypts are damaged more by the psyllids as the increase in nitrogen content increases the chance of survival of the psyllids and their populations grow rapidly (White, 1969). The requirements of insects for mineral salts is not well defined but it is known that they are very important for the ionic balance and cell membrane permeability of insects, acting as activators of enzymes (Panizzi & Parra, 1991). Thus, different types of fertilizers can affect insect populations positively or negatively. Some nutrients are very important in plant–insect relationships and have been thoroughly researched, such as nitrogen (White, 1969), magnesium (Santana et al., 1999), silicon (Camargo et al., 2011) and others.

To minimize plant stress, certain cultural practices have been recommended as a measure to strengthen the plant and provide higher resistance to psyllids. As excess nitrogen in the leaves leads to an increase in insect populations, a balanced fertilisation and irrigation in the dry season is recommended to avoid the concentration of this nutrient in the leaves (Garrison, 2001).

Although silicon is not considered an essential element for plants, it is absorbed and is involved in the formation of structures of defences such as trichomes and spines. It also contributes to greater leaf toughness by forming polymers (crystals) that are immobilised in the leaf tissue (Camargo et al., 2011). Thus, the application of industrial ashes rich in silicates has been recommended for commercial planting of eucalypts in regions poor in this element.

5.4 Chemical control

Adult psyllids should be monitored before damage becomes evident and the numbers of adults present should be recorded on a weekly basis. During subsequent seasons control action should be taken, if necessary, when populations or damage approach the levels that were previously identified to be intolerable. The foliar damage is primarily caused by larvae, but sprays are generally aimed at killing eggs or newly hatched larvae before the damage occurs, which is why the adults should be monitored. Therefore, a decision to spray should be based on the numbers of adults infesting the plants several weeks before larval damage becomes intolerable (Dreistadt et al., 2007).

The overlapping generations of the psyllids makes chemical control even harder because it requires successive spraying of insecticides. This increases production costs and requires additional work, which demmands the definition of a management strategy to be more critical. Usually the control is made through IPM, mainly biological control using predators or parasitoids. In most case the chemical control is not recommended because it is expensive, less efficient and may cause environmental damage (Santana & Burckhardt, 2007).

There are, however, instances when chemical control is necessary. When *C. eucalypti* was introduced into the USA, it caused up to 30% of production loss in commercial plantations of *E. pulverulenta* in California (Dahlsten et al., 1998b), making it necessary to apply chemical control. Nevertheless, considering the fast dispersal and establishment of this psyllid, eradication methods are not effective and chemical control is expensive and efficient for a short time. Its ready adaptation to tropical climatic conditions, its fast dispersal and the large areas planted with eucalypts suggest that this psyllid should be controlled by a programme of Integrated Pest Management, based on the management of the pest and its interaction with the ambient environment and other organisms.

6. Modelling Psylloidea dispersion

Invasive species can be introduced accidentally by humans or by natural dispersal. To provide the right conditions for the occurrence of a particular species, it is necessary to understand the factors that circumscribe their niche, such as abiotic conditions that define the physiological limits to the persistence of species, biological factors that influence the survival of populations (which may be negative in the case of competition, predation and parasitism, or positive, in the case of mutualism), and dispersal ability, which reflects what sites are accessible to individuals of a species (Soberón & Peterson, 2005).

Far from home and their environment and free of predatory and competitive processes, invasive alien species have favorable conditions for expansion and occupation, especially if its space or ecosystem has been altered by successive processes of human interventions. Possible areas of risk of introduction, spread and future distribution of invasive species are often estimated by bioclimatic modelling, also known as ecological niche modelling.

The process of modelling the potential geographic distribution of biological species can be summarized in the following steps: 1) a set of points of occurrence (georeferenced) is combined with a set of environmental variables, creating a niche group of points. Each point in this niche is formed by the values assumed by the environmental variables at each point of occurrence; (2) a modelling algorithm is used to create a niche model from the set of points niche; and (3) the model of niche created by the algorithm is applied on a certain geographic region, taking into account the same environmental variables used to create this niche model (Rodrigues, et al. 2010). Work is currently underway to model the potential distribution of introduced psyllids in Brazil so that plantations can be planned and managed in order to minimize the threat from these pests.

7. Conclusions

Jumping plant-lice are a major threat to large scale commercial eucalypt plantations. Of the approximately 350 described Australian species, over 250 are associated with eucalypts (Hollis, 2004). Of these, only seven species have been introduced into other continents and four have become economically important pests. However, there is large potential for other Australian spondyliaspidine species to be accidentally or intentionally exported into other continents. Improving the existing taxonomic base would help in the rapid recognition of new spodyliaspidine introductions. Incidentally, two of the four major psyllid pests on eucalypts have been described, in part, on the basis of non-Australian material (*Blastopsylla occidentalis* and *Ctenarytaina spatulata*). *Ctenarytaina peregrina*, which is almost certainly of Australian origin, is currently known only from Europe. A detailed taxonomic revision of *Ctenarytaina* is particularly important if we are to understand the threat from members of this genus.

Of the eucalypt species observed as hosts of *C. fiscella*, *E. grandis* and its hybrids are particularly important for Brazilian national forestry. Besides these, other species such as *E. camaldulensis* and *E. urophylla* are also susceptible and they are planted over large areas, or are the base for the production of clones. These species are already attacked by the red gum lerp psyllid (*G. brimblecombei*) and the introduction of one more pest can make the use of some genetic materials of great productivity unviable. For all these reasons, the acquisition of good taxonomic knowledge, and the amassing of control tools for incorporation into IPM programme are essential if Brazilian forestry is to be protected from this important group of insects.

8. References

ABRAF (2011). Anuário estatístico da ABRAF 2011 ano base 2010/ ABRAF. Brasília,130 pp.
Anonymous (2007). *Blastopsylla occidentalis*: another new *Eucalyptus* pest in South Africa. *Plant Protection News*, vol. 72, p. 2

Appleton, C. (2009). NZ FRI http://www.nzffa.org.nz/images/design/Pests/Cardiaspina-fiscella/Cardiaspina-fiscellaFHNews64.html, Access in November/30, 2009

Azevedo, F. & Figo, M.L. (1979). *Ctenarytaina eucalypti* Mask. (Homoptera, Psyllidae). *Boletino Servicio de Plagas Forestales*, vol. 5, pp. 41-46

Bekker-Migdiscova, E.E. (1973). Systematics of the Psyllomorpha and the position of the group within the order Homoptera. In: Narchuk, E.P. (ed.), Doklady na Dvadsat Chetvertom Ezheghodnom Chtenii Pamyati N.A. Kholodkovskogo, 1-2 aprelya 1971, p. 90-117. (British Lending Library Translation RTS 8526)

Berry, J. (2006). Brown lace lerp hyperparasitoid found in New Zealand. *Biosecurity*, No. 68, pp. 18-19

Berti Filho, E.; Costa, V.A.& Zuparko, R.L. & LaSalle, J. (2003). Ocorrência de *Psyllaephagus bliteus/quadricyclus* Riek (Hymenoptera: Encyrtidae) no Brasil. *Revista de Agricultura*, vol. 78, pp. 304-304

Bouvet, J.P.R.; Harrand, L. & Burckhardt, D. (2005). Primera cita de *Blastopsylla occidentalis* y *Glycaspis brimblecombei* (Hemiptera: Psyllidae) para la República Argentina. *Revista da la Sociedad Entomológica Argentina*, vol. 64, pp. 99-102

Brennan, E.B.; Gill, R.J.; Hrusa, G.F. & Weinbaum, S.A. (1999). First record of *Glycaspis brimblecombei* (Moore) (Homoptera: Psyllidae) in North America: Initial observations and predator associations of a potentially serious new pest of *Eucalyptus* in California. *Pan-Pacific Entomol*, vol. 75, pp. 55-57

Brennan, E.B.; Hrusa, G.F.; Weinbaum, S.A. & Levison, Jr. W. (2001). Resistance of *Eucalyptus* species to *Glycaspis brimblecombei* (Homoptera: Psyllidae) in the San Francisco by area. *Pan-Pacific Entomologist*, vol. 77, pp. 249-253

Brennan, E.B. & Weinbaum, S.A. (2001a). Performance of adult psyllids in no-choice experiments on juvenile and adult leaves of *Eucalyptus globulus*. *Entomologia Experimentalis et Applicata*, vol. 100, pp. 179–185

Brennan, E.B. & Weinbaum, S.A. (2001b). Effect of epicuticular wax on adhesion of psyllids to glaucous juvenile and glossy adult leaves of *Eucalyptus globulus* Labilliardière. *Australian Journal of Entomology*, vol. 40, pp. 270–277

Brennan, E.B. & Weinbaum, S.A. (2001c). Stylet penetration and survival of three psyllid species on adult leaves and 'waxy' and 'de-waxed' juvenile leaves of *Eucalyptus globulus*. *Entomologia Experimentalis et Applicata*, vol. 100, pp. 355–363

Brooker, M.I.H. (2000). A new classification of the genus *Eucalyptus* L'Hér. (Myrtaceae). *Australian Systematic Botany*, vol. 13, pp. 79-148

Burbidge, N. (1960). The phytogeography of Australia. *Australian Journal of Botany*, vol. 8, 75-209

Burckhardt, D. (1991). *Boreioglycaspis* and Spondyliaspidine classification (Homoptera: Psylloidea). *Raffles Bulletin of Zoology*, vol. 39, pp. 15-52

Burckhardt, D. (1994). Psylloid pests of temperate and subtropical crop and ornamental plants (Hemiptera, Psylloidea): a review. *Trends in Agricultural Sciences, Entomology*, vol. 2, pp. 173-186

Burckhardt, D. (1998). *Ctenarytaina eucalypti* (Maskell) (Hemiptera, Psylloidea) neu für Mitteleuropa mit Bemerkungen zur Blattflohfauna von *Eucalyptus*. *Mitteilungen der Entomologischen Gesellschaft Basel*, vol. 48, pp. 59-67

Burckhardt, D. (2005). Biology, ecology and evolution of gall-inducing psyllids (Hemiptera: Psylloidea), 143-157. In Raman, A.; Schaefer, C.A. & Withers, T.M. (eds) Biology, ecology and evolution of gall-inducing arthropods. Science Publishers, Inc., Enfield (NH), USA

Burckhardt, D. & Basset, Y. (2000). The jumping plant-lice (Hemiptera, Psylloidea) associated with *Schinus* (Anacardiaceae): systematics, biogeography and host plant relationships. *Journal of Natural History*, vol. 34, pp. 57-155

Burckhardt, D. & Elgueta, M. (2000). *Blastopsylla occidentalis* Taylor (Hemiptera: Psyllidae), a new introduced eucalypt pest in Chile. *Revista Chilena de Entomología*, vol. 26, pp. 57-61

Burckhardt, D.; Lozada, P.W. & Diaz, B.W. (2008). First record of the red gum lerp psyllid *Glycaspis brimblecombei* (Hemiptera: Psylloidea) from Peru. *Mitteilungen der Schweizerischen Entomologischen Gesellschaft*, vol. 81, pp. 83-85

Burckhardt, D. & Mühlethaler, R. (2003). Exotische Elemente der Schweizer Blattflohfauna (Hemiptera, Psyllidae) mit einer Liste weiterer potentieller Arten. *Mitteilungen der Entomologischen Gesellschaft Basel*, vol. 53, pp. 98-110

Burckhardt, D. & Ouvrard, D. (2007). The taxonomy, biogeography and host plant relationships of jumping plant-lice (Hemiptera: Psyllidae) associated with creosote bushes (*Larrea* spp., Zygophyllaceae). *Systematic Entomology*, vol. 32, pp. 136-155

Burckhardt, D. & Ouvrard, D. (2011). A revised classification of the jumping plant-lice (Hemiptera: Psylloidea). *Zootaxa*, in preparation

Burckhardt, D.; Santana, D.L.Q.; Terra, A.L.; Andrade, F.M.; Penteado, S.R.C.; Iede, E.T. & Morey, C.S. (1999). Psyllid pests (Hemiptera, Psylloidea): in South American eucalypt plantations. *Mitteilungen der Schweizerischen Entomologischen Gesellschaft*, vol. 72, pp. 1-10

Cadahia, D. (1980). Proximidad de dos nuevos enemigos de los *Eucalyptus* en España. *Boletino Servicio de Plagas Forestales*, vol. 6, pp. 165-192

Camargo, J.M.M.; Moraes, J.C.; Zanol, K.R.M. & Queiroz, D. L. (2011). Interação silício e insetos-praga: Defesa mecânica ou química? *Revista de Agricultura (Piracicaba)*, vol. 85, pp. 10-12

Camargo, J.M.M.; Santana, D.L.Q.; Dedecek, R.; Zanol, K.R.M. & Melido, R.C.N. (2009). Avaliação da resistência de clones de *Eucalyptus camaldulensis* Dehn. Ao psilideo de concha *Glycaspis brimblecombei* Moore (Hemiptera: Psyllidae). In: 1° Congresso Brasileiro sobre florestas energéticas, 2009, Belo Horizonte

Campbell, K.G. (1992). The biology and population ecology of two species of *Cardiaspina* (Hemiptera: Psyllidae) in plague numbers on *Eucalyptus grandis* in New South Wales. *Proceedings of the Linnean Society of New South Wales*, vol. 113, pp.135-150

Collet, N. (2001a). Biology and control of psyllids, and the possible causes for defoliation of *Eucalyptus camaldulensis* Dehnh. (river red gum) in south-eastern Australia – a review. *Australian Forestry*, vol. 64, pp. 88-95

Collett, N. (2001b). Insect pests of young eucalypt plantations. Forest Science Centre, Heidelberg. Note Number: AG0799. December, 2001

Cook, L.G. (2001). Extensive chromosomal variation associated with taxon divergence and host specificity in the gall-inducing scale insect *Apiomorpha munita* (Schrader) (Hemiptera: Sternorrhyncha: Coccoidea: Eriococcidae). *Biological Journal of the Linnean Society*, vol. 72, 265-278

Cook, L.G. & Gullen, P.J. (2004). The gall-inducing habit has evolved multiple times among the eriococcid scale insects (Sternorrhyncha: Coccoidea: Eriococcidae). *Biological Journal of the Linnean Society*, vol. 83, 441–452

Daane, K. & Paine, T.D. (2005). Biological Control of psyllids on lemon-scented and spotted gum in California. Latest news - updated graphs of pysllid populations through 2004. University of California. Accessed in:
http://nature.berkeley.edu/biocon/dahlsten/lemon_gum/

Dahlsten, D.L. (2001, modified 2002). Berkeley University. Biological control of psyllids on lemon-scented and spotted gum in California by).
http://www.cnr.berkeley.edu/biocon/dahlsten/lemon_gum/
Los Angeles County Agricultural Commissioner's Office New Agricultural Pest for Southern California - Spotted Gum Lerp Psyllid, *Eucalyptolyma maideni*.
http://acwm.co.la.ca.us/pdf/SpoGum.pdf

Dahlsten, D.L. (2003). Biological control of the red gum lerp psyllid, aspects of eucalyptus species in California. Center of biological control, University of California. 5 pp.
http://www.cnr.berkeley.edu/biocon/dahlsten/rglp/ Accessed in 02/Aug/2005.

Dahlsten, D.L.; Hansen, E.P.; Zuparko, R.L. & Norgaard, R.B. (1998a). Biological control of the blue gum psyllid proves economically beneficial. *California Agriculture*, vol. 52, pp. 35-40

Dahlsten, D.L.; Rowney, D.L.; Copper, W.A.; Tassan, R.L.; Chaney, W.E.; Robb, K.L.; Tjosvold, S.; Bianchi, M. & Lane, P. (1998b). Parasitoid wasp controls blue gum psyllid. *California Agriculture*, vol. 52, pp. 31-34

Dal Pogetto, M.H.F.A.; Wilcken, C.F.; Christovam, R.S.; Prado, E.P. & Gimenes, M.J. (2011). Effect of formulated entomopathogenic fungi on red gum lerp psyllid *Glycaspis brimblecombei*. *Research Journal of Forestry*, vol. 5, no. 2, pp. 99-106

Dreistadt, S.H. & Dahlsten, D.L. (2007). Pest Notes: Psyllids. UC ANR Publication 7423. Accessed April 10, 2007, in:
http://www.ipm.ucdavis.edu/PMG/PESTNOTES/pn7423.html

Durán, P.M. & Urrutia, A.B. (2001). Dos nuevos psilidos en Chile: *Ctenarytaina eucalypti* y *Blastopsylla occidentalis*. Santiago: CONAF, GEDEFF.

Garrison, R.W. (2001). New agricultural pest for Southern California. Spotted Gum Lerp Psyllid, *Eucalyptolyma maideni*. San Diego County Agricultural Commissioner's office.
http://www.sdcounty.ca.gov/reusable_components/images/awm/Docs/ipd_sp ottedpsyllid.pdf Accessed in: August/10/2011

Grimaldi, D.A. (2003). First amber fossils of the extinct family Protopsyllidiidae, and their phylogenetic significance among Hemiptera. *Insect Systematics & Evolution*, vol. 34, pp. 329–344

Hodkinson, I.D. (1974). The biology of the Psylloidea (Homoptera): a review. *Bulletin of Entomological Research*, vol. 64, pp. 325-338

Hodkinson, I.D. (1984). The biology and ecology of the gall-forming Psylloidea (Homoptera), 59-77. In: Ananthakrishnan (ed.), Biology of gall insects. Oxford and IBH Publishing, New Dehli, India.

Hodkinson, I.D. (1991). First record of the Australian psyllid *Blastopsylla occidentalis* Taylor (Homoptera; Psylloidea) on *Eucalyptus* (Myrtaceae) in Mexico. *Pan-Pacific Entomologist*, vol. 67, p. 72

Hodkinson, I.D. (1999). Biocontrol of eucalyptus psyllid *Ctenarytaina eucalypti* by the Australian parasitoid *Psyllaephagus pilosus*: a review of current programmes and their success. *Biocontrol News Information*, vol. 20, pp. 129-134

Hodkinson, I.D. (2007). A new introduced species of *Ctenarytaina* (Hemiptera: Psylloidea) damaging cultivated *Eucalyptus parvula* (=*parvifolia*) in Europe. *Deutsche Entomologische Zeitschrift*, vol. 54, pp. 27–33

Hodkinson, I.D. (2009). Life cycle variation and adaptation in jumping plant lice (Insecta: Hemiptera: Psylloidea): a global synthesis. *Journal of Natural History*, vol. 43, No. 1-2, pp. 65-179

Hollis, D. (1987). A review of the Malvales-feeding psyllid family Carsidaridae (Homoptera). *Bulletin of the British Museum (Natural History) (Entomology)*, vol. 56, No. 2, pp. 87–127

Hollis, D. (2004) Australian Psylloidea: jumping plantlice and lerp insects. CSIRO, Camberra, Australia

Hollis, D. & Broomfield, P. S. (1989). *Ficus*-feeding psyllids (Homoptera), with special reference to the Homotomidae. *Bulletin of the British Museum (Natural History) (Entomology)*, vol. 58, pp. 131–183

Hurtado A. & Reina I. (2008). Primera cita para Europa de *Glycaspis brimblecombei* Moore (Hemiptera: Psyllidae), una nueva plaga del eucalipto. *Boletín Sociedad Entomológica Aragonesa*, vol. 43, pp. 447-449

Iede, E.T.; Leite, M.S.P.; Penteado, S.R.C. & Maia, F. (1997). *Ctenarytaina* sp. (Homoptera: Psilidae) associada a plantios de *Eucalyptus* sp. em Arapoti, PR. In: CONGRESSO BRASILEIRO DE ENTOMOLOGIA, 16, Salvador. Resumos. p. 253

Klimaszewski, S.M. (1964). Studies on systematic of the suborder Psyllodea. Annales Zoologici, vol. 22, pp. 81-138

Laudonia, S. & Garonna, A.P. (2010). The red gum lerp psyllid, *Glycaspis brimblecombei* (Hem.: Psyllidae) a new exotic pest of *Eucalyptus camaldulensis* in Italy. *Bulletin of Insectology*, vol. 63, no. 2, pp. 233-236

Li, F.S. (2011). Psyllomorpha of China. 2 volumes Science Press, Beijing, China.

Ladiges, P.Y. (1997). Phylogenetic history and classification of eucalypts. In: (eds. J. E. Williams & J. H. Z. Woinarski) *Eucalyptus Biology*. Pp. 16-29. Cambridge University Press, Cambridge

Majer, J.D.; Recher, H.F. & Ganesh, S. (2000). Diversity patterns of eucalypt canopy arthropods in Eastern and Western Australia. *Ecological Entomology*, vol. 25, pp. 295-306

Majer, J.D.; Recher H.F.; Wellington, A.B.; Woinarski, J.C.Z. & Yen, A.L. (1997). Invertebrates of eucalypt formations. In: (eds. J.E. Williams & J.H.Z. Woinarski) *Eucalyptus Biology*. Pp. 278-302. Cambridge University Press, Cambridge.

Maschio, L.M.A.; Andrade, F.M.; Leite, M.S.P.; Bellote, A.F.J; Ferreira, C.A.; Iede, E.T.; Nardelli, A.M.B.; Auer, C.G.; Grigolleti, Jr.A. & Wiechetek, M. (1997). Seca dos ponteiros do eucalipto em Arapoti-PR./ Dryout of eucalipt buds in Arapoti, PR. In: *Conferência IUFRO sobre silvicultura de eucalipto*, pp. 353-359

Meza, P.A. & Baldini A.R. (2001). El psilido de los eucaliptos *Ctenarytaina eucalypti* Maskell (1890) (Hemiptera, Psyllidae). Chile, CONAF, vol. 21, no. 39

Moore, K.M. (1961). Observations on some Australian forest insects. 7. The significance of the *Glycaspis* species (Hemiptera: Homoptera: Psyllidae) association with their

Eucalyptus species hosts: erection of a new subgenus and descriptions of thirty eight new species of *Glycaspis*. *Proceedings of the Linnaean Society of New South Wales*, vol. 86, pp. 128-167

Moore, K.M. (1970). Observations on some Australian forest insects. 24. Results from a study of the genus *Glycaspis* (Homoptera: Psyllidae). *Australian Zoologist*, vol. 15, pp. 343-376

Morgan, F.D. (1984). Psylloidea of South Australia. D.J. Woolman, Government Printer, South Australia, 136 pp.

Olivares, T.S.; Burckhardt, D.H. & Cerda L.A. (2004). *Glycaspis brimblecombei* Moore, "Psyllido de los eucaliptos rojos" (Hemiptera: Psyllidae: Spondyliaspidinae): caracteres taxonómicos. *Revista Chilena de Entomología* vol. 30, no. 1, pp. 5-10

Onore, G. & Gara, R.L. (2007). First record of *Glycaspis brimblecombei* (Hemiptera: Psyllidae) in Ecuador, biological notes and associated fauna. *4th European Hemiptera Congress Ivrea, Turin, Italy*, extended abstracts, pp. 41-42

Ouvrard, D.; Burckhardt, D.; Azar, D. & Grimaldi, D. (2010). Non-jumping plant-lice in Cretaceous amber (Hemiptera: Sternorrhyncha: Psylloidea). *Systematic Entomology*, vol. 35, pp. 172-180

Paine, T.D. & Millar, J.G. (2002). Insect pests of eucalyptus in California: Implications of managing invasive species. *Bulletin of Entomological Research*, vol. 92, pp. 147-51

Paine, T.D.; Dreistadt, S.H.; Garrison, R.W. & Gill, R.J. (2007). *Eucalyptus* redgum lerp psyllid. http://www.ipm.ucdavis.edu/PMG/PESTNOTES/pn7460.html. Accessed April 10, 2007

Panizzi, A.R. & Parra, J.R.P. (1991). Ecologia nutricional de insetos e suas implicações no manejo de pragas. *São Paulo: Manole*, 259 pp.

Percy, D.M. (2003). Radiation, diversity, and host–plant interactions among island and continental legume-feeding psyllids. *Evolution*, vol. 57, pp. 2540–2556

Peris-Felipo F.J.; Mancusi G.; Turrisi, G.F. & Jiménez-Peydró, R. (2011). New corological and biological data of the Red Gum Lerp Psyllid, *Glycaspis brimblecombei* Moore, 1964 in Italy (Hemiptera, Psyllidae). *Biodiversity Journal*, vol. 2, pp. 13-17

Pinzón, F.O.P.; Guzmán, C.M. & Navas, N.F. (2002). Contribución al conocimiento de la biología, enemigos naturales y daños del pulgón del eucalipto *Ctenarytaina eucalypti* (Homoptera: Psyllidae). *Revista Colombiana de Entomología*, vol. 28, pp. 123–128

Prieto-Lillo, E.; Rueda, J.; Hernández, R. & Selfa, J. (2009). Primer registro del psílido rojo del eucalipto, *Glycaspis brimblecombei* (Homoptera: Psyllidae), en la Comunidad Valenciana. *Boletín Sanidad Vegetal Plagas*, vol. 35, pp. 277-281

Raman, A.; Burckhardt, D. & Harris, K.H. (2009). Biology and adaptive radiation in the gall-inducing Cecidomyiidae and Calophyidae (Hemiptera) on *Mangifera indica* (Anacardiaceae) in the Indian subcontinent. Tropical Zoology, vol. 22, no. 1, pp. 27–56

Rodrigues, F.A.; Rodrigues, E.S. da; Corrêa, P.L.P.; Rocha, R.L. de A. & Saraiva, A.M. (2010). Modelagem da biodiversidade utilizando redes neurais artificiais. www.inf.pucminas.br/sbc2010/anais/pdf/wcama/st02_04.pdf   Accessed   1 February 2011

Rosales, C.J.; Lobosque, O.; Carvalho, P.; Bermúdez, L. & Acosta, C. (2008). *Glycaspis brimblecombei* Moore (Hemiptera: Psyllidae). "Red Gum Lerp". Nueva plaga forestal en Venezuela. *Entomotropica*, vol. 23, pp. 103-104

Santana, D.L.Q. (2003). *Ctenarytaina spatulata* Taylor, 1997 (Hemiptera: Psyllidae): morfologia, biologia, dinâmica, resistência e danos em *Eucalyptus grandis* Hill. ex Maiden, Doctorate Thesis.

Santana, D.L.Q., Andrade, F.M.; Bellote, A.F.J. & Grigoletti, Jr.A. (1999). Associação de *Ctenarytaina spatulata* e de teores de Magnésio foliar com a seca dos ponteiros de *Eucayptus grandis*. *Boletim de Pesquisa Florestal, Colombo, PR*, vol. 39, pp. 41-49

Santana, D.L.Q., Bellote, A.F.J. & Dedecek, R.A. (2003a). *Ctenarytaina spatulata*, Taylor: Água no solo, Nutrientes minerais e suas interações com a seca dos ponteiros de eucalipto. *Boletim de Pesquisa Florestal, Colombo, PR*, vol. 46, pp. 57-67

Santana, D.L.Q. & Burckhardt, D. (2007). Introduced *Eucalyptus* psyllids in Brazil. *Journal of Forest Research*, vol. 12, pp. 337-344

Santana, D.L.Q., Carvalho, R.C.Z.; Favaro, R.M. & Almeida, L.M. (2004). *Glycaspis brimblecombei* (Hemiptera: Psyllidae) e seus inimigos naturais no Paraná. In: XX Congresso Brasileiro de Entomologia p. 450

Santana, D.L.Q., Menezes, A.O. & Bizzi, R.M. (2002). Ocorrência de *Psyllaephagus* sp. (Hymenoptera: Encyrtidae) parasitando *Ctenarytaina eucalypti* (Maskell) (Homoptera Psyllidae) no Brasil. In: 19° Congresso Brasileiro de Entomologia, Manaus, Resumos, p. 149

Santana, D.L.Q., Menezes, A.O.; Silva, H.D.; Bellote, A.F.J. & Favaro, R.M. (2003b). O psilídeo-de-concha (*Glycaspis brimblecombei*) em eucalipto. *Embrapa Florestas, Comunicado Técnico*, 3 pp.

Santana, D.L.Q. & Zanol, K.M.R. (2006). Biologia de *Ctenarytaina spatulata* Taylor, 1997 (Hemiptera, Psyllidae) em *Eucalyptus grandis* Hill ex Maiden. *Acta Biologica Paranaense*, vol. 35, pp. 47-62

Santana, D.L.Q., Zanol, K.M.R.; Botosso, P.P.C. & Mattos, P.P. (2005). Danos causadas por *Ctenarytaina spatulata* Taylor, 1977 (Hemiptera: Psyllidae) em *Eucalyptus grandis* Hill. Ex Maiden. *Boletim de Pesquisa Florestal, Colombo, PR*, vol. 50, pp. 11-24

Santana, D.L.Q. & Zanol, K.M.R. (2005). Morfologia externa das ninfas e adultos de *Ctenarytaina spatulata* Taylor (Hemiptera, Psyllidae). *Revista Brasileira de Entomol ogia*, vol. 40, pp. 340-346

Santana, D.L.; Zanol, K.M.R.; Oliveira, E.B.; Anjos, N. & Majer, J. (2010). Feeding and oviposition preferences of *Ctenarytaina spatulata* Taylor (Hemiptera: Psyllidae) on *Eucalyptus* spp and other Myrtaceae growing in Brazil. *Revista Brasileira de Entomologia*, vol. 54, pp. 149-153,

Soberon, J. M., Peterson A. T. (2005). Interpretation of models of fundamental ecological niches and species' distributional areas. *Biodiversity Informatics*, vol. 2, pp. 1-10

Tamesse, J.L; Laurentine, S.; Wenceslas, Y. & Joly, D.V. (2011). First record of *Blastopsylla occidentalis* Taylor, 1985 (Hemiptera: Psyllidae), a *Eucalyptus* psyllid in Cameroon, Central Africa. Entomological Research, vol. 40, no. 4, pp. 211–216

Taylor, K.L. (1985). Australian psyllids: A new genus of Ctenarytainini (Homoptera: Psylloidea) on *Eucalyptus*, with nine new species. *Journal of the Australian entomological Society*, vol. 24, pp. 17-30

Taylor, K.L. (1987). *Ctenarytaina longifolia* sp. n. (Homoptera: Psylloidea) from *Lophostemon confertus* (R. Brown) in Australia and California. *Australian Journal of Entomology,* vol., 26, no. 3, pp. 229–233

Taylor, K.L. (1990). The tribe Ctenarytainini (Hemiptera: Psylloidea): A key to known Australian genera, with new species and two new genera. *Invertebrate Taxonomy,* vol. 4, pp. 95-121

Taylor, K.L. (1997). A new Australian species of *Ctenarytaina* Ferris and Klyver (Hemiptera: Psyllidae: Spondyliaspidinae) established in three other countries. *Journal of the Australian entomological Society,* vol. 36, pp. 113-115

Thao, L.M.; Moran, N.A.; Abbot, P.; Brennan, E.B.; Burckhardt, D.H. & Bauman, P. (2000). Cospeciation of psyllids and their primary prokaryotic endosymbionts. *Applied and environmental microbiology,* July 2000, pp. 2898-2905.

Thao, L.M.; Clark M.A.; Burckhardt, D.H., Moran, N.A. & Bauman, P. (2001). Phylogenetic analysis of vertically transmitted psyllid endosymbionts (*Candidatus* Carsonella ruddii) based on atpAGD and rpoC: Comparisons with 16S-23S rDNA-derived phylogeny. *Current microbiology,* vol. 42, pp. 419-421.

Valente, C. & Hodkinson I.D. (2009). First record of the Red Gum Lerp Psyllid, *Glycaspis brimblecombei* Moore (Hem.: Psyllidae), in Europe. *Journal of Applied Entomology,* vol. 133, pp. 315-317

Valente, C.; Manta, A. & Vaz, A. (2004). First record of the Australian psyllid *Ctenarytaina spatulata* Taylor (Homoptera: Psyllidae) in Europe. *Journal of Applied Entomology,* vol. 128, pp. 369–370

White, I.M. & Hodkinson, I.D. (1985). Nymphal taxonomy and systematics of the Psylloidea (Homoptera). *Bulletin of the British Museum (Natural History) (Entomology),* vol. 50, pp. 153–301

White, T. C. R. (1968). Uptake of water by eggs of *Cardiaspina densitexta* (Homoptera: Psyllidae) from leaf of host plant. *Journal of Insect Physiology,* vol. 14, pp. 1669-1683

White, T.C.R. (1969). An index to measure weather-induced stress on trees associated with outbreaks of psyllids in Australia. *Ecology,* vol. 50, pp. 905-909

White, T. C. R. (1970). Some aspects of the life history, host selection, dispersal, and oviposition of adult *Cardiaspina densitexta* (Homoptera: Psyllidae). *Australian Journal of Zoology,* vol. 18, pp. 105-17

Wilcken, C.F.; Couto, E.B.; Orlato, C.; Ferreira-Filho, P.J. & Firmino, D.C. (2003). Ocorrência do psilídeo-de-concha (*Glycaspis brimblecombei*) (Hemiptera: Psyllidae) em florestas de eucalipto do Brasil. *Circular Técnica IPEF,* No. 201, pp. 1–11

Withers, T. 2000. Parasitoid on *Cardiaspina fiscella.* Access in 01/August/2011, in: http://www.nzffa.org.nz/farm-forestry-model/the-essentials/forest-health-pests-and-diseases/Pests/Cardiaspina-fiscella/Parasitoid-Cardiaspina-FHnews95. From Forest Health News No. 95, April 2000

Withers, T.M. (2001). Colonization of eucalypts in New Zealand by Australian insects. *Austral Ecology,* vol. 26, No. 5, pp. 467–476

Yang, M.M.; Mitter, C. & Miller, D.R. (2001). First incidence of inquilinism in gall-forming psyllids, with a description of the new inquiline species (Insecta, Hemiptera, Psylloidea, Psyllidae, Spondyliaspindinae). *Zoologica Scripta,* vol. 30, pp. 97-113

Yang, M.M. & Mitter, C. (1994). Biosystematics of hackberry psyllids (*Pachypsylla*) and the evolution of gall and lerp formation in psyllids (Homoptera: Psylloidea): a preliminary report. *In*: Price PW, Mattson WJ, Baranchikov YN, eds. The Ecology and Evolution of Gall-forming Insects. *U.S.D.A. Forest Service (North Central Forest Experiment Station) General Technical Report*, NC-174, St. Paul, Minnesota, pp. 172-185

Zondag, R. (1982) *Ctenarytaina eucalypti* (Maskell), (Hemiptera, Psyllidae). Blue-gum psyllid. *New Zealand Forest Service, Forest and Timber Insects in New Zealand*, No. 53, 4 pp.

3

Biological Studies and Pest Management of Phytophagous Mites in South America

Carlos Vásquez[1,*], José Morales-Sánchez[1], Fernando R. da Silva[2]
and María Fernanda Sandoval[3]
[1]*Universidad Centroccidental Lisandro Alvarado, Decanato de Agronomía
Departamento de Ciencias Biológicas. Barquisimeto, Estado Lara*
[2]*University of Amsterdam. Institute for Biodiversity and Ecosystem Dynamics (IBED)
Research Group of Population Biology, Amsterdam*
[3]*Instituto Nacional de Salud Agrícola Integral (INSAI). Av.
Principal Las Delicias. Edif. INIA – Maracay, Estado Aragua*
[1,3]*Venezuela*
[2]*The Netherlands*

1. Introduction

Mites are the most diverse representatives of an ancient lineage in phylum Arthropoda-subphylum Chelicerata- subclass Acari. Their body plan is strikingly different to that of other arthropods in not having a separate head, instead, an anterior region, the cephalothorax, combines the functions of sensing, feeding, and locomotion (Walter & Proctor 1999). Antennae, mandibles and maxillae are also absent; rather, a pair of often pincer-like mouthparts are present, the so-called chelicerae. Those members of the subclass Acari, which feed on plants, are known as phytophagous mites. Mites constitute one of the most heterogeneous cheliceran groups, since they are extremely diverse in their morphology, biology and ethnology, enabling them to colonize different environments. Their remarkable diversity in acarine morphology is reflected in the variety of ecological and behavioral patterns that mites have adopted (Krantz 2009). Thus, species can inhabit soil, litter (i.e. Cryptostigmata, Mesostigmata and Prostigmata), water (Hydrachnidia) or plants (Prostigmata or Mesostigmata). Phytophagy is widespread enough among the Trombidiform Acariformes so as to suggest that there was an early evolution commitment to plant feeding by several primitive predaceous and saprophagous trombidiform lineages (Krantz 2009). Some Prostigmatan mites, chiefly spider mites, false spider mites and eriophoid mites, use their specialized mouthparts to feed on the vascular tissues of higher plants and with their activity they can cause losses to field and protected crops (Evans 1992), becoming economically important pests.

This review summarizes more important phytophagous mites in tropical crops in South America, biological aspects, damage, and also main control strategies in tropical conditions.

---

*Corresponding Author

2. Economically important mite species

Species of agricultural importance can exhibit either phytophagous or predatory habits. Most important taxa including exclusively phytophagous mites are Eriophyoidea, Tetranychoidea (Tetranychidae, Tenuipalpidae). Tarsonemidae is also a family of mites which includes several pest species. All these taxa cover the most important crop pest species distributed worldwide, and several more geographically restricted species. The Eriophyoidea is a large superfamily of worldwide distribution. Over 3,000 species belonging to about 250 genera are known in the world. These worm-like or fusiform mites cause many forms of plant abnormalities such as galls, leaf blisters and rusts. Most eriophyid mite species are monophagous or are limited to plant species within a single genus. Some rust mites and gall mites are important pests on economic plants.

The Tetranychidae, also known as spider mites, is a large family including of some 1,200 species belonging to over 70 genera of worldwide distribution. Spider mites cause mechanical damage by sucking cell content from leaves. At first, it shows up as a stippling of light dots on the leaves; and sometimes leaves became bronze in color. As feeding continues, the leaves turn yellow and drop off. Often leaves, twigs, and fruit are covered with a large amount of webbing. In Tenuipalpidae, also known as false spider mites or flat mites, about 800 species have been described in over 25 genera. Only *Brevipalpus*, *Tenuipalpus* and *Raoiella* and few other genera become pests of economic plants, mainly on tropical fruit crops and ornamental plants. Virus transmission has only been documented in some Eriophyidae or Tenuipalpidae species.

Due to the economic losses caused by mite pests, management tactics need to be established to keep population levels under the economic threshold of infestation. This practice should be based on integrated pest management (IPM) including spraying chemical products, using biological control agents and/or resistant varieties. Currently, chemical control has to deal with serious control failures in mite populations, since the evolution of pesticide resistance in phytophagous mites is very common. Consequently, different chemical molecules are being currently developed to face this phenomenon, mainly in spider mites. However, the low level of immigration of susceptible individuals and the rapid reproductive rate associated to these mite groups have made it difficult to manage population in crops.

Biological control is an environmentally safe, cost-effective and energy efficient pest control, either on its own or as a component of integrated pest management. Although several mite species belonging to Bdellidae, Cheyletidae, Cunaxidae, Stigmaeidae and Tydeidae have shown predatory habits, Phytoseiidae mites have been more widely included in biological control programs, due to their capacity for surviving and reproducing on other arthropods. Additionally, some phytoseiid mites have shown to be resistant or less susceptible to chemical compounds commonly used to control pest mites in commercial crops, thus making them suitable for their use in integrated pest management programs.

3. Spider mites

This family consists of two subfamilies: Bryobinae and Tetranychinae. Spider mites occur on most of the major food crops and ornamental plants in almost all environments where mainly Tetranychinae species can potentially cause economic damage.

Spider mites are soft-bodied, medium-sized mites. They are often red, green, orange or yellow in color when alive. The gnathosoma has a capsule-like structure known as the stylophore, which is formed by the fusion of chelicerae. The movable digit of the chelicerae is very long, often whip-like and recurved proximally. A pair of stigmata is located near the base of the chelicerae, where the peritremes arise. The palps are five-segmented. The palptarsus and tibia often form a thumb-claw process. The tarsus often has an enlarged distal eupathidium (spinneret) in the Tetranychinae and this is used to spin webbing in many species. The size and shape of the spinneret is of taxonomic significance. The idiosoma is often covered with a striate cuticle. The pattern of the striation and the shape/density of lobes distributed on the striae are useful diagnostic characters.

There are three or four pairs of normal setae in two rows (v1-2, sc1-2) and two pairs of eyes on the dorsal propodosoma. On the opisthosomal dorsum, there are five rows of setae: c, d, e, f and h. The number, location, length and structure of dorsal setae are of taxonomic significance. Female genital pores are transverse and are bordered anteriorly by a genital flap and laterally by characteristic cuticular folds. The structures of the paired lateral claws and the medial empodium are of taxonomic importance. The empodium may be claw-like or pad-like with tenant hairs. Claws may bear dorsal or ventral hairs. The tarsi of legs I and II bear duplex setae (a long solenidion and a short normal tactile seta with their bases joined together. The number of duplex setae and their positions are of taxonomic significance.

Wedge-shaped males are smaller than ovoid females and have a tapering opisthosoma. Males have a protrudable aedeagus, the shape of which is very important in species identification.

Life history and biology: as other Prostigmata mites, the spider mite life cycle consists of egg, larva, protonymph, deutonymph and adult stages, except for some *Schizotetranychus* and *Eotetranychus* species, which may have one nymphal stage in males (Zhang 2003). Moulting takes place during the quiescent stages between each active stage. Development from egg to adult often takes one to two weeks or more, depending on mite species, temperature, host plants, humidity and other environmental factors. Males develop slightly faster than females and soon they search for and fight for quiescent deutonymph females. Unfertilized eggs produce only haploid males, while fertilized eggs produce diploid females.

3.1 Spider mite species of economic importance in South America

3.1.1 *Tetranychus urticae* Koch

The two-spotted spider mite (TSSM) is considered one of the most harmful tetranychid species in agriculture both in temperate or tropical countries (Cerna *et al.* 2009, González Zamora *et al.* 1993). Although a recent checklist includes some 920 host plant species in 70 genera (Bolland *et al.* 1998), only 150 of these host species have relevant economic value. The TSSM feeds on cell chloroplast on the underside of the leaves and symptoms become clearly visible on the upper side as characteristic whitish or yellowish punctures which, under high population levels, can join and become brownish o even cause leaf drop and plant death (Tomczyk & Krompczyńska 1985, Zhang 2003).

Due to its relevance and economic impact on tropical agriculture in South America, the TSSM has been the most extensively studied spider mite species. A number of papers related to biological aspects of the TSSM are available (Table 1).

Host Plants	Life cycle (days)	N° eggs/female/day	Longevity (days)	Reference
Sweet pepper	8.2 (27°C, 70% RH)	2.6	12.2	Gallardo *et al.* (2005)
Cotton	Non Bt 7.5 Bt 7.3 (25°C, 57.4% RH)		Non Bt: 16.7 Bt: 16.6	Esteves Filho *et al.* (2010)
Gerbera	21.6 (25°C, 70% RH)	3.75	8.83	Silva *et al.* (2009)

Table 1. Some biological parameters of the TSSM in various host plant and environmental conditions in South America.

Population management: For decades population management of tetranychid mites has been based mostly on chemical control in South America. However, biological control is being increasingly used under greenhouse conditions, mainly in Argentina, Brazil, and Colombia. In Colombia reductions of TSSM population on rose trees have been observed after periodical release of *Neoseiulus* sp. (Forero *et al.* 2008). Even though a lower number (19.4%) of phytophagous mites was found in plots chemically treated, a lower percentage (damage index 1 and 3, 8 and 13% less, respectively) of leaf damage was observed when *Amblyseius* sp. was released. On the other hand, the consumption rate (functional response) after a 24 h. period was 6.66 eggs, 18.06 larvae, and 19.15 nymphs under laboratory conditions and 4.56 eggs, 12.65 larvae, and 15.71 nymphs under greenhouse conditions (Forero *et al.* 2008). Greco *et al.* (2005) found that *Neoseiulus californicus* (McGregor) is a promising agent for successful TSSM control through conservation techniques, on strawberry crops in La Plata, Argentina. Accordingly, the authors showed that initial relative densities had an important effect on system dynamics. Thus, when pest/predator ratio was 5/1 (at initial pest densities from 5 to 15 females/leaflet) the final number of active *T. urticae*/leaflet was significantly lower than the economic threshold level (ETL), while at 20 females/leaflet this number did not differ from the ETL. At 7.5/1 ratio, the final number of active *T. urticae*/leaflet, at initial pest densities from 5 to 15 females/leaflet, reached the ETL without surpassing it. At 10/1 and 15/1 ratios, pest densities exceeded the ETL only at 15 initial *T. urticae*/leaflet. Since *N. californicus* showed to be very effective in limiting pest densities, conservation of this predator promoting favorable pest/predator ratios may result in early control of *T. urticae*.

Furthermore, *Phytoseiulus macropilis* (Banks) demonstrated a positive functional response (increases prey consumption). However, all indices evaluated showed that *P. macropilis* was unavailable to control the TSSM efficiently when the population numbers were low, possibly to keep itself in an environment with low populations (Ferla *et al.* 2011). Biological control by this predator is more efficient when five or more prey items are present at a given time.

3.1.2 *Tetranychus cinnabarinus* (Boisduval)

This species is commonly known as the carmine spider mite and it is associated with more than 120 host plant species such as cotton, strawberry, tomato, eggplant, and also ornamental species and fruit trees (Biswas *et al.* 2004). In Chile, the carmine spider mite is commonly found on carnation, strawberry, melon and beans from Arica, Parinacota and dessert areas where chemical control is frequently used (Klein & Waterhouse 2000).

Tello *et al.* (2009) found that the life cycle (egg-adult) of *T. cinnabarinus* on carnation var. Celta lasted 12.8 days (29.4 °C, 42.3 RH and 14:10 h (L:D) photo phase) and female longevity and mean daily oviposition rate were 24.28 days and 3.92 eggs/female/day, respectively. Under these conditions life table parameters were as follows: the intrinsic rate of increase (r_m) 0.183; the finite rate of increase (λ) 1.201 individuals/female/day; the mean generation time (T) was 20.24 days; and the net rate of reproduction (R_0) was 40.809.

Control strategies: most efforts to control pest population densities under threshold levels have relied on chemical tactics; however, particular attention has been paid to other management strategies in last decades, including resistant cultivars. So far, kidney bean has shown moderate resistance to other arthropod pests such as thrips (Cardona *et al.* 2002) and it seems to be based on antixenosis and antibiosis mechanisms (Frei *et al.* 2003). In general, these defense mechanisms are considered plant responses to stressing conditions either abiotic (drought, salinity) or biotic (herbivore or pathogens attacks), which induce the development of physical barriers to prevent feeding or secondary metabolites affecting oviposition or survival (Tomczyk & Krompczyńska 1985, Gardner & Agrawal 2002). Vásquez *et al.* (2007) observed that mean oviposition of *T. cinnabarinus* females was significantly lower on 22-day-old-leaf disks, ranging from 0.98 to 1.17 eggs/female/day on ICA-Pijao and Coche beans cultivars, respectively (Table 2), while females reared on 55 day-old leaf disks oviposition increased about 58 or 95%

		Number of eggs/female	
		22 day-old	55 day-old
Cultivar	n		
Tacarigua	25	1.16 a ± 0.7342 (34)	1.66 b ± 0.8119 (72)
Coche	25	1.17 a ± 0.5905 (30)	1.86 b ± 0.7126 (86)
ICA-Pijao	25	0.98 a ± 0.3852 (15)	1.92 b ± 0.8907 (99)

Values in a file followed by same letter did not show significant differences (Tukey's Test; P<0.05) Number in parenthesis represent total egg number during a five day period. From Vásquez *et al.* (2007).

Table 2. Oviposition (mean ± S.D.) of *T. cinnabarinus* on 22 or 55 days-old leaf disks from different kidney bean cultivars.

In general *T. cinnabarinus* females showed a higher rate of survival on 22 day-old disks, ranging from 5.67 to 6.63 living individuals in Tacarigua and Coche leaves; while on 55 day-old leaves survival ranged from 5.42 to 1.80 on Coche and ICA-Pijao, respectively. A significant survival reduction was observed on mites reared on 55 day old leaf disks from ICA-Pijao cultivar (Fig. 1), suggesting that this cultivar could dissuade to mite from feeding, affecting thus mite survival.

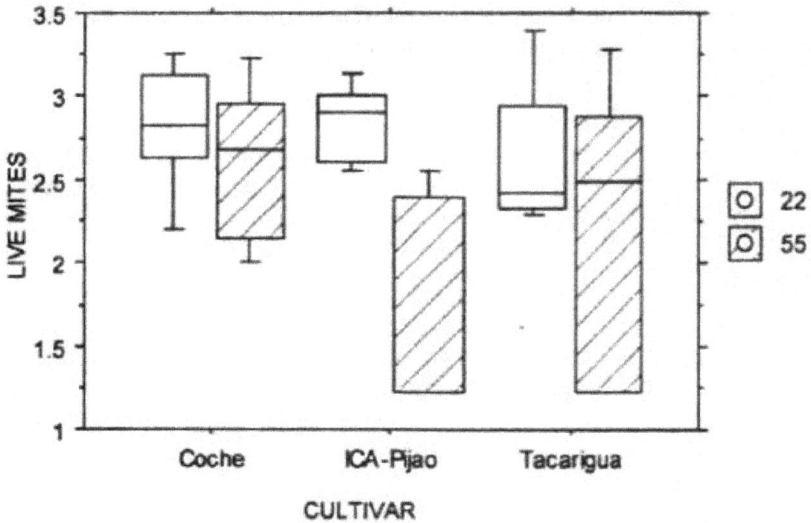

Fig. 1. Number of live mites on kidney bean cultivars on 22- or 55 day-old leaf disks. From Vásquez *et al.* (2007).

3.1.3 Red spider mites (*Tetranychus desertorum* Banks and *Tetranychus ludeni* Zacher)

The red spider mite, *T. desertorum* has been found on about 173 plant species in Argentina, Brazil, China, the United States of America, Mexico, Venezuela and some other countries (Bolland *et al.* 1998). This spider mite species might cause severe damage to bean plants, as *T. urticae* does (Moraes & Flechtmann 2008). In Venezuela, previous studies have demonstrated that bean cultivars are susceptible to *T. desertorum* attacks under irrigation conditions (Doreste 1984). Rivero & Vásquez (2009) showed that total developmental time of *T. desertorum* females on kidney bean 'Tacarigua' was 6.8 days, with partial duration of immature stages corresponding to 3.8, 1.4, 1.0 and 0.7 days for egg, larva, protonymph and deutonymph, respectively. Higher mean fecundity (6.93 eggs/female/day) was observed on day 4 and females lived during 10 days. The recorded life table parameters were as follows: net reproduction rate (Ro) = 41.10 individuals; generation time (T) = 11.15 days; intrinsic natural growth (rm) = 0.144 individuals/female/day, and finite natural increase rate (λ) = 1.155 individuals/female.

Furthermore, *T. ludeni* is a worldwide pest that attacks various plant species, such as *Phaseolus vulgaris* L., representing a severe threat in several countries, including Venezuela (Morros & Aponte 1994). According to these authors, the life cycle (egg-adult) of *T. ludeni* lasted 9.98 and 9.25 days for females and males, respectively (26.34± 3.92°C and 69.44± 19.44% HR). The life table also showed the following values: reproduction rate (Ro) 77.42; mean generation time (T) 19.63; intrinsic rate of natural increase (r) 0.2526 individuals/female/day; finite rate of natural increase (λ) 1.2874 individual/female/week. Morros & Aponte (1995a, b) evaluated the effect of two levels of mite infestation on vegetative and reproductive stages of the black bean *P. vulgaris* under greenhouse and field conditions. The authors observed greater reduction on number of leaves and internodes, leaf area, and dry weight of vegetative organs when mite infestation occurred in the early vegetative stage (Table 3).

	Greenhouse		Field	
	Development stage			
	Vegetative	Reproductive	Vegetative	Reproductive
Foliar area (cm²)	263.65	289.84	30.76	36.13
Number of leaves	22.83	25.00	34.54	45.72
Number of internodes	12.77	14.50	13.13	14.63
Dry leaf weight (g)	2.80	3.35	3.28	6.10
Dry plant weight (g)	1.61	1.80	2.41	3.83
Dry pod weight (g)	1.82	1.88	0.36	0.56
Total dry weight (g)	7.26	8.27	6.06	10.50

From Morros & Aponte (1995a, b).

Table 3. Reductions of plant performance as effect of *T. ludeni* feeding on black beans in vegetative or reproductive stages under greenhouse and field conditions.

3.1.4 The avocado brown mite, *Oligonychus punicae* Hirst

The avocado brown mite (ABM) is considered an important tetranychid pest in southern California (USA), as high population densities can cause severe defoliation on several avocado cultivars (McMurtry & Johnson 1966; McMurtry 1985). This tetranychid mite predominantly feeds on the upper leaf surface, although feeding could extend to the lower leaf surface at high population levels (Tomczyk & Kropczynska 1985). Damage to host plants is shown by a bronze tone on the leaves, and is associated with the rates of oviposition and female production (McMurtry 1970).

In tropical America, the ABM has been reported in more than 20 plant species, such as *Mangifera indica* L., *Musa sapientum* L., *Punica granatum* L. and *Vitis vinifera* L. (Ochoa *et al.* 1994; Bolland *et al.* 1998). In Venezuela, it was previously recorded on *Musa* spp. from Sur del Lago, Zulia State (Freitez & Alvarado 1978; Quirós 1978). More recently, in Lara State, it has been observed as an occasional pest whose feeding on grape delays fruit ripening (Vásquez *et al.* 2008a). These authors provided information about life cycle, fecundity and longevity of the ABM on six grapevine cultivars (Tucupita, Villanueva, Red Globe, Sirah, Sauvignon and Chenin Blanc), under laboratory conditions at 27 °C, 80% RH, and L12:D12 photoperiod (Tables 4 and 5).

Periods of pre-oviposition, oviposition and post-oviposition in *O. punicae* females varied among grape cultivars. Shorter pre- and post-ovipositional periods were found on Chenin Blanc leaves (1.2 and 0.9 days, respectively), while on others cultivars these parameters ranged from 1.6–2.0 to 1.2–2.5 days, respectively (Table 5). Oviposition periods lasted from 6.7 days (on Villanueva) to 16.1 days (on Sauvignon). Average daily egg production was highest on Tucupita (2.8 eggs/female/day) and lowest on Sirah (0.9 eggs/female/day). Daily oviposition rate ranged from 2.0 to 6.1 eggs/female up to day 7 in females feeding on Tucupita leaves, while on Chenin Blanc, Red Globe, Sauvignon, Sirah and Villanueva leaves it varied from 1.1–6.1, 0.2–5.3, 0.3–5.0, 0.2–2.7 to 0.5–3.5 eggs/female, respectively (Fig. 2). Also, female longevity of *O. punicae* was affected by grape cultivar: females lived longest on Sauvignon (17.5 days), and shortest on Villanueva (8.1 days)

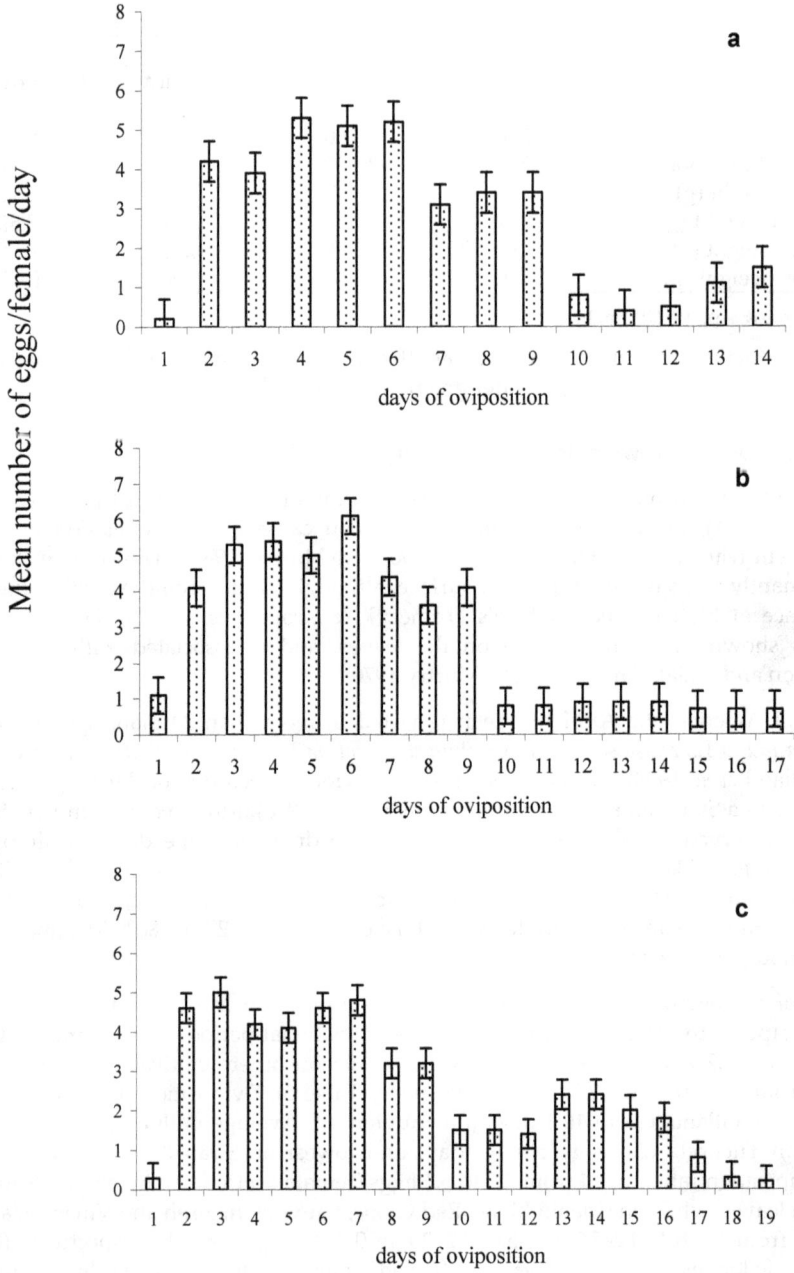

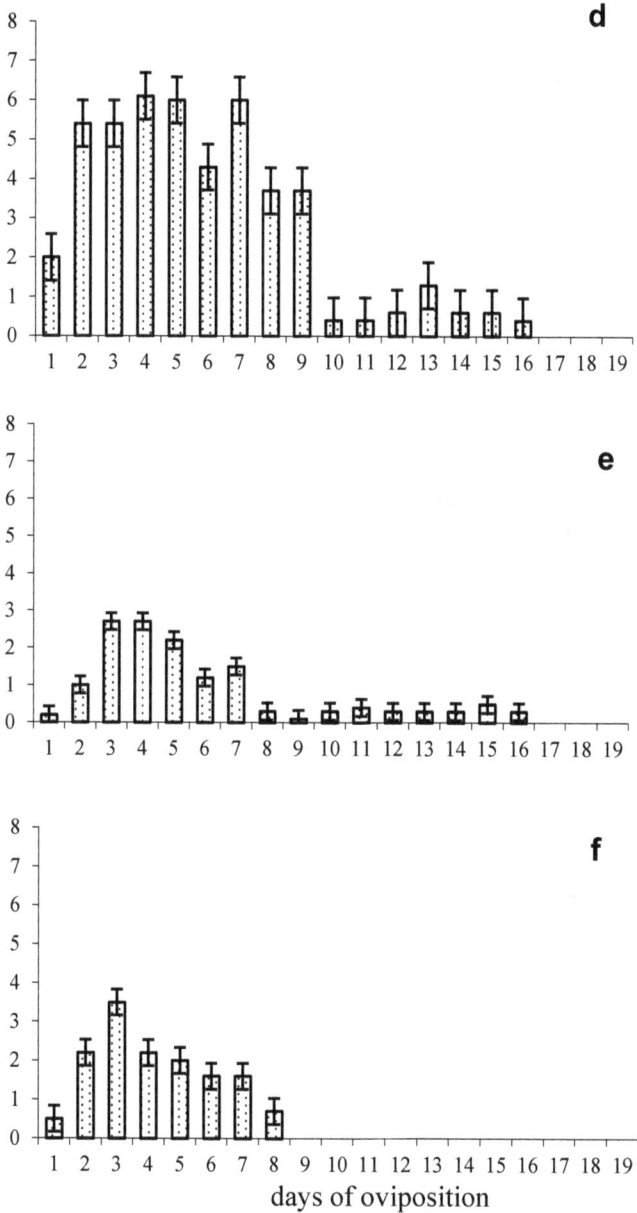

Fig. 2. Daily oviposition rate of *O. punicae* females feeding on grapevine leaves cultivars Red Globe (a), Chenin Blanc (b), Sauvignon (c), Tucupita (d), Sirah (e) and Villanueva (f).

Stages	Mean duration (hours) ± SD					
	Tucupita	Sauvignon	Villanueva	Red Globe	Chenin Blanc	Sirah
Egg	104.80±	106.50±	112.80±	105.20±	105.60±	111.20±
	5.9330a	7.5498a	5.215a	8.438a	6.066a	3.347a
Larva	22.40±	20.25±	22.40±	21.00±	29.60±	27.60±
	1.3416ab	4.1130b	1.949ab	6.042b	4.775a	4.506ab
Protochrysalis	11.80±	15.75±	10.80±	13.80±	13.40±	11.60±
	1.4832ab	3.403a	1.304b	3.115ab	1.517ab	1.517ab
Protonymph	13.00±	14.75±	15.60±	19.00±	21.40±	16.60±
	2.8284b	2.363ab	6.229ab	2.828ab	4.930a	5.413ab
Deutocrhysalis	11.80±	11.00±	11.80±	12.20±	13.50±	11.60±
	1.3038a	0.8165a	2.950a	3.347a	4.359a	2.074a
Deutonymph	22.60±	17.00±	23.60±	21.80±	20.00±	24.40±
	2.7928a	1.414a	3.578a	2.775a	6.733a	5.899a
Teliochrisalis	11.20±	9.25±	10.80±	15.00±	11.25±	10.80±
	1.9235ab	3.403b	1.924ab	1.414a	1.500ab	2.683ab
Life cycle[a]	197.6±	198.40±	208.80±	209.20±	215.00±	217.20±
	4.5056b	8.5391b	4.5497ab	12.6770ab	13.7840ab	8.8148a
	(8.16)	(8.27)	(8.70)	(8.72)	(8.96)	(9.05)

Values in a row followed by the same letter are not significantly different according to the Tukey's multiple comparison differences (P=0.0036, df=5, F=4.78).
[a]Values in brackets, days to complete development. From Vásquez et al. (2008a).

Table 4. Developmental time of different life stages of O. punicae in various grapevine cultivars.

Cultivars	N	Longevity	Preoviposition[1]	Oviposition [1]	Postoviposition [1]	Daily fecundity[2]
Tucupita	10	12.60 ± 5.1897abc	1.20 ± 0.4216b	11.40 ± 3.2042bc	1.20 ± 0.4216ab	2.82 ± 2.4186a
Sauvignon	10	17.50 ± 4.8028a	1.80 ± 0.4216ab	16.10 ± 2.8460a	1.40 ± 0.8433ab	2.15 ± 1.4130bc
Red Globe	10	13.20 ± 3.3417abc	1.80 ± 0.4216ab	11.10 ± 2.6013bc	1.30 ± 1.0593ab	2.72 ± 1.9132ab
Chenin Blanc	10	14.40 ± 3.8930ab	1.20 ± 0.4216b	13.50 ± 3.3417ab	0.90 ± 1.1005b	2.16 ± 1.9949ab
Villanueva	10	8.10 ± 1.1972c	1.60 ± 0.6992ab	6.70 ± 1.1595d	1.40 ± 1.2649ab	1.79 ± 0.9433cd
Sirah	10	10.30 ± 3.4383bc	2.00 ± 0.6667a	7.90± 4.9318cd	2.50 ± 1.3540a	0.94 ± 0.9532d

[1]Transformed values by $y = \sqrt{x + 0.5}$ [2] Transformed values by $y = \sqrt{x + 1.5}$

Values in a column followed by different letter showed significant differences. P<0.05 (preovip: F=4.21; ovip: F=12.8; and postovip: F=2.36; df=5); P<0.001 (longevity: F= 7.05; df=5). From Vásquez et al. (2008a).

Table 5. Adult longevity, preoviposition, oviposition and postoviposition times (days) and fecundity (eggs/female/day) of O. punicae on several grapevine cultivars.

An effect of host plant on mite reproduction has been established for several Tetranychid species (e.g., de Ponti 1977; Ribeiro *et al.* 1988; Hilker & Meiners 2002; Praslička & Huszár 2004). Previous studies have demonstrated that grape leaves and fruits synthesize phenolic compounds in response to fungal attacks or abiotic factors (Morrissey & Osbourn 1999). Furthermore, Harborne (1994) hypothesized that low molecular weight phenol compounds could act synergistically with tannins to provide plant resistance. Thus, reproductive parameters of *O. punicae* seemed to be associated with flavonoid content of grape cultivars; the higher the flavonoids content, the lower the mites' fecundity (Fig. 3). These findings could be considered an ecological approach for sustainable pest management programs in grapevine in tropical areas.

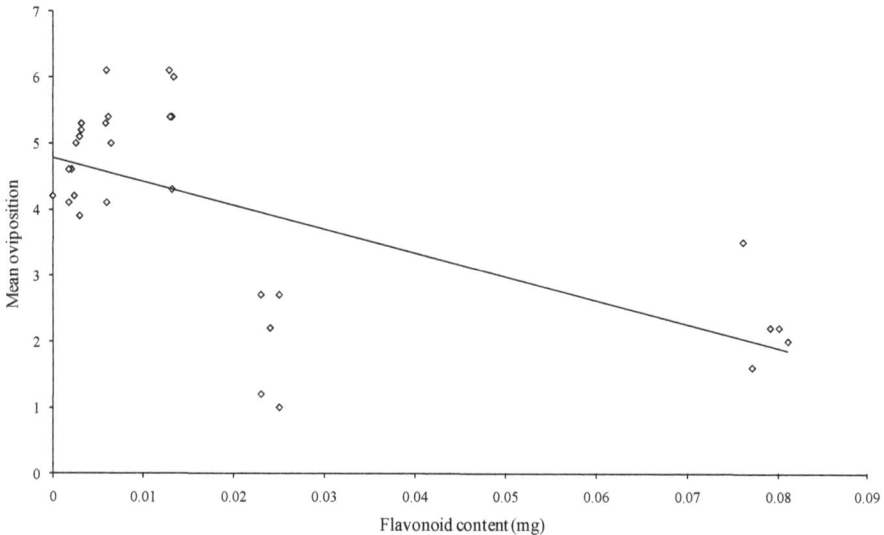

Fig. 3. Lineal regression between flavonoid content and mean oviposition of *O. punicae* on grapevine cultivars (y = 4.78 - 36.42x). From Vásquez *et al.* (2008a).

4. Tenuipalpidae mites

The false spider mites include about 800 described species belonging to over 25 genera, mostly found in tropical and subtropical areas (Jeppson *et al.* 1975, Baker & Tuttle 1987). This family consists of three subfamilies: Tegopalpinae, Brevipalpinae and Tenuipalpinae, the two latter ones including most of the described species: *Brevipalpus* and *Tenuipalpus* are the two largest genera and also the most economically important pest on citrus (Kitajima *et al.* 1972), coffee (Chagas *et al.* 2000), passion-fruit (Kitajima *et al.* 1997) and ornamentals (Smith-Meyer 1979). Since 2004, *Raoiella indica* Hirst is also becoming more important in the Caribbean islands and northern countries in South America (Vásquez *et al.* (2008b).

The false spider mites are also known as flat mites because most species are dorsoventrally flattened. They are slow-moving and are usually found on the lower surface of the leaves near the midrib or veins. Some species feed on the bark while others live in flower heads, under leaf sheaths or in galls. Only a small number of species belonging to a few genera have become pests of economic plants and they are most commonly found on tropical fruit crops and ornamental plants.

Life history and biology: Thelytoky is commonly observed in *Brevipalpus* mites, since female offspring consist in females and rarely males are found (Childers *et al.* 2001), being males and females haploid (Pijnacker *et al.* 1980). Life cycle of *Brevipalpus* is shown in figure 4, consisting in four active stages: larva, protonymph, deutonymph and adult; an inactive stage being observed between each active one. According to Goyal *et al.* (1985), developmental rate depends on temperature, relative humidity and host plant species.

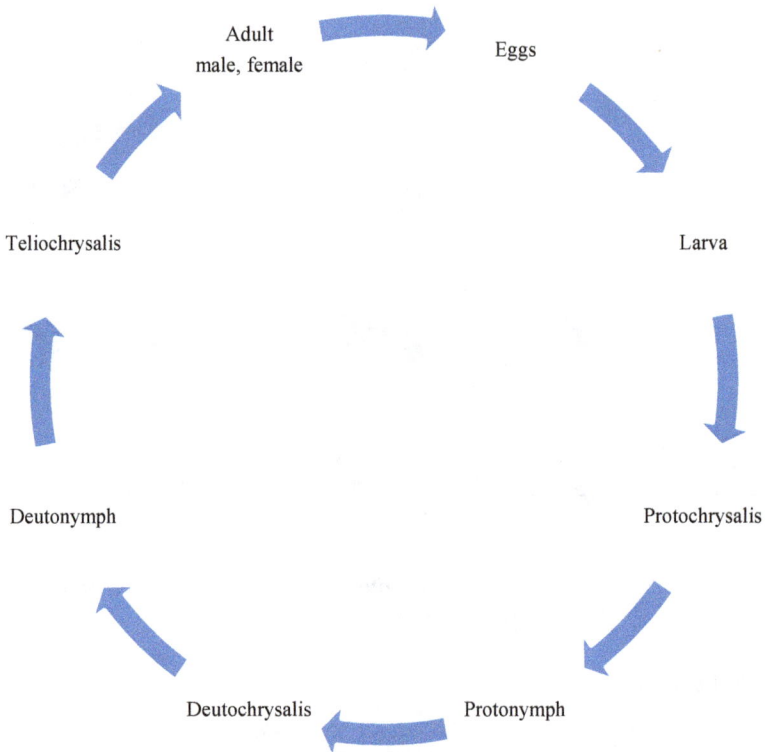

Fig. 4. Life cycle of *Brevipalpus* mites. From Childers *et al.* (2001).

4.1 Tenuipalpidae mite species of economic importance in South America

4.1.1 *Brevipalpus and Tenuipalpus* species

Brevipalpus californicus (Banks), *B. obovatus* (Donnadieu) and *B. phoenicis* (Geijskes) constitute the major economically important species in this genus. These mites usually feed on lower leaf surface and aggregate along the mid-vein or major lateral veins and they inject toxic saliva into fruits, leaves, stems, and bud tissues of their host plant, while *Tenuipalpus pacificus* Baker and

B. phoenicis have been observed feeding on the upper leaf surface of orchids (Childers *et al.* 2003). Ochoa *et al.* (1994) reported 177 host plant species in Central America, including 114 host plants for *B. phoenicis*, 29 for *B. californicus* and 34 for *B. obovatus*.

Probably the major threat related to these three *Brevipalpus* species is that concerning their ability to vectoring a virus borne disease called leprosis (Kitajima *et al.* 1996). Virus particles of citrus leprosis are short, bacilliform, 120–130 nm long (occasionally up to 300 nm), and 50–55 nm wide and occur mostly in parenchyma cells of the lesion area in affected orange leaves, fruits, or stems (Rodrigues *et al.* 2003). Leprosis is a serious disease on citrus from Brazil, Argentina, Paraguay and Venezuela, and probably Colombia and Uruguay (Childers *et al.* 2001). More recently, leprosis has been recorded in Panama (Dominguez *et al.* 2001) and it still represents a long term problem and causes severe damage on trees, yield reduction, and even death of trees if mite population control measurements are not taken into account (Rodrigues 2000).

4.1.2 The red palm mite, *Raoiella indica* Hirst

The red palm mite (RPM) was firstly reported in the Caribbean region in 2004 (Flechtmann & Etienne 2004). It is currently widely distributed in most of the Caribbean islands (Kane *et al.* 2005, Rodrigues *et al.* 2007). Subsequent mainland colonization was verified in Venezuela (Vásquez et al. 2008b) and Florida (Peña *et al.* 2008); and more recently, the pest has been reported to occur in Mexico (NAPPO 2009), Brazil (Návia *et al.* 2011) and Colombia (Carrillo *et al.* 2011). The RPM may cause severe damage to Arecaceae, especially coconut (*Cocos nucifera* L.), but also on Musaceae and other plant families (Flechtmann & Etienne 2004, Flechtmann & Etienne 2005, Etienne & Flechtmann 2006).

In coconut nurseries, mite feeding can provoke death of younger plants; meanwhile damage in adult plants is evidenced by a strong yellowish color in mature leaves (Peña *et al.* 2006, Welbourn 2005). After the red palm mite detection in the Americas, 70–75% fruit production decreasing has been estimated in Trinidad and Tobago and Venezuela (Návia 2008). Although the RPM primarily infests palms and bananas, some wild heliconia, gingers and other ornamental plant species growing under heavily infested coconut trees revealed small and sporadic colonies of the red palm mite (Roda 2008). However, according to Hoy *et al.* (2006), it is not clear whether these are valid host plants or whether the enormous mite populations on coconuts and other palms have temporarily moved on to substory plants under the palms.

Currently, the governments and researchers from those countries where the RPM has been recorded are making efforts to implement an alternative to pest management different from that of chemical control, since application of chemical products is not only hazardous to environment and public health, but also expensive. Thus, research is devoted to test the effectiveness of the predatory mite, *Amblyseius largoensis* Muma, the most frequent phytoseiid mite species found in association to the RPM in the Caribbean islands and Venezuela. Preliminary research has shown that this predator seems to be a promising agent of biological control (Carrillo *et al.* 2010, Rodríguez *et al.* 2010). However, when being considered in a population management program, other ecological strategies should also be included, such as using plant resistance from different genotypes which have been obtained in northern Venezuela by Dr. Antonio Ruiz (Instituto Nacional de Investigaciones Agrícolas [INIA-Irapa]), organic or inorganic fertilization programs in order to reestablish soil fertility and lastly, irrigation to avoid abiotic stress in old and young coconut plantations.

5. Eriophyid mites

The Eriophyoidea is a large superfamily of worldwide distribution. About 3,000 species belonging to over 250 genera are known in the world and a number of new species have been recently described from natural vegetation in South America (Table 6). The superfamily consists of three families: Phytoptidae, Diptilomiopidae and Eriophyidae. About three-quarters of the described species in Eriophyoidea belong to the Eriophyidae.

Species	Host plant	Country	Reference
Abacarus nectandrae *Aceria megalops*	*Nectandra membranacea* (Lauraceae) *Guapira opposita* (Nyctaginaceae)	Brazil	Flechtmann & Moraes (2002b)
Acalitus santibanezi	*Ipomoea murucoides* (Convolvulaceae)	Mexico	García-Valencia & Hoffmann (1997)
Aceria anisodorsum	*Caesalpinia peltophoroides* (Leguminosae)	Brazil	Flechtmann & Santana (2007)
Aceria coussapoae *Shevtchenkella desmodivagus* *Cosella ceratopudenda*	*Coussapoa microcarpa* (Cecropiaceae) *Desmodium barbatum* (Leguminosae) *Piptadenia gonoacantha* (Leguminosae).	Brazil	Flechtmann & Moraes (2002a)
Aceria inusitata	*Caesalpinia echinata* (Leguminosae)	Brazil	Britto *et al.* (2008)
Amrineus cocofolius	*Cocos nucifera* (Arecaceae)	Brazil	Flechtmann (1994)
Dichopelmus ibapitanga	*Eugenia uniflora* (Myrtaceae)	Brazil	Reis *et al.* (2010)
Epitrimerus torus *Epitrimerus angustisternalis*	*Acalypha reptans* (Euphorbiaceae) *Bougainvillea spectabilis* (Nyctaginaceae)	Brazil	Flechtmann (2010)
Juxtacolopodacus phalakros *Procalacarus perporosus* *Scolotosus centrolobii* *Scolotosus hartfordi* *Metaculus tanythrix*	*Mollinedia clavigera* (Monimiaceae) *Randia armata* (Rubiaceae) *Centrolobium robustum* (Leguminosae) *Centrolobium tomentosum* (Leguminosae) *Dicksonia selloviana* (Dicksoniaceae)	Brazil	Flechtmann & De Queiroz (2010)
Tetra tarabanensis	*Bulnesia arborea* (Zygophyllaceae)	Venezuela	Flechtmann & Vásquez (2007)

Table 6. Some eriophyoid mite species recently described from South America.

Eriophyoid mites are tiny worm-like or fusiform mites and they form galls or live freely on various host-plants (Royalty & Perring 1996; Westphal & Manson 1996). The wounding and injecting of specific salivary secretions into host-cells result in a specific response of the affected leaf, stem, or bud tissues; such as gall differentiation, hypersensitive reaction, or non-distortive feeding effects and in some cases complex symptoms, considered as syndromes (Petanović & Kielkiewicz 2010). Most species are monophagous and many species are limited to plant species within a single genus, with few exceptions (Zhang 2003). Most species cause little harm to their host plants, however, some rust mites and gall mites are important pests on economic plants.

Life cycle: Eriophyid mites passes through the egg, larva, nymph and adult stages. Both females and males complete their life cycle in about a week at an average temperature of around 25°C. The mating process is indirect, since male deposits spermatophores on host plants, and then the genital flap in the female presses the spermatophore into the body and crushes it. Females lay up to three eggs per day for up to a month, with a total of up to 87 eggs per female.

5.1 Eriophyid mite species of economic importance in South America

5.1.1 The coconut mite, *Aceria guerreronis* Keifer

The coconut mite is an invasive mite pest that has been disseminated and established rapidly in main coconut (*C. nucifera*) production areas (Návia *et al.* 2009) causing copra yield reductions, premature dropping of fruit (Moore & Howard 1996). *A. guerreronis* is mostly found in the meristematic region of fruit, under the perianth (Fernando *et al.* 2003). Development of colonies usually starts at earlier stages of fruit formation, resulting in discolored areas that frequently become larger and necrotic, longitudinal cracks and eventually producing an exudate (Haq *et al.* 2002). In addition to fruit damage, this mite attacks the growing tip of plantlets, turning it dark brown and often causing death of attacked plants in Brazil (Aquino *et al.*1968).

Despite the great economic importance of the coconut mite, information on basic aspects is scarce. Acording to Moraes & Zacarias (2002), the use of biological control agents based on the location of the *A. guerreronis* are crucial tools to manage this mite pest.

5.1.2 Tomato russet mite, *Aculops lycopersici* (Masse)

The tomato russet mite (TRM) is cosmopolitan in distribution and widespread in almost all areas where solanaceous crops are grown (Jeppson *et al.* 1975). This eriophyoid is one of the most common key pests of the commercially grown tomato, *Lycopersicon esculentum* Mill on a worldwide scale. In addition, the TRM host range includes tomatillo, potato, eggplant, poha (cape gooseberry), wild blackcurrant, popolo, wild gooseberry, blackberry, tobacco, bell pepper, cherry pepper, tolguacha, eggplant, Jerusalem cherry, hairy nightshade, black nightshade, horse nettle, morning glory, Jimson weed, Chinese thorn apple, petunia, nightshade, small flowered nightshade, amethyst, field bindweed, and Brinjal (Perring 1996). According to Duso *et al.* (2010), it is important to investigate: (a) TRM bioecological data useful to improve control strategies; (b) the different specialized intimate interactions that TRM establishes with different plants and/or different areas of the same plant; (c) the strength of the biochemical and physiological mechanisms/steps determining the intensity of closeness/dependence with the host plant.

6. Tarsonemid mites

The Tarsonemidae includes about 545 species of 45 genera widely distributed. This family consists of three subfamilies: Pseudotarsonemoidinae, Acarapinae and Tarsoneminae; the latter including most of the described species in the two large genera: *Tarsonemus* (over 270 species) and *Steneotarsonemus* (over 70 species). Tarsonemid mites exhibit various feeding habits, some species feed on fungus, algae, plants, and some of them can prey on parasite insects (Moraes & Flechtmann 2008). Some plant-feeding tarsonemid mites are pests of agricultural crops, most of them belonging to a few genera in the Tarsoneminae, except for the *Polyphagotarsonemus*. Since Tarsonemid mites feed on surface cells, more significant damage is observed in young tissues of the host plant. Symptoms are characterized by leaf discoloration with a silver aspect. Expanding leaves became shriveled or curled and eventually shed and plants severely attacked stop growing. Occasionally, plant tissue ontogeny is altered due to toxins injected.

Life history and biology: Tarsonemid mites are haplodiploid, being males produced by arrhenotoky and females by sexual reproduction; however, thelytoky has also been observed (Norton *et al*. 1993). Its life cycle consists of egg, larva, "pupa" and adult (male and female) stages, but "pupa" is an inactive stage in which the nymph stage takes place.

Polyphagotarsonemus latus (Banks) is undoubtedly the most important pest of many crops and ornamentals in field or greenhouses worldwide. This species disperses by wind, human transport of infested products and also through insects living on plants. *P. latus* females have been observed as phoront on *Bemisia tabaci* (Gennadius) on beans (*P. vulgaris*) in Colombia and on cucumber (*Cucumis sativus* L.) var. poinsett-76 and sesame (*Sesamun indicum* L.) var. INIA-1 in Venezuela (Bautista *et al*. 2005).

Life-history traits depend on temperature, host plants and even on varieties. On pepper, the developmental period (egg – adult) is, as an average, 4.1 days at 25°C for males and females, respectively. Adult female and male longevity is 11 and 15 days, respectively. The female/male sex ratio is 2.8 in the laboratory, and 2.3 on seedlings in a greenhouse. The intrinsic rate of increase is 0.359, the finite rate of increase is 1.43 individuals/female/day, the mean generation time 10.34 days and the net reproductive rate 41.0.

Tarsonemid mites of the genus *Steneotarsonemus* are phytophagous and specialized on monocotyledon plants (Lindquist 1986; Almaguel *et al*. 2000). In America, *Steneotarsonemus spinki* Smiley is considered the most destructive pest mite in rice ever (Rodríguez *et al*. 2009). In 1997, the rice mite was first reported in Cuba (Ramos & Rodríguez 1998), causing about 30 – 90% yield (Almaguel *et al*. 2000). Soon, *S. spinki* spread to the Caribbean (Haiti, Dominican Republic and Puerto Rico), Central America (Panama, Costa Rica, Nicaragua, Honduras and Guatemala), South America (Colombia and Venezuela) and North America (Mexico and USA) (Rodríguez *et al*. 2009, Sandoval *et al*. 2009). Although the rice mite has been detected in rice fields in Venezuela, important mite infestation focuses have not been observed to affect rice production in this country.

Pest control: Most information about control strategies of this species has been compiled in Cuba. In this country, nine predator species have been found, 4 Phytoseiid, 3 Ascid and 2 Laelapid species (Table 7).

Mite family	Species
Phytoseiidae	*Neoseiulus baraki*
	Neoseiulus paraibensis
	Proprioseiopsis asetus
	Neoseiulus paspalivorus
Ascidae	*Aceodromus asternalis*
	Asca sp.
	Proctolaelaps sp.
Laelapidae	*Aceodromus asternalis*
	Hypoaspis sp.

From Rodríguez *et al.* (2009).

Table 7. Predatory mites associated to *S. spinki* in rice fields in Cuba.

Another important tarsonemid species, *Steneotarsonemus furcatus* De Leon has been found on coconut, damaging fruit in Central America and in Brazil (Ochoa *et al.* 1994, Návia *et al.* 2005) and various gramineous species in Cuba (La Torre *et al.* 2005), and more recently, *Steneotarsonemus concavuscutum* Lofego & Gondim was examined on coconut in northeastern Brazil (Lofego & Gondim 2006).

7. Conclusions

In South America, there is a significant volume of literature dealing with taxonomy, biology and crop damage caused by phytophagous mites on agricultural crops. Research has provided a great deal of valuable information about their impact on several important crops in Brazil, Colombia, Mexico and Venezuela. At present, a multi-institutional and multi-disciplinary project on the biological control of the coconut mite in Tropical America is being developed. As a preliminary step, some explorations have been conducted in target countries to assess the abundance of the coconut mite, the diversity and prevalence of the natural enemies associated with this pest, and to evaluate the potential of these natural enemies as biological control agents for their use in the IPM programs. In addition, much work has been carried out to determine other host plants of the coconut mite and its natural enemies. Furthermore, some Government and Research Institutions from Brazil, Trinidad and Tobago and Venezuela are making efforts to find some ecological strategies to control the Red Palm Mite *Raoiella indica* Hirst (Acari: Tenuipalpidae), and thus minimize its impact on commercial coconut farms and other important crops in tropical areas.

8. References

Almaguel L, Hernández, J, Torre P de la, Santos A, Cabrera RI, García A, Rivero LE, Báez L, Cáceres I, Ginarte A. (2000) Evaluación del comportamiento del ácaro

Steneotarsonemus spinki (Acari: Tarsonemidae) en los estudios de regionalización desarrollados en Cuba. Fitosanidad, 4, 15 – 19.

Aquino MLN, Flechtmann CHW, Arruda GP (1968) Contribuição ao estudo do ácaro causador da necrose do olho do coqueiro em Pernambuco e seu controle. Recife, IPA, 17 pp. (Boletim Técnico, 34).

Baker AE, Tuttle M (1987) The false spider mites of Mexico (Tenuipalpidae: Acari) U.S. Department of Agriculture, Technical Bulletin No, 1706, 237 pp.

Bautista L, E Arnal, O Aponte (2005) Relación forética de *Polyphagotarsonemus latus* (Banks) (Acari: Tarsonemidae) y adultos de *Bemisia tabaci* (Gennadius) (Hemiptera: Aleyrodidae). Entomotropica, 20(1): 79 – 80.

Biswas GC, Islam W, Haque MM, Saha RK, Hoque KMF, Islam MS, Haque ME (2004) Some biological aspects of carmine mite, *Tetranychus cinnabarinus* Boisd. (Acari: Tetranychidae) infesting egg-plant from Rajshahi. Journal of Biological Sciences, 4 (5): 588 – 591.

Bolland HR, Gutierrez J, Flechtmann CHW (1998) World catalogue of the spider mite family (Acari: Tetranychidae). Leiden, Koninklijke Brill NV, 408 pp.

Britto EPJ, Gondim Jr. MGC, Návia D (2008) A new deuterogynous eriophyid mite (Acari: Eriophyidae) with dimorphic males from *Caesalpinia echinata* (Caesalpiniaceae) from Brazil: description and biological observations. International Journal of Acarology, 34 (3): 307 – 316.

Cardona C, Frei A, Bueno JM, Díaz J, Gu H, Dorn S (2002) Resistance to *Thrips palmi* (Thysanoptera: Thripidae) in beans. Journal of Economic Entomology, 95: 1066 – 1073.

Carrillo D, Návia D, Ferragut F, Peña JE (2011) First report of *Raoiella indica* (Acari: Tenuipalpidae) in Colombia. Florida Entomologist, 94(2): 370 – 371.

Carrillo D, Peña JE, Hoy MA, Frank JH (2010) Development and reproduction of *Amblyseius largoensis* (Acari: Phytoseiidae) feeding on pollen, *Raoiella indica* (Acari: Tenuipalpidae), and other microarthropods inhabiting coconuts in Florida, USA. Experimental and Applied Acarology, 52(2): 119 – 129.

Cerna E, Landeros J, Ochoa Fuentes YM, Luna Ruiz J, Vázquez Martínez O, Ventura López O (2009) Tolerancia del ácaro *Tetranychus urticae* Koch a cuatro acaricidas de diferente grupo toxicológico. Investigación y Ciencia, 44 (4-10): 4-10.

Chagas CM, Rossetti V, Colariccio A, Lovisolo O, Kitajima EW, Childers CC (2000) *Brevipalpus* mites (Acari: Tenuipalpidae) as vectors of plant viruses. *In:* Int. Congress Acarology (10, 2000, Melbourne). Halliday, RB; Walter, DE; Proctor, HC; Norton, RA; Colloff, MJ. (Eds) Proceedings. CSIRO Pub.

Childers CC, French JV, Rodrigues JCV (2003) *Brevipalpus californicus, B. obovatus, B. phoenicis,* and *B. lewisi* (Acari: Tenuipalpidae): a review of their biology, feeding injury and economic importance. Experimental and Applied Acarology, 30 (1-3): 5 – 28.

Childers CC, Kitajima EW, Welbourn WC, Rivera C, Ochoa R (2001) *Brevipalpus* como vectores de la leprosis de los cítricos. Manejo Integrado de Plagas, 60: 61 – 65.

de Ponti O (1977) Resistance in *Cucumis sativus* L. to *Tetranychus urticae* Koch. II. Designing a reliable laboratory test for resistance based on aspects of the host–parasite relationship. Euphytica, 26: 641 – 654.

Domínguez FS, Bernal A, Childers CC, Kitajima EW (2001) First report of the citrus leprosis virus in Panama. Plant Disease (Disease Notes), 85(2): 228.

Doreste E (1984) Acarología. San José, Instituto Interamericano de Cooperación para la Agricultura, 391pp.

Duso C, Castagnoli M, Simoni S, Angeli G (2010) The impact of eriophyoids on crops: recent issues on *Aculus schlechtendali*, *Calepitrimerus vitis* and *Aculops lycopersici*. Experimental and Applied Acarology, 51 (1-3): 151 – 168.

Esteves Filho, AB, Oliveira JV de, Torres, JB, Gondim Jr, MGC (2010) Biologia comparada e comportamento de *Tetranychus urticae* Koch (Acari: Tetranychidae) e *Phytoseiulus macropilis* (Banks) (Acari: Phytoseiidae) em algodoeiro bollgardTM e Isolinha não-Transgênica. Neotropical Entomology, 39 (3): 338 – 344.

Etienne J, Flechtmann CHW (2006) First record of *Raoiella indica* (Hirst, 1924) (Acari: Tenuipalpidae) in Guadalupe and Saint Martin, West Indies. International Journal of Acarology, 32: 331-332.

Evans GO (1992) Principles of acarology. CAB International. Wallindford, UK. 563 pp.

Ferla NJ, Marchetti M, Johann L, Haetinger C (2011) Functional response of *Phytoseiulus macropilis* under different *Tetranychus urticae* (Acari: Phytoseiidae, Tetranychidae) population density in laboratory. Zoologia (Curitiba), 28 (1): 17 – 22.

Fernando LCP, Aratchige NS, Peris TSG (2003) Distribution patterns of coconut mite, *Aceria guerreronis*, and its predator *Neoseiulus aff. paspalivorus* in coconut palms. Experimental and Applied Acarology, 31: 71 – 78.

Flechtmann CHW (1994) *Amrineus cocofolius* n.g., n.sp. (Acari: Eriophyidae) from Brazil. International Journal of Acarology, 20(1): 57 – 59.

Flechtmann CHW (2010) Two new *Epitrimerus* Nalepa (Eriophyidae, Prostigmata) species from Brazil and further remarks on the immature stages of a few eriophyoid species. Systematic & Applied Acarology, 15: 71 – 80.

Flechtmann CHW, de Queiroz DL (2010) New taxa in the Eriophyidae (Acari, Prostigmata) from forest trees in southern Brazil. Zootaxa, 2337: 18 – 30.

Flechtmann CHW, Etienne J (2004) The red palm mite, *Raoiella indica* Hirst, a threat to palms in the Americas (Acari: Prostigmata: Tenuipalpidae). Systematic and Applied Acarology, 9:109-110.

Flechtmann CHW, Moraes de, GJ (2002a) New Brazilian eriophyid mites (Acari: Eriophyidae). Zootaxa, 75: 1 – 12.

Flechtmann CHW, Moraes de, GJ (2002b) Three new species of eriophyid mites (Acari: Eriophyidae) from the State of São Paulo, Brazil. Zootaxa, 23: 1 – 8.

Flechtmann CHW, Santana DLQ (2007) A new deuterogynous eriophyid mite (Acari: Eriophyidae) from a semideciduous tree in Southern Brazil. International Journal of Acarology, 33(2): 129 – 132.

Flechtmann CHW, Vásquez C (2007) A new species and new records of plant mites of *Bulnesia arborea* (Zygophyllaceae) from Venezuela. Boletín del Centro de Investigaciones Biológicas, 40 (1): 57 – 67.

Flechtmann, CHW, Etienne J (2005) Un nouvel acarien ravageur des palmiers: En Martinique, premier signalement de *Raoiella indica* pour les Caraïbes. Phytoma, 548: 10-11.

Forero, G, Rodríguez M, Cantor F, Rodríguez D, Cure JR (2008) Criterios para el manejo de *Tetranychus urticae* Koch (Acari: Tetranychidae) con el ácaro depredador *Amblyseius*

(*Neoseiulus*) sp. (Acari: Phytoseiidae) en cultivos de rosas. Agronomía Colombiana, 26 (1): 78 – 86.

Frei A, Gu H, Bueno J, Cardona C, Dorn S (2003) Antixenosis and antibiosis of common beans to *Thrips palmi* Karny (Thysanoptera: Thripidae). Journal of Economic Entomology, 96: 1577 – 1584.

Freitez F, Alvarado G (1978) Contribución al conocimiento de ácaros en Musaceae de Venezuela (Acarina). Revista de la Facultad de Agronomía, 26:197–207.

Gallardo A, Vásquez C, Morales J, Gallardo S (2005) Biología y enemigos naturales de *Tetranychus urticae* en pimentón. Manejo Integrado de Plagas y Agroecología, 74: 34 – 40.

García-Valencia AS, Hoffmann A (1997) Especie nueva de ácaro eriófido en México (Prostigmata: Eriophyidae). Anuales del Instituto de Biología de la Universidad Nacional Autónoma de México, serie Zoológica, 6 (2): 253 – 260.

Gardner SN, Agrawal AA (2002) Induced plant defense and the evolution of counter-defense in herbivores. Evolutionary Ecology Research, 4: 1131 – 1151.

González Zamora JE, García-Marí F, Ribes A, SaquésJ, Masiello L, Orenga S (1993) Métodos de muestreo binomial y secuencial para *Tetranychus urticae* Koch (Acari: Tetranychidae) y *Amblyseius californicus* (McGregor) (Acari: Phytoseiidae) en fresón. Boletín de Sanidad Vegetal y Plagas, 19: 559-586.

Goyal M, Sadana GL, Sharma NK (1985) Influence of temperature on the development of *Brevipalpus obovatus* (Acarina:Tenuipalpidae). Entomon, 10: 125 – 129.

Greco NM, Sanchez NE, Liljesthrom GG (2005) *Neoseiulus californicus* (Acari: Phytoseiidae) as a potential control agent of *Tetranychus urticae* (Acari: Tetranychidae): effect of pest/predator ratio on pest abundance on strawberry. Experimental & Applied Acarology, 37:57 – 66.

Haq MA, Sumangala K, Ramani N (2002) Coconut mite invasion, injury and distribution, pp. 41 – 49. *In*: LCP Fernando, GJ Moraes & IR Wickramananda (Eds.), Proceedings of the International Workshop on Coconut Mite (*Aceria guerreronis*). Sri Lanka, Coconut Research Institute, 117pp.

Harborne J (1994) Do natural plant phenols play a role in ecology? Acta Horticulturae, 381:36 – 43.

Hilker M, Meiners T (2002) Induction of plant responses to oviposition and feeding by herbivorous arthropods: a comparison. Entomologia Experimentalis et Applicata, 104:181 – 192.

Hoy M, Peña J, Nguyen R (2006) Featured Creatures. University of Florida. Florida Department of Agriculture and Consumer Services, Division of Plant Industry. Available in: http://entomology.ifas.ufl.edu/creatures/orn/palms/red_palm_mite.htm#dist. July 19, 2010.

Jeppson H, Keifer H, Baker E (1975) Mites injurious to economic plants. Univ. of California. Press. Riverside. 614 pp.

Kane, E., R. Ochoa, G. Mathurin & E. Erbe. 2005. *Raoiella indica* Hirst (Acari: Tenuipalpidae): An island-hopping mite pest in the Caribbean. The ESA Annual Meeting and exhibition. Ft. Lauderdale, Florida, USA. Available in: http://www.sel.barc.usda.gov/acari/PDF/Raoiella%20indica-Kane%20et%20al.pdf.

Kitajima EW, Muller GW, Costa AS, Yuki VA (1972) Short, rodlike particles associated with citrus leprosis. Virology, 50:254-258.

Kitajima EW, Rezende JAM, Freitas JC (1996) Two types of particles associated with lesions induced by *Brevipalpus* mites in different plant hosts. VII. Encontro Nacional de Virologia. S. Lourenço, MG. 275 (abstract).

Kitajima EW, Rezende JAM, Rodrigues JCV, Chiavegato LG, Piza Jr CT, Morozini W (1997) Green spot of passion fruit, a possible viral disease associated with infestation by the mite *Brevipalpus phoenicis*. Fitopatologia Brasileira, 22: 555 – 559.

Klein C, Waterhouse DF (2000) The distribution and importance of arthropods associated with agriculture and forestry in Chile. ACIAR Monograph 68, Canberra, Australia. 234 p.

Krantz GW (2009) Habits and habitats. pp. 64 – 82. *In*: A manual of acarology. GW Krantz & DE Walter (Eds.) Texas Tech University Press. Lubbock, Texas, USA.

La Torre, P de; Almaguel L, Botta E, Cáceres I (2005) Plantas hospedantes de *Steneotarsonemus furcatus* de leon (Acari: Tarsonemidae) en Cuba. Neotropical Entomology, 34 (3): 517 – 519.

Lindquist EE (1986) The world genera of Tarsonemidae (Acari: Heterostigmata): a morphological, phylogenetic, and systematic revision, with a reclassification of family-group taxa in the Heterostigamata. Memoirs of the Entomological Society of Canada, 136, 1 – 517.

Lofego AC, Gondim Jr MGC (2006) A new species of *Steneotarsonemus* (Acari: Tarsonemidae) from Brazil. Systematic & Applied Acarology, 11: 195 – 203.

McMurtry JA (1970) Some factors of foliage condition limiting populations growth of *Oligonychus punicae* (Acarina: Tetranychidae). Annals of Entomological Society of America, 63:406–412.

McMurtry JA (1985) Avocado. pp. 327 – 338 In: Helle W, Sabelis M (Eds) Spider mites: their biology, natural enemies and control, vol. 1B. Elsevier Science Publishers B. V, Amsterdam, The Netherlands,

McMurtry JA, Johnson HG (1966) An ecological study of the spider mite *Oligonychus punicae* (Hirst) and its natural enemies. Hilgardia 37(11):363–402

Moore, D, Howard FW (1996) Coconuts. pp. 561 – 570. In: Lindquist EE, Sabelis MW & Bruin J (Eds.) Eriophyoid mites: their biology, natural enemies and control. Amsterdam, Elsevier.

Moraes GJ de, Flechtmann CHW (2008) Manual de acarologia: acarologia básica e ácaros de plantas cultivadas no Brasil. Ribeirão Preto, Holos, 288 pp.

Moraes GJ de, Zacarias MS (2002) Use of predatory mites for the control of eriophyid mites. pp. 78–88. *In*: Fernando LCP, Moraes GJ de, Wickramananda IR (Eds.), Proceedings of the International Workshop on Coconut Mite (*Aceria guerreronis*), 6–8 January 2000, Coconut Research Institute, Lunuvila, Sri Lanka.

Morrissey J, Osbourn AE (1999) Fungal resistance to plant antibiotics as a mechanism of pathogenesis. Microbiology and Molecular Biology Reviews, 63:708 – 724.

Morros ME, Aponte O (1994) Biología y tabla de vida de *Tetranychus ludeni* Zacher en caraota *Phaseolus vulgaris* L. Agronomía Tropical, 44(4): 667 – 677.

Morros ME, Aponte O (1995a) Efecto de dos niveles de infestación de *Tetranychus ludeni* Zacher sobre las fases de desarrollo de la caraota I: Nivel de campo. Agronomía Tropical, 54(2): 189 – 194.

Morros ME, Aponte O (1995b) Efecto de dos niveles de infestación de *Tetranychus ludeni* Zacher sobre las fases de desarrollo de la caraota II: Nivel de invernadero. Agronomía Tropical, 54(2): 195 – 202.

NAPPO (2009) Detecciones del ácaro rojo de la palma (*Raoiella indica*) en Cancún e Isla Mujeres, Quintana Roo, México. Notificación oficial de Plaga. Published on Nov 20, 2009.

Návia D (2008) Riesgo del "ácaro rojo de la palma", *Raoiella indica* Hirst, para Brasil. En: Acta del VI Seminario Científico Internacional de Sanidad Vegetal. La Habana, Cuba. 22-26 sept.

Návia D, Marsaro Jr AL, Silva FR da, Gondim Jr MGC, Moraes GJ de (2011) First report of the red palm mite, *Raoiella indica* Hirst (Acari: Tenuipalpidae), in Brazil. Neotropical Entomology, 40(3): 409 – 411.

Návia D, Moraes GJ de, Lofego AC, Flechtmann CHW (2005) Acarofauna associada a frutos de coqueiro (*Cocos nucifera* L.) de algumas localidades das Américas. Neotropical Entomology, 34(2), 349 – 354.

Návia D, Moraes, GJ, Querino, RB (2009) Geographic pattern of morphological variation of the coconut mite, *Aceria guerreronis* Keifer (Acari: Eriophyidae), using multivariate morphometry. Brazilian Journal of Biology, 69 (3): 773 – 783.

Norton RA, JB Kethley, DE Johnston & BM OConnor (1993) Phylogenetic perspectives on genetic systems and reproductive modes of mites. pp. 8-99. In: DL Wrensch & MA Ebbert (Eds.) Evolution and diversity of sex ratio in insects and mites. Chapman & Hall Publications, New York.

Ochoa R, Aguilar H, Vargas C (1994) Phytophagous mites of Central America: an illustrated guide. CATIE, Turrialba, Costa Rica, 234 pp.

Peña JE, Mannion C, Howard FW, Hoy MA (2008) Red palm mite, *Raoiella indica* Hirst (Acari: Prostigmata: Tenuipapidae). *In*: Encyclopedia of Entomology. Capinera J, Editor. Kluwer Academic Publishers, Dordrecht, The Netherlands.

Peña JE, Mannion CM, Howard FW, Hoy MA (2006) *Raoiella indica* (Prostigmata: Tenuipalpidae): The red palm mite: a potential invasive pest of palms and bananas and other tropical crops of Florida. University of Florida IFAS Extension, ENY-837. 2006 (July 12, 2007). Available in: http://edis.ifas.ufl.edu/BODY_IN681.

Perring TM (1996) Vegetables. pp. 593-606. *In*: Lindquist EE, Sabelis MW, Bruin J (Eds.) Eriophyoid Mites: their biology, natural enemies and control. Elsevier, Amsterdam, The Netherlands.

Petanović R, Kielkiewicz M (2010) Plant–eriophyoid mite interactions: specific and unspecific morphological alterations. Part II. Experimental and Applied Acarology, 51: 81 – 91.

Pijnacker LP, Ferwerda MA, Bolland HR, Helle W (1980) Haploid female parthenogenesis in the false spider mite *Brevipalpus obovatus* (Acari: Tenuipalpidae). Genetica, 51: 211 – 214.

Praslička J, Huszár J (2004) Influence of temperature and host plants on the development and fecundity of the spider mite *Tetranychus urticae* (Acarina: Tetranychidae). Plant Protection Science, 40(4):141-144

Quirós M (1978) Presencia del 'ácaro marrón del aguacate', *Oligonychus punicae* (Hirst) en hojas de plátano del Distrito Baralt, estado Zulia. Revista de la Facultad de Agronomía (LUZ), 5(1):422 – 424.

Ramos M, Rodríguez H (1998) *Steneotarsonemus spinki* Smiley (Acari: Tarsonemidae): Nuevo Informe para Cuba. Revista de Protección Vegetal, 13 (1): 25-30.

Reis AC, Gondim Jr. MGC, Návia D, Flechtmann CHW (2010) Eriophyoid mites (Acari: Prostigmata: Eriophyoidea) from fruit trees in Northeastern Brazil – a new genus, three new species and a redescription. Zootaxa, 2694: 43–56.

Ribeiro LG, Villacorta A, Foerster LA (1988) Life cycle of *Panonychus ulmi* (Koch, 1836) (Acari: Tetranychidae) in apple trees, cultivar Gala and Golden Delicious. Acta Horticulturae, 232:228

Rivero E, Vásquez C (2009) Biologia e tabela de vida de *Tetranychus desertorum* (Acari: Tetranychidae) sobre folhas de feijão (*Phaseolus vulgaris*). Zoologia (Curitiba), 26 (1): 38 – 42.

Roda A, Dowling A, Welbourn C, Peña JE, Rodrigues JCV, Hoy MA, Ochoa R, Duncan RA, De Chi W (2008) Red palm mite situation in the Caribbean and Florida. Proceedings of the Caribbean Food Crops Society, 44(1): 80-87.

Rodrigues JCV (2000) Relações patógeno-vetor-planta no sistema leprose dos citros. Thesis Ph.D. Piracicaba, Brasil,Centro de Energia Nuclear na Agricultura, da Universidade de São Paulo.

Rodrigues JCV, Kitajima EW, Childers CC, Chagas CM (2003) Citrus leprosis virus vectored by *Brevipalpus phoenicis* (Acari: Tenuipalpidae) on citrus in Brazil. Experimental and Applied Acarology, 30: 161–179.

Rodrigues JCV, Ochoa R, Kane EC (2007) First report of *Raoiella indica* Hirst (Acari: Tenuipalpidae) and its damage to coconut palms in Puerto Rico and Culebra Island. International Journal of Acarology, 33(1): 3–5.

Rodríguez H, Miranda I, Louis J, Hernández J (2009) Comportamiento poblacional de *Steneotarsonemus spinki* Smiley (Acari: Tarsonemidae) en el cultivo del arroz (*Oryza sativa* L.). Temas de Ciencia y Tecnología, 13 (39): 55 – 66.

Rodríguez H, Montoya A, Flores-Galano G (2010) Conducta alimentaria de *Amblyseius largoensis* (Muma) sobre *Raoiella indica* Hirst. Revista de Protección Vegetal, 25(1): 26 – 30.

Royalty RN, Perring TM (1996) Nature of damage and its assessment. pp. 493–512. In: Lindquist EE, Sabelis MW, Bruin J (Eds) Eriophyoid mites – their biology, natural enemies and control. Elsevier, Amsterdam.

Sandoval MF, Almaguel L, Fréitez F, Vásquez C (2009) Situación actual del ácaro del arroz, *Steneotarsonemus spinki* Smiley (Acari: Tarsonemidae) en Venezuela. Entomotropica, 24(3): 135 – 139.

Silva EA, Reis, PR, Carvalho, TMB, Altoé, BF (2009) *Tetranychus urticae* (Acari: Tetranychidae) on *Gerbera jamesonii* Bolus and Hook (Asteraceae). Brazilian Journal of Biology, 69(4): 1121 – 1125.

Smith-Meyer MKP. 1979. The Tenuipalpidae (Acari) of Africa with keys to the world fauna. Republic of South Africa. Dept.Agric.Tech.Serv. Entomology Memoir 135 pp.

Tello V, Vargas R, Araya J (2009) Parámetros de vida de *Tetranychus cinnabarinus* (Acari: Tetranychidae) sobre hojas de clavel, *Dianthus caryophyllus*. Revista Colombiana de Entomología, 35(1): 47 – 51.

Tomczyk A; D Krompczyńska (1985) Effects on the host plant. pp. 317 – 329. In: Spider mites: their biology, natural enemies and control, vol 1A. W. Helle and M. W. Sabelis. (Eds.) Elsevier, Amsterdam. 405 pp.

Vásquez C, Colmenárez M, Valera N, Díaz L (2007) Antibiosis of kidney bean cultivars to the carmine spider mite, *Tetranychus cinnabarinus* (Boisduval) (Acari: Tetranychidae). Integrated Control of Plant-Feeding Mites. IOBC wprs Bulletin, 30 (5): 133 – 138.

Vásquez C, Morales J, Aponte O, Sanabria ME, García G (2008a) Biological studies of *Oligonychus punicae* (Acari: Tetranychidae) on grapevine cultivars. Experimental and Applied Acarology, 45: 59 – 69.

Vásquez C, Quirós de G M, Aponte O, Sandoval MF (2008b) First report of *Raoiella indica* Hirst (Acari: Tenuipalpidae) in South America. Neotropical Entomology, 37 (6): 739 – 740.

Walter D, H Proctor (1999) Mites: ecology, evolution and behaviour. CABI Publishing. University of New South Wales Press, Sydney, New South Wales. 322 pp.

Welbourn C. (2005) Red palm mite *Raoiella indica* Hirst (Acari: Tenuipalpidae). Pest Alert, 2005. (on line). (Accessed on August 17th, 2011) Available in: htpp://www.daocs.state.fl.us/pi/enpp/ento/r.indica.html.

Westphal E, Manson DCM (1996) Feeding effects on host plants: gall formation and other distortion. pp. 231–241. In: Lindquist EE, Sabelis MW, Bruin J (Eds) Eriophyoid mites — their biology, natural enemies and control. Elsevier, Amsterdam.

Zhang ZQ (2003) Mites of greenhouses: identification, biology and control. CABI Publishing. Wallingford UK. 244 pp.

4

Biological and Ecological Studies on Land Snails and Their Control

Ahmed Sallam[1] and Nabil El-Wakeil[2,*]

[1]*Plant Protection Dept. Faculty of Agriculture, Sohag University, Sohag,*
[2]*Pests & Plant Protection Dept. National Research Centre, Dokki, Cairo*
Egypt

1. Introduction

Mollusca are the second largest phylum of the animal kingdom, forming a major part of the world fauna. The Gastropoda is the only class of molluscs which have successfully invaded land. They are one of the most diverse groups of animals, both in shape and habit. Among gastropods, land snails (subclass: Pulmonata) are one of the most numerous with almost 35,000 described species of the world. The present study dealt also with the chemical analysis of the mucus of three common land snails, *Eobania vermiculata* (Müller), *Theba pisana* (Müller) and *Monacha obstructa* (Montagu), and identification of the chemical compositions by using GC-MS. Results revealed that several variations in composition were observed between all species. Oxime, methoxy-phenyl and cyclotrisiloxane, hexamethyl were major components found that in three species, the total areas detected were 86.23, 76.83 and 70.83, respectively. This different composition of mucus may be due to differences from one species to another; different mechanical properties (function) are influenced by external factors such as temperature, humidity, light intensity, soil conditions and food supply. On reviewing literature conducted on land snails of Egypt, as far as can be ascertained, most studies were focused on Lower Egypt but Upper Egypt was neglected. So, the present investigation was designed to fulfil this gap and to promote and enhance the studies of land snails, especially those which have economic importance.

The Phylum Mollusca is probably the third most important animal group after the arthropods and vertebrates (South, 1992) Snails and slugs belong to the class Gastropoda. Snails and slugs are molluscs, a group of invertebrate animals with soft unsegmented bodies. Slugs are often described as snails without a shell, while snail bodies are enclosed in calcareous shells (Barker, 2001; Ramzy, 2009). The terrestrial mollusca including snails and slugs are destructive agricultural pests causing economic damage to a wide variety of plants including horticulture, field crops, and forestry. In addition they are of importance in medical and veterinary practice, since they serve as intermediate hosts for certain

Three common land snails, *E. vermiculata*, *T. pisana* and *M. obstructa*, are important crop pests and cause considerable damage in agriculture and horticulture, especially in areas where they find the conditions necessary for rapid multiplication, as shown in Fig. (1).

---

*Corresponding Author

Damage caused by snails depends not only on their activity and population density, but also on their feeding habits, which differ from one species to another. The damage involving considerable financial loss is inflicted on cereal, vegetables, Egyptian clover as well as other agricultural and field crops. The land snails feed on leaves, roots, tubers and ornamental plants (Bishara et al., 1968; El-Okda, 1981). In addition, during movement snails cause an undesirable smell which prevents men and even animals from feeding on these contaminated plants (El-Okda, 1984; Kassab & Daoud, 1964). Crops contaminated by snail slime lose their marketability and hence their export potential in many countries (Baker & Hawke, 1990; Ittah & Zisman, 1992). Land snails cause also a heavy damage to seed of oil plants and leaves of ornamental plants, as well as, citrus, peach, plam and vegetable, i.e. cabbage, carrot and bean. (El-Deeb et al., 1999; El-Okda, 1979, 1981; Ismail et al., 2003; Lokma, 2007; Shahawy et al., 2008).

The snail movement is rather slow and sluggish for a short distance depending upon temperature, food and natural of soil. They were active during optimum temperature, Humidity and moistened soil. They aestivate during the hot summer and hibernate during the cold winter (Kassab & Daoud, 1964). In Egypt, the land snails dispersing in northern Governorate, i.e. Alexandria, Kafr El-Shikh, Behera, and Domiate, Egypt (El-Okda, 1980; Hashem et al., 1993; Kassab & Daoud, 1964) At the present time these snails distribute in Ismaellia, Sharkia, Monofia, Gharbia, Minia, and Assiut Governorates, Egypt (El-Deeb et al., 2004; El-Massry, 1997; Metwally et al., 2002; Ramzy, 2009; Shoieb, 2008).

Gastropods such as slugs and snails secrete a trail of mucus from their pedal gland while traveling across a surface (Denny, 1983). The unique mechanical properties of snail pedal mucus enable the animal's locomotion while also causing the mucus to function as an adhesive to the substrate. The mucus trail performs a number of other functions, including the provision of mechanisms for re-tracing a path and for finding a mate of the same species by following a trail (Al-Sanabani, 2008). An understand ding of the functionality of trail mucus, including its interactions with water vapour, can therefore lead to a means of controlling the reproduction of snails and thereby limiting their impact on the environment, especially vegetable crops. When freshly deposited by terrestrial snails, trails of pedal mucus are reported to be in the range of 10 – 20 mm thick (Denny, 1989). But since the mucus typically consists of between 90 and 99.7% water by weight (Denny 1983), the trails dry to leave a much thinner solid film. It is generally believed that the fundamental structure of mucus gels consists of giant protein– polysaccharide complexes. This complex is usually classified into the broad categories of mucopolysaccharides and glycoproteins (Davies & Hawkins, 1998; Denny, 1983); mucus secretions can function as effective adhesives due to their viscoelasticity (Abd El-wakeil, 2005; Daoud, 2004; Grenon & Walker, 1980).

It is clearly known that successful control methods of terrestrial mollusca depend greatly on the broad base of knowledge of biological and ecological aspects of mollusca particularly in integrated approaches (El-Deeb et al., 2003a; Gabr et al., 2006; Ramzy, 2009; Shoieb, 2008). In Egypt, there is currently no information on the biochemical structure of *E. vermiculata, T. pisana* and *M. obstructa* mucus. A part of this work focuses on identifying the chemical composition of the mucus of three species of mollusks, i.e. *E. vermiculata, T. pisana* and *M. obstructa* under Egyptian conditions, and to compare the compositions between them. This book chapter is an attempt to gain information on land snails as agricultural pest and their control.

(A) Shell of *Eobania vermiculata* (Müller, 1770).

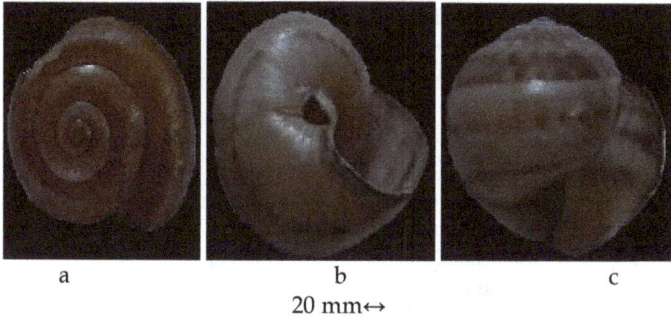

20 mm↔
(B) Shell of *Theba pisana* (Müller, 1774).

(C) Shell of *Monacha obstructa* (Montagu, 1803)
Fig. 1. Land snail types in Egyptian Agricultural fields (Cited from Ramzy, 2009).

2. Calssification of gastropods

The identification of terrestrial mollusca species could be classified according to the full description of Godan (1983) as follows:

Kingdom	:	Animalia
Sub Kingdom	:	Metazoa
Phylum	:	Mollusca
Class	:	Gastropoda
Sub class	:	Pulmonata

Order	:	Stylomatophora
Super Family	:	Helicoidae (Rafinesque, 1815)
Family	:	Helicidae (Rafinesque, 1815)
Genus	:	Cochlicella (Férussac, 1820)
Species	:	*Cochlicella acuta* (o.f. Müller, 1974)
	:	Cochlicella barbara (L., 1758)
	:	*Cochlicella ventricosa* (Draparnoud, 1801)
Genus	:	Helicella (Férussac, 1820)
Species	:	*Helicella obvia* (Hartman, 1840)
	:	Helicella bolenensis (Locard, 1884)
	:	*Helicella vestials* (Locard, 1882)
Sub family	:	Monacheae (Fitzinger, 1833)
Genus	:	Monacha (Fitzinger, 1833)
Species	:	*Monacha cantiana* (Montagu, 1803)
	:	*Monacha cartusiana* (o.f. Müller, 1774)
	:	*Monacha obstructa* (Pfeiffer, 1842)
Sub family	:	Helicinae (Rafinesque, 1815)
Genus	:	Theba (Riss, 1826)
Species	:	*Theba pisana* (o.f. Müller, 1774)
Genus	:	Cepaea (Held, 1837)
Species	:	*Cepaea hortensis* (o.f. Müller, 1774)
	:	*Cepaea silvatica* (Draparnoud, 1801)
	:	*Cepaea vindobonensis* (Férussac, 1821)
Genus	:	Eobania (Hesse, 1915)
Species	:	*Eobania Vermiculata* (o.f. Müller, 1774)
Genus	:	Helix (L., 1658)
Species	:	*Helix pomatia* (L., 1658)
	:	*Helix cantareus* (Risso, 1826)
	:	*Helix aperta* (Born, 1778)
	:	*Helix aspersa* (o.f. Müller, 1774)

Super family	:	Limacoidae (Rafinesque, 1815)
Family	:	Zonitidae (Mörch, 1864)
Sub family	:	Zonitinae (Mörch, 1864)
Genus	:	Oxychilus (Fitzinger, 1833)
Species	:	*Oxychilus alliarius* (Müller, 1822)
Family	:	Succinidae

Genus	:	Succinea
Species	:	Succinea ovalis

Family	:	Limacidae (Rafinesque, 1815)
Genus	:	Limax (L., 1758)
Species	:	*Limax maximus* (L., 1758)
	:	*Limax tenellus* (o.f. Müller, 1774)
	:	*Limax flavus* (L., 1758)
Genus	:	Deroceras (Rafinesque, 1820)
Species	:	*Deroceras reticulatum* (o.f. Müller, 1774)
	:	*Deroceras laeve* (o.f. Müller, 1774)
	:	*Deroceras caruanae* (Pollonera, 1891)

Family	:	Arionidae (Gray, 1840)
Sub family	:	Arioninae (Gray, 1840)
Genus	:	Arion (Férussac, 1819)
Species	:	*Arion ater* (L., 1758)
	:	*Arion rufus* (L., 1758)
	:	*Arion hortensis* (Ferussac, 1819)

3. Identification of snails

For identification of snails, the height and breadth of the shell as well as its shape and color are the main features (Fig. 2). The number of whards of the shell is established by observing the shell from above and counting downwards from the apex, which is clearly the beginning of the spire (Fig. 3). The dotted line indicates the extent of the first whorl. Compared with the shells of older snails, those of younger animals have a sharp edge, which is, moreover, neither thickened, folder back, nor enlarged. In those families in which these characteristics are not shown in the shells of older animals, the shell edge of younger snails is soft, flexible and without calcium deposits. The edge of the mouth (peristome) is the growing region of the shell and may be regarded as being composed of an outer lip (Godan, 1983).

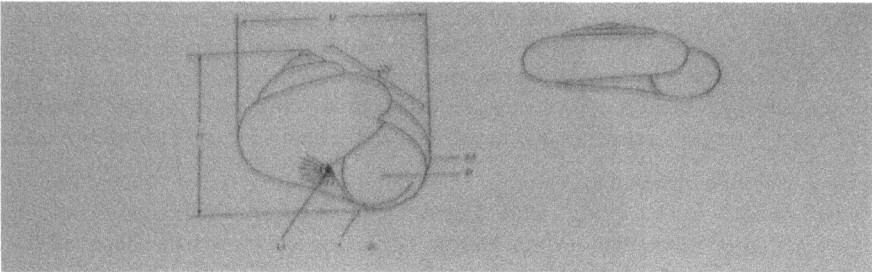

Fig. 2. Left shell concial to globular: right shell flat-conical. L. lip; *M. mouth* of shell (aperture), *P. peristome* (shell mouth edge); U. umbilicus; W. whorls; B. breadth; H. height (cited from Godan, 1983).

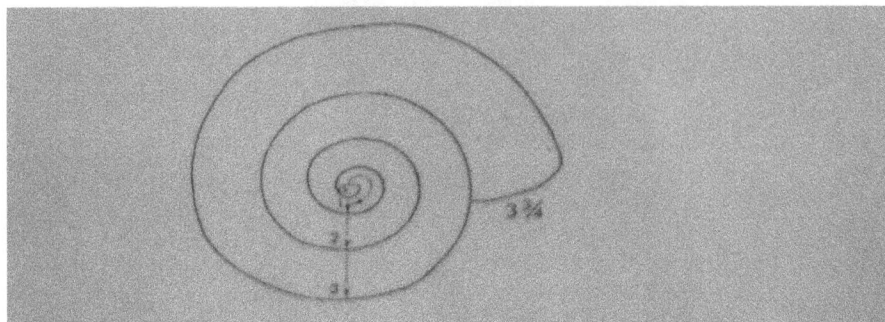

Fig. 3. Spire of the shell, Scheme, a apex; dotted line indicates extent of first whorl (cited from Godan, 1983).

4. Biology of terrestrial snails

Snails and slugs are simultaneously hermaphroditic, self incompatible. Five phases can be distinguished in the reproductive cycle of snails and slugs: Courtship, copulation, nest – building, egg – laying and embryonic development followed by egg hatching. Reproductive behaviour begins only when the humidity is high (80 – 85% for snail and 90 – 95% for slug).

Biological aspects of the land snails have been studied by many researches; for example, Kassab & Daoud (1964) found that the life cycle of *Helicella vestalis* was a relative simple. It laid the eggs in clutches, each contained from 25 to 30 eggs or more in the soft soil and were deposited in small cavities or holes in the soil. Eggs deposited at any time during spring season. The eggs were round and white in colour with calcareous or limy shell. Under normal conditions, the incubation period lasted from 12-15 days on the average. Soon after emerged from the eggs the young snails which were seen with their small and mucous shells began to move about in search for food. They were a little bit bigger than the head of a pin. Gradually, they added coils to the shell as they grew and rate of growth depended to some extent on the abundance of food and weather conditions. *H. lucorum* were sexually mature three years after hatching when the largest diameter of their shell was equal to greater than 25 mm as reported by Ramzy (2009) and Staikou et al. (1988).

Staikou & Lazaridou- Dimitriadou (1990) found that *M. cartusiana* (M.) reached maturity within one year at a size of 8 – 10 mm and could lay eggs immediately upon maturation. The reproductive period started in the beginning, middle or end of autumn depending on the weather condition which also effect on the growth of newly hatched individuals. While El-Massry (1997) revealed that *M. cartusiana* began to lay clutches from mid November to mid February. Number of clutches and clutch size were changed during breeding season.

Mohamed (1999) rearing snail *E. vermiculata* and slug *Limax flavus* (L.) at 20°C ± 1 and 80% R. H. for snail and 90% R. H. for slug and feeding on lettuce + cucumber and carrot. Investigated that some biological observation such as: Mating, egg laying and growth development.

Mating usually takes place at night, frequently on the soil surface. The snail *E. vermiculata* needs introductory behaviour (foreplay) with reciprocal lactile, oral contract and curving turns to reach on optimal position with respect to the genital opening of the partner (Fig. 4). This is followed by dart shooting, the pushing of a calcareous dart into the mating partner of body, which is assumed to facilitate meeting by increasing behavioural synchrony. Finally, the copulation is reciprocal; spermatophores are transferred after simultaneous intromission (Baur & Baur, 1999).

Fig. 4. Mating behaviour in snails (Cited from Godan, 1983).

Mating behaviour in slug *Limax flavous* is the two slugs twist around one another, like structure averted and intertwined to from a light spiral, and spermatophore are transferred. Courtship may last 3 – 4 hours in snail and slug (Fig. 4) (Godan, 1983).

Slug eggs are usually laid into soil holes and crevices into or on the soil, under stones and on decaying wood (Figs. 5, 6 and 7).

Fig. 5. The clutches of eggs of *Monacha obstructa* (A) and *Oxyloma elegans* (B) in soil (Cited from Ramzy, 2009).

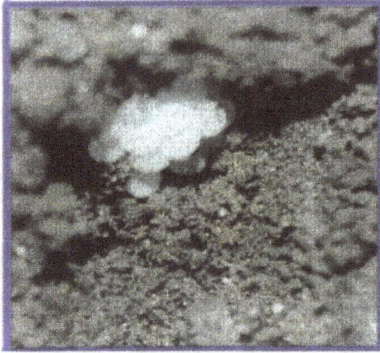

(1) Clutch of eggs of E. *vermiculata*

(2) Hatching

(3) Adult stage

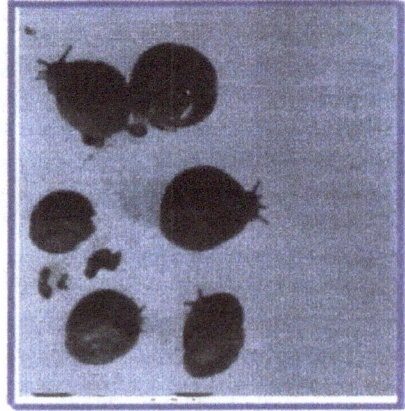

(4) The development of juveniles

Fig. 6. Life–cycle of E. *vermiculata* (cited from Mohamed, 1999).

Adult snail prepares to lay a clutch of eggs by excavating a deep hole in moist soil (Fig. 6). The animal pushes much effort in digging down with the anterior part to foot. Snails make a circular chamber about 5.7 – 6.1 cm. in deep. There is relationship between the body length and hole depth. This hole of egg – deposition is preparing after 10 hours of copulation.

During this time the snail is relatively conspicuous as the shell remains visible at the surface of nest. After that, egg deposition process takes place about 24 hours. Once egg – lying has finished the snail with draws, its foot and covers over the entrance of the hole with soil.

Snail eggs are whitish in colour and spherical shaped (2.9 – 3.0 mm diameter), while slug eggs are oval and strung along a thread (4.9 – 5.6 mm diameter). Snail and slug young does grow at a steady and relative fast. Growth continues at an almost constant rate even after animal maturation, by which time the slug has reached to weight 13 g. while the snail to 8.4 g. and the shell of snail is equal of 1/5 of a body weight. Animal snails and slugs arrive to adulthood after 321 to 364 days in average and their size and weight are more slowly increased (Figs. 6 and 7).

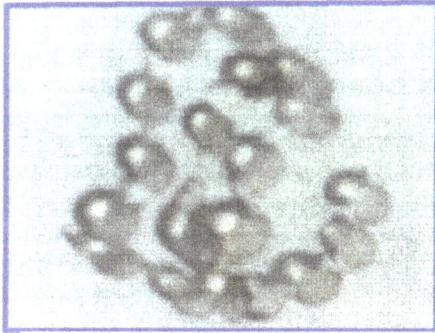

(1) Clutch of eggs of L. *flavus*   (2) Hatching

(3) The development of juveniles

(4) Adult stage

Fig. 7. Life-cycle of L. *flavus* (cited from Mohamed, 1999).

Generally, the highest number of eggs, hatchability percentage and longer life span were recorded in snail *E. vermiculata* and slug *L. flavus* in December followed by January, November, February and October, while these were decreased as temperature increased in March. The results are summarized in Table (1).

Type of land snail	Parameters			References
	No. of eggs/cultch	Hatchability	Life span(day)	
Eobnia vermiculata	80.92±5	91.43±3.2	1379±9.8	Mohamed, 1999
Limax flavus	62.23±2.7	91.882.1	677.87±10.4	Mohamed, 1999
Helicella vestalis	71.68	93.66	468.58	El-Massray, 1997
Helicella sp.	37.6	51.2	----------	Arafa, 1997
Eobania sp.	19	38.4	----------	Arafa, 1997
Monacha sp.	22.8	100	----------	Arafa, 1997
Monach cartus	22.55	99.1	----------	Ismail, 1997

Table 1. Some biological data of certain land snails.

5. Ecology of terrestrial snails

Terrestrial snails are mainly nocturnal, but following a rain may come out of hiding during the day. Temperature and moisture, rather than light, are the main factors to account for their nocturnal habits. Native snails may be found everywhere but prefer habitats offering shelter, adequate moisture, an abundant food supply and an available source of lime. Forested river valleys generally provide such habitats, and those with outcrops of limestone usually show the most abundant and varied mollusk faunas. Snails are very adaptable to times of drought and adverse climatic conditions. During these periods, the snails close the shell aperture with a mucus flap (epiphragm) which hardens and prevents desiccation. Snails can remain in this dormant state (aestivation) for years, breaking dormancy when climatic conditions are favorable again. Some ecological observation such as: Survey, population dynamic and movement, daily activity and dispersal of land snails, have been studied by many researchers (Bishara et al., 1968; Daoud 2004; El–Deeb et al., 1996, 2004; El–Okda, 1984; Metwally, et al., 2002; Ramzy, 2009).

5.1 Survey of land snails species

Bishara et al. (1986) found that *Euparypha pisana* (Müller), *Theba sp.*, *E. vermiculata, Rumina decolata, Helicella sp.*, and *Cochlicella acuta* were common species in field and orchards of the northern Delta Nile in Egypt.

El–Okda (1984) stated that *Monacha* sp. and *Oxychilus* sp. were found in Ismaellia Governorate on the Egyptian clover (*Trifolium alexandrium*), mango orchards, citrus and ornamental nurseries, in addition to wheat fields. He added that beans, watermelon, maize and tomato were attacked by land snails. El–Deeb et al. (1996) and (2003b) surveyed different terrestrial snails on the field crops, vegetables, ornamental plants and in orchards at Kafr El–Shekh and Dakahlia Governorate, Egypt. The obtained results showed that in Kafr El- Shekh Governorate, *E. vermiculata, Succinia putris* and *Cepaea nemoralis* snails were common. In Demietta Governorate *M. cartusiana, E. vermiculata, Cepaea nemoralis, C. acuta, Oxychilus aliavus* and *Helicella* sp. were recorded on different host plants. *M. cartusiana, Succinia putris, E. vermiculata, C. acuta* and *C. nemoralis* were recorded on different host plants at Dakahlia Governorate.

Metwally et al. (2002) found that six species of terrestrial mollusca belonging to families Helicidae and limacidae were recorded on different crops at 23 locatities representing 10 districts at Monofia and Gharbia Governorate. These species were *M. cartusiana*, (the glassy clover snail), *E. vermiculata*, (the Brown garden snail), *C. acuta*, (the conical snail), *O. alliarus*, the slugs, *Limax flavus* and *Deroceras reticulatum*, (the gray garden slug). *M. cartusiana* snail have the upper hand on snail incidence compared to other species; similar results were obtained by El-Deeb et al. (2003b). In Upper Egypt, Ramzy (2009) identified nine land snail species in Assiut governorate, Egypt. All the species recorded belong to order: Pulmonata from eight families. These species were *Pupoides coenopictus, Vallonia pulchella, Oxyloma elegans, Vitrea pygmaea, E. vermiculata, T. pisana, M. obstructa, Helicodiscus singleyanus inermis, and Cecilioides acicula*. The results are summarized in Table (2).

Species	Governorate	Referenceq
1- *Monacha sp.*; and *Oxychilus sp.*	Ismaellia	El- Okda, 1984
2- *E. vermiculata; Succinia putris;* and *Cepaea nemoralis*	Kafer El-Shekh	El-Deeb et al., 1996
3- *M. cartusina.; E. vermiculata.; C. nemoralis,; C. acuta.; Oxchilus aliavus.;* and *Helicella sp*	Demietta	El-Deeb et al., 1996
4- *M. cartusina.; Succinia putris.; E. Vermiculata.; C. acuta.; and C.nemoralis*	Dakahlia	El-Deeb et al., 1996
5- *M. Cartusina.; E. Vermiculata.; C. Acuta.; O. Aliavus.; Limax flavus* and *Deoceras reticulatum*	Monofia & Gharbia	Metwally et al., 2002
6- *Pupoides coenopictus, Vallonia pulchella, Oxyloma elegans, Vitrea pygmaea, Eobania vermiculata, Theba pisana, Monacha obstructa, Helicodiscus singleyanus inermis,* and *Cecilioides acicula*	Assiut	Ramzy, 2009

Table 2. Survey of land snail species in Egypt Governorates

5.2 Population dynamics of land snails

The population of *Monacha obstructa* in Egyptian clover field began to increase gradually from the end March to the middle of April, whenever the suitable temperature and humidity (Kady, 1983).

The life cycle, population dynamics and secondary production of the land snail *M. cartusina* in northern Greece was studied by Staikou & Lazaridou– Dimitriabou (1990). Demographic analysis of populations of *M. cartusina* revealed that (a). From two to three cohorts existed in the field throughout the year. (b). The reproductive period started in the beginning middle or end of autumn depending upon weather conditions. (c). Growth of the newly hatched individuals was also influenced by weather conditions. Population fluctuation of the land snails, *T. pisana, H. vestalis* and *C. acuta* in citrus orchards were noticed by Hashem et al. (1992). The main activity season collapsed from February to November for *T. pisana* and *C. acuta*. They added that the land snail *H. vestalis* proved to be the most abundant species reaching the peak during March to June (Shoieb, 2008).

5.3 Movement, daily activity and dispersal of land snails

Snail activity differs from one species to another; it is influenced by external factors such as temperature, humidity, light intensity, soil conditions and food supply. The migratory behavior of land snails is greatly affected by the microclimatic conditions in their habitat. Activity increased as a result fall in temperature below 21°C and arise till 30°C. Moisture also influenced the activity of land snails. The mucus of land snails consists of 98% water. At high temperature activity is inhibited by water lack. Light influences the activity of land snails. They remain in their hides during the day and only after dusk; they emerge to go in search of food. A decrease in light intensity with a fall in temperature below 21°C and a rise in humidity through dew fall at dusk, resulted in the animal moving on to vegetation, while in the morning and during the day the snails returned to the upper soil or between the earth clods where it is cool and shady (Godan, 1983). The dispersal of the land snail *T. pisana* in South Australia was influenced by variation in habitat; snails moved on free way average where grassy vegetation was scattered. Snails moved out of a well grazed permanent pasture to adjacent weedy roadside vegetation and trees in early summer. They returned to the pasture in autumn. Average movement varied between 0.1 m. and 1.1 m. per day, some snails moved > 55 m in a month in spring – autumn and 75 m in three month in autumn – winter (Baker, 1988; El-Deeb et al., 2003b).

The effect of moon light, temperature and relative humidity on daily activity of the small sand snail, *H. vestalis* was studied by El–Massry (1997). It was found that the activity of *H. vestalis* investing navel orange trees was significant different from time to time during the day and from season to another during the year. Also, moon light showed a significant effect on the daily activity of *H. vestalis*. The lowest number of active snails was recorded in summer, while the highest one was recorded in spring (Lokma et al., 2007). On the other hand, the time from 10 to 2 o'clock represented the time of the lowest activity, while the highest activity was recorded between the mid night and the six o'clock in the morning (Ismail et al., 2003).

6. Chemical analysis of mucus land snails

6.1 Sample collection

Mature snails, i.e. *E. vermiculata, T. pisana and M. obstructa* were collected in the field from different locations in Egypt during the winter and spring of 2005. These animals were transported in white cloth bags to the laboratory. Healthy individuals were kept in round plastic boxes (13 cm in diameter) containing moistened soil and feeding on cabbage paper from the market for one year under laboratory conditions (25±5°C temperature and 70±5% R.H.). Mucus (5 ml) was collected from roughly 100 individuals by stimulating the surface of live snails by small plastic syringe (5 ml). The samples were stored at -20°C in a deep freezer until analysis according to Sallam et al. (2009).

6.2 Chemical analysis

Mucus from *E. vermiculata, T. pisana,* and *M. obstructa* were analysed by GC-MS which was performed with an agilent 6890 gas chromatograph equipped with a mass spectrometric detector (MSD) model agilen 5973. A fused silica capillary column (HP-5MS), 5% phenyl polysiloxane as non-polar stationary phase (30 m x 0.25 mm x i.d) and 0.25 μm film thickness was used. Operating conditions were as follows: injector port temperature, 2508C. Helium was used as a carrier gas at a flow rate of 1.0 ml/min pulsed splitless mode programmed at 8°C/min to 260°C, and held for 18 min. The total analysis time was 41 min. A 1 ml volume was injected splitless. The mass spectrometric detector (MSD) was operated in electron impact ionization mode with an ionizing energy of 70 eV, scanning from m/z 50 – 500. The ion source temperature was 230°C and the quadruple temperature was 150°C. The electron multiplier voltage (EM voltage) was maintained at 1100 V above autotune, and a solvent delay of 3 min was employed. The instrument was manually tuned using perfluorotributylamine (PFTBA). Identification was based on comparison with the MS computer library (NIST Software Package, Finnigan) and on the respective retention indices. The separated components were identified by matching data with those of the data published by Wiley7n.1.

6.3 Chemical components of mucus

Chemical constituents of mucus from *E. vermiculata*, *T. pisana* and *M. obstructa* GC-MS analysis of mucus from *E. vermiculata* detected the presence of the following compounds: iso-valeric acid, methyl, 3-methoxyamino-propanoate, oxime, methoxy-phenyl, pantanolic acid, 4-methyl, cyclotrisiloxane, hexamethyl, tetradecanal, furan, 2-isobutenyl-4vinyl, and di-n-octylphthalate (Table 3). Data showed that eight compounds were identified in this mucus after comparison with library data by Wiley7n.1. Both oxime, methoxy-phenyl, and iso-valeric acid were mainly characterized by a high concentration of total compounds (81.58 and 8.76%, respectively), while methyl, 3-methoxyaminopropanoate, and di-n-octylphthalate were characterized by low concentration of total compounds (0.63 and 0.56%), respectively.

Peak No.	Name of compounds	Chemical formula	Molecular weight (MW)	Area %	Retention time (RT)
1	Iso-valeric acid	$C_5H_{10}O_2$	102.07	8.76	3.45
2	Methyl, 3-methoxyamino-propanoate	$C_5H_{11}NO_3$	133.07	0.63	3.51
3	Oxime-, methoxy-phenyl	$C_8H_9NO_2$	151.06	81.58	4.01
4	Pantanolic acid, 4-methyl	$C_6H_{12}O_2$	116.08	1.26	4.36
5	Cyclotrisiloxane, hexamethyl	$C_6H_{18}O_3Si_3$	222.06	4.65	6.63
6	Tetradecanal	$C_{14}H_{28}O$	212.21	1.01	16.45
7	Furan, 2-isobutenyl-4-vinyl	$C_{10}H_{16}O$	152.12	1.55	16.85
8	Di-*n*-octylphthalate	$C_{24}H_{38}O_4$	390.28	0.56	26.64
9	Total			100.00	

Table 3. Chemical composition of mucus from *E. vermiculata* (cited from Sallam et al., 2009).

Data in Table 4 show the chemical constituents of mucus from *T. pisana*. Ten compounds were identified in this mucus after comparing with library data from Wiley7n.1. Both oxime, methoxy-phenyl, and cyclotrisiloxane, hexamethyl were mainly characterized by a high concentration of total compounds (56.28 and 20.55%, respectively). In contrast, 1,2,4-trichloroacetophenone and pyridine, 1-acetyl-5-(3.4.-dihydro-2H-pyrrol-5-yl) 1,2,3.4-tetrahydro were characterized by a low concentration of total compounds (1.36 and 1.10%), respectively.

Peak No.	Name of compounds	Chemical formula	Molecular weight (MW)	Area %	Retention time (RT)
1	Butanoic acid, 2-methyl	$C_5H_{10}O_2$	102.07	2.89	3.62
2	Oxime-, methoxy-phenyl	$C_8H_9NO_2$	151.06	56.28	4.11
3	Cyclotrisiloxane, hexamethyl	$C_6H_{18}O_3Si_3$	222.06	20.55	6.69
4	1-Butyl-2,4,6-trimethyl benzene	$C_{13}H_2O$	176.16	5.81	7.60
5	Thymyl methyl ether	$C_{11}H_{16}O$	164.12	2.98	8.02
6	Pyridine,1-acetyl-5-(3,4-dihydro-2H-pyrrol-5-yl) 1.2.3.4-tetrahydro	$C_{11}H_{16}N_2O$	192.13	1.10	8.41
7	Cyclotetrasiloxaneoctamethyl	$C_8H_{24}O_4Si_4$	296.07	2.41	9.38
8	Benzene, 1,4-Bis(trimethylsilyl)	$C_{12}H_{22}Si_2$	222.13	3.87	10.22
9	Thiosulfuric acid, S-(2-aminoethyl) ester	$C_2H_7NO_3S_2$	156.99	2.74	20.18
10	1,2,4-trichloroacetophenone	$C_8H_5Cl_{30}$	221.94	1.36	26.45
11	Total			100.00	

Table 4. Chemical composition of mucus from *T. pisana* (cited from Sallam et al., 2009).

Also, data in Table 5 show that the chemical constituents of mucus from *M. obstructa*. Seven compounds were identified in this mucus, bothoxime, methoxy-phenyl, and diethyl phthalate were mainly characterized by a high concentration of total compounds (57.95 and 14.20%, respectively), while pentadecane and hexatriacontane were characterized by a low concentration (3.55 and 2.22%), respectively.

Peak No.	Name of compounds	Chemical formula	Molecular weight (MW)	Area %	Retention time (RT)
1	Oxime-, methoxy-phenyl	$C_8H_9NO_2$	151.06	57.95	3.97
2	Cyclotrisiloxane, hexamethyl	$C_6H_{18}O_3Si_3$	222.06	12.88	6.62
3	Pentadecane	$C_{15}H_{32}$	212.25	3.55	13.19
4	Diethyl phthalate	$C_{12}H_{14}O_4$	222.09	14.20	15.04
5	10-Methylnonadecane	C_2OH_{42}	282.33	4.51	16.62
6	Hexatriacontane	$C_{36}H_{74}$	506.58	2.22	18.04
7	1,2-Benzene dicarboxylic acid, dibutyl ester	$C_{16}H_{22}O_4$	278.15	4.70	20.23
8	Total			100.00	

Table 5. Chemical composition of mucus from *M. obstructa* (cited from Sallam et al., 2009).

It is obvious that the different composition of mucus from different species of land snails, oxime, methoxy-phenyl, and cyclotrisiloxane, hexamethyl were major components found in all species. This difference in composition of mucus may be due to differences from one species to another, different mechanical properties (function) and are influenced by external factors such as temperature, humidity, light intensity, soil conditions and food supply. These data agreement with Meikle et al. (1988) found substantial differences between the mucus of six coral species. It should not be surprising that different forms of mucus have different compositions and different mechanical properties. There is wide variation in the gross composition of mucus secretions (Davies & Hawkins, 1998) and in their function (Denny, 1989). As Davies & Hawkins (1998) point out, relatively little is known about the structure of invertebrate mucus secretions. In the study by Smith & Morin (2002), the composition of the adhesive form of march periwinkle mucus was compared to the trail mucus used during locomotion. They found that the trail mucus consists primarily of large, carbohydrate-rich molecules with some relatively small proteins. In contrast, the adhesive mucus has 2.7 times as much protein with no significant difference in carbohydrate concentration. This change in composition corresponds to an order of magnitude increase in tenacity with little clear change in overall concentration.

Previous research on marine mucus secretion found that roughly 50% of the dry weight was inorganic residue (Connor, 1986; Davies et al., 1990). This value can also be estimated from the percentage of inorganic salts in seawater (3.56%) (Schmidt-Nielsen 1990); for typical marine mucus containing 96 – 98% water, we would predict that inorganic material would make up 46 – 64% of the dry weight (Sallam et al., 2009).

7. Damage and feeding behaviour of land snails

Land mollusca pests are serious problem, every year; damage involving considerable financial losses is inflicted on cereal, potatoes, vegetables, lettuce, carrots, cabbage, maize, clover as well as other agricultural and horticultural crops. They eat leaves, root and tuber of nearly all vegetables, field crops, ornamental plants as well as fruits in field, garden and green house. Land snails cause heavy damage, especially to seeds and seedlings of cereals and seeds of oil plants. Damage was manifested in chewing soft vegetative growth, flowers and fruits, beside eating seeds, roots and tubers after sowing or during repining. On the other hand, land molluses left viscous liquids upon the plant on which they had been fed giving bright trace films. Moreover, unpleasant garlic odour was smelt on Egyptian clover which infested with *M. abstracta*, making farm animals refuse it. (El-Okda, 1980; Imevore & Ajayi, 1993; Ismail et al., 2003; Ramzy, 2009).

El-Okda (1980) mentioned that land mollusca attacked raw Succulent vegetables and proffered soggy parts. These pests attacked seeds, seedlings, roots and tuber crops. The more succulent raw leaf vegetables, fruits and buds were extra ordinarily attacked in addition to flower damage when land mollusca become abundant. Also, these land mollusca leave unpleasant slimy tracks on the injured parts.

Imevbore & Ajayi (1993) reported that when mature snails fed once daily at 2 % of body weight on diet containing 20 different feeds categorized as leaves, fruits, and household waste. Feed intake data indicated that the African giant snail a definite preference for fresh fruits and low preference for household waste.

The food preference and consumption of certain vegetable plants and field crops leaves for three land snails: *M. Cantiana, Succina Putris* and *T. pisana* was studied by El- Deeb et al. (2001). Results showed that Egyptian clover was the most preferred crop for *M. contiana* followed by lettuce, cucumber, carrot, cabbage and squash, while carrot fruits were the most preferable for *S. putris* followed by Egyptian clover, lettuce, cabbage, cucumber and squash. On the other hand, the lettuce was the most preferable for *T. pisana* followed by cabbage, carrot, cucumber, clover and squash, see Fig (8). In the same time, the results indicated that the bran was the most preferred bait for the three snails species followed by crushed wheat, crushed bread, crushed rice and crushed maize. The relative susceptibility of five fruits species (apple, orange, pear, plum trees and banana plants) to the four land snail species infestation was studied under laboratory conditions by identification of fruit leaf cells in their excrement. Results indicated that *T. pisana, H. vestalis* and *C. acuta* highly prefer pear, while *E. vermiculata* prefers banana. Orange ranks second in their favorability. On the contrary, plum, and apple are not preferred for the snail species, beside banana for *C. acuta*

Moreover, pear and orange are mostly attacked by *T. pisana* and banana by *E. vermiculata* (Shahawy et al., 2008; Tadros et al., 2001).

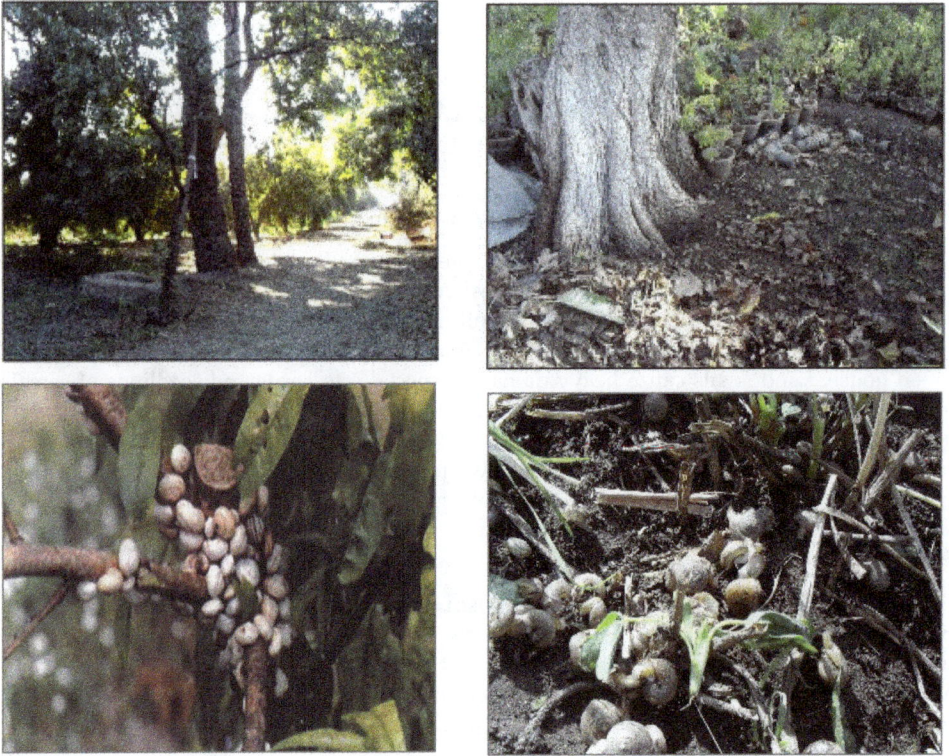

Fig. 8. Feeding damage by terrestrial snails on different plants (cited from Ramzy 2009).

8. Control of terrestrial mollusca

Molluscan pests can be destroyed in several different ways: by chemical means (Molluscicides), through the use of biotechnical measures, by mechanical methods and by interference with the environment. These different methods can, and sometimes have to be used simultaneously.

The control of pulmonates by whatever method is not simple, and each of the above-mentioned methods has its associated problems. A prerequisite from species to species is even in the same area.

8.1 Legislative control

Different ways of dispersal of economic pest mollusca may occur by man, plants, seeds, soil, ships, trains, airplane, cars, cargo, baggage, postal packages and food stuffs (Herzberg & Herzberg, 1962). For example, most of gastropds entering the USA with baggage come from Europe, Asia, Africa, Hawaii and the Philippines (Godan, 1983). Most Countries have now established quarantine regulations allowing important of plant and baggage under certain strictly specified conditions. Land mollusca among the economic pest species are listed in the rules and regulations of plant pest control in USA, the plant quarantine treatments contains dosages for fumigation of containers and ships to control gastropods.

Several investigators used the fumigants to combat the quarantinable snails. Richardson & Roth (1963) used ethyleue oxide 10 %, carbondioxide 90 % mixture (carboxide) instead of methyl bromide for fumigation against the very resistant aestivated *C. barbara* (L.) and *T. pisana* on military cargo from Mediterranean areas, and it showed good efficiency, penetration, and stability as a suitable substitute for methyl bromide.

Ittah & Zisman (1992) evaluated the volatile allylalcohol derivatives for control of *T. pisana* snails on cut roses for export. They found that allyl alcohol esters (propionate, formate, acetate) to be very effective in the fumigation of snails infesting flowers of roses. Concentration of allyl acetate of 0.022 mM at 25°C for 3 hrs was found to cause 100 % snail mortality without any fear of phytotoxicity towards roses.

8.2 Mechanical control

1. Hand collection with subsequent squashing of the slugs and snails is the oldest mechanical methods and was the only control measure used up to the middle ages (Carman, 1965; Mahrous et al., 2002; Shah, 1992). Common salt was then employed this causes an over reproduction of mucus by snails. Later, traps and protective barriers were used. The flower pots tiles, cabbage leaves, flat stones towels and also drain pipes, these hand to be searched for snails in early morning such dark shelters are especially effective as traps when they are baited with raw fresh potato slices of fresh lettuce leaves. It is necessary, however, to collect the snails daily and to kill them with a strong solution of common salt or in boiling water.
2. Burning –over- burning vegetation on which aestivating snails attach will reduce throwers or commercial weed- burners is effective in reducing snail populations along fence rows and in areas where other measures may not be practical. (Joubert & walters, 1951)
3. Protective barriers of dehydrating substance such as cattle salt, caustic soda, kainite or completely dry quick lime can act as barrier. Nakhla (1995) studied a mechanical method to protect orchard trees from land snails by using a band of metal sheets around the tree trunk. The rings tested were in the shape of cornea and were made of copper sheet, aluminum sheet, wire screen gauze (14 mesh) and a fiber cord, (Rape). The results revealed that wire screen rings gave the highest percentage of production against the different snails, i.e. *E. vermiculota, C. acuta, H. vestalis* and *T. pisana* followed by the copper sheet rings. Aluminum sheet rings gave intermediate protection while the rape rings gave poor protection

8.3 Agricultural control

Plough of the soil before sowing seeds of wheat, in early spring, is the most effective means of ecological control, that protect wheat from damage caused by slugs. This way results in exposure of the slugs and their eggs to predators and the weather (Wouters, 1970). Slugs need moist soil for egg-laying and cool shelter for protection during the day opening up the garden to more light and air movement will reduce the amount of favorable habitat (Mahrous et al., 2002; Shetlar, 1995). There are many methods which are effective in the agricultural practices:

1. Plowing-In open fields, adjacent to outside storage and on base housing areas, plowing the soil twice a year has been found to reduce both snails and slugs populations. Cultivating the soil in late autumn destroys many of the immature and adult snails, as well as eggs that have been deposited in the soil.
2. Disking and culti-packing. This is helpful in reducing land snail populations in areas where plowing may be not practical because of thin top soil or where erosion may be a serious problem. The mechanical action of the disc and culti-packr will eliminate many adult snails, while stirring the soil will destroy many eggs.
3. Improvement of soil structure, results in exposure of the sensitive egg and juvenile slugs to direct sunlight and they die of desiccation.
4. Clearing of the edges of fields and irrigation ditches after the harvest can reduce the level of infestation, but does not eliminate it altogether.

8.4 Biological control

In the following section the possibilities are discussed of using such environmentally harmless methods of control of molluscs, whether by diseases, parasites, predators or biotechnical procedures (Baker, 1989; Godan, 1983).

A. Fungi

Fungi attack mainly eggs of gastropods, for examples those, of *Deroceras reticulatum* by *verticullium chlamydosorium*. This fungus is not suitable for use in biological control, despite a high infection rat of the eggs (Mead, 1961). While fungal infections are important when rearing molluscs in the laboratory, since the may destroy the whole stock, in the field, fungi seldom cause total elimination or even a reduction in gastropod number.

B. Bacteria

Bacteria of the genera Escherichia, Alcaligenes and Bacillius, in France an aerobacter infection occurred on a *Helix fram* (Mead, 1961).

The combination between "victoback12" as (*Bacillus thuringiensis*) and the parasitic nematoda *Rhabditis* sp. can play an effective role in controlling *E. vermiculuta* as well as the other terrestrial snail and slugs in Egypt (Azzam & Belal, 2002)

C. Viruses

Very little is known about the effect of viruses on molluscs. This is a most neglected area of research.

D. Nematodes

The nematodes are the third large group of worms which parasitize molluscs; they are mostly of interest to veterinary medicine, but in a few cases also affect man. A comprehensive review of snails considered them as intermediate hosts of nematodes (Azzam, 1999; Azzam & Hegab, 2000; Wilson et al., 1996).

The slug nematodae *Phasmarhabditis hermaphrodita* (Schiender) has been successfully used control slugs in field experiments (Wilson et al., 1996). The snail parasitic nematode *Rhabditis* sp. was recorded for the first time in Egypt and showed high infectivity different snails, slugs and insects in laboratory (Azzam, 1998). Production of this nematode from different pests was investigated and it was found that *L. flavus* slug and *E. vermiculata* snail was the most adequate host producing high numbers of this nematode, as shown in Table (6) (Azzam, 1999).

Azzam & Hegab (2000) exposed the terrestrial snail, *E. vermiculata* to different concentrations (10-300 I.S/ Snail) of the snail parasitic nematode, *Rhabditis* sp. under laboratory conditions. They are found that the period needed for snail death decreased by increasing concentration.

E. Parasites and predators

E.1. Parasites

The most important of these parasitic include protozoa, flat worms, lung worms, carabid beetles, the glowworm larvae of lampyrid bettles as well as the larvae of Sciomyzidae (Diptera), as shown in tables (6 and 7). Protozoa associated with molluscs both parasitically and symbiotically (Baker, 1989; Godan, 1983). The sprrozoan *klossia* sp. has been found in the kidney in various slugs and snails, and an amoeba is found as on endo-parasite in *Arion rufus*

Tetrhymena rostrata (Kahl) was isolated from the soil in which the slugs live, showing that contamination occurs through the soil. Slug is probably infected by contact with cysts in the soil during the winter. It is also possible that the parasites are transferred by way of the eggs; the some eggs laid by *Deroceras reticulatum* were found to be infested with ciliates.

Field slugs infested with *T. rostrata* show the following symptoms: mantle shield swollen, posterior end of the body some what elongated and laterally compressed; after death the body shortened, and a tumor-like swelling of the mantle shield.

E. 2. Predators

Attempts have understandably been made to use predators for control of pest gastropods, especially for the biological control of land snails. For example *Byfo melanosticus, Rantigran tigrian* used in control *Lavicaulis alte* (Muthumani et al., 1992). Insects from order diptera, family sciomyzidae including 84 species as predators of Molluscs (Knutson, 1966; Neff, 1964; Verbeke, 1964). Beetles from order coleoptera, family carabidae, species, i.e. *Thermophilum hexaasticum* (Gerstaeker), *Carabus violaceus* (L.), and *Scophinorus striatopunctatus* (Choudoir) were used as biological control agents for *Achatina fulica*.

Fouad et al. (2000) illustrated the predation potential of some vertebrates, i.e. rodents, (*Rattus rattus rattus, Rattus norvegicus, Arvicanthes niloticus* and *Acomys cahirinus*) and birds (*Ardola ibis ibis* and *Bubulcus ibis*) for the different stages of three land snail species, *M. abstructa, T. pisana* and *E. vermiculata* under laboratory conditions.

It is concluded that the predation potential for the vertebrate animals markedly differed according to predator species and size of the prey or tested snail species as the rodent *R. norvegicus* exhibited a comparatively high potential for all stages of the tested gastropod species followed by *R. rattus* when compared with the *A. niloticus*, while *A. cahirinus* showed the lowest effect against all tested gastropod species. On the other hand, *A. ibis ibis* occupied the first order between bird species when compared with *B. ibis* birds. In the same time, the obtained data revealed that all tested predators exhibited a high predacious effect against all land snails particularly their juvenile stages as animals characterized by soft shell were more vulnerable for predator attacking, while the adults of snails were enable to protect themselves either by disappearing inside the hard shell or by releasing extensive mucus, as *M. obstructa* more acceptance to both vertebrate species which have small size than *T. pisana* and *E. vermiculata*.

Species	Host	
	Snail	**Slug**
Plathelmiuthes		
Trematoda		
Brachycoelium obesum (Nicolli)		*Deroceras agreslo* (L.)
Brachylaema helices (Pomaliae)	*Cepaea hortensis* ((Müller)	
Brachyiaema nicolli (Witenberg)		*Deroceras agreslo* (L.)
Brachyiaema obesum (Nicolli)		*Deroceras agreslo* (L.)
Dicrocoelium dendriticum (Rudolphi)	*Helicella obvia* (Hartmann)	*Arion fasciatus* (Nilsson)
(D. lanceolatum (Rudolphi)	*Helicella itala* (L.)	*Arion subfuscus* (Draparneud)
(Lancet liver fluke)(Grazing stock)	*Monacha cartisicne* (Müller)	*Deroceras reticulatum* (Müller)
	Cochlicelle acuta (Müller)	*Limax tenellus* (Müller)
	Cochlicopa lubrica (Müller)	
Leucochloridlum paradoxum	*Succinea putris* (L.)	
Monocerus Sp.		*Arion* spp.
Cestods		
Anomotaenia arionis (Siebold)		*Arion ater* (L.)
Aporina delafondi Railliet		*Arion rufus* (L.)
Choanotaenia crassiscolex (Linstow)	*Oxychilus cellarius* (Müller)	*Deroceras* spp., *Arion* spp.
Cyslicercoidos dukae (Holland)	*Succinea Putris* (L.)	
Davainea proglottina (Davaine)	*Helicella itala* (L.)	*Arion hortensis* (Ferussac)
	Helicella obvia (Hartmann)	*Arion intermedius* (Normand)
	Succinnea putris (L.)	*Deroceras caruanae* (Pollonera)
	Succineidae	*Deroceras leave, D. reticulatum* (Müller)
		Lehmannia marginata (Müller)
		Limax flaves (L.), *Limax maximus* (L.)
Hemenolepis multiformis (Creplin)		*Arion* spp.
Raillietina bonini (Megnin)	*Cepaea hortensis* (Müller)	*Arion* sp., *D. reticulatum* (L.)
		Marginata
Teania bothrioplitis (Filippi)	*Monacha cartusiane* (Müller)	
Nemathelminthes		
Nematoda		
Aelurostrongytus abstrusus (Railliet) (Lungworm of catts)	*Helix* spp.	*Arion circumscri Deroceras agreste., D. reticulatum*
Alloionema appendiculatum (Schneider)		*Arionater* (L.) *A. tufus* (L.) *D. agreste*
		D. agreste
Angiostoma helices (Conte & Bonnet)	*Helix asperea* (Müller)	
Angiostoma limacis (Dujardin) *Leptodera angiostoma* (Oujardin)		*Arion* spp., *Limax* spp.
Angiostrongylus cntonensis (Chen)	*Achatina fulica*	*Deroceras leave* (Müller), *Limax* spp.
Crenosoma vulpis (Dujardin) (Cat, Dog, Fox)	*Cepaea* spp., *Helix* spp., *Succinea* spp.	*Arion* spp., *D. reticulatum*
Cystocaulus ocreatus (Davtian)	*Cochilicella acuta* -Helicidae	*D. reticulatum* (Müller)
Hexamermis albicans (Siebold)	*Succinea patrus* (L.)	*D. agreste., D. reticulatum*
Mermis nigrescens (Dujardin)	*S. patrus* (L.)	*D. agreste., D. reticulatum*
Mullerius capillaries (Müller) (Lungworm of Sheep & Cattle)	*Cepaea vindobonensis, C. acuta H. obvia, H. Pomatia, M. cartusiane, T. pisana*	*A. ater, A.hortensis, D. laeve D. reticulatum, Limax* sp.
Protostrongylux nufescens (Longworm of Sheep)	*Cepaea* sp., *Helicella* sp.,*Helix* sp., *Monacha* spp.	

Table 6. Terrestrial gastropods intermediate hosts of worm parasites.

Molluscan predators have two ways of seizing their prey. If the prey is a snail, it is reached through the mouth of the shell, the predator penetrating deeper and devouring its prey as it does so: Ganaxis feed on terrestrial snails in this way. *Edentuline ovoidea* (Brugier) predatory snails used for control of *Achatina fulica* (Bowdied). Other predators on terrestrial gastropods, *Edentulina affiris*, and *Haplotrema minimum* (Ancey) are used to control the *H.*

aspersa (Müller). The latter species reduced the population and succeeds in keeping is number down. (Godan, 1983; Zeidan, 2001).

Species	Host	
	Snail	Slug
Amoeba		
Acanehamoebe sp		*Arion fasciatus* (Nilsson)
Amoeba sp		*Arion rufus* (L.)
Ryplobia hellcis (Trypanoplasma)	*Cepaea hortensis* (Müller), *C. nemorallis* (L.) *Helix pomatia* (L.) (in Receptaculum semnis)	
Trichomonas liracis	*Cernuella Virgata* (Dacosta)	*Limax* sp.
Sporozoa		
Isopora incerta (Schneider)		*Limex cinereoniger* (Wolf)
Isopora rera Aime (Schneider)		*Limax* sp.
Klossia loosi (Nabin)	*Helix, Cepaea and Succinea sp.*	*Arion* spp., *Limax* sp.
Pfeifferirclla impudica (Leger & Hollande)		*Lehmannia marginata* (Müller)
Plistiphora husseyi (Michaud)	*Achating zobra* (Sganzin)	
Trichodina echauinae	*Achatina zebra* (Sganzin)	
Ciliata		
Colpoda aspersa (Kahl)		*Deroceras agreste* (L.)
Colopoda steini (Maupas)		*Deroceras agreste* (L.) *Deroceras reticulatum* (Müller) *Lehmannia marainata* (Müller)
Concophthirus steenstrupi (Stein)	*Helix* spp.	*Arion ater* (L.) *Deroceras agreste* (L.)
Semitricholina sphaeronuclea	Zonitidae	-------------------------------
Tetrahymena limacis (Warren)	*Trichia lubomirskii* (Slosarski)	*Arion hortensis* (Ferussac), *Deroceras reticulatum* (Müller), *D. leave* (Müller), *Lehmannia marginata* (Müller), *Limax flavusl, L. maximus* (L.) *Milax gagates* (Dr.)
Tetrahymena rostrata	*Zonitoides nitidus* (Müller)	*Arion intermedius* (Normand) *Deroceras reticulatum* (Müller)

Table 7. Parasitic protozoan in terrestrial gastropods (cited from Zeidan, 2001).

Predation by the carnivorous snail, *oxychilis* sp. on the newly hatched and youngsters of *Monacha* sp. snails was observed by El-Okda (1984). Another predatory decollates snails (*Rumina decolata*). These snails are used very successfully in commercial citrus groves in California and provide excellent control to the brown garden snails (Fisher & Orth, 1985) and to slugs (Allikas, 1997).

8.5 Chemical control

Chemical control of exotic snails typically employs metaldehyde, methiocarb (Mesurol), salt, or combinations of these chemicals with other molluscicides in a myriad of bait formulations or foliar sprays.

8.5.1 Methods of application

Spraying, Dusting and use of pellet bait. Today the use of bait is the most common method of gastropod control in both agriculture and horticulture, where as spraying and dusting are less frequently used. El-Massry (1997) studied the effectiveness of certain pesticides namely, i.e. methomyl (Lannate); paraquate (Garamoxone); oxyfluorfen (Goal); Glyphosate (Lansar) and pendimethalin (Stomp) against adult stage of three species of land snails (*H. vestalis, M. contiana* and *E. vermiculata*) under laboratory conditions using three methods for testing, i.e. direct spray, dipping and poisonous bait technique. He found that toxicity of any tested compound was varied according to the method of application. Mothomyl was the most effective toxicant against adults of the three tested species (Gabr et al., 2006).

8.5.2 Chemical components as snailcides

A. Metaldehyde treatments

They are applied during dry climatic conditions are usually more successful than the degree of control achieved during damp, high humidity conditions, at which time snails are likely to be more active (Dax1, 1970; Moens, 1970). The principal toxic effect of metaldehyde is through stimulation of the mucous gland which cause excessive sliming, leading to death by dehydration. Metaldehyde is broken down into acetaldehyde by sunlight, so where possible the pellet should be put in shady places, particularly under the leaves of the affected plants (Henderson, 1970; Henderson & Triebskorn, 2002).

B. The pesticidal properties of methiocarb

They are similar to the toxic action of other carbamates which prevent effective nerve transmission by inhibiting the enzyme acetyl cholinesterase. The methyl carbamate most widely used as amolluscicide is methiocarb. This compound is more poisonous than metaldehyde pellets and less active as a contact killer, acting more as stomach poison when ingestion the baits for controlling slugs *D. reticulatum* (Getzin & Cole, 1964). The effectiveness of methiocarb is compromised less by low temperatures and high humidity than that of metaldehyde which is amagor advantage (Mallet & Bougran, 1971).

C. Differences in the symptoms of poisoning by metaldehyde and methiocarb

The symptoms shown by gastropods poisoned with metaldehyde differ markedly from those following carbamate poisoning (Godan, 1965). The metaldehyde can affect molluscs either by contact, with absorption through the skin, or through the gut when eaten. The main effect is that of an irritant, causing the molluscs to produce masses of mucus, leading to dehydration and sometimes death. Loss of mucus also means that the animals can no longer move around, so that death and dying animals are found close to the baiting site. Molluscs that have been poisoned by methiocarb can however, move around for a while, but then swell up with fluid and become immobile dying shortly afterwards. In dry conditions this swelling can be reduced, and some animals may recover, although generally recovery rates are lower than with metaldehyde (Abd El-Wakeil, 2005).

D. In addition to these molluscicides, sodium chloride (common table salt)

It is an effective dehydrating agent. It may be applied as ainch barrier application on the perimeter of known/ suspected snail infested area. During periods of rain or high relative humidity, salt are aestivating.

E. Further chemical compounds

They are being tested for their possible molluscicidal effects, even substance with other uses, such as fertilizer, fungicides, insecticides and herbicides (Van der Gulk & Springett, 1980). Fox (1964) used herbicides among the substance used for plant protection in agriculture and horticultural plants. The results indicated that herbicides were effective against land snails. Mode of action was in contact over long period of time with snail or as direct feeding on treated plants.

El-Massry et al. (1998) tested fertilizers, urea, ferrous sulfate and calcium super phosphate against many species of land snails; *H. vestalis, M. Contiana* and *E. vermiculata* in the laboratory. He found that urea the highest toxic effect against the three species of land snails, respectively, followed by ferrous sulfate, while calcium super phosphate showed the last toxic.

The evaluation of a molluscicide is based on testes of its toxicity, persistence of effectiveness, attractiveness to gastropods and also the chances of recovery of affected slugs or snails studies by many researchers (i.e. Arafa, 1997; El-Okda et al., 1989; El-Sebae et al., 1982; Ghamry et al., 1994; Mahrous et al., 2002; Radwan & El-Zemity, 2001).

El-Sebae et al. (1982) tested locally formulated bran baits containing aldicarb, methomyl or Dupont- 1642 against land snails, *H. vestalis, E. vermiculata* and *T. pisana*. Different wheat and rice brains are containing 0.5% aldicarb or methomyl showed high attractant action and toxicity for land snails, represented by their high mortality percentages.

El-Okda et al. (1989) evaluated the efficacy of the formulated local 0.5 %, aldicarb, oxamyl, methiocarb, Lannat and metaldhyde in controlling the land molluscs ; *H. aspersa, Eobania sp., Theba sp., Rumina* sp. and *oxychilus* sp. The results indicated that, aldicarb, oxamyl and Lannat gave the highest toxicity against the most snails and slugs species, while methiocarb and metaldehyde were less toxic.

Ghamry et al. (1994) evaluated fourteen insecticides against tow land snails; *M. contiana and E. vermiculata*. Results from bait testes revealed that, methomyl, dithiocarb, carbaryl, chlorpyrifos and dimethoate were effective for killing snails after 12 days under laboratory conditions. On the other hand, the same trend was observed with those insecticides under field conditions.

Arafa (1997) tested sucmate- granules, Mesural and Nuvacron against the land snail *E. vermiculata* under field conditions. The results revealed that sucmate- granules gave the highest mortality (100%) when used as poison baits while Novacron gave 46% mortality when it used as direct spray after two weeks. Mesural gave 48.5 % mortality percentage when it used as poison baits.

Radwan & El-Zemity (2001) synthesized a new series of 1,2,4- triazol derivatives and screened for their molluscicidal activity against two type of terrestrial snail, *H. aspersa* and *T. pisana*, by two methods of application, either as contact or as bran baits. Several of the tested compounds exhibited good molluscicidal activity, and *T. pisana* was more sensitive than *H. aspersa*. Substitution at the 0 and or P- positions of the phenyl ring with chlorine or bromine gave higher molluscicidal activity than unsubstituted compound, with O, P- dichloro substitution being optimum. In addition compounds containing two triazole moieties showed higher molluscicidal activity, particulary as stomach poisons, that the contact toxic effect of the corresponding compound with one triazole ring. In general, carbamate derivatives were more active than their corresponding 1,2,4- triazol derivatives.

Mahrous et al. (2002) tested seven pesticides to evaluate their molluscicidal activity as poisonous baits against *M. Cartusiana* in Sharkia Governorate, Egypt. The obtained results that the molluscicidal efficiency of the tested pesticides after 15 day- treatment could be arranged as follows: fenamiphos> sethoxydim> oxamyl> monocrotophos> butachlor> biofly and seed grad. On the other hand, carrier or attractive materials usually used in poisonous baits showed insignificant effect on the molluscicidal activity of fenamiphos in controlling *M. cartusiana*.

8.6 Botanical pesticides against molluscan pests

Plant products known to possess molluscicidal activity against the snails of agricultural importance are presented in Table (8) along with details of their forgeneric names, plant parts tested to possess biological activity and their formulations (Prakash & Rao, 1997). Azardirachtin, and active component isolated from neem kernel extract, was also reported to show mulluscicidal activity against *Lymnea luteda* (Ramesh, 1983). The molluscicidal activity of saponins isolated from saponaria roots was investigated (Kadey et al., 1982). Also, Kady et al. (1986) attributed the molluscicidal action of the wild herb, *Peganum harmala* (L.) (seeds) to its alkaloidal constituents which affect the respiration and/ or the nervous sytem of the snails. Kishor & Sati (1990) reported "spirostanol glycoside" from an ornamental plant, *yacca aloifolia* to be 100 % toxic at 10 ppm when tested against the snails like *Biumphalaria glabrata*. El- Hwashy et al. (1996) showed that the ethanolic leave extracts of Cauliflower, Oshar and pergularia were most effective against *E. vermiculata* snails when tested as residue film technique with mortality percentage of 88.8, 88.8 and 77.7 respectively. Molluscicidal activity and repellent properties of thirteen monoterpenoidal compounds were studied against the snail, *H. aspersa*. Camphor, thymol, (R)-carvone and carvacrol proved to be potent molluscicides of the compounds tested only citronellol, geraniol, (±) methanol and thymol were highly effective as repellents (El-Zemity et al., 2001). Repellency effect of 28 plant extracts obtained from different parts of 13 indigenous plants, i.e. Damsisa, Halfa barr, Colocynth, khella, Harmala, Datura, Santonica, Sucalyptus (L.), Eucalyptus (S), Enab eddhib, Calotropis, Alocasia, Halouk and Geranium, was investigated against *M. cantiana* land snails, using one and two choice feeding methods (Abd El- All et al., 2002). All plant materials either extracted with hexane and/ or ethanol showed a considerable snail repellency effect, when their crude extracts were tested using one choice feeding methods. In contrast results of two choice feeding test method Indicated that all the tested plant extracts showed snail repellant effect except Damsisa, Enabeddip, Calotropis and Geranium hexane crude extracts, which failed to achieve 50 % or more repellency level.

Generic name	Part of plant and its formulations	Biological activity	Reference
Azardirachta indica A. Juss. (Meliaceae)	Kernel extract of neem and neem cake extract	Toxicity to *Lymnaea matalensis* and *L.auricularea*.	Reynaud, 1986; Sasmal, 1991
Balantisa egyptica (L.) Delite (Simaroubaceae)	Nut and leaf powders	Toxicity to the snails	Belen, 1982
Citrus mitis (Linn.) Macf. (Rutaceae)	Dried fruit powder and its aqueous extract	Toxicity to the snails	Belen, 1982
Coryzadenia balsamifera Griff. (Hernandiaceae)	Fruit and leaf aqueous extracts	Toxicity to the snails	Belen, 1982
Croton figluim (L.) (Euphorbiaceae)	Leaf extract in water	Toxicity to the snails	Belen, 1982
Entade gigas(L.) Fawc and Rendle (Mimosaceae)	Leaf seed and bark extracts in water	Toxicity to the snails	Belen, 1982
Jatropha curcas (L.) (Euphorbiaceae)	Fruit and seed aqueous extract	Toxicity to the snails	Belen, 1982
Menispermum coculus (L.) (Menispermaceae)	Leaf, branch, fruit and seed extracts in water	Toxicity to the snails	Belen, 1982
Prago pabularia Hybrid (Cactaceae)	Aqueous leaf extract	Toxicity to the snails	Belen, 1982
Thevetia nerifolia Juss.ex Steud (Apocynaceae)	Alcoholic fruit extract	Toxicity to the garden snail	Johri et al., 1993

Table 8. Molluscicidal properties of plant products (Cited from Parakash & Rao, 1997).

8.7 Integrated mollusca control

Integrated pest management (IPM) is an economic necessity, and is vital for our modern agriculture, this approach requires a good understanding of all biological and ecological aspects of the mollusca in question (Snails or slugs). These methods include using all control procedures to suppress moluosca populations to non-damaging levels as follows:

1. Plough of the soil before sowing seeds, results in exposure of the sensitive egg and Juvenile slugs and snails to direct sunlight and they die of desiccation.
2. Hand collection of the snails and slugs daily and to kill them with a strong solution of common salt or in boiling water (Mahrous et al., 2002)

3. Burning over is a quick method of clearing land of pests before sowing crops.
4. Beer – baited traps have been used to trap and down slugs and snails, and scrape off the accumulated snails and slugs daily and destroy them by crushing (Olhendorf, 1996)
5. Protective barriers of dehydrating substance, will keep snails and slugs out of planting beds.
6. The use of poisonous bait is the most common method of gastropod control.

9. References

Abd El-A11, S.M., Zedan, H.A., El-Deeb, H.I., & Soliman, A.M. (2002). Repellent and toxic effect of various indigeneous plant extracts and pesticides against *Monacha cantiana* land snails. *Journal of Agricultural Science Mansoura University* 27: 4167-4181.

Abd El-Wakeil, K. F. (2005): Ecotoxicolgical studies on terrestrial isopods (Crustacea) in Assiut, Egypt. *Ph. D. Thesis, Assiut University Egypt,* 271pages.

Allikas, G. (1997). Orchid Online. *AOS magazine orchids,* Austuralia, pp. 32-38.

Al-Sanabani, A. S. M. (2008): Biological studies on terrestrial slugs (Gastropoda) in Assiut, Egypt with special reference of their ecology. *Ph. D. Thesis, Fac. of Sci., Thamar Univ., Yemen* 174 pages.

Arafa, A.A.I. (1997). Studies on some land molluscs at Sharkia Governorate. *M.Sc. Thesis Faculty Agricultural Al-Azhar University* 137pages.

Azzam K.M. (1998). First record of the snail parasitic nematode *Rhabditis* sp. isolated from Egyptian terrestrial snails and its compability to infect other pests. *Egyptian Journal of Biological Pest control* 8: 27-29.

Azzam, K.M. (1999). Production of the snail parasitic nematode *Rhabiditis* sp. from different pests. *Proceedings of 1st regional symposium for Applied Biological Control in Mediterranean Countries Cairo,* Egypt, 25-29.

Azzam, K.M. & Belal, M.H. (2002). Molluscicidal effect of a bacterial exotoxin, victoback12 as *Bacillus thuringiensis,* and the snail parasitic nematode, *Rhabditis* sp. as biocontrol agents against *Eobania vermiculuta* (Müller) snail. *Proceedings of 1st Conference of the central Agricultural Pesticide Laboratory,*pp. 499-501.

Azzam, K. M. & Hegab, A.M. (2000). Potential of the snail parasitic nematode, *Rhabditis* sp. in controlling the snail *Eobania vermiculata* (Müller) and its effect on Albino rat, *Rattus norvegicus. Journal of Agricultural Science Mansoura Unversity* 25: 1825-1829.

Baker, G.H. (1988). Dispersal of *Theba pisana* (Mollusca: Helicidea). *Journal of Applied Ecology* 25: 889-900.

Baker, G.H. (1989). Damage, population dynamics, movement and control of pest helicid snails in Southern Australia. *BCPC Mono. No. 41* slugs and snails in world Agric.

Barker, G. M. (2001). The biology of terrestrial Molluscs. *CABI publishing, New Yourk, USA:* 558pp.

Baker, G.H. & Hawke, B.G. (1990). Life history and population dynamics of *Theba pisana* (mollusca: Helicidae), in a cereal pasture rotation. *Journal of Applied Ecology* 27: 16 – 29.

Baur, B. & Baur A. (1991). Effect of courtship and repeated copulation on egg production in the simultaneously hermaphroditic land snail *Arianta abustorum* (L.). *Invertebrate Reproduction and Development* 21: 201-206.

Belen, E.M (1982). Plant extracts possessing rodenticidal and molluscicidal activities. *Monitor* 10: 6-7

Bishara, S.I., Hassan, M.S. & Kalliny, A.S. (1968). Studies on some land snails injurious to agriculture in U.A.R. Revue de zoologie et de botanique africaines Lxx VII (3-4): 239-252.

Carman, G.E. (1965). Electrical trapping device for land snails. *Journal of Economic Entomology* 58: 786-787.

Connor VM. 1986. The use of mucous trails by intertidal limpets to enhance food resources. Biol Bull 171: 548 – 564.

Daoud, M. I. A. (2004). Ecological and morphological studies on some terrestrial snails in Dakahlia governorate. *M. Sc. Thesis Agriculture Faculty Al-Azhar University* 177pages.

Davies, M.S. & Hawkins S.J. (1998). Mucus from marine mollusks. Advances in Marine Biology 34: 101 – 112.

Davies MS, Jones HD, Hawkins SJ. 1990. Seasonal variation in the composition of pedal mucus from *Patella vulgate* L. *Journal* of Experimental Marine Biology and Ecology 144: 101 – 112.

Daxl, R. (1970). Der Einflub von Temperature and relativer Luftfeuchte die Molluskizide Wirkung die Metaldehyde. Isolan and Ioxynil auf *Limax flavus* und dessen Eier. *Zeitschrift Angewandte Entomologie* 67: 57-87.

Denny, M.W. (1983). Molecular biomechanics of mollucan mucous secretions. *In*: Wilbur, K., Simkiss, K., Hochka, P.W., (eds.) Mollusca, Vol. 1, New York: Academic Press. pp 341 – 465.

Denny, M.W. (1989). Invertebrate mucous secretions: functional alternatives to vertebrate paradigms. *In*: Chantler, E., Ratcliffe, N.A. (eds.) Mucus and Related Topics. Symposia of the Society for Experimental Biology 43: 337 – 366.

El-Deeb, H. I., Abdel-Halim, A., Koutb, I., Khidr, F. K. & Edress, N. M. (2004): Studying some ecological aspects associated with the prevalent land snails at Kafr El-Sheikh governorate. *Journal of Agricultural Science Mansoura University* 29: 2847-2853.

El-Deeb, H. I., Ewies, E. A., Kandil, M. A., Gabr, W. M. & Mobarak, S. A. (2003a). Toxicity and biochemical studies of methomyl and diazinon on different ages of the land gastropod species *Monacha obstructa*. *Journal of Agricultural Science Mansoura University* 28: 7011-7023.

El-Deeb, H. I., Zidan, Z. H., Fouad, M. M. & Asran, F. D. (2003b). Survey Of Terrestrial Snails And Their Malacophagous Insects At Three Governorates In Egypt. *Egyptian Journal of Applied Science* 18: 355-361.

El-Deeb, H.I., Foda, M.E., Zedan, H. & Edress, N.M. (2001). Bait preference and food consumption of some land snails in Egyptian new reclaimed lands. *Journal of Environment Science* 3: 90-104.

El-Deeb, H.I., Wilson, M. & Eshara, E.H. (1999). Ecological studies on certain land snails infest some Economic crops at Beheira Governorate, Egypt. *Proceedings of 2nd Intentional Conference Pest Control*, Mansoura, Egypt, 19-28.

El-Deeb, H.I; El-Ghamry, E.M; El-Hwashy, N. & Essa, N. (1996). Relative abundance of some land snails in certain governorate of Egypt. *Journal of Agricultural Science Mansoura University* 21(8): 2977-2983.

El-Hawashy, N.M., Zedan, H.A. & Abd-All, S.M. (1996). Toxicity effect of certain indigenous plant extracts on *Eobania vermiculata* land snails in Egypt. *Journal of Agricultural Science Mansoura University* 21: 4133-4138.

El-Massry, S.A.A. (1997). Studies on the control of some land snails infesting certain fruit trees. *Ph.D. Thesis, Agriculture Faculty Zagazig University* 123-131 pp.

El-Massry, S.A.A., Ghamry, El.M., Hegab, A.M. & Hassan, A.I. (1998). Effect of certain fertilizers against some snails infesting navel orange trees. *Egyptian Journal of Applied Science* 13: 654-662.

El-Okda, M.K. (1979). Land snails of economic importance at Alexandria region with some notes on the morphological features classification, economic damage & population on ornamental plants. *Agricultural Research Review* 57: 125-131.

El-Okda, M.K. (1980). Land snails of economic importance on vegetable at Alex&ria & neighboring region. *Agricultural Research Review* 58: 79-85.

El-Okda, M.M.K. (1981). Locomation activity and infestation abundance of certain terrestrial mollusca in fruit orchard, Alexandria province, ARE. *Proceedings of 4th Arab Pesticides Conference*, Tanta University, Egypt 2: 279 – 287.

El-Okda, M.K. (1984). Land mollusca infestation & chemical control in El-Ismaelia Governorate. *Agricultural Research Review*, Egypt 62: 87-92.

El-Okda, M.M.K., Emara, M.M., & Selim, A.M. (1989). The response of the harmful & useful terrestrial mollusca towards several toxicant. Efficacy of six toxicants under laboratory conditions. *Alexandria Science Exchage* 10: 375 –385.

El-Sebae, A.H., El-Okda, M.M.K. & Marei, A.S. (1982). Local formation of three carbamated oxims as terrestrial molluscicides. *Agricultural Research Review* 60: 85-91.

El-Zemity, S.R., Mohamed, Sh. A., Radwan, M.A. & Sherby, Sh.M. (2001). Molluscicidal efficacy & repellency of some naturally occurring monoterpenoids against the land snail, *Helix aspersa*, Muller (Mollusca: Pulmonata). *Annals Agricultural Science Ain Shams University* Cairo, 46: 339-349.

Fisher, T.W. & Orth, R.E. (1985). Biological control of snails: observations of the snails, *Rumina decollate* (L.), 1758 (Subulinidae) with particular reference to its effectiveness in the biological control of *Helix aspersa* (Muller), 1974 in Californai occasional paper/ *Department Entomology, California- Riverside University* (1): 111 p., USA.

Fouad, M.M., Ibrahim, I.K., Khidr, F.K. & El-Deeb, H.I. (2002). The predational potential of some vertebrates for three land snail species. *Proceedings of 2nd International Conference, Plant Protection Research Institute*, Cairo, Egypt, 85-87.

Fox, C.J. (1964). The effect of five herbicides on the numbers of certain invertebrate in grass land soil control. *Journal of plants Science* 44: 405-409.

Gabr, W. M., Youssef, A. S. & Khidr, F. K. (2006): Molluscicidal effect of certain compound against two land snail species, *Monacha obstructa* and *Eobania vermiculata* under laboratory and field conditions. *Egyptian Journal Agricultural Research* 4: 43-50.

Getzin, L.W. & Cole, S.G. (1964). Evaluation of potential molluscicides for slug control. *Washington Agricultural Experiment Washington State University Bulltein* 658: 1-9.

Ghamry, E.M., Amer, k.Y. & Wilson, B.M. (1994). Screening test of some insecticides against two land snails *Monacha cantiana* & *Eobania vermiculata* in Sharkia Governorate. *Zagazig Journal of Agricultural Research* 21: 1539-1545.

Godan, D. (1983). Pest slugs & snails –Biology & control. *Springer Verlag Berlin Heidelberg*, 73-74: 287-293.

Godan, D. (1965). Untersuchungen über die Molluskizide wirkung der Carbamate. I. Teil: Ihre Toxizitat auf Nacktschnecken. *Zeitschrift Pflanzenkrank- und Schutz*, 72: 398-410.

Grenon, J.F., Walker, G. (1980). Biomechanical and rheological properties of the pedal mucus of the limpet, *Patella vulgata* L. *Comparative Biochemistry and Physiology* B 66: 451 – 458.

Hashem, A. G., Nakhla, J.M., Tadros, A.W., & Korashy, M.A. (1993). Monitoring land snails on sweet orange trees in Behera governorate. *Egyptian Journal of Agricultural Research* 20: 699-707.

Hashem, A.G., Nakhla, J.M. & Tadros, A.W. (1992). The seasonal fluctuation in the land snails population on citrus trees in the northern reclaimed lands. *Al-Azhar Journal of Agricultural Research* Cairo, Egypt 16: 325-337.

Henderson, I.F. (1970). The Fumigant effect of metaldehyde on slugs. *Annals Applied of Biology* 56: 507-510.

Henderson, H. & Triebskorn, R. (2002). Chemical control of Terrestrial gastropods. CAB International, Molluscs as crop pests Barker, G.M. (ed.), 31 pages.

Herzberg, F. & Herzberg, A. (1962). Observations on reproduction in *Helix aspersa*. *American Midland Naturalist Journal*, 68: 297-306.

Imevbore, E.A. & Ajayi S.S. (1993). Food preference of the African giant snail, Archachatina *marginata* in captivity. African J. Ecol., 31 (3): 265-267. C.F. Review Agricultural Entomology, 1994, 82: 891pp.

Ismail, S.A.A. (1997). Ecology, biology & control of certain terrestrial snails infesting some vegetables & field crops in Sharkia Governorate. *Ph.D. Thesis, Agriculture Faculty, Zagazig University* 107-116 pp.

Ismail, Sh. A. A., El-Massry, S. A. A., Khattab, M. M. & Hassan, A. Sh. (2003). Daily activity and damage caused by *Eobania vermiculata* Müller (Gastropoda) in citrus orchards. *Egypttain Journal of Agricultural Research* 18: 1-6.

Ittah, Y. & Zisman, U. (1992). Evaluation of volatile allyl alcohol derivatives for control of snails on cut roses for export. *Pest Science* 35: 183-186.

Johri, R., Katiyar, A.k. & Johri, P.K. (1983). Indigenous plant as molluscicide. *Journal of Applied Zoological Research* 4: 48-50.

Joubert, C.J. & Walters, S.S. (1951). Control of snails & slugs. *Farming in South Africa*, 26: 379-380.

Kady, M.M., El-Sebae, A.H., Abd El-Ghany, A.A. & Zedan, H.A.(1986). Alkaloids with molluscicidal activity from the wild herb *peganum harmala* (L.). *Journal of Agricultural Science Mansoura University* 11: 1246-1251.

Kady, M.M., Ghanim, A.A., El-Adl,A.A. & Nassar, O.A. (1983). Seasonal abundance of the terrestrial snail *Monacha obstructa* & its chemical control. *Journal of Agricultural Science Mansoura University* 8: 377-383.

Kady, M.M., Hssan, R.A. & Said, A.A. (1982). Chemical & toxicological studies on saponins isolated from saponaria roots. *Journal of Agricultural Science Mansoura University* 7: 370pp.

Kassab, A. & Daoud, H. (1964). Notes on the biology & control of land snail of economic importance in the U.A.R. *Jouranl of Agricultural Research Review*, Cairo, 42: 66-98.

Kishor, N. & Sati, O.P. (1990). A new molluscicidal spirostanol glycocide of *Yacca aloifolia*. *Journal of Natural Products* 53: 1557-1559.

Knutson, L.V. (1966). Biology & immature stages of malacophagous flies: *Antichaeta analis*, *A. brevipennis*, *A. obliviosa* (Sciomyzidae) *Transactions of the American Entomological Society* 92: 67-191.

Lokma, M. H. E. (2007). Studies on some terrestrial gastropods inguious to field crops at Sharkia governorate. *M. Sc. Thesis Faculty of Agriculture Zagazig University*, Egypt 147pp.

Mahrous, M.E., Ibrahim, M.H. & Abd-El-Aal, E.M. (2002). Control of certain land snails under field conditions in Sharkia Governorate. *Egypttain Journal of Agricultural Research* 29: 1041-1045.

Mallet., C. & Bougaran, H. (1971). Action molluscicide de divers carbamates. *In: XXième Symposium International de Phytopharmacie et de Phytiatrie*, 12: 207-215.

Mead, A.R. (1961). The Giant African snail: A problem in economic malacology 1-257. *Chicago University press, Chicago, Illinosi, USA*. C.F.Godan, 1983.

Meikle P., Richards G.N. & Yellowless, D. (1988). Structural investigations on the mucus from six species of coral. Marine Biology 99:187–193.

Metwally, A.M., Zedan,H.A., El-Saeid, A.B. & El-Akra, T.M.M. (2002). Ecological studies on certain land snails in Monofia & Gharbia Governorate. *Proceedings of 2nd International Conference, Plant Protection Research Institute*, Cairo, Egypt: 65-79.

Moens, R. (1970). Techniques chimiques de destruction de gites a limnees dans les prairies. Colloque information Scient., L" Assainissement des Prairies.par les Triatements molluscicides" *Gembloux centre de Recherches agronomiques de l'Etat*, 232-55, Gembloux.

Mohamed, M.F.M. (1999). Ecological & biological studies on land snails & slugs in Egypt. *Ph.D. Thesis, Faculty Agriculture. Cairo, University* 166-170 pp.

Muthumani, K., Jeyakumar, E., Mani, A. & Mathavan, S. (1992). Biological control of veronicellid slug, *Lavicaullius alte* (stylomatophorn :veronicollidae) using frog as a predator. *Environment and Ecology* 10: 649-652.

Nakhla, J.M. (1995). A mechanical method for controlling land snails on pear trees in Egypt. *Journal of Agricultural Research* 73: 357-363.

Neff, S.E. (1964). Snail-Killing Sciomyzid flies: Application in biological control. *Verhandlungen des Internationalen Verein Limnologie* 15: 933-939.

Prakash, A. & Rao, J. (1997). Botanical pesticides in agriculture. *CRC press, Inc. Lewis publishers*, 345- 354 pp.

Radwan, M.A. & El-Zemity, S.R. (2001). Synthesis & molluscicidal structure- activity relationships of some novel 1,2,4-triazole N-methyl carbamates. *Pest Management Science* 57: 707-712.

Ramesh, B. G. (1983). Neem extract destroys snails. *Science Reporter* (India) Dec., (Issue):23

Ramzy, R.R. (2009). Biological and ecological studies on land snails at Assiut, Egypt. *M. Sc. Thesis, Faculty of Science, Assiut University*, Egypt, 164 pages.

Reynaud, P.A. (1986). Control of the azolla pest, *Lymnaea natalensis* with molluscicides of plant origin. *International Rice News* 11: 27-28.

Richardson, H.H. & Roth, H. (1963). Ethylene oxide Fumigants to eliminate Quarantinable snails Cochlicelle or Theba in Cargo. *Journal of Economic Entomology*, 56: 836-839.

Sallam A.A.A., El-Massry S.A. & Nasr I.N. (2009). Chemical analysis of mucus from certain land snails under Egyptian conditions. *Archives of Phytopathology and Plant Protection* 42: 874–881.

Sasmal, S. (1991). Bionomics & control of pests of *Azolla pinnata* R.Br. *Ph.D.Thesis Utkal University Bhubaneswar, India* p168. C.F. prakash & Rao, 1997.

Schmidt-Nielsen K. (1990). Animal physiology: adaptation and environment. *Cambridge University Press* p 607.

Shah, S. (1992). Management of the African snail. *Indian, farming*, 41 (11): 21 pp, C.F. Rev. Agric. Entomol., 81 (7): 744 pp.

Shahawy, W. A., Hendawy, A. S., Abada, A. E. & Kassem, A. A. (2008). Land snails infesting rice plants and their accompanied parasitoids and predators at Kafr El-Sheikh governorate, Egypt. *Egyptian Journal of Agricultural Research* 86: 971-980.

Shetlar, D.J. (1995). Slugs & their management, Ohio State Univ. Ext. Factsheet HYG-2010-95.

Shoieb, M.A. (2008). Occurrence and distribution of terrestrial mollusks in Suez canal Governorates and south of Sinai, *Egyptian Journal of Agricultural Research* 86: 989-994.

Smith, A.M. & Morin, M.C. (2002). Biochemical differences between trail mucus and adhesive mucus from marsh periwinkle snails. *Biological Bulletin* 203: 338 – 346.

South, A. (1992). Terrestrial slugs. Biology, & control. Dept. of Biological Sciences, City of London polytechnic Chapman & Hall London New York, *Tokoy of Melbourne, Marras*, C.F. Zedan, *et al.* (2002)

Staikou, A. & Lazaridou- Dimitriadou, M. (1990). Aspects of the life cycle, population dynamics, growth & secondary production of the snail *Monacha cartusiana* (Muller) (Gastropoda: pulmonata) in Greece. *Malacologia* 31: 353-362.

Staikou, A., Lazaridou-Dimitriadou, M. & Farmakis, N. (1988). Aspect of the life cycle, population dynamics, growth & secondary production of the edible snail *Helix Lucorum*, (L.) (Gastropoda, pulmonta) in Greeca. *Journal of molluscan studies* 54: 139-155.

Tadros, A.W., Nakhla, J.M. & Rizk, M.A. (2001). The relative susceptibility of fruit tree species to land snail infestation as indicated by cells in their excrement. *Egyptian Journal of Agricultural Research* 79: 57-68.

Van der Gulik, J. & Springette, J.A. (1980). The effect of commonly used biocides on slugs. *Proceedings of 3rd N.Z. weed & pest control Conference*, pp. 225-229.

Verbeke, J. (1964). Contribution a etude des dipteres malacophages. II. Donnecs nouvelles sur la taxonomic et la repartition geographique des sciomyzidae palearctiques. *Bulletin de l'Institut Royal des Sciences Naturelles de Belgique Entomologie* 36: 1-15. C.F. Godan, 1983

Wilson, M.J., Hughes, L.A., Hamacher, G.M., Barahona, L.D. & Glen, D.M. (1996). Effect of soil incorporation on the efficacy of the Rhabditia nematode, *Phasmarhabditis hermaphrodita*, as a biological control agent for slugs. *Annals of Applied Biology* 128: 117-126.

Wouters, L. (1970). Schneckenbekampfungi in Oost-Flevoland. *Pflanzenschutz Nachrichten*, Bayer, 23, 173-177. C.F. Godan, 1983.

Zeidan, H.A. (2001). Biological control of land snails. *Proccedings of 3rd scientific conference on role of natural enemies in the biological control of agricultural pests*. Faculty of Agriculture, Mansoura University, Egypt pp. 46-60.

5

Research on One Kind of Essential Oil Against Drugstore Beetle *Stegobium paniceum* (L.)

Can Li

*Department of Biology and Engineering of
Environment, Guiyang University, Guiyang
People's Republic of China*

1. Introduction

Insect pest management in stored products is facing a crisis due to several serious drawbacks of using insecticides, such as the development of resistance in the treated pest, toxic residues, and the increasing cost of application (Tapondjou et al 2002). Hence it is necessary to develop alternative pest control techniques for protecting stored commodities (Gunasekaran and Rajendran 2005). Toward this end, intensified efforts have lead to an increasing number of research studies to find safe, effective and viable alternatives (Tapondjou et al. 2002). Carbon dioxide gas (CO_2) and essential oils from plants have received considerable attention for the control of stored products insects, because of their relative safety to the non-target organisms (AliNiazee 1972; Bekele et al 1996; Juliana and Su 1983; Sudesh et al 1996; Bouda et al 2001).

Chinese medicinal materials (CMM) are widely available in China. While most of these products are stored prior to use for health protection or disease treatment, great losses often occur during storage due to infestation by insect pests. Drugstore beetle *Stegobium paniceum* (L.) is the most widely encountered insect causing serious damages to stored products. It is one of the dominant species found on stored CMM in Hubei, Guizhou and Shandong, causing huge economic losses (Can et al 2004; Guilin and Wangxi 1996; Zhaohui and Fangqiang 2001). A series of measures have been taken to control infestation (Nielsen 2001; Hashem 2000, Toh 1998; Platt et al 1998). However, concerns over health and safety associated with traditional synthetic insecticides have prompted the development of plant-based insecticides.

The present study explores the efficacy of essential oil from *Z. bungeanum* Maxim against larvae and adults of *S. paniceum*. The contact toxicity and fumigant toxicity were investigated, and the subsequent development of treated insects was recorded.

2. Materials and methods

2.1 Extraction of essential oil

The fruits of *Z. bungeanum* were collected from the mountains in the west of China. After cleaning and pounding into powder, the materials were soaked in clear water (1:5, w:v) at

60℃ for 24h. After a 14 hour hydro-distillation, the crude oil was collected, dehydrated with anhydrous sodium sulfate and stored at 4℃ until use.

2.2 Collecting and rearing insects

The insects were obtained from a local CMM storehouse. They were mass reared on *Euphorbia* kansui Liou in a growth chamber under controlled conditions (29±1℃; 75±5% r.h.; light for 14 h).

2.3 Fumigant toxicity of crude oil to drugstore beetle

The oil was introduced onto a filter paper (7cm diameter, surface 38.5cm^2) placed in the center of a 1L glass jar. To prevent insects from contacting the oil, the insects and cultural medium were separated by a piece of gauze placed around the filter.

Larvae (25 to 30 days-old, n=20 in each of the three replicates) were exposed to concentrations of 12, 24, 48, 96 and 192 ul crude oil in the glass jars. Clear water was used as control. The jars were covered with fine gauze to prevent the insects from fleeing. The insects were incubated in the growth chamber at 29±1℃ and 75±5% r.h.. During the exposure, the test insects were fed on *E. kansui* Liou. After exposure of 6 to 144 h, the insect mortalities were determined respectively. Mortality was corrected by Abbotts (1925) formula when the mortality of control insects is above 10%.

Similarly, adult insects (2 to 3 days-old, 20 per replicate) were exposed to concentrations of 12, 24, 48, 96 and 196 ul in 1L glass jars respectively. The glass jars were covered by gauze to prevent the insects from flying away and they were kept in the same conditions as the larvae. Insect mortalities were determined after exposure to oil for 3 to 72 h.

2.4 Contact toxicity of oil to drugstore beetle

The crude oil was diluted with clear water to 200, 400, 667, 1000, and 2000 ppm respectively for this test. One ul of each of these solutions was introduced onto the back of one insect. The glass jars were placed into growth chamber at 29±1℃ and 75±5% r.h.. Three replicates of each dose and three controls treated with water were tested (n=20 per replicate). The toxicity to insects was assessed after treating for 12 to 144 h.

2.5 Observation of subsequent development

After LT$_{50}$ (The median mortality time) treatment, the living insect samples were transferred to similar sized jars and were incubated in the growth chamber, and untreated insects were transferred as a control group. They were under regular observation to assess their subsequent development in another day. The development time of the larvae and the number of eggs laid by each adult were recorded.

3. Results

3.1 Fumigant toxicity of essential oil to drugstore beetle

The percentages of mortality for insects exposed to different doses of oil are shown in Figures 1 and 2. In general, the higher mortality was achieved when the insects were exposed to a higher dose of oil or a longer exposure period. The dose of 192 ul/L of oil achieved complete mortality of larvae after the pests were exposed for 96 h and the dose of 96 ul/L achieved the same result after exposure of 120 h. The dose of 48 ul/l needed 144 h or longer to achieve complete mortality. The lowest dosage of 12ul/l induced little mortality (<10%) when the pests were exposed for 6h or shorter.

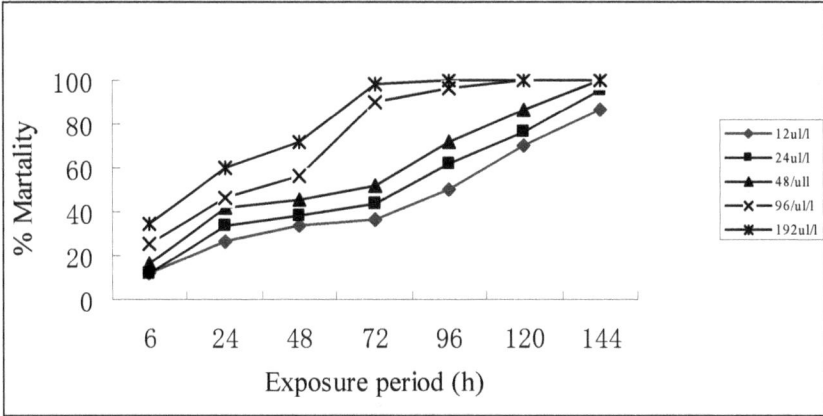

Fig. 1. Fumigant toxicity of essential oil to the larvae of drugstore beetle *S. paniceum*. Mortality was NOT corrected by Abbotts (1925) formula in all calculations because the mortality of control insect is <5%.

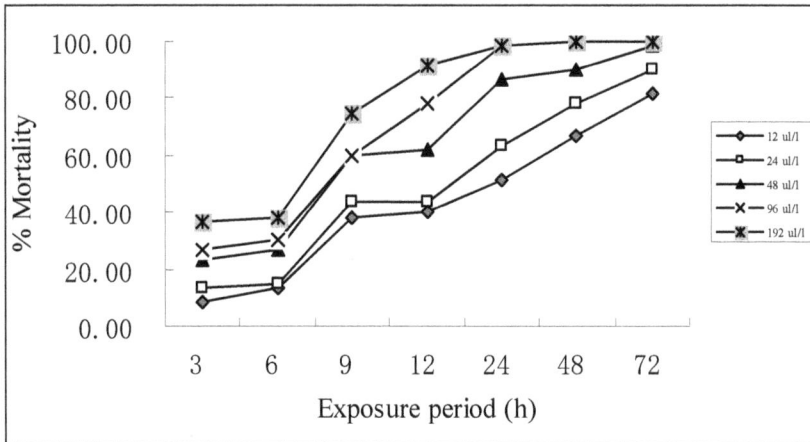

Fig. 2. Fumigant toxicity of essential oil to the adults of drugstore beetle *S. paniceum*.

Figure 2 illustrates the fumigant toxicity of essential oil on the adults. The trend is similar to that shown in Figure 1. With the increase of concentration and exposure time, the percentage of mortality improved. It reached 100% at the concentration of 96 or 192 ul/L of oil with the exposure time of 48 h. After 6 h treatment, the mortality rate was significantly higher than control group at all doses.

Significant differences of susceptibility were noted between adults and larvae under the same conditions. In fumigant test for 12h the calculated regression line equation was $Y=3.24+0.66X$ for larva and $Y=3.05+1.40X$ for adult, respectively (Probit analysis). Similar analysis of 24h treatment generated an equation $Y=3.61+0.70X$ for larvae and $Y=2.80+1.99X$ for adult. Comparison of LD_{50} values for the two life stages showed that the adult was more susceptible ($LD_{50}=24.57$ ul/l for 12h; $LD_{50}=12.64$ ul/l for 24h) than larva ($LD_{50}=485.84$ ul/l for 12h; $LD_{50}=98.92$ ul/l for 24h) (Table 1).

Life stages	Fumigant toxicity 24h		Fumigant toxicity 12h	
	LD_{50}(ul/l)	Regression model	LD_{50}(ul/l)	Regression model
Larva	98.92	y=3.61+0.70x,r=0.996	485.84	y=3.24+0.66x,r=0.988
Adult	12.64	y=2.80+1.99x, r=0.968	24.57	y=3.05+1.40x,r=0.976

Table 1. LD_{50} and regression model calculated for mortality within exposure period of12 and 24 h to fumigant toxicity.

3.2 Contact toxicity of essential oil to drugstore beetle

Figure 3 represents contact toxicity of the larvae of different concentrations of essential oil solution. The dose of 2000 ppm of solution achieved complete mortality of larvae after the pests were exposed for 120 h. The dose of 1000 ppm or 667 ppm of the solution also achieved complete mortality after 144 h exposure (Fig 3). The percentage of mortality of adult reached 100% at the concentration of 2000 ppm and exposure of 144 h. It also achieved complete mortality at 1000 ppm of the solution for 144h (Fig. 4).

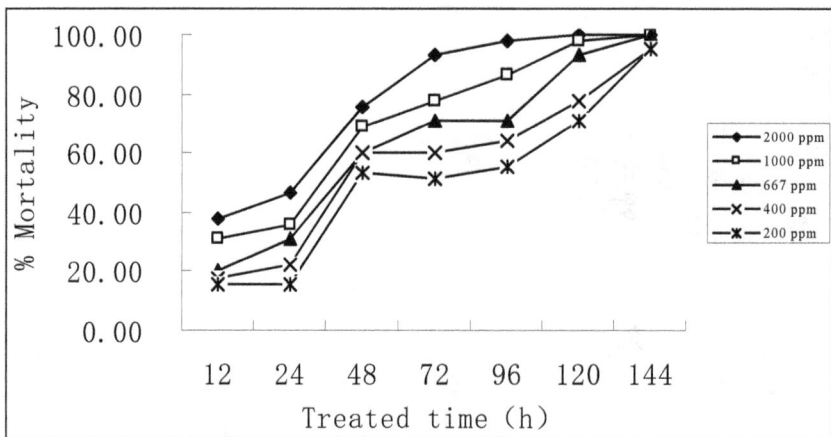

Fig. 3. Contact toxicity of essential oil to the larvae of drugstore beetle S. paniceum.

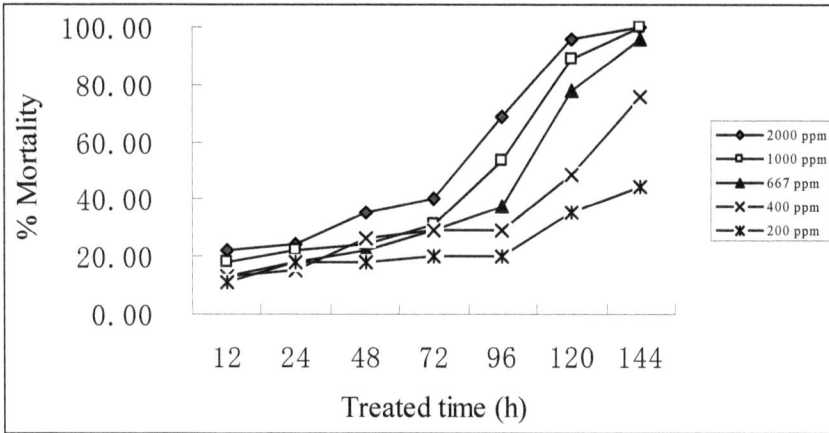

Fig. 4. Contact toxicity of essential oil to the adults of drugstore beetle *S. paniceum*.

The mortality of larvae was significantly higher than that of adults (Fig 3 and Fig 4) when treated for the same period and concentration. According to Probit analysis, the calculated regression line equation of 24 h for larva was $Y=1.82+0.91X$ and for adult was $Y=2.35+0.60X$. Similar trend was found with the 48 h data. Comparison of LD_{50} values for life stages showed that the larva was more susceptible ($LD_{50}=3175.91$ ppm for 24h; $LD_{50}=292.13$ppm for 48h) than adult ($LD_{50}=23610.74$ ppm for 24h; $LD_{50}=6784.18$ ppm for 48 h) (Table 2).

Life stages	Contact toxicity 24h			Contact toxicity 48h		
	LD_{50}(ppm)	Regression model		LD_{50}(ppm)	Regression model	
Larva	3175.91	y=1.82+0.91x	r=0.948	292.13	y=2.92+0.85x	r=9.981
Adult	23610.74	y=2.35+0.60x	r=0.967	6784.18	y=2.09+0.76x	r=0.986

Table 2. LD_{50} and regression model calculated for mortality within treated time of 24 and 48 h for contact toxicity.

3.3 The subsequent development of test insects

Subsequent observation on the adult pests revealed that the essential oil also prevented adults from laying eggs normally. Significant differences were observed on the number of eggs and development time of larvae between the treated and untreated pests (p<0.05). The development time of the treated larvae was 29.77±6.27 (d) (n=13), and untreated was 16.77±2.17 (d) (n=13) (mean ±SE). The mean number of eggs laid by treated adult was 14.21±3.28 (n=27), comparing with 63.50±23.94 (n=27) (mean ±SE) by the untreated insects.

4. Discussion

Contact and fumigant insecticidal actions of plant essential oils have been well demonstrated against pests in stored products (Murray B. Isman 2000). Several investigations have demonstrated contact, fumigant and antifeedant effects of a range of essential oil constituents against the red flour beetle *Tribolium castaneum* and the maize weevil *Sitophilus zeamais* (Ho et al 1994, 1995, 1997; Huang and Ho 1998). Essential oils from many plants have been developed as pest control agents. Perhaps the most attractive benefit of using essential oils (or their constituents) as protectants is their low mammalian toxicity (Murray B. Isman 2000).

Insecticidal activity of *Z. bungeanum* Maxim volatile oil was reported before . The results showed that *Z. bungeanum* Maxim volatile oil had high repellent and insecticidal activity against *Sitophilus* zeamais and *Tribolium* castaneum. The volatile constituents of the essential oil were identified by their retention index and mass spectrum in comparison with those of standard synthetic compounds (Zhi'an et al 2001). The major chemical constituents are C_9, C_{10} alcohols and alkenes. The most abundant constituent was 3-cyclohexen-1-ol, 4-methyl-1-(1-methylethyl) (33.578%), followed by Eucalyptol (15.656%) and Benzene,1-methoxy-4-(1-propenyl) (8.33%). Among those chemical components, beta-Phellandrence (23.2%), ailha-Pinene (1.89%) (Jun et al 2001) could be the key factors to control the pests.

The infestation of stored Chinese medicinal materials by drug store beetle is a serious problem. The use of insecticides is avoided because of toxicity. In this study, we investigated the effects of the essential oil from *Z. bungeanum* on the control of drug store beetle and found that it not only prolonged the development time of larvae but can also prevented adults from laying eggs successfully. Effective contact and fumigant insecticidal actions of such oil were demonstrated, and the essential oil may help control expansion of drugstore beetle populations. Adult drugstore beetles were more susceptible than larvae to fumigant action of the oil. Adults showed higher mortalities at the same concentration than larvae, likely cause by suffocation. However, larvae were more susceptible than adults to contact treatment. These results are likely because the respiration of adults is higher than that of larvae and the adults' hard elytra may have prevented the infiltration of oil. The distinct responses to oil at these two life stages may be attributed to the morphological and behavioral differences (Tapondjou 2002). It is important to note that the susceptibility varies during the life cycle when using these oils as pest control regime.

Essential oil from *Z. bungeanum*, as a plant product, is unlikely to affect the nutritional or medicinal composition of stored CMM. Its use can be beneficial for human health and environment safety (Murray B. Isma 2000). The effectiveness and safety of the essential oil supported its use as a natural pest control agent.

5. Acknowledgements

This research was sponsored in part by Guizhou provincial Modernization of Chinese Material Medical project (No.QKHZYZ[2011]5049),Organization Department of CPC Guizhou Committee [TZJF-2009-02] and Guiyang city science and technology bureau [No. 2010(1-4)].

6. References

AliNiazee, M.T. (1972). Susceptibility of the confused and red flour beetles to anoxia produced by helium and nitrogen at various temperatures. *Journal of Economic Entomology*, Vol.65,No1, pp. 60-64, ISSN 0022-0493

Bekele, J. A.; Obeng-Ofori, D.; Hassanali, A. (1996). Evaluation of Ocimum suave (Wieed) as a source of repellents toxicants and protectants in storage against three stored product insect pests. *International Journal of Pest Management* ,Vol.42, pp.139-142, ISSN 0967-0874

Bouda, H.; Tapondjou, L.A.; Fontem, D.A. & Gumedzoe, M.Y.D.(2001), Effect of essential oils from leaves of Ageratum conyzoides, Lantana sitophilus, Chromolaena odorata on the mortality of Sitophilus zeamais (Coleoptera, Curculionidae), *Journal of Stored Products Research*, Vol.37, No. 2, (April 2001) , pp. 103-109, ISSN 0022-474X

Can, L.; Zizhong, L.; & Youlian, Y.(2004). Analysis of the structure of insect community on the stored Chinese Medicinal Materials in Guiyang. *Journal of mountain agriculture and biology*, Vol.23,No.1 (February 2004),pp. 41-45 ISSN 1008-0457

Guilin, L. & Wangxi, D.(1996). The community quantitative characters of insects on the stored TCM materials. *Journal of huazhong agriculture university* Vol.15,No.4, (December 1996), pp.338-345. ISSN 1000-2421

Gunasekaran, N. & Rajendran, S. (2005). Toxicity of carbon dioxide to drugstore beetle Stegobium paniceum and cigarette beetle Lasioderma serricorne. *Journal of Stored Products Research*, Vol.41,No.3,pp.283-294. (September 2005),ISSN: 0022-474X

Hashem, M.Y. (2000). Suggested procedures for applying carbon dioxide (CO_2) to control stored medicinal plant products from insect pests. *zeitschrift fur pflanzenkrankheiten und pflanzenschutz-journal of plant diseases and protection*,Vol. 107,No.2,(June 2000),pp. 212-217. ISSN 0340-8159

Ho, S.H.; Cheng, L.P.L.; Sim, K.Y. & Tan, H.T.W.(1994). Potential of cloves (*Syzygium aromaticum* [L.] Merr. and Perry) as a grain protectant against *Tribolium castaneum* (Herbst) and *Sitophilus zeamais* Motsch. *Postharvest Biology and Technology*,Vol.4, No.1-2.(April 1994), pp.179-183,ISSN 0925-5214

Ho, S.H.; Ma,Y. & Huang, Y. (1997). Anethole, a potential insecticide from Illicium verum Hook F., against two stored product insects. *International Pest Control*, Vol.39, No.2, (June 1997), pp.50-51, ISSN 0020-8256

Ho, S.H.; Ma, Y.; Goh, P.M. & Sim, K.Y. (1995). Star anise, Illicium verum Hook F. as a potential grain protectant against *Tribolium castaneum* (Herbst) and *Sitophilus zeamais* Motsch. *Postharvest Biology and Technology*, Vol.6, No.3-4 (October,1995), pp 341-347. ISSN 0925-5214

Huang, M.T.; Ferraro, T. & Ho, C.T.(1994). Cancer chemoprevention by phytochemicals in fruits and vegetables. Food Phytochemicals for Cancer Prevention. *IACS Symposium Series*,Vol.546, 2-16(December 1993) ISBN13: 9780841227682eISBN: 9780841214101

Huang, Y.; Hee, S.K.; & Ho, S.H. (1998). Antifeedant and growth inhibitory effects of a-pinene on the stored-product insects, Tribolium castaneum (Herbst) and Sitophilus zeamais Motsch. *International Pest Control* .Vol.40, No.1, (Appil 1998),pp. 18-20, ISSN 0020-8256

Huang, Y. & Ho, S.H. (1998). Toxicity and antifeedant activities of cinnamaldehyde against the grain storage insects, *Tribolium castaneum* (Herbst) and *Sitophilus zeamais* Motsch. Journal of Stored Prodducts Research. Vol.34, No.1,(January 1998)pp.11-17 ISSN 0022-474X

Juliana, G. & Su, H.C.F. (1983). Laboratory studies on several plant materials as insect repellents for protection of cereal grains. *Journal of Economic Entomology*. Vol.76, pp. 154-157.ISSN 0022-0493

Jun,Q.;Tong, C. & Qing, L. (2001). Simultaneous Distillation and Solvent Extraction and GC/ MS Analysis of Volatile Oils of *Zanthoxylum Bungeagum* Maxim. *Journal of Guizhou University of technology* (Natural Science Edition) Vol.30,No.6,(November 2001) pp.1-6 ISSN : 1009-0193

Murray,B. Isman (2000). Plant essential oils for pest and disease management. *Crop Protection*, Vol.19, No.8-10, (September 2000) pp.603-608. **ISSN:** 0261-2194

Nielsen, P.S. (2001).The effect of carbon dioxide under pressure against eggs of Ephestia kuehniella Zeller and the adults of S. paniceum (L.) and Oryzaephilus surinamensis (L.). *Anzeiger fur Schadlingskunde* Vol.74, No.3,(June 2002),pp.85-88. ISSN 1436-5693

Platt, R.R.; Cuperus, G.W.; Payton, M.E. Bonjour, E.L. & Pinkston, K.N. (1998). Integrated pest management perceptions and practices and insect populations in grocery stores in south-central United States. *Journal of stored products research*, Vol.34, No.1 (January 1998), pp.1-10 ISSN 0022-474X

Sudesh, J.; Kapoor, A.C. & Ram, S. (1996). Evaluation of Some Plant Products Against Trogoderma granarium Everts in Sorghum and Their Effects on Nutritional Composition and Organoleptic Characteristics. *Journal of stored products research*, Vol.32,No.4(October 1996,),pp.345-392 ISSN 0022-474X

Tapondjou, L.A.; Adler, C.; Bouda, H. & Fontem, D.A. (2002). Efficacy of powder and essential oil from Chenopodium ambrosioides leaves as post-harvest grain protectants against six-stored product beetle. *Journal of stored products research*, Vol.38,No.4.(November 2002), pp.395-402 ISSN 0022-474X

Toh, C. A. (1998). Studies on pest insects of crude drugs. *Korean Journal of Pharmacognosy*, Vol.29,pp.365-373 ISSN, 0253-3073

Zhaohui, L. & Fangqiang, Z. (2001). Research on numerical characteristic of the structure of insect community on the stored Chinese Medicinal Materials in Shangdong province. *Stored Food stuff* Vol.30, No.3, (June 2001),pp.12-16. ISSN 1000-6958

Zhi'an G.; Jingchuan. Z. & Zhihai, X. (2001) Study on Chemical Constituents of the Essential Oil from *Zanthoxylum bungeanum* Maxim by Gas Chromatography-Mass Spectrometry. *Chinese journal of chromatography*, Vol.19, No.6,(November 2001) pp.567-568. ISSN 1000-8713

6

Baculoviruses: Members of Integrated Pest Management Strategies

Vanina Andréa Rodriguez,
Mariano Nicolás Belaich and Pablo Daniel Ghiringhelli
*LIGBCM-AVI (Laboratorio de Ingeniería Genética y Biología
Celular y Molecular Area Virosis de Insectos)
Universidad Nacional de Quilmes/Departamento de Ciencia y Tecnología
Argentina*

1. Introduction

Integrated Pest Management (IPM) can be defined as "an ecologically based pest control strategy that relies heavily on natural mortality factors and seeks out control tactics that disrupt these factors as little as possible" (Flint & Bosch, 1981). IPM began to be applied because the extensive use of chemical insecticides show different types of environmental damages, as development of resistant insects, the appearance of new pests, injury to bird and mammals populations and human health damage due to the release of toxic waste on environment and food contamination. The aims of IPM are to protect crops, with minimum cost and risk for humans, animals and ecosystems. The development and application of IPM requires the knowledge of how the ecosystem influences on pest insects and its natural control agents, and how to modify this environment to control particularly pest insects and avoid the related chemical agent's problems.

2. Integrated pest management (general overview)

IPM applies different tactics, like pest resistant plants, use of entomopathogens such as bacteria and viruses, and strategies that involves cultural, physical, mechanical, biological and chemical control. The use of these combined tactics reduces the chances of generating resistance and insect survival.

2.1 Pest resistant plants (transgenesis using plant genes)

Plants have a vast metabolic capability and produce many secondary chemicals which are toxic, anti-nutritional, or aversive to species might otherwise be potential predators (Norris & Kogan, 1980). Examples include the pyrethrins from chrysanthemums and alkaloids like nicotine from tobacco. Other compounds implicated in protection from insect attack include the terpenoids, steroids, flavonoids, phenolic, glucosinolates, cyanogenic glycosides, rotenoids, saponins and non protein amino acids (Gatehouse et al., 1991). As secondary compounds are the products of multi-enzyme pathways which involve the interaction of many gene products, such defense system are in most cases too complex to be used in plant genetic engineering (Dawson *et al.*, 1989; Hallahan *et al.*, 1992). However, a few plant defense mechanisms are based on the product of a single gene, and the target site of many is the insect digestive system. Most of these types of single genes are suitable for gene transfer. Being of plant origin, they have the advantage that they are likely to have a high degree of compatibility with the metabolic system of the transgenic host plant.

Crops resistant to insect attack offer an alternative strategy of pest control upon chemical pesticides. Transgenic plant technology can be a useful tool in producing resistant crops, by introducing novel resistance genes into a plant species. Several different classes of plant proteins have been shown to be insecticidal towards a range of economically important insect pests. Genes encoding insecticidal proteins have been isolated from various plant species and transferred to crops by genetic engineering. Amongst these genes are those that encode:

1. Protease inhibitors (serine and cysteine). The damage of leaves of certain solanaceous plants, either by insect feeding or mechanical wounding, induced the synthesis of protease inhibitors (Green & Ryan, 1972; Shumway *et al.*, 1976; Walker-Simmons & Ryan, 1977; Brown *et al.*, 1985). The first gene of a plant successfully used to be transferred to another plant was a trypsin inhibitor (cowpea trypsin inhibitor, CpTI) (Pusztai *et al.*, 1992; Graham *et al.*, 1995; Xu *et al.*, 1996). As an example of a commercial deployment of a proteinase inhibitor transgene to date, could be mentioned the culture of genetically engineered cotton varieties in China. These varieties express two transgenes to improve cotton protection, Bt toxins against lepidopteran larvae and CpTI. In 2005, Bt/CpTI cotton was grown on over 0.5 million hectares (Gatehouse, 2011).
2. Inhibitors of α-amylase. This enzyme is able to catalyse the hydrolysis of α-1-4 glycosidic bonds, transforming polysaccharides into mono and disaccharides (Grossi-de-Sá & Chrispells, 1997; Franco *et al.*, 2002; Pelegrini *et al.*, 2006). Interference of the carbohydrate absorption could be a way to reduce the insect pest feeding (Yamada *et al.*, 2001). Leguminous seeds are known as rich sources of proteinaceous α-amylase inhibitors (α-AIs) (Payan, 2004). Expression of α-amylase inhibitors (*a*-AIs) from both scarlet runner bean (*Phaseolus coccineus*) and common bean (*Phaseolus vulgaris*) has been shown to be effective in transgenic plants, showing high protection against seed weevils in pea (Shade *et al.*, 1994; Schroeder *et al.*, 1995), azuki bean (Sarmah *et al.*, 2004), chickpea (Sarmah *et al.*, 2004; Ignacimuthu & Prakash., 2006), cowpea (Solleti *et al.*, 2008) and coffee (Barbosa *et al.*, 2010).
3. Lectins. This group of carbohydrate-binding proteins constitutes entomotoxic factors present in many plant species. During the last decade a lot of progress was made in the study of the properties of a few lectins that are expressed in response to phytophagous insects. Based on their activity towards pest insects, these proteins have a high potential for use in pest control strategies. For example, the use of plant lectins has been applied to control numerous pests: melon fruit fly larvae (Kaur *et al.*, 2009); *Aedes aegypti* larvae, which has developed tolerance to many other insecticides (Coelho *et al.*, 2009); and the cotton leafworm *Spodoptera littoralis*, an economically important caterpillar in agriculture and horticulture (Hamshou *et al.*, 2010).

2.2 Entomopathogen bacteria (*Bacillus thuringiensis*)

Bacillus thuringiensis is a Gram positive bacteria belonging to Eubacteria. It was isolated in the early twentieth century in Japan from dead larvae of the silkworm. This bacteria, produces spores and crystalline bodies composed of one or more proteins with insecticidal activity

(Schnepf *et al.*, 1998, Sedlak *et al.*, 2000). The crystalline toxins, named Cry δ-endotoxins, exist in a variety of forms: bipyramidal, spherical, rhomboidal, cuboidal and irregular, among others, and are active against a large number of insect groups as well as nematodes and protozoa. Today there are over 40 groups of Cry proteins (Crickmore *et al.*, 2009). The δ-endotoxins are synthesized as an inactive pro-Cry-toxin and when they are ingested by the larvae feeding on plant debris or soil, the inclusions are solubilized in the alkaline conditions of the digestive tract of the larvae and are converted by the action of the insect proteases in active peptides (Feitelson *et al.*, 1992; Schnepf *et al.*, 1998). Active toxin is recognized by a specific receptor; it binds to microvilli of intestinal cells (Gazit *et al.*, 1998; Gerber & Shai, 2000), and generates ion channels. The natural ion imbalance dissipates and the pH of medium diminishes causing osmotic cell lysis and larva ceases to feed (Schnepf *et al.*, 1998). Moreover, tissue destruction allows mixing the contents of gastrointestinal tract with hemolymph. Both phenomena favor the germination of bacterial spores, resulting in the death of the larva few days after ingestion crystals (Aranda *et al.*, 1996; Schnepf *et al.* 1998; Crickmore *et al.*, 2009).

There are two main approaches to use Bt as a pest control agent:

1. Preparations based on living or dead Bt containing spores and crystals are sprayed on crops, as if it were a conventional insecticide. This strategy are currently used in the United States, Europe, Argentina and Mexico as a biological control for insect and other invertebrate pests (mites, nematodes, flatworms and protozoa) that affect crops of corn, potato, tomato, sorghum, rice, coffee, beans, sugar cane, among others (Neppl, 2000). The application of Bt in insect control is not exempt from the emergence of resistance. Using of a combination of toxins reduce resistance to individual toxins, maintains a populational balance and prevents the prevalence of resistant variants (Georghiou & Wirth, 1997; Ives *et al.*, 2011; Yang *et al.*, 2011). Research and years of use have shown that the employment of Bt products is not hazardous to non-target arthropods, birds, fish, mammals, or environment (EPA, 2008).

2. The genes that encode different Cry proteins have been used to generate genetically modified (GM) plants. Transgenic plants are resistant to the attack by insect pests. The most widely GM crops commercialized so far include mainly maize, cotton and rice. The use of this modified crops has been approved in several countries, including United States, Brazil, Argentina, Canada, Australia, Spain, South Africa, among other (James, 2011). Considering that insecticidal crystal proteins can be released continuously into the soil in different forms during the growing period of Bt-plants (Zhou *et al.*, 2011), biosafety of the use of genetically modified plants is always questioned. However, data regarding the development and commercial use of transgenic Bt varieties have shown that the currently available Bt crops have no direct detrimental effects on non target organism due to their narrow spectrum of activity. In addition, the use of these modified crops, such as Bt maize and Bt cotton, results in significant reductions of insecticide application and has clear benefits on the environment and farmer health. Consequently, Bt crops can be a useful component of IPM systems to protect the crops from targeted pests (Yu *et al.*, 2011).

2.3 Entomopathogen viruses (*Baculoviruses*)

The insect viruses are intracellular parasites that can only reproduce inside a susceptible insect host. They are valuable natural control agents, providing a secure control, effective and sustainable in a variety of insect pests. The virus particles are present in the environment and usually can be found on the surface of plants or in the soil. Insects become infected by consuming plant material contaminated with viral particles on the surface or by contact with the soil. Baculoviruses are the most common type of insect viruses. It has been reported that infect over 600 species of insects worldwide. Most baculovirus infect caterpillars, the immature form of moths and butterflies. Naturally, these viruses are potent regulators of the population of many caterpillar pests, but the use as a tool for biological control in agriculture is limited by biological or technical reasons.

1. Advantages. Insect viruses are very safe to handle, since they are not infectious to organisms other than their natural hosts. Moreover, most insect viruses have a high specificity, so that the risk of affecting non-target beneficial insects is very low.
2. Disadvantages. Most insect viruses have a low speed of action on their insect host, during this time the plague is still eating and damaging. Insect death is also dose dependent, and very high doses are often necessary for adequate control. Usually, viruses are very effective against early larval stages; the late larval stages are less susceptible to virus infection. Virus particles exposed to sunlight or high temperatures are rapidly inactivated. In addition, some cultural practices can affect viral persistence, hiding the viral particles in the soil.

3. Baculovirus

3.1 Baculovirus biology

Baculoviridae is a viral family that infects insects. Their genomic material is composed by double strand, circular, super coiled DNA, with a size ranging from 80 to 180 kpb. These viruses are classified in four genera: *Alphabaculovirus* (nucleopolyhedroviruses –NPVs- that specifically infect insects of lepidopteran order), *Betabaculovirus* (granuloviruses -GVs- that specifically infect insects of lepidopteran order), *Gammabaculovirus* (NPVs that specifically infect insects of hymenoptera order) and *Deltabaculovirus* (NPVs that specifically infect insects of diptera order) (Jehle *et al.* 2006). The virus genome is packaged within a rod-shaped nucleocapsid which is further surrounded by a lipoprotein envelope to form the virus particle. This structure is then occluded at very late stages by a crystalline matrix (Occlusion Body or OB) largely comprising for a single occlusion protein (about 28 kDa), which serves to protect it in the environment. During the infection cycle two different phenotypes are generated: budded virus (BVs, responsible for the systemic cell to cell infection) and occluded-derived virus (ODVs, responsible for the host to host infection) (Figure 1).

The most common way of insect host primary infection is by ingestion during larval feeding of contaminated foliage (Figure 2). Following virus ingestion, the OBs are dissolved in the high pH conditions (pH 8.5 to 11) of the insect midgut, releasing the virus particles (ODVs) into the gut lumen (Granados & Lawler, 1981; Pritchett *et al.*, 1981; Pritchett *et al.*, 1984; Rohrmann, 2008). Released viruses bind to the columnar epithelial cells and enter the tips of the microvilli on the apical brush border of cells (Kuzio *et al.*, 1989; Faulkner *et al.*, 1997; Haas-Stapleton *et al.*, 2004). Following fusion between cell and virus membranes, the nucleocapsids are released into the cytoplasm and are transported to the nucleus, where viral DNA transcription and replication occurs. Into the cytoplasm, nucleocapsids are transported from the basal membrane to the haemocoel, acquiring the host derived membrane and virus encoded proteins (Washburn *et al.*, 2003). Secondary infection is achieved by BVs produced from the midgut cells. Thus, the viruses spread to other insect tissues including the fat body, endodermis, muscle sarcolemma and nerve ganglia (Harrap, 1970; Washburn *et al.*, 2003). Prior to death larvae become creamy in colour, cease feeding and show limited movement. In most baculovirus infections, the host tissues break down as a result of the expression of virus-encoded chitinase and cathepsin proteins (Ohkawa *et al.*, 1994; Hawtin *et al.*, 1995; Slack *et al.*, 1995). Then, OBs are released into the environment following the rupture of the insect cuticle; 10^9 OBs may be released from a single larva, and may remain viable in the environment for several years, until ingestion by another host larva resumes replication cycle (Evans & Harrap, 1982) (Figure 2).

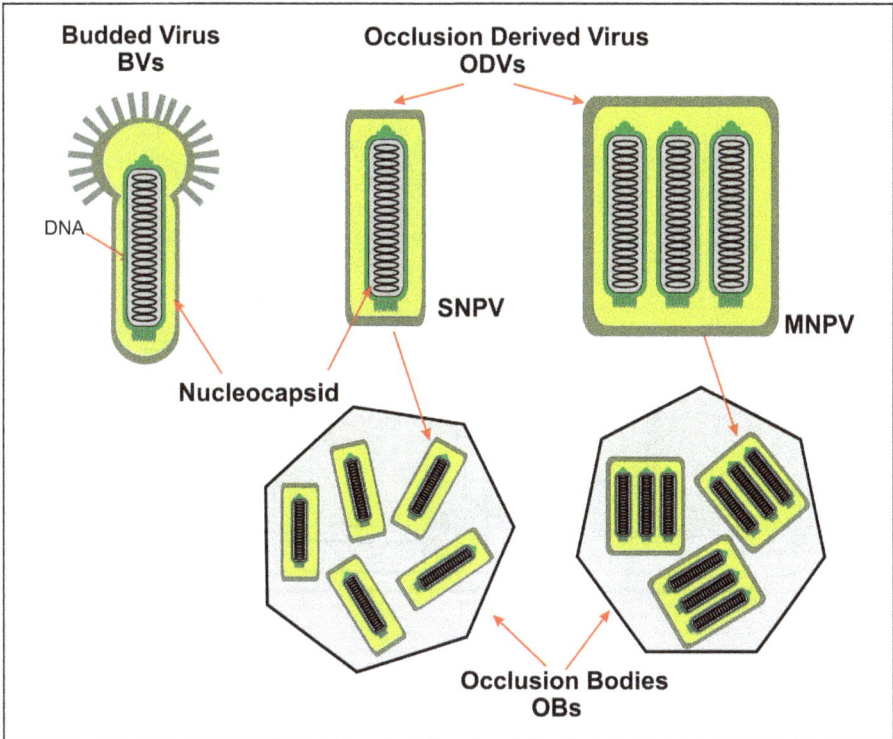

Fig. 1. **Alphabaculovirus phenotypes**. Occlusion bodies (OBs) containing occlusion derived virus (ODVs) responsibles for the primary infection in the midgut cells and Budded virus (BVs) that spread the infection to other larval tissues.

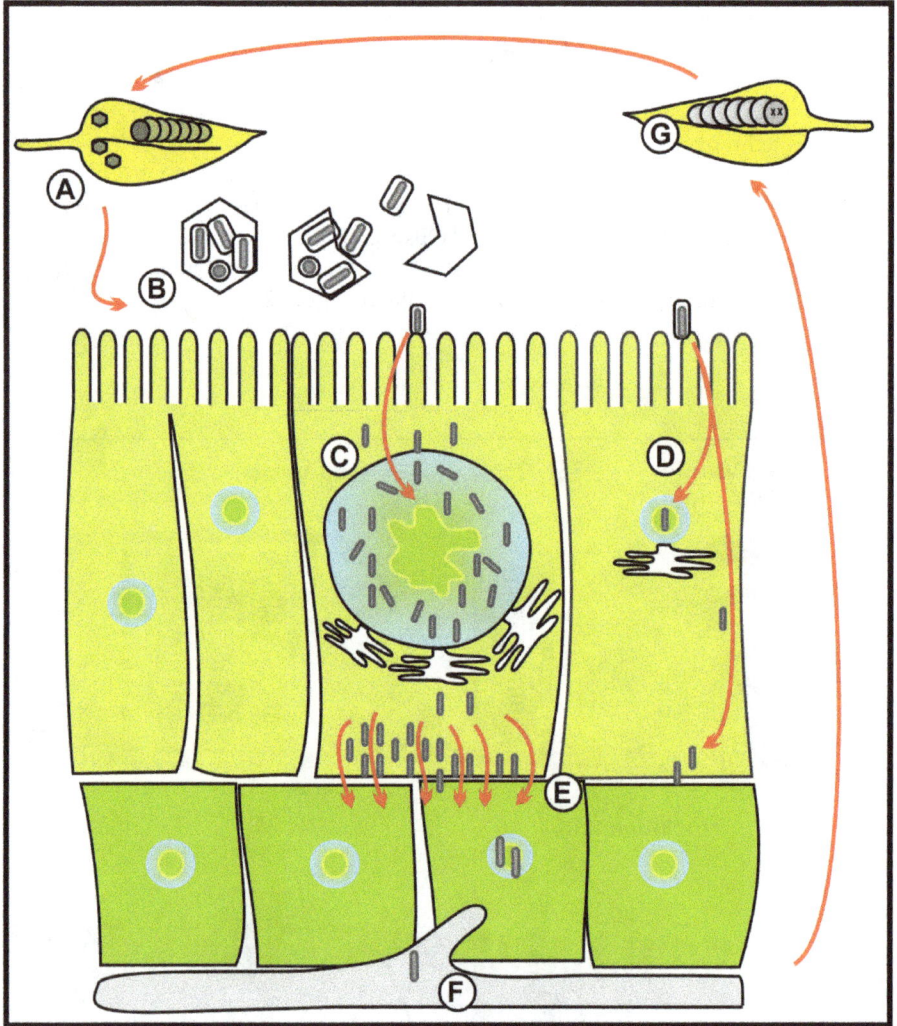

Fig. 2. **Infection cycle.** NPV biological cycle. **A.** Larva ingests a baculovirus-contaminated leaves. **B.** The virions are released and bind to midgut cell microvilli. **C.** Into columnar cells the nucleocapsids are transported to the nucleus, where the viral DNA is released and the BVs are generated. **D.** Nucleocapsids can pass through the cell to the haemocele. **E.** Viral particles enter into susceptible tissue cells and initiate a replication cycle to produce a new BV generation. **F.** Later, the OBs are generated into each infected cell. **G.** Cell lysis occurs, larval integument breaks, and the OBs are released to be eaten by other larva and thus restart the infection.

3.2 Baculovirus as control agents (Figure 3)

Wild type baculoviruses

A large number of Lepidoptera are pests of economically important crops such as soybean, sunflower and cotton, among others. Alpha- and Betabaculoviruses are very useful tools for control of lepidoteran pests. These viruses have a narrow host range, are safe for other insects and organisms in the environment (Szewczyk et al, 2006), and have strong pathogenicity and virulence. Other advantages are the stability in the environment for long periods and that can be applied using simple methods. Because of the great advantages of baculoviruses as biological control tools and their safety for the ecosystem, their use in IPM is accepted and is steadily increasing. Currently, different baculovirus are used in pest control management worldwide.

In North America, the use of baculoviruses as a pest control agent started as early as 1930 with the protection of pine trees with *Diprion hercyniae* NPV (Bird & Buek, 1961). Subsequently, this strategy was use to protect numerous commercial crops, including alfalfa, cabbage, corn, cotton, lettuce, soybean, tobacco and tomato (Granados & Federici, 1986). Another baculovirus that is currently used to control pest infection is *Helicoverpa zea* Simple Nucleopolyhedrovirus (HzSNPV). This virus infects several species belonging to the genera *Heliothis* and *Helicoverpa*, which include a wide range of crop pests (Chakrabarti et al., 1999). HzSNPV provides a tool for control against the bollworm, soybean, sorghum, corn, tomatoes and beans. In the 90´s of the latest century the virus was registered under the name GemStar™ and marketed by Thermo Trilogy Company. This virus produced in United States is also marketed in Australia by Aventis Crop Science and it is now used to protect crops against *Helicoverpa armigera* (Mettenmeyer, 2002), which attack near 200 crops including cotton, soybeans, chickpeas, sunflower, snuff, pepper, and corn, among others. HaSNPV has been adopted for large-scale viral pesticide production in China and has been used extensively to protect crops of cotton (Zhang et al., 1995).

In South America, probably the most successful program of using baculovirus in pest control is the program implemented in Brazil since 1980 to control the velvet bean caterpillar *Anticarsia gemmatalis* (pest of soybean) (Moscardi, 1989, 1999). In this program, the *Anticarsia gemmatalis* Multiple Nucleopolyhedrovirus (AgMNPV) is used as a biological insecticide against *Anticarsia gemmatalis* larvae, and by 2005 the treated area reached 2 million hectares (Szewczyk et al., 2006). One advantage of using this virus is that it is highly virulent and only needs to be applied once, whereas chemical insecticides should be applied twice. In addition, the use of this virus is 20 to 30% less expensive than chemical insecticides. Recently, a pilot plant was constructed in Embrapa Soja, Londrina, in order to improve the laboratory process and train people in the production of virus. This laboratory will be able to inoculate 20,000 to 30,000 larvae per day (Szewczyk et al., 2006).

Moreover, the *Cydia pomonella* Granulovirus (CpGV) has been used as pest control in crops of apples and pears. Its use has increased in Europe and North America since 2000 and is used in about 100,000 ha in those continents. Currently there are several commercial preparations that include Cyd-X, Virosoft CP4 (North America), Carpovirusine™ (France), Madex™ y Granupon™ (Switzerland), Granusal™ (Germany) and Virin-CyAP (Russia) (Rohrman, 2008). In Argentina, *Cydia pomonella* is a pest of pear, apple and walnut crops. The National Institute of Agricultural Technology (INTA; Argentina) in agreement with the

Natural Plant Protection company (France) developed a bio-insecticide called Carpovirus to control that pest, using a formulation based on CpGV.

For an additional example of baculovirus employed as tools of biological control and pest management, can be mentioned *Spodoptera exigua* Multiple Nucleopolyhedrovirus (SeMNPV). Formulations of this virus are being used to protect crops of sweet peppers in Spain against *Spodoptera exigua*. The company Biocolo SRL is responsible for the commercial production of this bio-insecticide with the capacity to treat 50,000 larvae per day. The introduction of this biological insecticide has helped to multiply more than 15 times the area of biologically protected crops grown, reaching 23,000 ha in 2008 (Caballero *et al.*, 2009).

Modified baculoviruses

Despite the advantages that baculovirus have, there are also some limitations for their use as pest control agents. Some of these limitations are the high costs of in vivo production and the low persistence in very sunny conditions (Ignoffo *et al*, 1977). Also, the virus must be used in early insect development stages, because in the late stages the insects are more resistant to infection (Washburn *et al.*, 2003). Another limitation is their slow speed to kill the pest (Ignoffo *et al*, 1992). To avoid this, recombinant baculovirus have been developed and offer attractive alternatives to broad-spectrum chemical control. These recombinant viruses can express specific toxins (Inceoglu *et al*, 2001), hormones (Elvira *et al*, 2010) or enzymes (Gramkow *et al*, 2010) and are much more efficient than the wild-type virus in speed to kill.

1. **Expression of insect-selective toxin.** The *aait* gene obtained from *Androctonus australis* scorpion is one of the most promising toxins to be expressed in baculovirus. A recombinant virus containing this gene showed to be 40% faster in killing larvae than the wild type and a reduction of host feeding by 60% (Cory *et al.*, 1994; Inceoglu *et al.*, 2001). The site of action of this neurotoxic polypeptide is one insect sodium channel. Lepidopterous larvae infected with an AaIT-expressing baculovirus reveal symptoms of paralysis identical to those induced by injection of the native toxin (Elasar *et al.*, 2001) and many of the physiological effects are very similar to those of pyrethroid insecticides which also act at the same target (Gordon *et al.*, 1992). Other useful insect-selective neurotoxins are SFI1 (obtained from a European spider *Segestria florentina*) and ButaIT (derived from the South Indian red scorpion *Mesobuthus tamulus* (Wudayagiri *et al.*, 2001). Some toxins could exert a cooperative effect when they are co-expressed, such as LqhIT1 and LqhIt2, obtained from *Leiurus quinquestriatus* scorpion (Regev *et al.*, 2003).

2. **Expression of insect hormone genes.** Disruption, over expression or inactivation of one or more insect hormones results in abnormal growth, feeding cessation and/or death. So, the insertion of genes that encode insect hormones were the first strategies used to generate genetically modified baculovirus. A recombinant virus of *Bombyx mori* MNPV (BmNPV) that encodes an active diuretic hormone (DH) showed to be 20% faster in killing larvae than wild type virus (Maeda *et al.*, 1989). Later, by the deletion of *egt* gene, which prevents the larval molt, the mutant virus resulted to be 30% faster in killing larvae and in a considerable reduction in food intake than wild type virus (O'Reilly & Miller, 1991). Also, this gene may be replaced by an exogenous gene and enhance the insecticidal activity (Arif, 1997; Popham *et al.*, 1997; Sun *et al.*, 2004).

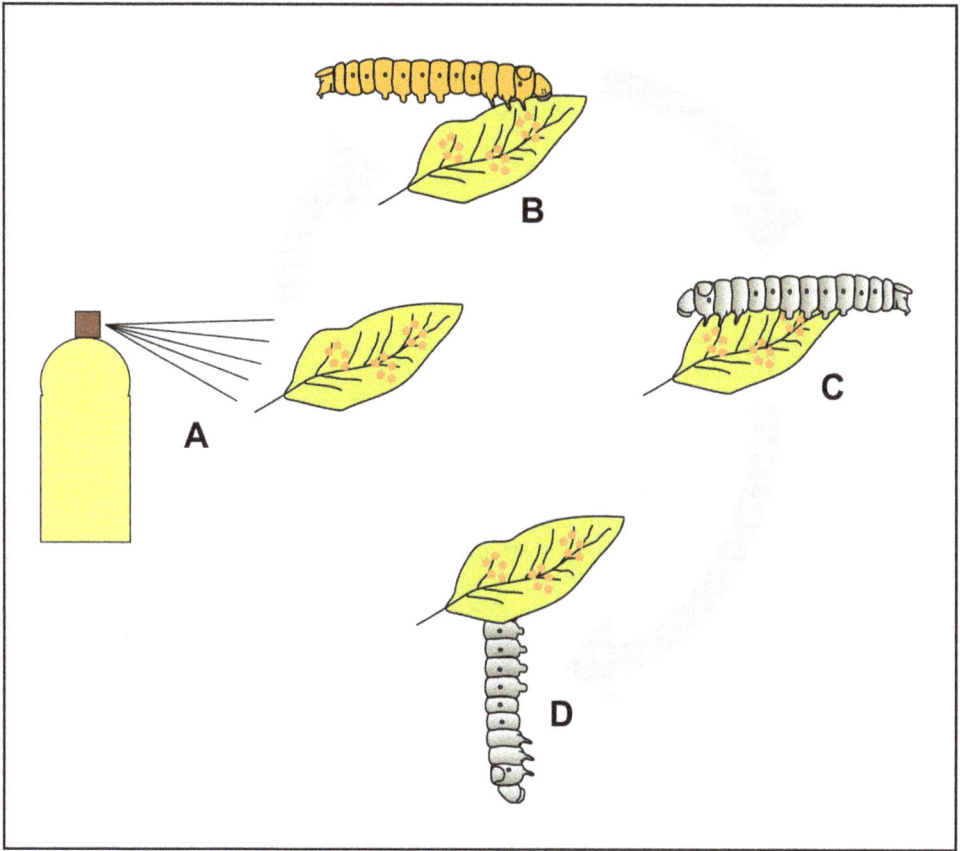

Fig. 3. **Baculoviral bioinsecticide. A.** Bioinsecticide spreading on leaves. **B.** Larvae feeding a contaminated leaf. **C.** Infected larvae. **D.** Dead larvae. Released virus will permit the resume of the infection cyle.

Baculovirus safety

Alpha- and Betabaculoviruses are not infectious for predatory or beneficial insects outside of the order Lepidoptera, or toward other non-targeted organisms (Black *et al.*, 1997; Szewczyk et al., 2006). Baculoviruses do not replicate in mammals, birds, fish, amphibians, reptiles and other vertebrates (Barsoum *et al.*, 1997). Moreover, different studies have concluded that the transgenes that encode toxins or other foreign proteins in recombinant baculovirus increase the speed to kill the specific target, but do not confer any selective ecological advantage compared with the wild-type baculovirus. (Cory *et al.*, 1997; Lee *et al.*, 2001). Another concern associated with GM organisms is the possibility of genetic recombination that results in the "jump" of the recombinant baculovirus foreign gene to another organism. If the pesticide based on a recombinant baculovirus is used long enough and in sufficiently high concentrations in the field, it is expected that genetic recombination could occur. In both the field and in laboratory conditions, this phenomenon is expected to be higher among highly homologous baculoviruses infectious to the same host (Hajos *et al.*, 2000). However, it is important to note that the recombinant virus with a phenotype more virulent than the wild type has disadvantages, since they produces fewer progeny and are rapidly out-competed in the ecosystem (Inceoglu *et al.*, 2001).

Future prospects

In order to improve the use of baculovirus in pest control management, various methodologies are being studied and developed. Among these we can mention:

1. Expression of fusion proteins to expand the host range. Because ButaIT toxin alone exhibits weak oral toxicity, and as an alternative to the use of recombinant baculovirus different strategies have been developed based on the generation of fusion proteins. One of the proteins used is a lectin, which functions as a carrier (GNA, *Galanthus nivalis* agglutinin) and the other protein is a toxin. A fusion protein comprising GNA and SFI1 has been shown to have insecticidal effects on both lepidopteran and homopteran plant pests (Fitches *et al.*, 2004; Down *et al.*, 2006) while a fusion protein combining GNA with ButaIT has also been shown to have insecticidal activity against lepidopteran larvae. The injection data reported show that fusion proteins containing SFI1 and ButaIT are insecticidal towards a range of insects including lepidopteran, dipteran, coleopteran and dictyopteran pests (Fitches *et al.*, 2010).

2. Antisense strategies. A new approach to the development of integrated pest control is the use of technologies based on iRNA (Gordon & Waterhouse, 2007; Price & Gatehouse, 2008). For efficient pest control methods using iRNA in the field, the major challenge is the development of easy and reliable methods for production and delivery of dsRNA. The dsRNA can be produced *in vivo* using a bacteria system or can be synthesized *in vitro*. Other way to produce dsRNA is using transgenic plants that express the desired antisense. The delivery of dsRNA can be done by spraying dsRNA on the crop plants or by feeding insects. In a recent study feeding heat-killed bacteria that produced dsRNA or *in vitro* synthesized dsRNA was used to silence five target genes in Colorado potato beetle, *Leptinotarsa decemlineata*. It was observed that the loss of function of these target genes caused larval mortality and significantly decreased insect growth (Zhu *et al.*, 2011). In a similar study, it was used a synthetic iRNA to inhibit a mitochondrial electron transport of *Plutella xylostela*, causing prominent insect mortality (Gong *et al.*, 2011). The use of different target genes could minimize the risk of resistance development. Offtarget effects and species specificity of dsRNA are two major potential issues that need to be addressed. However, the same way as was observed with *Caenorhabditis elegans* and other plant nematodes (Price & Gatehouse, 2008), in insects, described methods to feed dsRNA to a range of different insect species demonstrated that even highly conserved genes can be exploited to trigger species-specific iRNA without affecting non-target species (Whyard *et al.* 2009).

3. Combined control methods. The use of synergism between two different pathogens is a strategy that could be useful to conduct pest control strategies. To evaluate this possibility it have been conducted studies using *Spodoptera exigua* larvae and assessed the effect of using a baculovirus (*Spodoptera exigua* NPV) and a parasitoid (*Microplitis pallidipes*) to control this pest (Jiang *et al.*, 2011). The results obtained indicate that when *M. pallidipes*-parasitised *S. exigua* larvae were infected by SeMNPV, the virus did not affect the developmental period of *M. pallidipes* in the host, and most parasitoids

completed development, possibly because parasitism by *M. pallidipes* reduced larval sensitivity to the virus. The results of this study also indicate that *M. pallidipes* is an important vector of SeMNPV and contributes to natural epizootics of the virus. Female parasitoids that had developed or oviposited in virus-infected hosts, or that emerged from cocoons contaminated with virus, were able to transmit infective doses of virus to healthy host larvae (Jiang *et al.*, 2011).

4. Conclusions

Considering the amount of baculovirus species that have been isolated so far, its development as bio-pesticides has not been commensurate with all its potential. Most of the viruses found in commercial phase are produced by small or medium companies or by the users themselves, as is the case of USDA Forest Service (US Department of Agriculture), CIP (International Potato Center) or EMBRAPA (Empresa Brasilera de Pesquisa Agropecuaria). Among the factors limiting their commercialization it has been noted that they are too specific, they present a slow speed of action, they have a low persistence in the field, and are costly to produce infecting larva. Also, the approaches that involve *in vitro* production processes in insect cell cultures are still in development stage.

An entomopathogenic product will be considered as a viable alternative in pest control if it meets control with the same speed, ease of use, at the same cost than a chemical insecticide. This way, do not take into account their unique capabilities: the ability to replicate in their host and be dispersed in culture, the ability to synergistically act with natural enemies, and availability to be produced locally or regionally. Faced with this situation have been conducted many studies to increase the speed of action, to extend the host range and maintain their safety to non-target organisms. This involves the generation of genetically modified baculoviruses and the use of different viral formulations (Cherry & Williams, 2001).

Baculovirus survival in the environment can be affected by temperature, pH, moisture, the presence of additives, exposure to UV light, and by the action of some plant metabolites such as peroxidases, that generate free radicals (Hoover *et al.*, 1998; Zhou *et al.*, 2004).

Actually, there have been developed some protective agents against UV that have been included into the viral formulations. Some commercial bleach such as Phorwite AR, Blankophor, and Tinopal C1101 are very effective to protect *Lymantria dispar* Nucleopolyhedrovirus -LdNPV-, *Helicoverpa zea* Nucleopolyhedrovirus -HzNPV-, and *Spodoptera frugiperda* Multiple Nucleopolyhedrovirus -SfMNPV- (Shapiro *et al.*, 1994; Zou & Young, 1994, 1996; Mondragón *et al.*, 2007). Recent studies have shown that the addition of 1% (wt:vol) aqueous extracts of cocoa (*Theobroma cacao* L.) (Malvales: Malvaceae), coffee (*Coffea arabica* L.) (Gentianales: Rubiaceae), and green and black tea (*Camellia sinensis* L.) (Ericales: Theaceae) provided excellent UV radiation protection for the beet armyworm, *Spodoptera exigua* Multiple Nucleopolyhedrovirus under laboratory conditions (SeMNPV) (El-Salamouny *et al.*, 2009).

Regarding the safety of genetically modified baculovirus, a recombinant HzNPV carrying the *aait* gene was not pathogenic for bees, birds, fish, and other vertebrates (Szewczyk *et al.*, 2006). Natural enemies of the larvae, like parasitoids and predators, were not adversely affected by ingesting individuals infected with recombinant viruses (Li *et al.*, 1999; Smith *et al.*, 2000; Boughton *et al.*, 2003). So far, neither has been shown that the transgene be transferred from baculovirus to other organism or virus (Inceoglu *et al.*, 2001). On this basis, there is no evidence that a recombinant baculovirus is more dangerous than the corresponding wild type.

Finally, knowledge about the biology of baculoviruses suggests that bioinsecticides based on formulations containing these viruses have much lower risk to the environment than the classic chemical insecticides (Szewczyk *et al.*, 2006).

5. Acknowledgment

This work was supported by research funds from *Agencia Nacional de Promoción Científica y Técnica* (ANPCyT) and *Universidad Nacional de Quilmes*. PDG is member of the Research Career of CONICET (*Consejo Nacional de Ciencia y Tecnología*); MNB holds a postdoctoral fellowship of CONICET.

6. References

Aranda, E.; Sánchez, J.; Peferoen, M.; Güereca, L.; Bravo, A. (1996). "Interactions of the *Bacillus thuringiensis* crystal proteins with the midgut epithelial cells of *Spodoptera frugiperda* (Lepidoptera: Noctuidae)". J. Invertbr. Pathol. 68, pp. 203-212.

Barbosa, A.E.A.D.; Albuquerque1, E.V.S.; Silva1, M.C.M.; Souza1, D.S.L.; Oliveira-Neto1, O.B.; Valencia, A.; Rocha1, T.L.; Grossi-de-Sa, M.F. (2010). "α-amylases inhibitor-1 gene from *Phaseolus vulgaris* expressed in *Coffea arabica* plants inhibits α-amylases from the coffee berry borer pest". BMC Biotechnology. 10, pp. 44-52.

Barsoum, J.; Brown, R.; McKee, M.; Boyce, F.M. (1997). "Efficient transduction of mammalian cells by a recombinant baculovirus having the vesicular stomatitis virus G glycoprotein". Hum. Gene Ther., 8, pp. 2011-2018.

Bird, F.T.; Burk, J.M. (1961). "Artificial disseminated virus as a factor controlling the European spruce sawfly, *Diprion hercyniar* (Htg) in the absence of introduced parasites". Canad. Entomol. 96, pp. 228-238.

Black, B.C.; Brennan, L.A.; Dierks, P.M.; Gard, I.E. (1997). "Commercialization of baculoviral insecticides". In The baculovirus, ed by Miller LK, Plenum Press, New York, pp. 341-387.

Boughton, A.J., Obrycki, J.J.; Bonning, B.C. (2003). "Effects of a protease-expressing recombinant baculovirus on nontarget insect predators of *Heliothis virescens*". Biol. Control. 28, pp. 101-110.

Brown, W.E.; Takio, K.; Titani, K.; Ryan, C.A. (1985). "Wound-induced trypsin inhibitor in alfalfa leaves: identity as a member of the Bowman-Birk inhibitor family". Biochemistry 24, pp. 2195-2108.

Caballero, P.; Murillo, R.; Muñoz, D.; Williams, T. (2009). "El nucleopoliedrovirus de *Spodoptera exigua* (Lepidoptera: Noctuidae) como bioplaguicida: análisis de avances recientes en España". Revista Colombiana de Entomología 35, pp. 105-115.

Chakrabarty, S.; Monsour, C.; Teakle, R.; Reid, S. (1999). "Yield, biological activity, and field performance of a wild-type Helicoverpa nucleopolyhedrovirus produced in H. zea cell cultures". J. Invertebr. Pathol. 73, pp. 199-205.

Cherry, A.; Williams, T. (2001). "Control de insectos plaga mediante los baculovirus". En: Los baculovirus y sus aplicaciones como bioinsecticidas en el control biológico de plagas. P. Caballero, M. López-Ferber. & T. Williams Eds. Phytoma S.A., Valencia, España. pp. 389-450.

Coelho, J.S.; Santos, N.D.; Napoleão, T.H.; Gomes, F.S.; Ferreira, R.S.; Zingali, R.B.; Coelho, L.C.; Leite, S.P.; Navarro, D.M.; Paiva, P.M. (2009). "Effect of *Moringa oleifera* lectin on development and mortality of *Aedes aegypti* larvae". Chemosphere 77, pp. 934-938.

Cory, J.S.; Hirst, M.L.; Williams, T.; Halls, R.S.; Goulson, D.; Green, B.M.; Carty, T.M.; Possee, R.D.; Cayley, P.J.; Bishop, D.H.L. (1994). "Field trial of a genetically improved baculovirus insecticide". Nature (London). 360, pp. 138-140.

Crickmore, N.; Zeigler, D.; Feitelson, J.; Schnepf, E.; Van Rie, J.; Lereclus D. (2008). "*Bacillus thuringiensis* toxin database". Online:
http://www.lifesci.sussex.ac.uk/home/Neil_Crickmore/Bt/index.html.

Dawson, G.W.; Hallahan, D.L., Mudd, A.; Patel, M.M.; Pickett, J.A.; Wadhams, L.J.; Wallsgrove, R.M. (1989). "Secondary plant metabolites as targets for genetic modification of crop plants for pest resistance". Pestic. Sci. 27. pp 191-201.

Down, R.E.; Fitches, E.C.; Wiles, D.P.; Corti, P.; Bell, H.A.; Gatehouse, J.A. et al.. (2006). "Insecticidal spider venom toxin fused to snowdrop lectin is toxicto the peach-potato aphid, *Myzus persicae* (Hemiptera: *Aphididae*) and the rice brown planthopper, *Nilaparvata lugens* (Hemiptera:Delphacidae)". Pest Manag. Sci. 62, pp. 77–85.

Elazar, M., Levi, R.; Zlotkin, E.J. (2001). "Targeting of an expressed neurotoxin by its recombinant baculovirus". Exp. Biol. 204, pp. 2637-2645.

El-Salamouny, S.; Ranwala, D.; Shapiro, M.; Shepard, B.M.; Farrar, R.R. Jr. (2009). "Tea, coffee, and cocoa as ultraviolet radiation protectants for the beet armyworm nucleopolyhedrovirus". J. Econ. Entomol. 102, pp. 1767-1773.

Elvira, S.; Williams, T.; Caballero, P. (2010). "Juvenile hormone analog technology: effects on larval cannibalism and the production of *Spodoptera exigua* (Lepidoptera: Noctuidae) nucleopolyhedrovirus". J. Econ. Entomol. 103, pp. 577-582.

EPA (Environmental Protection Agency) (2008). "Pesticide Registration eligibility decision (RED): *Bacillus thuringiensis*. Online:
http://www.epa.gov/oppsrrd1/REDs/0247.pdf.

Evans, H.F.; Harrap, K.A. (1982). "Persistence of insect viruses". In Virus persistence. Vol 33, ed. BWJ Mahy, AC Minson & GK Darby. SGM Symposium Cambridge University Press, pp. 57-96.

Faulkner, P.; Kuzio, J.; Williams, G.V.; Wilson, J.A. (1997). "Analysis of p74, a PDV envelope protein of *Autographa californica* nucleopolyhedrovirus required for occlusion body infectivity in vivo". J. Gen. Virol. 78, pp. 3091-3100.

Feitelson, J.S.; Payne, J.; Kim, L. (1992). "*Bacillus thuringiensis*: insects and beyond". Nat. Biotech. 10, pp. 271-275.

Flint, M.L.; van den Bosch, R. (1981). "Introduction to Integrated Pest Management". Plenum Press. Nueva York, EEUU, pp. 240.

Fitches, E.C.; Bell, H.A.; Powell, M.E.; Back, E.; Sargiotti, C.; Weaver, R.J.; Gatehouse, J.A. (2010). "Insecticidal activity of scorpion toxin (ButaIT) and snowdrop lectin (GNA) containing fusion proteins towards pest species of different orders". Pest Manag. Sci. 66, pp. 74–83.

Fitches, E.; Edwards, M.G.; Mee, C.; Grishin, E.; Gatehouse, A.M.R.; Edwards, J.P.; Gatehouse, J.A. (2004). "Fusion proteins containing insect-specific toxins as pest control agents: snowdrop lectin delivers fused insecticidal spider venom toxin to insect haemolymph following oral ingestion". J. Insect Physiol. 50, pp. 61–71.

Franco, O.L.; Rigden, D.J.; Melo, F.R.; Grossi-de-Sá, M.F. (2002). "Plant α-amylase inhibitors and their interaction with insect α-amylases. Structure, function and potential crop protection". Eur. J. Biochem. 269, pp. 397-412.

Gatehouse, J.A.; Hilder, V.A.; Gatehouse, A.M.R. (1991). "Genetic engineering of plant for insect resistance". In Plant Genetic Engineering, Plant Biotechnology Series, Vol 1, ed. D. Grierson. Blackie & Sons Ltd, London/Chapman and Hall, New York, pp. 105-135.

Gatehouse, J.A. (2011). "Prospects for using proteinase inhibitors to protect transgenic plants against attack by herbivorous insects". Curr. Protein Pept. Sci. 12, pp. 409-416.

Gazit, E.; La Rocca, P.; Samson, M.S.; Shai, Y. (1998). "The structure and organization within the membrane of the helices composing the pore-forming domain of Bacillus thuringiensis δ-endotoxin are consistent with an "umbrella-like" structure of the pore". Proc. Natl. Acad. Sci. USA 95, pp. 12289-12294.

Georghiou, G.P.; Wirth, M.C. (1997). "Influence of exposure to single versus multiple toxins of Bacillus thuringiensis subsp. israelensis on development of resistance in the mosquito Culex quinquefasciatus (Diptera: Culicidae)". Appl. Environ. Microbiol. 63, pp. 1095-1101.

Gerber, D.; Shai, Y. (2000). "Insertion and organization within membranes of the δ-endotoxin pore-forming domain, helix 4-loop-helix 5, and inhibition of its activity by a mutant helix 4 peptide". J. Biol. Chem. 275, pp. 23602-23607.

Gong, L.; Yang, X.; Zhang, B.; Zhong, G.; Hu, M.Y. (2011). "Silencing of Rieske iron–sulfur protein using chemically synthesised siRNA as a potential biopesticide against Plutella xylostella". Pest Manag. Sci. DOI 10.1002/ps.2086

Gordon, D.; Moskowitz, H.; Eitan, M.; Warner, C.; Catterall, W.A.; Zlotkin, E. (1992). "Localization of receptor sites for insect-selective toxins on sodium channels by site-directed antibodies". Biochemistry 31, pp. 7622-7628.

Gordon, K.H.; Waterhouse, P.M. (2007). "RNAi for insect-proof plants". Nature Biotechnol. 25, pp. 1231-1232.

Graham, J.; McNicol, R.J.; Greig, K. (1995). "Towards genetic based insect resistance in strawberry using the cowpea trypsin inhibitor". Ann. Appl. Biol. 127, pp. 163-173.

Gramkow, A.W.; Perecmanis, S.; Sousa, R.L.; Noronha, E.F.; Felix, C.R.; Nagata, T.; Ribeiro, B.M. (2010). "Insecticidal activity of two proteases against Spodoptera frugiperda larvae infected with recombinant baculoviruses". Virol. J. 7, doi:10.1186/1743-422X-7-143.

Granados, R.R.; Federici, B.A. (1986). "Biology of baculoviruses". CRC, Boca Ratón, Florida.

Granados, R.R.; Lawler, K.A. (1981). "In vivo pathway of Autographa californica baculovirus invasión and infection". Virology 108, pp. 297-308.

Green, T.R. & Ryan, C.A. (1972). "Wound-induced proteinase inhibitor in plant leaves: a possible defence mechanism against insects". Science 175, pp. 776-777.

Grossi-de-Sá, M.F.; Chrispeels, M.J. (1997). "Molecular cloning of bruchid (Zabrotes subfasciatus) α-amylase cDNA and interaction of the expressed enzyme with bean α-amylase inhibitors". Insect Biochem. Mol. Biol. 27, pp. 271-281.

Haas-Stapleton, E.J.; Washburn, J.O.; Volkman, L.E. (2004). "P74 mediates specific binding of *Autographa californica* M nucleopolyhedrovirus occlusion-derived virus to primary cellular targets in the midgut epithelia of *Heliothis virescens* Larvae". J. Virol. 78, pp. 6786-6791.

Hajos, J.P.; Pijnenburg, J.; Usmany, M.; Zuidema, D.; Zavodszki, P.; Vlak, J.M. (2000). "High frequency recombination between homologous baculoviruses in cell culture". Arch. Virol. 145, pp. 159-164.

Hallahan, D.L., Pickett, J.A., Wadhams, L.J., Wallsgrove, R.M.; Woodcock, C.M. (1992). "Potential of secondary metabolites in genetic engineering of crops for resistance". In Plant genetic manipulation for crop protection-Biotechnology in agriculture No 7, ed. A.M.R. Gatehuse, V.A. Hilder and D. Boulter. CAB international, Wallingford, Oxon, UK, pp. 215-248.

Hamshou, M.; Van Damme, E.J.; Smagghe, G. (2010). "Entomotoxic effects of fungal lectin from *Rhizoctonia solani* towards *Spodoptera littoralis*". Fungal Biol. 114, pp. 34-40.

Harrap, K.A. (1970). "Cell infection by a nuclear polyhedrosis virus". Virology 42, pp. 311-318.

Hawtin, R.E.; Arnold, K.; Ayres, M.D.; Zanotto, P.M.; Howard, S.C.; Gooday, G.A.; Chappell, L.H.; Kitts, P.A.; King, L.A.; Posse, R.D. (1995). "Identification and preliminary characterization of a chitinase gene in the *Autographa californica* nuclear polyhedrosis virus genome". Virology 212, pp. 673-685.

Hoover, K.; Kishida, K.T.; DiGiorgio, L.A.; Workman, J.; Alaniz, S.A.; Hammon, B.D.; Duffey, S.S. (1998). "Inhibition of baculoviral disease by plant-mediated peroxidase activity and free radical generation". J. Chem. Ecol. 24, pp. 1949-2001.

Ignacimuthu, S.; Prakash, S. (2006). "Agrobacterium-mediated transformation of chickpea with alpha-amylase inhibitor gene for insect resistance". Journal of Biosciences 31, pp. 339–345.

Ignoffo, C.M. (1992). "Environmental factors affecting persistence of entomolpathogens". Florida Entomol. 75, pp. 516-525.

Inceoglu, A.B.; Kamita, S.G.; Hinton, A.C.; Huang, Q.; Severson, T.F.; Kang, K.; Hammock, B.D. (2001). "Recombinant baculoviruses for insect control". Pest Manag. Sci. 57, pp. 981-987.

Ives, A.R.; Glaum, P.R.; Ziebarth, N.L.; Andow, D.A. (2011). "The evolution of resistance to two-toxin pyramid transgenic crops". Ecol. Appl. 21, pp. 503-515.

James, C. (2011). "Global Status of Commercialized Biotech/GM Crops: 2010". Ithaca, NY, ISAAA www.isaaa.org.

Jehle, J.A.; Blissard, G.W.; Bonning, V.; Cory, J.S.; Herniou, E.A.; Rohrmann, G.F.; Theilmann, D.A.; Thiem, S.M.; Vlak, J.M. (2006). "On the classification and nomenclature of baculoviruses: A proposal for revision". Arch. Virol. 151, pp. 1257-1266.

Jiexian Jiang, Aiping Zeng, Xiangyun Ji, Nianfeng Wan and Xiaoqin Chen. (2011). "Combined effect of nucleopolyhedrovirus and *Microplitis pallidipes* for the control of the beet armyworm, *Spodoptera exigua*". Pest Manag. Sci. 67, pp. 705-713.

Kaur, M.; Singh, K.; Rup, P.J.; Kamboj, S.S.; Singh, J.J. (2009). "Anti-insect potential of lectins from Arisaema species towards *Bactrocera cucurbitae*". Environ. Biol. 30, pp. 1019-1023.

Kuzio, J.; Jaques, R.; Faulkner, P. (1989). "Identification of p74 a gene essential for virulence of baculovirus occlusion bodies". Virology 173, pp. 759-763.

Lee, Y.; Fuxa, J.R.; Inceoglu, A.B.; Alaniz, S.A.; Richter, A.R.; Reilly, L.M.; Hammock, B.D. (2001). "Competition between wild type and recombinant nucleopolyhedroviruses in a greenhouse microcosm". Biol. Control 20, pp. 84-93.

Maeda, S. (1989). "Increased insecticidal effect by a recombinant baculovirus carrying a synthetic diuretic hormone gene". Biochem. Biophys. Res. Commun. 165, pp. 1177-1183.

Mettenmeyer, A. (2002). "Viral insecticidas hold promise for bio-control". Farming Ahead 124, pp. 50-51.

Mondragón, G.; Pineda, S.; Martínez, A.; Martínez, A.M. (2007). "Optical brightener Tinopal C1101 as an ultraviolet protectant for a nucleopolyhedrovirus". Commun. Agric. Appl. Biol. 72, pp. 543-547.

Moscardi, F. (1989). "Use of viruses for pest control in Brazil: the case of the nuclear polyhedrosis virus of the soybean caterpillar, Anticarsia gemmatalis". Memorias do Instituto Oswakdo Cruz, Rio de Janeiro, 84, pp. 51-56.

Moscardi, F. (1999). "Assessment of the application of baculoviruses for control of Lepidoptera". Annu. Rev. Entomol. 44, pp. 257-289.

Neppl, C. (2000). "Management of resistance to Bacillus thuringiensis toxins". The Environmental Studies Program Education- University of Chicago (Chicago, U.S.A.).

Norris, D.M. & Kogan, M. (1980). "Biochemical and morphological bases of resistance". In Breeding Plants Resistant to Insects, ed. F.G. Maxwell and P.R. Jennings. John Wiley & Sons, London, pp. 23-60.

O'Reilly, D.R.; Miller, L.K. (1991). "A baculovirus blocks insecto molting by producing ecdysteroid UDP-Glucosyl Transferase". Science (Washington) 245, pp. 1110-1112.

Payan, F. (2004). "Structural basis for the inhibition of mammalian and insect α-amylases by plant protein inhibitors". Biochim. Biophys. Acta 1696, pp. 171-180.

Pelegrini, P.B.; Murad, A.M.; Grossi-de-Sa, M.F.; Mello, L.V.; Romeiro, L.A.; Noronha, E.F.; Caldas, R.A.; Franco, O.L. (2006). "Structure and enzyme properties of Zabrotes subfasciatus α-amylase". Arch. Insect Biochem. Physiol. 61, pp. 77-86

Popham, H.J.R.; Li, Y.; Miller, L.K. (1997). "Genetic improvement of Helicoverpa zea polyhedrosis virus as a biopesticide". Biol. Control 10, pp. 83-91.

Price, D.R.G.; Gatehouse, J.A. (2008). "RNAi-mediated crop protection against insects". Trends Biotechnol. 26, pp. 393-400.

Pritchett, D.W.; Young, S.Y.; Geren, C.R. (1981). "Proteolytic activity in the digestive fluid of larvae of Trichoplusia ni". Insect Biochem. 11, pp. 523-526.

Pritchett, D.W.; Young, S.Y.; Yearian, W.C. (1984). "Some factors involved in the dissolution of Autographa californica NPV by digestive fluids of Trichoplusia ni larvae". J. Invert. Pathol. 43, pp. 160-168.

Pusztai, A.; Grant, G.; Steward, J.C.; Bardocz, S.; Ewen, S.W.B.; Gatehouse, A.M.R. (1992). "Nutritional evaluation of the trypsin inhibitor from cowpea". British J. Nutr. 68, pp. 783-791.

Regev, A.; Rivkin, H.; Inceoglu, B.; Gershburg, E.; Hammock, B.D.; Gurevitz, M.; Chejanovsky, N. (2003). "Further enhancement of baculovirus insecticidal efficacy with scorpion toxins that interact cooperatively". FEBS Lett. 537, pp. 106-110.

Rohrmann, G. (2008). "Baculovirus Molecular Biology". Bethesda (MD): National Library of Medicine (US), NCBI.

Sarmah, B.K.; Moore, A.; Tate, W.; Molvig, L.; Morton, R.L.; Rees, D.P.; Chiaiese, P.; Chrispeels, M.J.; Tabe, L.M.; Higgins, T.J.V. (2004). "Transgenic chickpea seeds expressing high levels of a bean alpha-amylase inhibitor". Mol. Breed. 14, pp. 73–82.

Schnepf, E.; Crickmore, N.; Van Rie, J.; Lereclus, D.; Baum, J.; Feitelson, J.; Zeigler, D.R.; Dean, D.H. (1998). "*Bacillus thuringiensis* and its pesticidal crystal proteins". Microbiol. Mol. Biol. 62, pp. 775-806.

Schroeder, H.E.; Gollasch, S.; Moore, A.; Tabe, L.M.; Craig, S.; Hardie, D.C.; Chrispeels, M.J.; Spencer, D.; Higgins, T.J.V. (1995). "Bean alpha-Amylase Inhibitor Confers Resistance to the Pea Weevil (*Bruchus pisorum*) in Transgenic Peas (*Pisum sativum*)". Plant Physiol. 107, pp. 1233-1231. 1239.

Sedlak, M.; Walter, T; Aronson, A. (2000). "Regulation by overlapping promoters of the rate of synthesis and deposition into crystalline inclusions of Bacillus thuringiensis δ-endotoxins". J. Bacteriol. 182, pp. 734-741.

Shade, R.E.; Schroeder, H.E.; Pueyo, J.J.; Tabe, L.M.; Murdock, L.L.; Higgins, T.J.V.; Chrispeels, M.J. (1994). "Transgenic Pea-seeds expressing the alpha-amylase inhibitor of the common bean are resistant to bruchid beetles". Bio-Technology 12, pp. 793-796.

Shapiro, M.; Daugherty, E.M. (1994). "Enhacement in activity of homologous and heterologous viruses against gypsy moth (Lepidoptera, Lymantridae) by an optical brightener". J. Econ. Entomol. 87, pp. 361-365.

Shumway, L.K.; Yang, V.V.; Ryan, C.A. (1976). "Evidence for the presence of proteinase inhibitors I in vacuolar protein bodies of plant cell". Planta 129, pp. 161-165.

Slack, J.M.; Kuzio, J.; Faulkner, P. (1995). "Characterization of v-cath, a cathepsin, L-like proteinase expressed by the baculovirus *Autographa californica* multiple nuclear polyhedrosis virus". J. Gen. Virol. 76, pp. 1091-1098.

Smith, C.R.; Heinz, K.M.; Sansone, C.G.; Flexner, J.L. (2000). "Impact of recombinant baculovirus field applications on non-target heliothine parasitoid *Microplitis croce*ipes (Hymenoptera: Braconidae)". J. Econ. Entomol. 93, pp. 1109-1117.

Solleti, S.K.; Bakshi, S.; Purkayastha, J.; Panda, S.K.; Sahoo, L. (2008). "Transgenic cowpea (*Vigna unguiculata*) seeds expressing a bean alpha-amylase inhibitor 1 confer resistance to storage pests, bruchid beetles". Plant Cell Rep. 27, pp. 1841–1850.

Sun, X.; Sun, X.; van der Werf, W.; Vlak, J.M.; Hu, Z. (2004). "Field inactivation of wild type and genetically modified *Helicoverpa armigera* single nucleocapsid nucleopolyhedrovirus in cotton". Biocontrol Sci. Technol. 14, pp. 185-192.

Szewczyk, B., Hoyos-Carvajal, L.; Paluszek, M,; Skrzecz, I.; Lobo de Souza, M. (2006). "Baculoviruses-re-emerging biopesticides". Biotechnol. Adv. 24, pp. 143-160.

Walker-Simmons, M. & Ryan, C.A. (1977). "Immunological identification of proteinase inhibitor I and II in isolated tomato leaf vacuoles". Plant Physiol. 60, pp. 61-63.

Washburn, J.O.; Chan, E.Y.; Volkman, L.E.; Aumiller, J.J.; Jarvis, D.L. (2003). "Early synthesis of budded virus envelope fusion protein GP64 enhances *Autographa californica* multicapsid nucleopolyhedrovirus virulence in orally infected *Heliothis virescens*". J. Virol. 77, pp. 280-290.

Whyard, S.; Singh, A.D.; Wong, S. (2009). "Ingested double-stranded RNAs can act as species-specific insecticides". Insect Biochem. Mol. Biol. 39, pp. 824-832.

Wudayagiri, R.; Inceoglu, B.; Herrmann, R.; Derbel, M.; Choudary, P.V.; Hammock, B.D. (2001). "Isolation and characterization of a novel lepidopteran-selective toxin from the venom of South Indian red scorpion, *Mesobuthus tamulus*". BMC Biochem. 2, pp. 11-16.

Xu, D.P.; Xue, Q.Z.; McElroy, D.; Mawal, Y.; Hilder, V.A.; Wu, R. (1996). "Constitutive expression of a cowpea trypsin-inhibitor gene, CpTI, in transgenic rice plants confers resistance of two major rice insect pest". Mol. Breed. 2, pp. 167-173.

Yamada, T.; Hattori, K.; Ishimoto, M. (2001). "Purification and characterization of two α-amylase inhibitors from seeds of tepary bean (*Phaseolus acutifolius* A. Gray)". Phytochem. 58, pp. 59-66.

Yang, Z.; Chen, H.; Tang, W.; Hua, H.; Lin, Y. (2011). "Development and characterisation of transgenic rice expressing two *Bacillus thuringiensis* genes". Pest Manag. Sci. 67, pp. 414-422.

Yu, H.L.; Li, Y.H.; Wu, K.M. (2011). "Risk Assessment and Ecological Effects of Transgenic *Bacillus thuringiensis* Crops on Non-Target Organisms(F)". J. Integr. Plant Biol. 53, pp. 520-538.

Zhang, G.Y.; Sun, X.L.; Zhang, Z.X.; Zhang, Z.F.; Wan, F.F. (1995). "Production and effectiveness of the new formulation of Helicoverpa virus pesticide-emulsifiable suspension". Virologica Sinica 10, pp. 242-247.

Zhou, M.Z.; Sun, H.C.; Hu, Z.H.; Sun, X.L. (2004). "SOD enhances infectivity of *Helicoverpa armigera* single nucleocapsid nucleopolyhedrosis virus against *H. armigera* larvae". Virologica Sinica 18, pp. 506-507.

Zhou, X.Y.; Liu, N.; Zhao, M.; Li, H.; Zhou, L.; Tang, Z.W.; Cao, F.; Li, W. (2011). "Advances in effects of insecticidal crystal proteins released from transgenic Bt crops on soil ecology". Yi Chuan. 33, pp. 443-448.

Zhu, F.; Xu, J.; Palli, R.; Ferguson, J.; Palli, S.R. (2011). "Ingested RNA interference for managing the populations of the Colorado potato beetle, *Leptinotarsa decemlineata*". Pest Manag. Sci. 67, pp. 175-182.

Zou, Y.; Young, Y.S. (1994). "Enhancement of nuclear polyhedrosis virus activity in larval pest of Lepidoptera by a stilbene fluorescent brightener". J. Entomol. Sci. 29, pp. 130-133.

Zou, Y.; Young, Y.S. (1996). "Use of fluorescent brightener to improve *Pseudoplusia includens* (Lepidoptera: Noctuidae) nuclear polyhedrosis virus activity in the laboratory and field". J. Econ. Entomol. 89, pp. 92-96.

7

Bacillus thuringiensis – Based Biopesticides Against Agricultural Pests in Latin America

Ricardo Antonio Polanczyk, Sergio Antonio De Bortoli
and Caroline Placidi De Bortoli
Universidade Estadual Paulista (UNESP)
Brazil

1. Introduction

Twenty years after its discovery, the first attempts to use the entomopathogenic bacterium *Bacillus thuringiensis* (*Bt*) to control pests were made in Europe. Due to the successes achieved in some laboratory trials, commercial production of this pathogen began in France, where commercial formulations entered the market in 1938 as the product Sporeine. In the U.S., the use of this pathogen increased after 1950, mainly for the control of lepidopteran pests, resulting in the production of a formulation called Thuricide (Beegle & Yamamoto, 1992; Peferoen & Lambert, 1992; Lord, 2005).

Bt-based biopesticides have stimulated pathology and microbial control studies, and *Bt* is now one of the main pathogens used in agricultural pest control (Lord, 2005; Brar et al., 2006; Bravo et al., 2011). More than 200 *Bt*-based biopesticides account for 53% of the worldwide market for biopesticides, generating annual revenues of 120–140 million dollars (CAB, 2010). The Americas are responsible for almost 50% of this consumption, particularly the US and Canada, with Latin America accounting for only 8–10% of the total consumption (Guerra et al., 2001).

It is noteworthy that the public's concern of pesticide residues in food and the effect of pesticides on the environment has encouraged the use of microbial products in the production of vegetables and fruits of high commercial value in particular. The production of Dipel (*Bt* kurstaki-based biopesticide) began in 1970. This product proved to be 20–200-fold more potent than other *Bt*-based biopesticides (Beegle & Yamamoto, 1992). Dipel is currently used to control more than 167 lepidopteran pests (Glare & O'Callagham, 2000).

The insecticidal action of *Bt* is mainly due to crystalline inclusions (crystals or "Cry") that contain proteins synthesized during the second phase of the growth cycle, coinciding with the formation of spores (Bravo et al., 2001; Copping & Menn, 2000). These crystals become protoxins when dissolved in alkaline medium in the digestive system of insects. In the presence of digestive enzymes, protoxines are converted into 4 or more toxic polypeptides (δ endotoxins). The hydrolyzed Cry toxins cross the peritrophic membrane and bind to specific receptors present in the apical microvilli membrane of the midgut columnar cells, forming pores that increase membrane permeability, thereby affecting ion gradients and osmotic balance in the apical membrane. The increase in water absorption leads to the lysis and eventual rupture and disintegration of the midgut cells, which represses feeding and results in death due to starvation. The insect may also die due to an environment favorable for spore germination after cell lysis in the midgut. In this new environment, because of the mixing of the hemolymph and midgut contents, the pH becomes acidic, spores multiply, and eventually septicemia and insect death occur (Hofte & Whiteley, 1989; Knowles, 1994; Copping & Menn, 2000).

Although the detection of parasporal inclusions (Cry) in *Bt* cells and speculation about their importance in pathogenicity date back to the 1950s, their formulations were based on the number of spores, with inconsistent results, until the 1970s. However, since then, new formulations of *Bt* have considered the presence of δ-endotoxins. This fact, together with the potential of rapid growth and sporulation in relatively cheap substrates, increased the potential of *Bt* to control insects and enhanced market success.

The purpose of this chapter is to report the use of *Bt* in biological control programs in several countries of Latin America as well as to discuss the ongoing research to emphasize the potential efficacy of these pathogens in pest management. Other issues such as characterization, toxins, isolation, production, formulation effects on natural enemies, plants expressing *Bt* genes, and the evolution of resistance are discussed using previous publications (Brar et al., 2006; Estruch et al., 1997; Glare & O'Callagham, 2000; Charles et al., 2000; Valicenti & Zanasi, 2005).

2. Argentina

Despite the restricted use of *Bt* to control pests in this country, a number of projects have been developed by research institutes (Botto, 1996). *Bt*-based biopesticides were first used in the early 1950s. The target insect was *Colias lesbia pyrrothea* (alfalfa caterpillar/clouded yellow), and the use of biopesticides increased crop yields by approximately 72%. However, efforts aimed at increasing the use of this pathogen were not resumed until 1984 (Sosa-Gómez & Moscardi, 1991).

The Centro de Investigaciones Biologicas, belonging to the Universidad Nacional de Mar del Plata, performs the isolation, selection, morphological and biochemical characterization, and genetic toxicologic investigation of new *Bt* strains. Studies conducted at this center are aimed at developing and characterizing the efficacy of *Bt*-based agents against *Anticarsia gemmatalis*, *Spodoptera frugiperda* (Lepidoptera), *Anthonomus grandis*, *Diabrotica speciosa*, and *Tenebrio molitor* (Coleoptera) (Beron et al., 1999; Beron & Salerno, 2000).

Currently, imported products based on *Bt* are used in Argentina to control *Rachiplusia nu* and *A. gemmatalis* (Bac-Tur, Dipel, and Gale BT-PM). However, their use is not extensive since the market is dominated by chemical pesticides, particularly pyrethroids, and research of these products is still hampered by the severe economic crisis faced in recent years. In view of this, in the short term, it is unlikely that the use of entomopathogens, including *Bt*, will increase. The lack of policies at the national level and the isolated or discontinuous nature of the developed projects are factors that restrict the advancement of biological control in this country (Botto, 1996).

3. Brazil

Early studies in Brazil regarding the use of *Bt*-based products were conducted by Figueiredo et al. (1960) and Pigatti et al. (1960). These authors highlighted the high potential of this pathogen to control many pests such as *Ascia monuste orseis*, *Sylepta silicalis*, *Dirphia sabina*, *Azochis gripusalis*, *Alabama argillacea*, *Mocis repanda*, *Xanthopastis timais*, *Musca domestica*, and *Erinnyis ello*. The first results in relation to the use of *Bt*-based products were considered promising and spurred further research on *Bt*. The first project that aimed to control an important agricultural pest in Brazil, the fall armyworm (*S. frugiperda*), with *Bt* was started in 1993 by Embrapa Milho and Sorgo (CNPMS in Sete Lagoas - MG).

Until the early 1990s, only 3 commercial products were available in the Brazilian market, all based on *Bt kurstaki* (Dipel, Thuricide, and Bactospeine) (Habib & Andrade, 1991). The introduction of *Bt*-based products has encountered problems related to use strategies, marketing, as well as negative opinions advertised by the manufacturers and retailers of chemical insecticides, which diffused the idea that the insects should be rapidly controlled (Alves et al., 1998). Despite these issues, the number of *Bt*-based products on the market has recently increased.

This increase occurred mainly due to changes in the marketing strategies of the companies that sell these products, which emphasized some advantages, such as the maintenance of populations of parasitoids, predators, and pollinators (Alves et al., 1998). However, despite the environmental and social advantages of microbial products, the area treated with *Bt*-based biopesticides is approximately 150,000 hectares (Souza, 2001). This low use is due to a number of factors such as competition with chemicals (higher cost), specificity (reduced action spectrum), and low persistence of most commercially available formulations in the field (inactivation by ultraviolet radiation). New products based on *Bt* should be available in the Brazilian market in the coming years, increasing the availability of biopesticides. The cost of treating 1 hectare ranges from US$7.5 to US$15.0, depending of the crop (Alves et al., 1998).

The incidence of *Ecdytolopha aurantiana* (citrus borer) parasitism has risen sharply in citrus orchards in Brazil since the late 1990s (Gallo et al., 2002). This is mainly due to the incidence of variegated chlorosis, the insect vector, and its chemical control with pyrethroid insecticides through the fog application of non-selective insecticides, which in most cases have high shock power or contact, causing high mortality among natural enemies. The citrus borer limits production in many citrus growing regions of São Paulo State, causing losses of up to 60%, which translates to losses of up to $50 million per year; moreover, citrus borer infestation creates export barriers for some countries. *Bt*-based biopesticides effectively control this pest (Alves et al., 2001b) when the first application is made with more than 6 females caught per trap; the second application occurs 20–30 days later. The bacterium has been used in over 50,000 hectares of citrus orchards in Brazil, mainly in São Paulo State. Among the most important restrictions in the use of this method is the difficulty in synchronizing the short period of larvae exposure (just 4 h on the surface before it penetrates the fruit) and product application and its low persistence in the field due to the action of ultraviolet radiation.

The persistence of the product in the field can be increased by adding vegetable or mineral oil (Gallo et al., 2002) or using formulations containing microencapsulated factors that protect against ultraviolet rays, as reported by Dunkle & Shasha (1989). These authors reported that spores and crystals microencapsulated with Congo red or folic acid exhibited persistent viability of at least 50% of its toxic activity for 12 days. To reduce the impact of this pest on citrus, the Group of Advisers on Citrus recommends, among others measures, the use of products based on *Bt* plus a half dose of insecticide (pyrethroid).

Lepidopteran larvae that cause major defoliation in soybean include *A. gemmatalis* (velvetbean caterpillar), which consumes approximately 90 cm^2 of leaves over the course of its development (Gallo et al., 2002). Despite the high efficiency of *Bt* against this pest (Habib & Amaral, 1985; Bobrovski et al., 2002), its use is restricted mainly due to the widespread use of *Baculovirus anticarsia* (nuclear polyhedrosis virus). This biopesticide has a lower cost, ranging from US$1.3 to US$2.0 per hectare (Alves et al., 1998) and high efficiency, controlling over 80% of the pest population (Moscardi, 1998). Its fast growth can be attributed to the simple strategies used, including the amenability of its multiplication in the field by producers themselves, and the private sector's interest in producing and marketing the product (Moscardi, 1998).

The "Associação Riograndense de Empreendimentos de Assistência Técnica e Extensão Rural" (EMATER-RS), the rural extension agency of the Rio Grande do Sul State, one of the main producers of soybean, led a program aimed at the biological control of *A. gemmatalis* that sprayed approximately 13,000 hectares with Dipel in 2001 and 2002. However, the total area treated with this biopesticide is possibly higher because the data refer only to cities where EMATER-RS works, and do not represent the entire state.

In Brazil, a major pest of pastures is *Mocis latipes* (striped grass looper). This pest occurs cyclically (of some years); when in large numbers, it can significantly reduce the amount of forage available by completely consuming plant leaves. *Bt*-based products are used against small caterpillars in initial outbreaks at concentrations of 0.6–1 kg/ha. An important aspect is that these products are selective, being harmless to humans, natural enemies, and other animals (Gallo et al., 2002), thus indicating its usefulness in pastures and forage crops.

An economic viability study of *Thyrinteina arnobia*, the most important pest of *Eucalyptus*, was performed using *Bt* in approximately 4,000 hectares of *Eucalyptus grandis* and *Eucalyptus urophylla* (White, 1995). The cost of control was US$12.50/ha (equivalent to 1% of the timber produced), and this remained uniform even after pest infestation. A return on investment occurred in the first year after the control was instituted, because the measures adopted led to an income of US$34.20/ha/year. By contrast, a lack of control measures can result in losses of US$460.70–878.91/ha, or 37–70 times the cost of control, depending of the infestation level. Although studies comparing *Bt*-based products with chemicals have not been conducted for this pest, the economic feasibility of using *Bt* to control *T. arnobia* is obvious.

The pickleworm moths *Diaphania nitidalis* and *Diaphania hyalinata* are pests of cultivated cucurbits, and their larvae feed on all plant parts, preferring the fruit . The new leaves and branches become dry after being attacked. In fruits, the larvae open galleries and destroy the pulp, causing decay. *D. nitidalis* attacks fruits of any age, whereas *D. hyalinata* attacks the leaves and fruit skin. When using Dipel to control *D. nitidalis*, it is essential to determine the age of larvae because Dipel must be applied to early-stage larvae, i.e., before the larvae enter the fruit, as it is ineffective once larvae enter the fruit.

Other pests controlled with *Bt*-based produced are the imperial moth (*Eacles imperialis magnifica*) and cassava hornworm (*E. ello*), which can completely defoliate coffee and cassava plants, respectively. For the tomato pinkworm (*Tuta absoluta*), applications should be made weekly, in the evening or at night, simultaneously with the use of the egg parasitoid *Trichogramma pretiosum*. This pest can attack all parts of the tomato plant at any stage of development, consuming the leaves, branches, and mainly the apical buds and destroying new shoots, resulting in depreciation of the commercial value of the fruit (Gallo et al., 2002).

The fall armyworm in Brazil can reduce corn yield by 20%, and under dry conditions, it cuts off the plants close the ground, causing serious crop damage (Gallo et al., 2002). Although Thuricide is cited as a product registered for *S. frugiperda* control (Compêndio de Defensivos Agrícolas, 2010), its use is restricted due to low efficacy in most regions. Preliminary results from laboratory experiments conducted in the Laboratory of Pathology and Microbial Control of Insects, Piracicaba, SP, Brazil (ESALQ-USP) indicate that populations of fall armyworm from different geographic regions have different levels of susceptibility to the same strain of *Bt* (Polanczyk et al., 2005). These initial results confirm the need for comprehensive monitoring to verify variation among populations, which would increase the effectiveness of this strategy.

Bt-based biopesticides can be used in combination with other substances such as molasses and corn bran (Gravena et al., 1980a) to reduce *Diatraea saccharalis* (sugarcane borer) infestation in sugarcane crops by more than 50%. Parasitoids such as *Trichogramma pretiosum* enhance the effectiveness of *Bt kurstaki* in the control of *T. absoluta* (Marques, 1993). Similarly, pesticides, including insecticides (Assunção et al., 1980; Gravena et al., 1980b; Gravena et al., 1983; Zanuncio et al., 1992), have the same effect on biopesticides in controlling certain pests.

Published research by EMBRAPA indicates the importance of this *Bt* in agricultural pest control (Batista et al., 2005; Praça et al., 2004). In addition to EMBRAPA reports, several other federal and state institutions have performed research related to *Bt*. Among them, the Laboratory of Microbiology and Genetic of the UNISINOS University investigated the selection of *Bt* strains against coleopteran and lepidopteran pests (Fiuza et al., 2002, Pinto & Fiuza, 2002; Schünemann et al., 2005). ESALQ-USP studied the production, formulation, and selection of *Bt* strains for insect pests such as *T. absoluta*, *S. frugiperda*, and *Sitophilus oryzae* (Marques & Alves, 1996; Alves et al., 1997; Alves et al., 1999; Alves et al., 2001; Giustolin et al., 2001; Polanczyk et al., 2005). The sublethal effects of this pathogen on target insects and its interaction with other biological control agents have been studied (Polanczyk & Ahmed, 2005a, b). In the Laboratory of Entomology and Plant Pathology, projects are underway to select *Bt* strains with efficacy against lepidopteran pests and investigate their effects on natural enemies (Grecco et al., 2006; Pratissoli et al., 2006).

However, according to IBGE (2006), the 19 major crops of economic importance (cotton, peanuts, rice, oats, potatoes, cocoa, coffee, sugarcane, onions, barley, beans, oranges, castor oil plant, cassava, corn, soybeans, sorghum, wheat, and triticale) occupy an area of 146.6 million hectares. However, only 150,000 hectares are treated with *Bt*-based biopesticides for pest control, although their potential use may be extended up to 6 million hectares.

4. Colombia

Despite the prevalence of insect parasitoids in biological control programs in Colombia, some *Bt*-based products, albeit on a small scale, are used in pest control (Botto, 1996). However, a few decades ago, Revelo (1965) emphasized the potential of this pathogen in the control of *S. frugiperda*, *Agrotis ipsilon*, and *D. saccharalis*, the 3 major pests of maize in Columbia. In 1990, US$5.8 million was spent to control *S. frugiperda* in cotton crops, and US$4.2 million was spent to control the pest in corn and sorghum. In area, these amounts represent 430,000 and 440,000 treated hectares, respectively (Bosa & Cotes, 1997).

Important pests of many vegetable crops such as *Leptophobia aripa*, *Ascia monuste*, and *Plutella xylostella* have been well controlled by *Bt*-based biopesticides; thus, the damage caused by these insects is economically insignificant (Ruiz, 1998). However, Ruiz also states that in most cases, control measures involving the use of biological control must be accompanied by other tactics to maintain the pest population level below the economic injury level.

In 1989, the Laverlam Company began research of biopesticides to protect the environment and integrate tactics of integrated pest management in Colombia. These efforts resulted in the production of Turilav (*Bt kurstaki*), which is used to control *Heliothis* spp., *A. argillacea*, *Agrotis* spp., and *Spodoptera* spp., particularly in cotton crops.

In Valle del Cauca, the use of *Bt* to control *T. absoluta* is part of an integrated pest management program that has successfully reduced the costs of control by more than 54% to approximately US$650 per hectare (Belotti et al., 1990; Garcia, 1992). In the same region, *Bt*-based biopesticides are used against *Caligo ilioneus*, an important sugarcane pest that reduces the weight of the plant by 26–56% and its sugar content by 8–18%. In cotton crops, *Bt* is applied to control occasional outbreaks of *A. argillaceae* when defoliation exceeds 30%.

Sánchez et al. (1999) verified the high efficacy of *Bt kurstaki* and *Bt aizawai* to control *D. hyalinata* and *D. nitidalis*. These pests are the most destructive parasites of melon crops, in which they defoliate the crop and feed on the branches and fruits, causing annual losses of up to 25% of the yield. The use of *Bt* control has reduced the amount of insecticides used, which were sprayed up to 24 times in just 1 season in some cases.

Bosa & Cotes (1997) reported the high efficacy of 2 *Bt* strains (*Bt*-127 and HD-137) against *S. frugiperda*. Arango et al. (2001) used 1,100 native *Bt* isolates to study their selectivity to control the fall armyworm. Among the tested material, 32 isolates exhibited activity against this pest. The most powerful strains expressed *cry1Aa*, *cry1Ab*, *cry1Ac*, *cry1B*, *cry1C*, and *cry1D*. Serotyping revealed the isolates to be *Bt kurstaki*, *Bt thuringiensis*, *Bt canadensis*, and *Bt indiana*.

Maduell et al. (2002) isolated *Bt* from 13 species of the *Piper* genus in Colombian forests, detecting *Bt* in 74% of the samples. Regarding the presence of *cry* genes, *cry1* was amplified from 70% of the material (usually toxic to lepidopterans), and 60% exhibited some toxicity (12–100% mortality) to *S. frugiperda* larvae. In similar research, Yaro et al. (1999) isolated and characterized this bacterium in different parts of the country.

Recent studies by several research institutes revealed that 3 native isolates (IBUN6.4, IBUN3.3, and IBUN2.6) exhibit promising efficacy against *Heliothis virescens*, and 2 others (IBUN4.0 and IBUN3.8) were selected for *S. frugiperda*. For *H. virescens*, Cry 1F is the most effective toxin, and Cry1B and Cry1E are effective against *S. frugiperda*.

δ-Endotoxin expression is directly associated with sigma factors responsible for different sporulation stages. Studies with mutagenic agents found a direct relationship between the mutation of σ^E and σ^K factors with changes in *cry1Aa*, *cry1Ba*, *cry2Aa*, and *cyt1Aa* expression and changes in the specificity of insecticidal toxins. Research undertaken by the IWC sought to develop asporogenic *Bt* strains via chemical agent-induced mutagenesis. A strain was obtained that is currently being characterized in bioassays against *S. frugiperda* to determine the effect of the mutations on the integrity of genes, proteins, and toxin efficiency. The great interest in asporogenic *Bt* strains is justified even though the generation of these strains is difficult. However, because they are mutants produced in the laboratory, these strains should be utilized only after careful assessment of their environmental impact.

Corporación Nacional para Investigaciones Biologicas, Corporación Colombiana de Investigación Agropecuária, and Microbiological Research Center da Universidad de los Andes e Biotechnology Institute da Universidad Nacional de Colombia have conducted *Bt* studies with the following objectives: pathogen isolation, δ-endotoxin toxic activity determination against lepidopterans, toxic protein characterization, and DNA segment amplification of *cry* genes.

5. Cuba

Since 1990, due to the drastic reduction in the assistance provided by the former Soviet Union, Cuba has instituted necessary reforms to meet the basic needs of its citizens. The country made several transformations in many areas, including agriculture, leading to the development of more sustainable technologies in integrated pest management programs (Rosset & Moore, 1999). In the 1960s, the importation of some products based on *Bt* to control *H. virescens* and *M. latipes* stimulated the search for native pathogen strains (Pérez & Vasquéz, 2001). The use of chemical insecticides, which were mainly provided by the former Soviet Union, prevailed in Cuban agriculture, with methyl parathion being widely used in pest control before 1990.

The Ministry of Agriculture of Cuba has accelerated and significantly expanded plans to increase the production biological control agents to replace imported chemical insecticides. Since 1990, there has been an 89% reduction in the import of chemical insecticides and fertilizers. At the end of 1991, approximately 56% of insecticides used in Cuban agriculture were organic, representing a savings of US$15.6 million per year. In 1994, with 222 labs, the Centros de Producción de Entomófagos y Entomopatógenos (CREE) began operations, producing insects, nematodes, and pathogens in 15 provinces of Cuba. Currently, CREE operates 280 labs (53 in cultivated sugarcane areas and 227 in fruit areas and other crops), and *Bt* is primarily used (Pérez & Vasquéz, 2001).

Due to the prevalence of these bacteria-based products in Cuba, researchers seek to discover new native strains with high potential for pest control. Several institutions such as Instituto de Investigaciones de Sanidad Vegetal, Instituto de Investigaciones Fundamentales de la Agricultura Tropical, Centro National de Sanidad Agropecuária, Instituto de Investigaciones de Cítricos y Frutales, Instituto Cubano de Investigaciones Derivados de la Caña de Azúcar, and Instituto de Ecología y Systemática have contributed to the development of biopesticides in several areas.

The first biopesticides produced in Cuba were based on *Bt*, and they have been used to control *H. virescens*, *M. latipes*, *P. xylostella*, *S. frugiperda*, *E. ello*, and *D. hyalinata* (Table 3).

Among the registered products, those belonging to the line Thurisav are used to control *M. latipes, P. xylostella, Heliothis* spp., *Trichoplusia ni, Spodoptera* spp., and mites.

M. latipes is one of the most important agricultural pests in Cuba, causing losses of approximately 86,000 tons per year (equivalent to 7,800 hectares). Pérez et al. (1991) tested several products based on different subspecies of *Bt* (*Bt thuringiensis, Bt kurstaki*, and *Bt dendrolimus*) against this pest in pastures, and *Bt dendrolimus* was the most effective, with mortality rates of 86–90%.

According to Blanco (2006), *Brassica oleracea capitata* is one of the most important vegetable crops in Cuban agriculture, and among the factors limiting its production is *P. xylostella* (diamondback moth), which reduces its yields by 75–95%. To avoid these losses, many governmental and nongovernmental organizations have worked together to develop control strategies focusing the context of integrated pest management (IPM). The tactics employed should include *Bt*-based biopesticides to manage this pest and others such as *Ascia monuste eubotea* and *T. ni* to significantly reduce the control costs. Due to the success of these products in Cuba, the area treated with *Bt* to control *P. xylostella* increased from 17,400 hectares in 1988–89 to 53,000 hectares in 1993–94. Initial entomopathogen studies to control pest mites (*Phyllocoptruta oleivora, Tetranychus tumidus, Polyphagotarsonemus latus*) began in 1980. For products based on *Bt*, the mortality varies from 70% to 100% depending on the species, preventing the unnecessary use of chemical insecticides (Pérez, 1996). Between 1994 and 1996, almost 1,000 hectares were sprayed with *Bt* (LBT-13) to control *P. latus* (broad mite) in potato crops, with an efficacy exceeding 85% at a dose of 3–5 L/ha. The same strain caused between 83% and 90% mortality in adults and 100% in nymphs of *P. oleivora* (citrus rust mite) in citrus crops 72 h after application (20 L of fermented solution/ha). Moreover, it is selective to natural enemies such as predatory *Cycloneda sanguinea* in all its stages of development (Rojas, 2006).

The LBT-24 strain is one of the most effective biological agents for *S. frugiperda* control, and the mortality observed in the field ranged from 70 to 90% (Montesbravo, 2006), resulting in an increase of 15% in crop yield. The use of *Bt* formulations to control *S. frugiperda* in Cuba is very important because this pest can reduce corn yield by up to 40%, and important results have been obtained using this tactic together with others, such as the nuclear polyhedrosis virus and parasitoids (*Telenomus* sp. and *Chelonus insularis*).

The efficacy of LBT and LBT-3-21 *Bt kurstaki* isolates has exceeded 84% in potato, tomato, and tobacco crops. In Cuba, more than 20 species of insects are potato pests, and some of them such as *P. latus, Spodoptera* spp. and *T. ni* are controlled by *Bt*. In potato and tomato, the use of *Bt* favors the preservation of the natural enemies of *Liriomyza trifolii*. The utilization of this pathogen allows the action of *Opius* spp. and *Heteroschema* sp., which induce greater than 70% mortality (Morales, 2006).

The lepidopteran *H. virescens* is the most important pest of cotton in Cuba, and the efficacy of *Bt*-based products is similar to that of chemicals (14% for chemicals vs. 15.13% for biopesticides) (Jiménez, 1996).

Forestry is an important component of the Cuban economy, and 18 species of harmful insects greatly limit production. The most important pest of pine forests is *Rhyacionia frustrana*, the larvae of which invade the sprouts and buds of pine. *Bt*-based insecticides are widely used to control this pest with satisfactory results (Vásquez et al., 1999).

6. México

The use of entomopathogenic agents to control pests in Mexico began in the 1950s. In 1968–69, tests demonstrated the effectiveness of products formulated with *Bt* (Thuricide 90T and Thuricide 90TS) in controlling *P. xylostella* (Carrillo, 1971). However, the use of these insecticides increased sharply after 1990 due to the great interest of the public and private institutions in the use of pathogens in pest control.

In 1999, the use of *Bt*-based insecticides in Mexico increased from 15% to 20% and became the most commonly used and accepted biopesticide in that country. It was employed in 100,000 hectares of corn, 174,000 acres of cotton, and 200,000 hectares of vegetables and other crops. In the Bajio region of Guanajuato, 100 tons of *Bt*-based biopesticides were applied in 2001. In the Aguascalientes State, products made from *Bt kurstaki* and *Bt aizawai* are used against pests of vegetables, spinach, and potatoes, treating approximately 1,000 hectares (Vallés, 1998). In the entire country, it was estimated that 4–10% of the insecticides used contain *Bt* bacterium as the active ingredient. The leading *Bt*-producing companies are multinational, and the cost of control is US$19/ha, which is competitive with the chemicals in the market (Guerra et al., 2001).

Among the pests reported in Table 5, some of them are very important in several crops. For example, *S. frugiperda* can cause losses of up to 58% in plants and reduce corn yield by 1,148 kg/ha at a plant density of 45,000/ha (Castro-Franco et al., 1995). This species, together with *Helicoverpa zea*, *Diatraea grandiosella*, and *D. saccharalis*, may be responsible for losses up to 30% in Mexican corn crops.

In addition to the aforementioned pests, Rodríguez et al. (1991) reported the effectiveness of *Bt* (DL$_{100}$ 75 mg/kg diet) against *Galleria mellonella*, an apiculture pest that causes losses of honey, pollen, and wax production of up to 10%. In addition, Rodriguez & Trumble (1993) reported that this pathogen was useful in controlling tomato pests in the IPM context without harmful effect on natural enemies such as *Trichogramma* spp.

Edwards et al. (1999) obtained different CL50 estimates of the same isolate for several S. frugiperda populations collected in 5 regions of Mexico. The authors emphasized that variation may be due to geographical isolation, which results in reproductive isolation and physiologically different populations showing differential susceptibility to Bt.

The Centro de Investigación y de Estudios Avanzados, the International Maize and Wheat Improvement Center, and the Institute of Biotechnology, National Autonomous University of Mexico (UNAM) work together to locate new *Bt* toxins to control corn pests. Other research obtained promising results against *D. grandiosella*, *D. saccharalis*, *S. frugiperda*, *H. zea*, and *T. ni* (Bohorova et al. 1996; Bohorova et al. 1997; Del Rincón-Castro et al., 2006; Magdalena et al., 2001).

Researchers at the University of Guanajuato and Centro de Investigación de Estudios Avanzados del IPN investigated the characterization and selection of *Bt* against *Manduca sexta* and *T. ni* (Rosas et al., 1994; Corona et al., 1998; Rosales-Reyes et al., 2003).

At the Institute of Biotechnology UNAM, research has being conducted on the interaction between receptors and Cry1Ab toxins (lepidoptera-specific) to identify the regions responsible for toxicity. At the same institute, research for new insecticidal proteins that can potentially control insect pests has been performed using *Bt* (Bravo et al., 1998, Guerra et al., 2007). This project constructed a database composed of 500 *Bt* strains isolated from Mexican soil. Some of these were effective in controlling pests such as *Epilachna varivestis*, *Tapinoma melanocephalum*, *Phyllophaga* spp., *Anomala* spp., and *Bemisia tabaci*.

Another approach taken by researchers is to study the interaction of the toxins Cry1C and Cry1D with the intestinal membranes of insects such as *S. frugiperda*, *Rhopalosiphum maidis*, and *D. grandiosella*. More recently, work has been conducted to analyze pore formation in the insect midgut apical membrane and the subsequent effects on *Manduca sexta* gut cells.

7. Concluding remarks

The entomopathogenic bacterium *B. thuringiensis* is used against a wide range of pests in several crops in Latin America. Its widespread use against *P. xylostella* is noteworthy, even in countries not covered in this paper due to a lack of accurate information, as well as the use of *Bt*-based products produced in Cuba by small labs funded by the federal government.

Regional production is restricted to Cuba, and more recently, to Mexico, whereas imported products are used in other countries, which directly increases the cost of pest control and indirectly increases the final cost of production. The ability to formulate these products on a commercial scale should be considered by the governments of Latin American countries and/or private industries. The decrease in cost to the farmer would boost the use of these microbial products and increase the producers' profit, as the "products of biological origin," or those grown without pesticides, have better acceptance and price on the market in several countries.

The use of *Bt*-based products in Latin American countries depends more on political and economic aspects than technological approaches. Considering the basic economic differences in Brazil and Cuba, it would be logical to assume that the use of microbial products in Brazil, including those formulated with *Bt*, would be much greater than that in Cuba. However, the Cubans, under U.S. blockade, produced technology and developed strategies to use entomopathogens that exemplify how it is possible to make microbial control one of the pillars of sustainable agriculture. Brazil created a program that represents the most successful microbial control in the world — the control of *A. gemmatalis* with nuclear polyhedrosis virus. Brazil also undertook programs using bioinsecticides based on *Bt* frequent to control dipteran vectors. However, in Brazil, the availability of large numbers of pesticide formulations, supported by an intensive propaganda financed by multinational companies, together with the lack of a structure to implement advancements of alternative methods of control in production, has limited the use of entomopathogens. The solution to this impasse involves greater investment in technical assistance and research, principally in the production and formulation of general entomopathogens, aiming to convince the farmer that the biological method may in many cases be an important tactic of control that can minimize or even replace some pesticides to control certain pests.

Apparently, the use of plants expressing *Bt* genes has a negative impact on the use of products formulated with this bacterium, but this is negligible and unlikely to affect the national economy significantly. These crops, in addition to performing a vital role in economic development, have great potential to increase the use of bioinsecticides to control pests.

Other aspects such as knowledge of the biology and/or behavior of the target insects are very important, as they are directly related to their field efficiency. This is true for *E. aurantiana* and *T. absoluta*, 2 very important pests of citrus and tomato crops, respectively. Larvae of *E. aurantiana*, after emerging, spend little time on the fruit surface, and the larvae of *T. absoluta* remain inside the leaf for the most part of the larval stage. In both cases, the immature insects are exposed to *Bt* quickly. Therefore, although the efficacy in relation to some pests depends on the formulation, pathogenicity, and virulence of the pathogen, other aspects should be considered and studied to better enable the control of pests.

8. References

Alves, S.B.; Moino Jr., A.; Almeida, J.E.M. 1998. Desenvolvimento, potencial de uso e comercialização de produtos microbianos. In: *Controle microbiano de insetos*, S.B. Alves (ed.), Piracicaba: FEALQ, p.1143-1163.

Alves, L.F.A.; Alves, S.B.; Pereira, R.M.; Capalbo, D.M.F. 1997. Production of *Bacillus thuringiensis* var. *kurstaki* Berliner grow in alternative media. *Biocontrol Science and Technology*, 7: 377-383.

Alves, L.F.A.; Alves, S.B.; Augusto, N.T. 1999. Seleção de inertes e adjuvantes visando a elaboração de formulação granulada de *Bacillus sphaericus* Neide. *Arquivos do Instituto Biológico*, 66(1): 77-81.

Alves, L.F.A.; Alves, S.B.; Lopes, R.B.; Augusto, N.T. 2001a. Estabilidade de uma formulação de *Bacillus sphaericus* armazenada sob diferentes temperaturas. *Scientia Agricola*, 58(1): 21-26.

Alves, S.B.; Medeiros, M.B.; Tamai, M.A.; Lopes, R.B. 2001b. Trofobiose e microrganismos na proteção de plantas. *Biotecnologia Ciência e Desenvolvimento*, 21: 16-21.

Altieri, M.A. 1999. *Agroecologia: bases científicas para uma agricultura sustenable.* Montevidéu: Nordan-Comunidad. 325p.

Arango, J.A.; Romero, M.; Orduz, S. 2001. Diversity of *Bacillus thuringiensis* strains from Colombia with insecticidal activity against *Spodoptera frugiperda* (Lepidoptera: Noctuidae). *Journal of Applied Microbiology*, 91:1-10.

Assunção, M.S.; Campos, A.R.; Gravena, S. 1980. Efeito de Amitraz, Amitraz + *Bacillus thuringiensis* e Amitraz + inseticidas no controle de *Heliothis* spp. em algodão. *Anais da Sociedade Entomológica do Brasil*, 9(2): 249-251.

Batista, A.C.; Melatti, V.M.; Demo, C.; Martins, E.S.; Praça, L.B.; Gomes, A.C.M.M.; Falcão, R.; Brod, C.S.; Monnerat, R.G. 2005. *Prospecção de estirpes de Bacillus thuringiensis efetivas para o controle de Anticarsia gemmatalis.* Brasília: EMBRAPA, 19p. (Boletim de Pesquisa e Desenvolvimento, 82).

Bravo, A.; Likitvivatanavong, S.; Gill, S. S.; Soberón, M. 2011. *Bacillus thuringiensis*: a story of a successful bioinseticide. *Insect Biochemistry and Molecular Biology*, 41: 1-9.

Beegle, C.C.; Yamamoto, T. 1992. Invitation paper (C.P. Alexander Fund): history of *Bacillus thuringiensis* Berliner reseach and development. *Canadian Entomologist*, 124: 587-616.

Belloti, A.C.; Cardona, C.; Lapointe, S.L. 1990. Trends in pesticide use in Colombia and Brazil. *Journal of Agricultural Entomology*, 7: 191-201.

Béron, C.; Arcas, J.; Salerno, G.L. 1999. Uso de *Bacillus thuringiensis* en el control de insectos plaga de cultivos de interés agronômico. In: *IV Simpósio Nacional de Biotecnologia Vegetal*, Resumos, Argentina, Buenos Aires, p.77.

Béron, C.; Arcas, J.; Salerno, G.L. 2000. Characterization of a *Bacillus thuringiensis* strain toxic against lepidopterans and coleopterans. In: *XXI International Congress of Entomology*, Abstratcs, Brasil, Rio de Janeiro, p.515.

Blanco, E. 2006. Control biologico de la polilla de la col *Plutella xylostella* (L.). INISAV, Cuba. 13 p. (http://www.aguascalientes.gob.mx/agro/produce/PLUT-BIO.htm). Acessado em 25/09/2006.

Bohorova, N.; Maciel, A.M.; Brito, R.M. 1996. Selection and characterization of mexican strains of *Bacillus thuringiensis* active against four major lepidopteran maize pests. *Entomophaga*, 41(2):153-165.

Bohorova, N.; Cabrera, M.; Abarca, C. et al. 1997. Susceptibility of four tropical lepidopteran maize pests to *Bacillus thuringiensis* Cry1-type insecticidal toxins. *Journal of Economic Entomology*, 90(2): 412-415.

Bobrowski, V.L.; Pasquali, G.; Bodanese-Zanettini, M.H.; Pinto, L.M.F.; Fiuza, L.M. 2002. Characterization of two *Bacillus thuringiensis* isolates from South Brazil and their toxicity against *Anticarsia gemmatalis* (Lepidoptera: Noctuidae). *Biological Control*, 25(2): 129-135.

Bosa, C. F.; Cotes, A. M. 1997. Evaluacion de la actividad insecticida de isolados nativas de *Bacillus thuringiensis* contra *Spodoptera frugiperda* (Smith). *Revista Colombiana de Entomologia*, 23(3-4): 107-112,

Botto, E.N. 1996. Controle biológico de plagas en la Argentina: informe de la situación actual. In: *El Control Biológico en America Latina*, M.C. Zapater (ed.), Argentina, Buenos Aires, p.1-8.

Branco, F.E. 1995. *Aspectos Econômicos do Controle de Thyrinteina arnobia (Stoll, 1782) (Lep.:Geometridae) com Bacillus thuringiensis (Berliner) em Povoamentos de Eucalyptus spp*. Dissertação: Mestrado. Universidade Federal do Paraná. 130p.

Brar, S.K.; Tyagi, V.R.D.; Valéro, J.R. 2006. Recent advances in downstream processes and formulations of *Bacillus thuringiensis* based biopectides. *Process Biochemestry*, 41: 323-342.

Bravo, A.; Sarabia, S.; Lopez, L. 1998. Characterization of *cry* genes in a mexican *Bacillus thuringiensis* strain collection. *Applied and Environmental Microbiology*, 64(12): 4965-4972.

CAB International. The 2010 Worldwide Biopesticides Market Summary. London: CPL Business Consultant, 2010. 40p.

Carbajal, J. 1995. Uso de *Bacillus thuringiensis* cepa LBT 13 en el control de *Phylocoptruta oleivora* en el cultivo de los cítricos. In: *III Encontro Nacional Científico-Técnico de Bioplaguicidas*, Anais, Cuba, Havana, p.18.

Carozzi, N.B.; Kramer, V.C.; Warren, G.W.; Evola, S.; Koziel, M.G. 1991. Prediction of insecticidal of *Bacillus thuringiensis* strains by polymerase chain reaction product profiles. *Applied and Environmental Microbiology*, 57: 3057-3061.

Carrillo, J.L. 1971. Pruebas de Thuricide (Bacillus thuringiensis) para combatir gusanos de la col en Chapingo, Mex. Agricultura Técnica em México, 3(2):58-60.

Castro-Franco, R.; García-Alvarado, J.S.; Galán-Wong, L.J.; Gallegos-Del Tejo, A. 1995. Efecto sinérgico de extractos de bagazo de Agave lechuguilla Torr. en la actividad insecticida de Bacillus thuringiensis contra Spodoptera frugiperda Smith. Phyton, 57(2): 113-119.

Cerón, J.; Ortiz; A.; Quintero; R.; Güereca; L.; Bravo, A. 1995. Specific PCR primers directed to identify cryI and cryIII genes within Bacillus thuringiensis strain collection. Applied and Environmental Microbiology, 61: 3826-3831.

Cerón, J.; Covarrubias L.; Quintero, R.; Ortiz, A.; Ortiz, M.; Aranda, E.; Lina, L.; Bravo, A. 1994. PCR analysis of the cryI insecticidal crystal family genes from Bacillus thuringiensis. Applied and Environmental Microbiology, 60: 353-356.

Charles, J.F.; Delécluse, A.; LeRoux, C.N. 2000. Entomopathogenic bacteria: from laboratory to field application. Dordrecht: Kluwer Academic Publishers, 524 p.

COMPÊNDIO DE DEFENSIVOS AGRÍCOLAS: Guia prático de produtos fitossanitáros para uso agrícola. 2005. 7 ed. São Paulo: Andrei. 1141p.

Copping, L. G.; Menn, J.J. 2000. Review biopesticides: a review of their action, applications and efficacy. Pest Management Science, 56: 651-676.

Corona, J.E.B.; Meza, J.E.L.; Ibarra, J.E. 1998. Caracterización de una cepa mexicana de Bacillus thuringiensis ssp. kenyae: un análises de sua baja toxicidad hacia lepidópteros. Vedalia, 5(1): 3-12.

Cruz, P.F.N. 1977. Resultados preliminares sobre a eficácia do Bacillus thuringiensis Berliner no controle do "mandorová" da seringueira (Erinnyis ello L.) (Lepidoptera: Sphingidae), na Bahia. Revista Theobroma, 7: 93-98.

Del Rincón-Castro, M.C.; Méndez-Lozano, J.; Ibarra, J.E. 2006. Caracterización de isolados nativas de Bacillus thuringiensis con actividad inseticida hacia del gusano cogollero del maíz Spodoptera frugiperda (Lepidoptera: Noctuidae). Folia Entomologica, 45(2): 157-164.

Dunkle, R.L.; Shasha, B.S. 1989. Response of starch-encapsulated Bacillus thuringiensis containing ultraviolet screens to sunlight. Environmental Entomology, 18(6): 1035-1041.

Edwards, M.L.; Hernández, M.J.L.; Rubio, A.P. 1999. Biological differences between five populations of fall armyworm (Lepidoptera: Noctuidae) collected from corn in Mexico. Florida Entomologist, 82(2): 254, 262.

Estruch, J.J.; Carozzi, N.B.; Desai, N.; Duck, N.B.; Warren, G.W.; Koziel, M.G. 1997. Transgenic plants: an emerging approach to pest control Nature Biotechnology, 15: 137-141.

Fernández-Larrea, O. 1999. A review of Bacillus thuringiensis (Bt) production and use in Cuba. Biocontrol News and Information, 20(1): 47-48.

Ferrer, F. 2001. Biological control of agricultural insect pest in Venzuela; advances, achievements and future perspectives. Biocontrol News and Information, 22(3): 67-74.

Figueiredo, M.B.; Coutinho, J.M.; Orlando. A. 1960. Novas perspectivas para o controle biológico de algumas pragas com Bacillus thuringiensis. Arquivos do Instituto Biológico, 27: 77-88.

Fiuza, L.M.; Pinto, L.M.N.; Azambuja, A.O.; Steffens, C.; Menezes, V.G.; Vargas, J.O. 2002. Pathogenic effect of Bacillus thuringiensis isolate from south Brazil against Oryzophagus oryzae (Coleoptera, Curculionidae). In: XXXV Annual Meeting of the SIP, VIII International Colloquim on Invertebrated Pathology and Microbial Control, VI

International Conference on Bacillus thuringiensis, Program and Abstracts, Brasil, Foz do Iguaçu, p.60.

Gallo, D.; Nakano, O.; Silveira Neto, S.; Carvalho, R.P.L.; Baptista, G.C.; Berti Filho, E.; Parra, J.R.P.; Zucchi, R.A.; Alves, S.B.; Vendramim, J.D.; Marchini, L.C.; Lopes, J.R.S.; Omoto, C. 2002. Entomologia Agrícola, Piracicaba: FEALQ, 920p.

Garcia, F.R. 1992. Manejo integrado de plagas en cultivos del Valle del Cauca. *ICA Informa,* 26: 11-12.

Glare, T.R.; O'Callaghan, M. 2000. *Bacillus thuringiensis: biology, ecology and safety.* Chichester: John Wiley & Sons, 350 p.

Gravena, S.; Sanguino, J.R.; Bara, J.R. 1980a. Controle biológico da broca da cana *Diatraea saccharalis* (Fabricius, 1794) por predadores de ovos e *Bacillus thuringiensis* Berliner. *Anais da Sociedade Entomológica do Brasil,* 9(1): 87-95.

Gravena, S.; Campos, A.R.; Maia, O.S.; Paulaneto, G. T. 1980b. Eficiência de *Bacillus thuringiensis* Berliner e *Bacillus thuringiensis* + Methomyl no controle de lepidópteros no tomateiro. *Anais da Sociedade Entomológica do Brasil,* 9(2): 243-248.

Gravena, S.; Araújo, C.A.M.; Campos, A.R.; Villani, H.C.; Yotsumoto, T. 1983. Estratégias de manejo integrado de pragas do algodoeiro em Jaboticabal, SP, com *Bacillus thuringiensis* Berliner e artrópodos benéficos. *Anais da Sociedade Entomológica do Brasil,* 12(1): 243-248.

Grecco, E.D.; Polanczyk, R.A.; Pratissoli, D.; Andrade, G.S.; Malacarne, A.M. 2006. Potencial de *Bacillus thuringiensis* Berliner no controle de *Spodoptera eridania* (Lep.: Noctuidae). In: *XXI Congresso Brasileiro de Entomologia,* Resumos, Brasil, Recife, ID 186-1.

Guerra, P.T.; Wong, L.J.G.; Roldán, H.M. 2001. Bioinseticidas: su empleo, producción y comercialización en México. *Ciencia UANL,* 4(2): 143-152.

Guerra, P.T.; Iracheta, M.M.; Pereyra, R.A.; Galán, L.J.W.; Gómez, R.G.; Tamez, R.S.G.; Rodriguez, C.R. 2007. Caracterización de isolados mexicanas de *Bacillus thuringiensis* tóxicas para larvas de lepidópteros e coleópteros. *Ciencia UANL,* 3(4): 477-482.

Giustolin, T.A.; Vendramin, J.D.; Alves, S.B.; Vieira, S.A. 2001. Efeito associado de genótipo de tomateiro resistente a *Bacillus thuringiensis* var. *kurstaki* sobre o desenvolvimento de *Tuta absoluta* Meyrich (Lep., Gelechiidae). *Neotroptical Entomology,* 30(3): 461 - 465.

Habib, M.E.M.; Amaral, M.E.C. 1985. Aerial application of *Bacillus thuringiensis* against the velvetbean caterpillar *Anticarsia gemmatalis* Huebner, in soybean fields. *Revista de Agricultura,* 60(2): 141-149.

Habib, M.E.M.; Andrade, C.F.S. de. 1991. Controle microbiano de insetos com o uso de bactérias. *Informe Agropecuário,* 15(167): 21-26.

Hofte, H.; Whiteley, H.R. 1989. Insecticidal crystal proteins of *Bacillus thuringiensis. Microbiological Reviews,* 53: 242-255.

Huerta, A.; Cogollor, G. 1995. Control de la polilla del brote del pino (*Rhyacionia buoliana* Den. *et* Schiff.) mediante isolados del bacterio *Bacillus thuringiensis* var. *kurstaki. Ciencias Florestales,* 10(1-2): 135-141.

IBGE. 2006. *Levantamento sistemático da produção agrícola.* http://www.ibge.gov.br/. Acessado em 27/09/2010.

Jiménez, J. 1996. Control de *Heliothis virescens* (Fabr.) com medios biológicos en el cultivo del tabaco. In: *Curso International de Sanidad Vegetal en Tabaco*, Cuba, INISAV, 14 p.

Jiménez, J.; Fernández, R.; Canzado, A.; Vásquez, M. 1997. Efectividad de biopreparados nacionales de *Bacillus thuringiensis* para la lucha contra *Heliothis virescens* en tabaco. In: *V Encuentro Nacional Técnico-Científico de Bioplaguicidas*, Resumos, Cuba, Havana, p.58.

Knowles, B.H. 1994. Mechanism of action of *Bacillus thuringiensis* insecticidal δ-endotoxins. *Advances in Insect Physiology*, 24: 275-308.

Lambert, B.; Peferoen, M. 1992. Insecticidal promise of *Bacillus thuringiensis*. *BioScience*, 42: 112-122.

Leréclus, D. 1988. Génétique et biologie moléculaire de *Bacillus thuringiensis*. *Bulletin de l'Institut Pasteur*, 86: 337-371.

Licor, L.; Almaguel, L.; López, A., 1995. Control de ácaros en toronja y nananja com *Bacillus thuringiensis* (LBT-13) en Ciego de Avila, in: *Encuentro Nacional Técnico-Científico de Bioplaguicidas*, 3, Resumos, Havana, p. 2-3.

Lord, J.C. 2005. From Metchnikoff to Monsanto and beyond: The path of microbial control. *Journal of Invertebrate Pathology*, 89:19-29.

Maduell, P.; Callejas, R.; Cabrera, K.R.; Armengol, G.; Orduz, S. 2002. Distribution and characterization of *Bacillus thruringiensis* on the phylloplene of species of *Piper* (Piperaceae) in three altitudinal levels. *Microbial Ecology*, 44: 144-153.

Iracheta, M.C.; Galán, L.W.;Ferré, J.M.;Pereyda, B.A. 2001. Selección de toxinas Cry contra *Trichoplusia ni*. *Ciência UANL*, 4(1): 55-62.

Marques, I.M.R. 1993. *Ação de Bacillus thuringiensis Berliner var. kurstaki sobre Scrobipalpuloides absoluta (Meyrick, 1917) (Lepidoptera: Gelechiidae) e sua interação com o parasitóide Trichogramma pretiosum Riley, 1879 (Hymenoptera: Trichogrammatidae)*. Piracicaba, SP. Tese Doutorado, ESALQ/USP, 75 p.

Marques, I.M.R.; Alves, S.B. 1996. Efeito de *Bacillus thuringiensis* Berl. var. *kurstaki* sobre *Scrobipalpuloides absoluta* Meyer (Lepidoptera: Gelechiidae). *Anais da Sociedade Entomológica do Brasil*, 25(1): 39-45.

Martínez, Z.; Tamara, P.; Lérida, A.; Brito, R.; Hernández, A.; Peñate, E.; Lluvides, J.; Marín, R. 1995. Utilización de la cepa LBT-13 y de *Phytoseiulus macropilis* para el control de *Tetranychus tumidus*. In: *III Encuentro Nacional Técnico-Científico de Bioplaguicidas*, Resumos, Cuba, Havana, p.7-8.

Monnerat R.; Martins, E.; Queiroz, P.; Ordúz, S.; Jaramillo, G.; Benintende, G.; Cozzi, J.; Dolores Real, M.; Martinez-Ramirez, A.; Rausell, C.; Cerón, J.; Ibarra, J.E.; Del Rincon-Castro, M. C.; Espinoza, A.M.; Meza-Basso, L.; Cabrera, L.; Sánchez, J.; Soberon, M.; Bravo, A. Genetic variability of Spodoptera frugiperda Smith (Lepidoptera: Noctuidae) populations from Latin America is associated with variations in susceptibility to Bacillus thuringiensis Cry toxins. Applied and Environmental Microbiology, no prelo.

Montero, J.F. 1991. Uso de *Bacillus thuringiensis* var. *kurstaki* para el control de plagas en cucúrbitas. In: *III Taller Centroamericano en Fitoprotección del Melón*, Memoria, Honduras, p.71-72.

Montesbravo, E.P. 2006. Control biologico de *Spodoptera frugiperda* Smith en maiz. INISAV, Cuba. 10 p. http://www.aguascalientes.gob.mx/agro/produce/SPODOPTE.htm. Acessado em 25/09/2006.

Morales, C.A.M. 2006. Control biologico dentro del MIP en solanaceas. INISAV, Cuba. 12p. http://www.aguascalientes.gob.mx/agro/produce/SOLAN-BI.htm. Acessado em 25/09/2006.

Moscardi. 1998. Utilização de vírus entomopatogênicos em campo. In: Controle Microbiano de Insetos, S.B. Alves (ed.), Piracicaba: FEALQ, p. 509-540.

Pérez A.R. 1996. *Elementos para el manejo integrado del ácaro rojo,* Tetranichus tumidus *Banks (Acari: Tetranychidae) en plátano y banano.* Tese de Doutorado, INISAV, La Habana, Cuba, 89p.

Pérez, N.; Vasquéz, L.L. 2001. Manejo ecológico de plagas. In: *Transformando el campo cubano. Avances de la agricultura sostenible,* F. Funes et al. (eds.,) Cuba, La Habana, p.191-226.

Pérez, T., Jimenez , J.; Pazos, R. 1991. Efectividad de *Bacillus thuringiensis* obtenido en medio liquido estatico contra *Mocis* spp. en areas de pastos. *Protección de Plantas,* 1(1): 15-20.

Pigatti, A.; Figueiredo, M.B.; Orlando, A. 1960. Experiências de laboratório sôbre a atividade de novos inseticidas contra o mandorová da mandioca. *Biológico,* 26: 47-51.

Pinto, L.M.N.; Fiuza, L.M. 2002. Toxicity of *Bacillus thuringiensis kurstaki* HD68 ICPs to the *Spodoptera frugiperda* (Lepidoptera, Noctuidae) larvae from south Brazil. In: *XXXV Annual Meeting of the SIP, VIII International Colloquim on Invertebrated Pathology and Microbial Control,VI. International Conference on Bacillus thuringiensis,* Program and Abstracts, Brasil, Foz do Iguaçu, p.90.

Polanczyk, R. A. 1999. *Isolamento e seleção de Bacillus thuringiensis Berliner para controle de Spodoptera frugiperda (Smith & Abbot, 1797) (Lepidoptera: Noctuidae).* Dissertação: Mestrado. Universidade Federal do Rio Grande do Sul. 64p.

Polanczyk, R.A.; Alves, S.B. 2005. Interação entre *Bacillus thuringiensis* e outros entomopatógenos no controle de *Spodoptera frugiperda. Manejo Integrado de Plagas Y Agroecologia,* 74: 24-33.

Polanczyk, R.A.; Alves, S.B. 2005. Interação entre *Bacillus thuringiensis* e outros entomopatógenos no controle de *Spodoptera frugiperda. Scientia Agricola,* 62(5): 464-468.

Polanczyk, R.A.; Alves, S.B.; Padulla, L.F. 2005. Screening of *Bacillus thuringiensis* against three brazilian populations of *Spodoptera frugiperda* (Lepidoptera: Noctuidae). *Biopesticides International,* 1(1,2): 114-124.

Pratissoli. D.; Polanczyk, R.A.; Vianna, U.R.; Andrade, G.S.; Oliveira, R.G.S. 2006. Desempenho de *Trichogramma pratissoli* Querino & Zucchi (Hymenoptera, Trichogrammatidae) em ovos de *Anagasta kuehniella* (Zeller) (Lepidoptera, Pyralidae) sob efeito de *Bacillus thuringiensis* Berliner. *Ciência Rural,* 36(2): 369-377.

Praça. L.B.; Batista, A.C.; Martins, E.S.; Siqueira, C.B.; Dias, D.G.S.; Gomes, A.C.; Falcão, R.; Monnerat, R.G. 2004. *Pesquisa Agropecuária Brasileira,* 39(1): 11-16.

Revelo, M.A. 1965. Efectos del *Bacillus thuringiensis* sobre algunas plagas Lepidópteras del maíz bajo condiciones tropicales. *Agricultura Tropical,* 27(1): 392-395.

Rodrigues, B.A.; Trumble, J.T. 1993. A successful IPM system for processed tomatoes in Sinaloa, Mexico using *Bacillus thuringiensis* as a key component. In: *XXVI Annual Meeting of the Society for Invertebrate Pathology,* Program and Abstracts, Asheville, p.67.

Rodríguez, M. C.; Trujillo, H. S. 1991. Control de la polilla mayor de la cera *Galleria melonella* por isolados de *Bacillus thuringiensis* el municipio de Tecomán, Colima, México. *Revista Latino Americana del Microbiologia*, 33: 203-207.

Rojas, L. A. 2006. Control biologico de acaros fitofagos em diferentes cultivos. INISAV, Cuba. 16 p. (http://www.aguascalientes.gob.mx/agro/produce/acarbio.htm). Acessado em 25/09/2006.

Rosales-Reyes, T.; Salcedo-Hernandez, R.; Ibarra, J.E.; Barobza-Corona, J.E. 2003. Identificación de los genes cry en isolados mexicanas de *Bacillus thuringiensis* con potencial inseticida. *Acta Universitaria*, 13:39-46.

Rosas, A.M.M.; Perez, V.M.J.; Henonin, L.A.; Ibarra, J. 1994. Aislamiento y caracterizacion de una cepa nueva de *Bacillus thuringiensis*, esencialmente inmovil y altamente toxica a *Manduca Sexta* (Lepidoptera: Sphingidae). *Vedalia*, 1(1): 3-12.

Rosset, P.; Moore, M. 1999. Food security and local production of biopesticides in Cuba. *ILEIA Newsletter*, 13(4): 18-23.

Ruiz, R.A.V. 1998. El control biológico de insectos plagas en hortalizas. In: *Nuevos aportes del control biológico en la agriocultura sostenible*, Lizárraga, T. A., Barreto, C.U., Hollands, J. (eds.), Bogotá: Editora Nacional, p.269-294.

Sáenz, C.E.; Salazar, J.D.; Rodriguez, A.; Alfaro, D.; Oviedo, R. 2002. Resultados de vinte anos da direção de investigação de extensão de cana-de-açúcar (DIECA) no manejo integrado de pragas em Costa Rosa. In: *VIII Congresso Nacioanl STAB*, Brasil, Pernambuco, p.58-66.

Sánchez, R.M.Y.; Vargas, B.H.L.; Sánchez, G.G. 1999. Evaluación de variedades de *Bacillus thuringiensis* para el control de *Diaphania* spp. (Lepidoptera: Pyralidae) en melón, *Revista Colombiana de Entomologia*, 25(1-2): 17-21.

Santos, G.P.; Zanuncio, T.V.; Neto, H.F.; Zanuncio, J.C. 1995. Suscetibilidade de *Eustema sericiea* (Lepidoptera: Notodontidae) ao *Bacillus thuringiensis* var. *kurstaki*. *Revista Ceres*, 42(242): 423-430.

Schnepf, E.; Crickmore, N.; Van Rie, J.; Lereclus, D.; Baum, J.; Feitelson, J.; Zeigler, D.R.; Dean, D.H. 1998. *Bacillus thuringiensis* and its pesticidal crystal proteins. *Microbiology and Molecular Biology Reviews*, 62: 775-806.

Schünemann, R.; Barbieri, C.; Fiuza, L.M. 2005. Toxicidade de delta-endotoxinas de *Bacillus thuringiensis* à *Anticarsia gemmatalis* (Lepidoptera, Noctuidae). In: *IX Simpósio de Controle Biológico*, Resumos, Brasil, Recife, p.80.

Souza, M.L. de. 2001. Utilização de microrganismos na agricultura. *Biotecnologia Ciência e Desenvolvimento*, 21: 28-31.

Sosa-Gómez, D.R.; Moscardi, F. 1991. Microbial control and insect pathology in Argentina. *Ciência e Cultura*, 43(5): 375-379.

Valicente, F.H.; Zanasi, R.F. 2005. *Uso de meio alternativos para produção de bioinseticida à base de* Bacillus thuringiensis. Sete Lagoas: EMBRAPA, 4p. (Circular Técnica, 60).

Vallés, S.S. 1998. Avances en el uso del control biologico en el estado de Aguascalientes. *Cuaderno de Trabajo*, 85: 1-22.

Vásquez, L.L.; Menéndez, J.M.; López, R. 1999. Manejo de insectos de importância forestal en Cuba. *Revista Manejo Integrado de Plagas*, 54: 4-14.

Zanuncio, J.C.; Guedes, R.N.C.; Cruz, A.P. da; Moreira, A.M. 1992. Eficiência de *Bacillus thuringiensis* e de Deltametrina, em aplicação aérea, para o controle de *Thyrinteina arnobia* Stoll, 1782 (Lepidoptera: Geometridae) em eucalipital no Pará. *Acta Amazonica*, 22(4): 485-492.

Yaro, E.; Cerón, J.; Uribe, D. 1999. Caracterización de aislamientos nativos de *Bacillus thuringiensis* provenientes de bosque natural y zonas de cultivos agrícolas Colombianos. *Revista Colombiana de Entomologia*, 25(3-4): 185-190.

8

Recent Advances in Our Knowledge of Baculovirus Molecular Biology and Its Relevance for the Registration of Baculovirus-Based Products for Insect Pest Population Control

Renée Lapointe[1], David Thumbi[1] and Christopher J. Lucarotti[2]
[1]Sylvar Technologies Inc., New Brunswick,
[2]Natural Resources Canada, Canadian Forest
Service-Atlantic Forestry Centre, New Brunswick
[1,2]Canada

1. Introduction

Public demand for safer, environmentally-benign alternatives to synthetic chemical pesticides and more stringent barriers put in place by regulatory agencies worldwide has led to increased interest in microbial pest control agents (MPCA) based on viruses, bacteria, fungi, protozoa and nematodes as the active ingredient. The MPCA market has recently experienced an increase of 47% between 2004/2007 with sales worth $396 million in 2007/2008 (CPL Consultants, 2010). Despite this increase, the microbial biopesticide market still only represents about 1% of the sales of chemical pesticides (CPL Consultants, 2010; Marrone, 2007). Factors impeding the establishment of strong MPCA markets are complex (Chandler et al., 2011; Marrone, 2007; Ravensberg, 2011) but include the burdensome costs associated with the registration of commercial products that are aimed at relatively small niche markets (Chandler et al., 2011; Ehlers, 2011). The main priority of regulatory agencies is to protect human health and safety and the environment from potential risks associated with the use of pest control products. A common feature of MPCA registration processes around the world is that they grew out of registration processes designed for chemical pesticides with adjustments allowing for the reduced risks of MCPAs (Chandler et al., 2011; Ehlers, 2011). Even though public attitudes to the use of biological control agents has been favourable (63%) a large proportion of the public (46%) has expressed concerns about the consumption of food treated with microbial pesticides (Cuddleford, 2006).

Baculoviruses only infect insects, are ubiquitous in the environment and are known to be important in the regulation of many insect populations. Baculoviruses are host specific, infecting only one or a few closely-related species, helping to make them good candidates for management of crop and forest insect pests with minimal off-target impacts (Cory, 2003; England et al., 2004; Hewson et al., 2011; Raymond et al., 2005). In fact, baculoviruses have been recognized as being amongst the safest of pesticides (Black et al., 1997; Gröner, 1986) and have been included on lists of "low risk" biocontrol agents by the European Union (Leuschner, 2010; Regulation of Biological Control Agents in Europe; http://www.rebeca-net.de). Since the start of their commercial use, baculoviruses have been tested extensively to assess their safety in order to meet registration requirements (reviewed in Burges et al., 1980a, 1980b; Gröner, 1986; Ignoffo, 1973). As of 2010, over 24 baculovirus species have been reported to be registered for use in insect pest management throughout the world (Kabaluk et al., 2010; Quinlan & Gill, 2006). Market share of baculoviruses is 6% of all microbial pesticides (Marrone, 2007; Quinlan & Gill, 2006) and millions of hectares have been treated with registered baculovirus products over the years (Kabaluk et al., 2010; Moscardi, 2011; Szewczyk et al., 2009). Despite many years of use and testing against nontarget organisms, no adverse effects have ever been attributed to baculoviruses (McWilliam, 2007; OECD, 2002). In this review of baculoviruses, we discuss how baculovirus evolution, host range determination and pathogenesis have contributed to their inherent safety for nontarget organisms including humans.

2. Classification and origins of baculoviruses

The virus family *Baculoviridae* is divided into four genera that are restricted to three insect orders: *Alphabaculovirus* (nucleopolyhedrovirus or NPV) in Lepidoptera, *Betabaculovirus* (granulovirus or GV) also in Lepidoptera, *Gammabaculovirus* (NPV) in Hymenoptera, and *Deltabaculovirus* (NPV) in Diptera (Jehle et al., 2006). The large, covalently-closed, double-stranded DNA genome is packaged into an enveloped, rod-shaped capsid. The virions of NPVs are enveloped either singly (SNPV) or in groups of multiple virions (MNPV) which are then occluded in a protein called polyhedrin to form the occlusion body (OB). Virions of GVs are enveloped and occluded singly in an ovicylindrical granule (also an OB) formed from the protein granulin. Baculoviruses are normally named for the initial host from which they were first isolated. The International Committee on Taxonomy of Viruses (ICTV) lists the designated members of the Baculoviridae (http://www.ictvdb.org/Ictv/index.htm).

To date, 58 baculovirus genomes have been sequenced; 41 alphabaculoviruses, 13 betabaculoviruses, three gammabaculoviruses and one deltabaculovirus. Baculovirus genome sizes range from the smallest gammabaculovirus, *Neodiprion lecontei* nucleopolyhedrovirus (NeleNPV), at 81,755 bp (Lauzon et al., 2004) to the largest betabaculovirus, *Xestia c-nigrum* granulovirus (XcGV) at 178,733 bp (Hayakawa et al., 1999). No matter the genus or genome size, all baculoviruses share 31 core genes in common (Miele et al., 2011). These are essential genes involved in oral infection (*pif-0 (p74), pif-1, pif-2, pif-3, pif-4/19kd/odv-e28, pif-5/ odv-e56*), cell cycle arrest (*odv-ec27, ac81*), replication (*dnapol, helicase, lef-1, lef-2*), late gene transcription (*lef-4, lef-5, lef-8, lef-9, p47*) and virus assembly, packaging and release (*38k/ac98, alk-exo, desmoplakin, gp41, odv-e18, odv-nc42, odv-ec43, p6.9, p33/ac92, p49, vlf-1, vp39, vp91, vp1054*) (Miele *et al.* 2011). Twenty of these core genes (*p47, lef-4, lef-5, lef-8, lef-9, vlf-1, pif-0, pif-1, pif-2, pif-3, pif-4, pif-5, vp91, vp39, 38k, ac68, ac81, p33, dnapol, helicase*) are also found in insect dsDNA viruses belonging to the genus *Nudivirus* (Wang et al., 2011) and a number (e.g., *lef-4, lef-5, lef-8, p47, 38k/ac98, vp91, pif-0, pif-1, pif-2, pif-3*) are also found in bracoviruses (polydnaviruses) associated with parasitic wasps belonging to the family Braconidae (Bézier et al., 2009). It has been suggested that bracoviruses arose from the insertion of a nudivirus ancestor into braconid wasps about 100 million years ago (mya) (Bézier et al., 2009). Nudiviruses and baculoviruses, however, are thought to have shared a common ancestral virus (Wang & Jehle, 2009). Deltabaculoviruses and gammabaculoviruses are thought to be more primitive than the alphabaculoviruses and betabaculoviruses because of their smaller genomes and tissue tropism which is limited to midgut epithelial cells (Lauzon et al., 2004) and, in the case of deltabaculoviruses, cells of the posterior midgut and gastic caeca (Moser et al., 2001). Gammabaculoviruses, however, are thought to be more closely (although still distantly) related to the alphabaculoviruses and betabaculoviruses than are the deltabaculoviruses (Herniou et al., 2004). The virions of *Culex nigripalpus* deltabaculovirus (CuniNPV) are occluded in a 90-kDa protein that bears no similarity to polyhedrin/granulin proteins or any other protein in available sequence databases (Perera et al., 2006).

The occlusion derived virions (ODVs) that emerge from OBs are the universal virion phenotype for all baculoviruses as they are responsible for the initial oral (*per os*) infection of host insect gut cells. In lepidopteran hosts, the initial, primary infection of midgut cells by ODVs is followed by secondary infection of tissues within the insect hemocoel that is effected by the budded virion (BV) phenotype. The genome content of ODVs and BVs is identical but differences in virion morphology, structural proteins, envelopes, antigenicity, and cellular site of maturation are the basis for their respective patterns of infectivity. In mosquito hosts, following the primary infection of the gastric caeca and posterior midgut by ODVs, deltabaculovirus BVs also spread the infection further from cell to cell but only to these same tissues (Moser et al., 2001). Sawfly gammabaculoviruses do not appear to have a BV phenotype (Duffy et al., 2006; Garcia-Maruniak et al., 2004; Lauzon et al., 2004) and OBs are only produced in the nuclei of midgut epithelial cells (Federici, 1997).

In Lepidoptera, NPVs have been reported from 28 families and GVs from 19 families (Martignoni & Iwai, 1981). In the Diptera, NPVs have been reported from the Calliphoridae, Chironomidae, Culicidae, Sciaridae, Tachinidae and Tipulidae (Martignoni & Iwai, 1981). Fewer families of sawflies (Argidae, Diprionidae, Pamphiliidae and Tenthredinidae) are reported to be infected by NPVs (Martignoni & Iwai, 1981). However, due to a general lack of viral isolates, sequence data and other information, most of the baculoviruses listed by Martignoni & Iwai (1981) are not yet considered as valid species by ICTV (http://www.ictvdb.org/Ictv/index.htm). For example, the NPV of the pamphiliidid sawfly, *Acantholyda erythrocephala*, has been reported to occur in the fat body (Jahn, 1967), something which is not characteristic of gammabaculoviruses.

Recent phylogenetic analyses have indicated that the Hymenoptera, not the Coleoptera, are basal to the holometabolous insects that also include the Lepidoptera and Diptera (Savard et al., 2006). When and how the four genera of baculoviruses came to infect their different insect hosts is not known but selection pressure and co-evolution with their respective hosts appears to have constrained each baculovirus genus to a single insect order (Herniou et al., 2004). Historically, the Hymenoptera have been subdivided into the more advanced Apocrita, including ants, bees and wasps, and the basal Symphyta that includes the sawflies. It now appears, however, that the evolution from the ancestral hymenopteran to the Euhymenoptera (Apocrita and Orussoidea) was monophylletic and that the different superfamilies of sawflies constitute separate branches off the lower end of this lineage (Farris & Schulmeister, 2011). In this light, gammabaculoviruses have only been confirmed in the Diprionidae and, considering the paraphyletic origins of the different groups of sawflies, it may be the case that gammabaculoviruses are restricted to the Diprionidae or closely related families within the superfamily, Tenthredinoidea.

3. Baculovirus pathogenesis and potential blocks to infection

As is the case for all viral pathogens, baculovirus replication is dependent upon the availability of permissive host cells. The accessiblity and susceptibility of host cells to viral invasion and replication is classified into three categories; permissive, semi-permissive and non-permissive. A permissive infection results in successful viral replication and subsequent production of infectious virions that can transmit the infection to other permissive cells and individuals. Semi-permissive infections result in limited viral progeny resulting from defects in some replication events, such as gene expression or viral DNA replication. In non-permissive infections, cells do not support viral replication and the process does not yield infectious progeny. Determining what factors influence the level of permissiveness of an insect cell to a particular baculovirus has proven to be challenging because baculovirus host range is affected not only by the interactions between the baculovirus and the host cell at the molecular level (reviewed in Miller & Lu, 1997; Thiem & Cheng, 2009) but also by aspects of insect behaviour and physiology (reviewed in Cory & Hoover, 2006).

As hosts for viruses, insects can present challenges because of their sporadic and/or episodic availability and their relatively short life spans. Long periodicity of population fluctuations, for example the spruce budworm (*Choristoneura fumiferana*) which can span over 30 years (Royama, 1992), indicates that baculoviruses must be able to persist in the environment for long periods while waiting for permissive hosts to become available. The OB and its surrounding polyhedral envelope (PE) (a protein/carbohydrate matrix) (Gross & Rohrmann, 1993; Gross et al., 1994; Russell & Rohrmann, 1990) help protect the ODVs from degradation by such environmental factors as dessication and ultraviolet radiation (UV) (reviewed in Slack & Arif, 2007). OBs and ODVs can be further protected from UV radiation by establishing natural reservoirs in sheltered environments such as those in and on plants and in soil (Raymond et al., 2005; Witt, 1984).

3.1 Midgut lumen and pH

Baculoviruses are predominantly diseases of the larval stages of insects. When a larval host consumes foliage or water that is contaminated with OBs, the alkaline pH (8-11) of the larval midgut (Fig. 1) dissolve the PE and OB matrix within minutes (Adams & McClintock, 1991) releasing ODVs into the midgut lumen. The gut environment, into which OBs enter, is a first deciding specificity factor as OBs will only dissolve in an alkaline environment. The dissolution of OBs is further facilitated by OB-associated alkaline proteases. While the PE lattice is sufficiently narrow to restrict access by large digestive enzymes of vertebrates, it does allow infiltration by anions from the alkaline midgut of insects. Midgut pH of Lepidoptera averages 10.5, while within the Diptera, only in the Culicidae does the gut pH reach 10 (Terra et al., 1996). Within the Hymenoptera, only sawflies are known to harbor baculoviruses and their midgut pH, although lower than those of lepidopteran and culicid mosquitoes, is between 6.7 and 8.7 (Heimpel, 1955), which is more alkaline than that of bees and wasps (Apidae) at pH 5.7 (Terra et al., 1996). When compared to the midgut of other coleopteran families, such as the Coccinelidae, Chrysomelidae and Cerambycidae (midgut pH 5.4 to 6.9), the midgut pH of Scarabeidae is 10.4 (Terra et al., 1996). Although not a baculovirus *per se*, the nudivirus of the scarab *Oryctes rhinoceros* (OrN) shares homologies with baculoviruses *per os* factors and other core genes (Wang et al., 2011). Although the nudiviruses have established more complex tissue tropisms and transmission routes than baculoviruses, their primary route of infection is also *per os* (Wang & Jehle, 2009).

In addition to protecting the ODVs against environmental factors, the stability of the crystalline structure of the OBs has been shown to assist in the dispersal of the virus by vertebrates. The acidic pH of the stomachs (from pH 1 to 7) of vertebrates (Fig. 1) helps to preserve the integrity of the OBs. Excreted OBs, recovered from the digestive tracts of non-host invertebrate and vertebrate animals (Lautenschlager et al., 1980; Vasconcelos et al., 1996) were found to remain infectious to their insect larval hosts, leading to the suggestion that the consumption of baculovirus-infected larvae by various non-target animals plays a role in the dissemination of OBs (Entwistle et al., 1977; Lautenschlager et al., 1980).

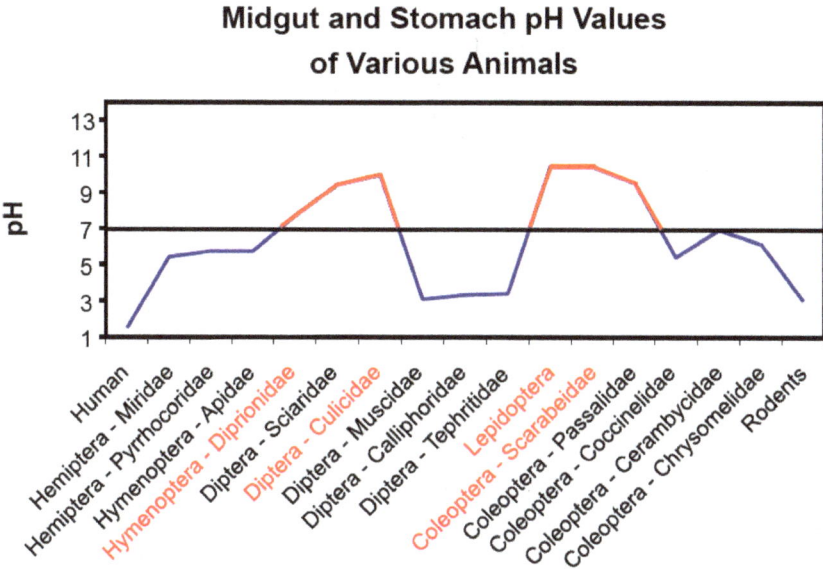

Fig. 1. Midgut and stomach pH values of various organisms. Red line shows alkaline values and blue line, acidic values. Insect orders and families that have been identified as baculovirus or nudivirus hosts are labeled in red. pH value of midguts of Hemiptera, Hymenoptera (Apidae), Diptera, Lepidoptera and Coleoptera (Terra et al., 1996), pH value of Hymenoptera (Diprionidae) (Heimpel, 1955) and pH values for stomach of human (Guyton & Hall, 2006) and rodents (McConnell et al., 2008).

3.2 Peritrophic membrane

The peritrophic membrane (PM) of insects is an acellular sleeve-like structure that lines and protects the gut. In insects, the PM consists of chitin sheets, proteoglycans and chitin-binding proteins such as peritrophins and intestinal mucins, that cross-link to form a thick, three-dimensional mesh (Barbehenn & Martin, 1995; Lehane, 1997; Peters, 1992; Tellam et al.,

1999). The chitin sheets provide the PM with strength and flexibility and the hydrating capacity of the proteoglycans is implicated in determining the permeability of the PM (Lehane, 1997). The pores, ranging from 21 to 36 nm (Barbehenn & Martin, 1995), allow bidirectional trafficking of small molecules such as digestive enzymes and the regulation of the flow of hydrolytic products (water and nutrients) between endoperitrophic and ectoperitrophic spaces (reviewed in Lehane, 1997). Insect intestinal mucins (IIM) consist of potentially-glycosylated, mucin-like domains and cysteine-rich chitin-binding domains (CBD) (Sarauer et al., 2003; Wang & Granados, 1997a), which are thought to have a similar lubricating and protective role as vertebrate mucins (Sarauer et al., 2003; Tellam et al., 1999; Wang & Granados, 1997a).

The small pore size of the PM limits the passage of larger materials such as pathogens, toxins and noxious phytochemicals (e.g. tannins) that are also ingested into the endoperitrophic space during feeding (Barbehenn, 2001). The size of baculovirus nucleocapsids, being between 40–70 nm in diameter and 250–400 nm in length (Boucias & Pendland, 1998), precludes them from crossing the PM. Disruptions in the links between PM chitin sheets and proteins can result in a decrease in the ability of the PM to act as a barrier to pathogens (Plymale et al., 2008; Wang & Granados, 1997b; 2000). ODVs passively cross the PM through physical breaches caused by mechanical abrasions, chemical means (Wang & Granados, 2000) or possibly by natural degradation of the PM. Diet has also been shown to affect the PM thickness. When fed cotton foliage rather than artificial diet, tobacco budworm larvae (*Heliothis virescens*) were shown to form a thicker PM that, by decreasing the number of primary infection foci on the midgut epithelium (Plymale et al., 2008), led to lower susceptibility to *Autographa californica* MNPV (AcMNPV).

An active mechanism of disruption of the PM has evolved in many GVs and a few lepidopteran group II NPVs whose genomes encode a metalloprotease called enhancin. Enhancins are occluded in the OB matrix either inside or on the surface of ODVs (Hashimoto et al., 1991; Lepore et al., 1996; Slavicek & Popham, 2005; Wang et al., 1994). In *Trichoplusia ni*, TnGV ODV enhancin results in disruption of the PM through the degradation of IIM structures (Derksen & Granados, 1988; Hashimoto et al., 1991; Wang & Granados, 1997b). LdMNPV enhancins are located on the ODV surface and were found to be essential for *per os* infectivity of *Lymantria dispar* larvae even when the PM was absent suggesting that enhancins may play a role in addition to PM breaching (Hoover et al., 2010).

Homologues of 11K protein are conserved in most baculoviruses and can be present in multiple copies (Dall et al., 2001). These proteins are characterized by the presence of C_6 chitin-binding motifs or peritrophin-A domains that are also present in chitin binding proteins such as mucins, peritrophins and chitinases of insect guts and basal laminae of trachea (Dall et al., 2001; Tellam et al., 1999). The AcMNPV 11K proteins were found to associate with BVs and OBs (Lapointe et al., 2004) but not with the ODVs themselves (Braunagel et al., 2003; Zhang et al., 2005) and were shown to enhance oral infection (Lapointe et al., 2004; Zang et al., 2005) but not hemocoelic infection. Although the function of Ac150 and Ac145 is not yet known (Lapointe et al., 2004; Zhang et al., 2005), they are important mediators of primary infection where they impact infectivity differentially in two different hosts of AcMNPV, *T. ni* and *H. virescens* (Lapointe et al., 2004).

3.3 Entry in midgut cells

The midgut tissue consists principally of midgut epithelial cells that are prone to apoptosis (Uwo et al., 2002) and slough off regularly (Engelhard & Volkman, 1995) which, when considering the obligate nature of a viral infection, makes them a sub-optimal environment for viral replication. The process of midgut epithelial cell infection is carried out by ODVs. Comparatively, BVs are 10,000 fold less efficient than ODVs at infecting midgut cells (Volkman & Summers, 1977; Volkman et al., 1976). On the other hand, AcMNPV ODVs have been shown not to infect or even penetrate insect tissue culture cells (Volkman & Summers, 1977; Volkman et al., 1976). In a cell culture assays, 2.3×10^5 ODVs were shown to be required to infect a single *T. ni* cell (TN-368) making them 1700 to 1900 times less infectious than BVs (Volkman et al., 1976). ODV infectivity to TN-368 cells was improved, however, in the presence of midgut juices from *Heliothis zea* and *Estigmene acrea* larvae (Elam et al., 1990), indicating that midgut lumen factors are important to the primary infection process.

Due to the complexity of *in vivo* studies and the lack of midgut cell lines that would simplify studies of ODV host cell entry, the exact mechanisms directing the primary infection processes are still relatively unclear. Once across the PM, ODVs bind to ODV-specific receptors and fuse to the brush border microvilli of the columnar epithelial cells (Haas-Stapleton et al., 2004; Horton & Burand, 1993). Nucleocapsids are then released into the cell cytoplasm to initiate primary infection of the midgut epithelium (reviewed in Slack & Arif, 2007). ODV-associated proteins of both NPVs and GVs are numerous (Braunagel et al., 2003; Wang et al., 2011) and include structural proteins responsible for encapsidation, unique ODV-envelope proteins involved in tissue tropism and proteins involved in viral gene expression and DNA replication (Braunagel et al., 2003). The ODVs unique association with proteins involved in viral gene expression and DNA replication is thought to optimize viral replication in midgut cells. Over 40 proteins were found to be associated with the ODVs of the alphabaculovirus AcMNPV (Braunagel et al., 2003), the betabaculovirus *Pieris rapae* GV (Wang et al., 2011) and the deltabaculovirus CuniNPV (Perera et al., 2007). Most of the proteins found in both AcMNPV ODVs and BVs are nucleocapsid proteins while most of the tissue-specific envelope proteins are different for each phenotype (Braunagel et al., 2003; Wang et al., 2010).

ODV-envelope proteins are highly complex and over 10 envelope proteins have been found to be associated with the ODVs of AcMNPV (reviewed in Slack & Arif, 2007; Braunagel & Summers, 2007). Only few of these ODV proteins, termed the *per os* infectivity factors (PIFs), have been shown to be essential for the AcMNPV *per os* infection process (Braunagel et al, 1996; Fang et al., 2009; Faulkner et al., 1997; Kikhno et al., 2002; Pijlman et al., 2003; Ohkawa et al., 2005; Sparks et al., 2011; Xiang et al., 2011) but more are expected to be identified by *in vivo* infectivity assays. Recently, three of these PIFs have been shown to form a very stable core complex (PIF-1, PIF-2 and PIF-3), held together by non-covalent bonds, that contributes to efficient entry and nucleocapsid delivery into midgut cells (Peng et al., 2010). The lepidopteran midgut is rich in trypsin and chymotrypsin (Johnston et al., 1995; Terra et al., 1996) and it has been suggested that the tight conformation of this PIF-complex might protect the active, internal PIF domains from the harsh chemical conditions of the midgut. In contrast, P74 (PIF-0) is only loosely associated with the core PIF-complex and actually requires the midgut environment to undergo functional activation by protease cleavage (Slack et al., 2008) by an ODV-associated host alkaline protease and by a host midgut trypsin

(Peng et al., 2011). Binding of ODVs to the tip of the midgut microvilli most likely occurs through a protein receptor binding mechanism (Horton & Burand, 1993; Yao et al., 2004) that relies upon PIF-0, PIF-1 and PIF-2 (Haas-Stapleton et al., 2004; Ohkawa et al., 2005). Supporting ODV adaptation to the midgut environment, binding efficiency has been shown to be optimal at alkaline pH (Horton & Burand, 1993).

Studies suggest that ODV binding proteins from different viruses interact with cell receptors that are specific to different virus/host systems (Haas-Stapleton et al., 2003, 2004, 2005, Horton & Burand, 1993; Ohkawa et al., 2005). Larvae of the fall armyworm *Spodoptera frugiperda* are highly resistant to oral infection by AcMNPV but are susceptible to infection by BVs when injected directly into the hemocoel (Haas-Stapleton et al., 2003). The resistance to *per os* infection is due to a lower level of ODV binding to the midgut cells (Haas-Stapleton et al., 2005) indicating that ODV interaction with specific midgut receptors is necessary for productive primary infection of midgut columnar epithelial cells. In addition, co-inoculation studies demonstrated that AcMNPV did not compete with SfMNPV for receptor binding indicating further that midgut receptor specificity is important for baculovirus host range determination.

As early as 30 min after BVs enter the cell, AcMNPV infection stimulates the formation of filamentous F-actin cables (Charlton & Volkman, 1993) and nucleocapsids are transferred to the nucleus by active actin polymerization (Lanier & Volkman, 1998; Ohkawa et al., 2010). The motile nucleocapsids encounter the nucleus and quickly enter through nuclear pores (Ohkawa et al., 2010) where the viral DNA genome is released. The nucleus is the site of replication for all baculoviruses and viral assembly for all NPVs is also carried out fully in the nucleus. In GVs, however, the occlusion step occurs following the disintegration of the nuclear envelope and the merging of the nucleoplasm and the cytoplasm (reviewed in Federici, 1997; Rohrmann, 2011; Winstanley and Crook, 1993). *Spodoptera frugiperda* and *T. ni* cells lines, Sf9 and High Five™, respectively, are permissive to AcMNPV but not to *Bombyx mori* NPV (BmNPV). Infection by BmNPV BVs was shown to be restricted, in part, by defective nuclear import of BVs where the virions entered the cell cytoplasm but were not able to enter nuclei of the cells (Katou et al., 2006). Mammalian cells lines are non-permissive to baculoviruses. AcMNPV ODVs were shown to bind to human carcinoma cells lines, A549 and HepG2, at the very low efficacy of 3×10^6 and 6×10^6 ODVs per cell, respectively (Mäkelä et al., 2008). Where binding occurred, the ODVs were inefficiently internalized into cytoplasmic vacuoles and were not released into the cytoplasm for intracellular trafficking and nuclear entry (Mäkelä et al., 2008). Thus, the non-permissive infection of HepG2 and A549 cells by AcMNPV ODVs is caused by ineffective binding to and internalization in the cell.

3.4 Disease progression in gammabaculoviruses

In hymenopteran sawfly NPVs, virions are singly enveloped and the host range of each virus is restricted to a single species (Federici, 1997). ODVs initiate midgut cell infection and once infected, the midgut cell nuclei become the site of viral gene expression and DNA replication in a manner that is consistent with that described for alphabaculoviruses (Duffy et al., 2007). In infected balsam fir sawfly larvae, *Neodiprion abietis* NPV (NeabNPV) viral DNA increased over 200% within the first 2 hours post infection (hpi) (Duffy et al., 2007). Progeny virions were occluded directly without the production of BVs (reviewed in

Federici, 1997; Slack & Arif, 2007; Rohrmann, 2011). Sequence data from gammabaculoviruses failed to identify homologues to the alphabaculovirus fusion proteins, GP64 or F-protein (Duffy et al., 2006; Lauzon et al., 2006), indicating the absence of the BV phenotype. Also lacking in gammabaculoviruses are viral fibroblast growth factor (vfgf) homologues (Jehle et al., 2006), that have been shown to accelerate the establishment of systemic infections in alphabaculoviruses (Detvisitsakun et al., 2007; Passarelli, 2011). (See section 3.7.) The NeabNPV replication cycle is rapid and efficient, with over 100 nucleocapsids being occluded in each OB (Duffy et al., 2007). Rather than using the BV phenotype to propagate the initial infection to other midgut cells, the released OBs have been suggested to serve as inoculum for other gut cells (Rohrmann, 2011). Midgut cell lysis in sawflies, rather than being a hindrance to gammabaculoviruses, is utilized as a mechanism for dispersal and only hymenopteran baculoviruses encode a trypsin-like protein (Duffy et al., 2006; Lauzon et al., 2006) that is thought to aid this process (Rohrmann, 2011). Much of the horizontal transmission occurs while infected larvae are still alive and feeding because OBs are excreted in a virus-laden diarrhea that is infectious to other larvae (Federici, 1997).

3.5 Disease progression in deltabaculoviruses

The replication of deltabaculoviruses is restricted to the cells of the posterior midgut and gastric caeca (Moser et al., 2001). Virions are singly enveloped prior to occlusion and the host range is restricted within suborders (Andreadis et al., 2003). While the primary infection is initiated by ODVs, the amplification of virus progeny within larvae occurs by cell-to-cell transmission that is carried out by BVs (Moser et al., 2001) and sequence data from CuniNPV identified an homologue to the F-protein of LdMNPV (Afonso et al., 2001). Restriction to the midgut is further corroborated by the lack of vfgf homologues in CuniNPV (Jehle et al., 2006). The production of OBs is rapid (14-48 hpi) and occluded progeny are released upon larval death (Becnel, 2007) for horizontal transmission to permissive hosts.

3.6 Disease progression in betabaculoviruses

Betabaculoviruses are only known to infect lepidopteran larvae. Primary infection is by OBs called granules and is initiated in midgut cells. The granules occlude a single virion and the host range of GVs is very specific (Cory, 2003). Some GV infections are restricted to the midgut (Type III), others cause systemic infections that progress either only to fat body tissues (Type I) or extend to most organs and tissues (Type II) (reviewed in Federici, 1997; Rohrmann, 2011). The time to host death caused by Type II GVs is similar to that of alphabaculoviruses (Lacey et al., 2002) and Type I GVs take the longest to kill the host larva (Federici, 1997). The transmission of Type II GVs occurs following the death and liquefaction of the infected larvae (Federici, 1997). The only identified Type III GV, *Harrisina brillians* GV (HabrGV) is transmitted horizontally by the release of OBs into the gut lumen of the infected larvae and out of the larva in the frass (Federici & Stern, 1990), similar to gammabaculoviruses. With the exception of the Type III midgut tropic HabrGV, the infection of the midgut epithelium is transient and does not produce OBs (Federici, 1997; Hess & Falcon, 1987). The cytopathology of GVs differs from that of other baculoviruses in that, although viral replication occurs and BV particles are produced in the nucleus of infected cells, OBs are formed in an area where the nucleoplasm and cytoplasm merge following the dissolution of the nuclear membrane (Hess & Falcon, 1987; Winstanley & Crook, 1993). Sequence data from GVs identified homologues of the F-protein and fgf orthologues (Liang et al., 2011; Miele et al., 2011) that are conserved in baculoviruses that cause systemic infections (Rohrmann, 2011).

3.7 Disease progression in alphabaculoviruses

The most studied baculovirus systems are those of lepidopteran alphabaculoviruses where ODVs enter midgut cells and BVs distribute the infection to other tissues in the hemocoel. In the well characterized system, AcMNPV-infected *T. ni* larvae, the infection progresses sequentially from midgut epithelial cells (4 - 12 hpi) to midgut-associated tracheoblasts and tracheal epithelial cells (12 - 24 hpi) (Engelhard et al., 1994). The infection then proceeds to hemocytes and tracheoblasts (36 hpi), later to fat body tissues (48 hpi) and finally to the majority of the remaining larval tissues by 70 hpi (e.g. cuticular epidermis, gonads, Malpighian tubules, midgut epithelia and salivary glands) (Engelhard et al., 1994). Although alphabaculovirus-infected midgut cells do not typically yield OBs (Granados & Lawler, 1981), they are crucial to the establishment of the systemic infection either by re-channeling ODV nucleocapsids into BV (without transiting through the nucleus) or by producing *de novo* BVs after a cycle of replication in infected midgut cell nuclei (reviewed in Rohrmann, 2011). While all other infected organs remain infected and produce viral particles, the infected midgut recovers by 48 hpi by sloughing off infected cells (Engelhard et al., 1995) and replacing them with new, healthy midgut cells (Keddie et al., 1989). Normally permissive to AcMNPV, *T. ni* larvae become resistant to AcMNPV infection as the larval development progresses. Active midgut cell sloughing is thought to play an important role in the developmental resistance that occurs in fourth-instar *T. ni* larvae (Engelhard & Volkman, 1995).

Once in the nucleus, the baculovirus genome utilizes the host transcriptional machinery to initiate the regulatory events that will result in the initial production of nucleocapsids that bud out of the cell by interacting with the cell plasma membrane and acquiring envelope fusion protein (EFP). Group I alphabaculoviruses (e.g. AcMNPV) encode for two main EFPs, GP64, and F-protein while group II alphabaculoviruses (e.g. LdMNPV) encodes only F-protein (Pearson et al., 2000). Deltabaculoviruses encode F-protein while gammabaculoviruses do not encode a discernable EFP (Miele et al., 2011). These EFPs are essential for BVs to exit infected cells and for cell-to-cell transmission (Monsma et al., 1996; Oomens & Blissard, 1999). An alternate mode for exiting the midgut cells rapidly has been described for AcMNPV (Washburn et al., 1999). In infected cells, some of the co-enveloped nucleocapsids enter the nucleus to initiate a round of replication while others, initially, remain in the cytoplasm. GP64 is produced early and modifies the basal membrane of the cell to mediate exit of BVs (Keddie et al., 1989). Even before *de novo* BVs are produced, ODV nucleocapsids in the cytoplasm bypass replication by reaching GP64-modified basal membranes and budding directly to initiate systemic infection of non-midgut tissues (Washburn et al., 2003a; Zhang et al., 2004). The fast shuttle of nucleocapsids through midgut cells is thought to accelerate the establishment of secondary infections before the midgut cells are sloughed off thereby potentially contributing to the wider host range of MNPVs (Washburn et al., 1999; Washburn et al., 2003b) and counteracting host defense mechanisms. This process of direct transformation of ODV to the BV phenotype has not been observed in SNPVs (Rohrmann, 2011).

ODVs must cross the PM to establish primary infections but BVs need to breach the basal lamina (BL) at the base of the midgut epithelium to initiate secondary infections. The BL is an extracellular layer of protein sheets that are secreted by epithelial cells lining the midgut trachea and other organs (Rohrbach & Timpl, 1993). The BL serves as structural support for regenerating epidermal cells that replace senescing cells that were sloughed off during development or physical assaults to the gut and as a separation between the sterile hemocoelic organs and the midgut, thus preventing the passage of natural microbiota and pathogens acquired during feeding. A model for AcMNPV breaching of the BL has recently been proposed (Means & Passarelli, 2010; Passarelli, 2011). The tracheal system is the insect respiratory system and the first cells to be infected by BVs are the tracheoblasts (Engerhard et al., 1994; Kirkpatrick et al., 1994; Washburn et al., 1995) which are highly motile, single-cell tracheal projections that respond to signaling from oxygen deficient cells and organs. One essential component to this response are the fibroblast growth factors (FGF) that, through a cascade of activation involving fibroblast growth factors receptors (FGFR), trigger tracheal cell motility. To branch to other cells and tissues, tracheoblasts are thought to degrade the BL by secreting enzymes. Baculoviruses are the only viruses known to encode FGF signaling molecules (Passarelli, 2011). Conserved only in alphabaculoviruses and betabaculoviruses, but absent in midgut-restricted gammabaculoviruses and deltabaculoviruses, viral FGF (vFGF) has been shown to accelerate BV exit from midgut cells and secondary infection by rerouting the host respiratory system to the midgut epithelium by mimicking host FGF. Although not essential for host infection *per se*, the difference in speed of establishment of systemic infection and speed of kill (Detvisitsakun et al., 2007; Katsuma et al., 2006) is thought to impact the capacity of any given virus to infect different hosts (Passarelli, 2011).

BV entry into insect cells is effected by GP64 via a clathrin-mediated, low-pH dependent, endocytic process (Long et al., 2006). Insect cells have different receptors for GP64 and F-protein and these two proteins appear to act separately (Hefferon et al., 1999; Westenberg et al., 2007; Wickham et al., 1999). GP64 tropism is so broad in fact, that BVs from AcMNPV and other baculovirus species such as BmNPV have been shown to be taken up by numerous non-lepidopteran cell lines including mammalian and dipteran cell lines (Carbonell et al., 1985; Shoji et al., 1997). Given the broad cellular tropism of GP64, receptors are thought to be common molecules present in invertebrates and vertebrates such as phospholipids (Tani et al., 2001). Therefore, GP64-mediated BV entry into cells is most likely not a restricting event. Receptor specificity for F-protein, however, is more restricted. Mammalian cells were shown not to possess F-protein receptors and could not to be transduced with a gp64-null AcMNPV pseudotyped with baculovirus F-protein (Westenberg et al., 2007). The lack of the *gp64* gene in LdMNPV, might contribute to its narrow host range (Barber et al., 1993; Glare et al., 1995).

Through their distribution in the hemolymph and systemic infection of a variety of tissues, BVs are largely responsible for the amplification of the virus within infected host larvae. Although the primary infection process is essential to the infectivity of baculoviruses in lepidopteran baculoviruses the systemic infection and ultimate death of larvae are dependent upon production of BVs. Horizontal transmission to neighboring larvae is dependent upon the release of OBs that occurs following larval death. Expression of viral proteases (cathepsin) and chitinase, late post infection, ensures that, in most alphabaculoviruses and in some betabaculoviruses, progeny OBs are released in the environment by lysing larval tissues and the exoskeleton of larvae following death (Hawtin et al., 1997; Hom et al., 2002). In addition, in silkmoth and gypsy moth, virally-produced proteins, tyrosine phosphatase (ptp) (Kamita et al., 2005) and ecdysteroid uridine 5′-diphosphate (UDP)–glucosyltransferase (egt) (Hoover et al., 2011), are responsible for behavioral changes that occur during the infection process where infected larvae leave their normal sheltered habitats and climb to exposed surfaces. This alteration in larval behaviour is thought to assist in virus distribution by facilitating predation by animals and by increasing exposure to elements.

4. Replication cycle

For all baculoviruses, once inside the nucleus, genome replication events follow a strictly-controlled cascade of temporal and sequential events (reviewed in Friesen, 2007; Rohrmann, 2011). Baculovirus genes are transcribed in three temporal phases (early, late and very late) where later steps of each phase are dependent on occurrence of earlier molecular events (Carstens et al., 1979; Guarino & Summers, 1986; Guarino & Summers, 1987).

4.1 Early phase

Upon viral DNA release into the cell nucleus, cellular transcription is harnessed for the expression of viral immediate early genes (Carstens et al., 1979; Guarino & Summers, 1986, 1987) and host-cell transcripts are decreased progressively from 12 - 18 hpi until complete shut down by 36 hpi (Nobiron et al., 2003). Baculoviruses depend on the cellular RNA polymerase II, to initiate the cycle of replication (Fuchs et al., 1983; Huh & Weaver, 1990) by recognizing and initiating transcription of viral early promoter sequences (Fuchs et al., 1983; Hoopes & Rohrmann, 1991; Huh & Weaver, 1990). Early viral transcripts can be detected as early as 0.5 hpi and through to 6 - 9 hpi in AcMNPV-infected susceptible cells (Chisholm & Henner, 1988; Guarino & Summers, 1986). The early promoter regions are conserved throughout baculoviruses and mimic those of the host RNA polymerase II with a consensus TATA element at about 30 bp upstream from the RNA start site (Pullen & Friesen, 1995) and a CAGT motif that acts as an initiator element (Blissard et al., 1992; Pullen & Friesen, 1995). Early genes encode mainly for polypeptides (IE1, IE0, IE2 and PE38) that have regulatory functions which are responsible for the transcriptional regulation of other viral genes, for the initiation of viral DNA replication (Lu & Carstens, 1991; Stewart et al., 2005; Todd et al., 1995), or to take control of the host cell for the purpose of viral multiplication (Possee & Rohrman, 1997). Early baculovirus gene expression is mostly regulated at the level of transcription and transfected viral DNA is infectious to permissive cells (Burand et al., 1980; Cartens et al., 1980) indicating that initial transcription of early genes does not require viral factors to be present at the start of the infection. The lack of requirements for viral proteins has been substantiated recently by proteomic analysis of AcMNPV BV particles that have been shown not to contain regulatory proteins (Wang et al., 2010). Larvae of *Anticarsia gemmatalis* (velvetbean caterpillar) are highly resistant to AcMNPV. Infection was shown to be blocked at early stages of replication where, even though ODVs successfully entered the midgut cells and were rechanelled to tracheal cells, the immediate early gene, *ie-1*, was not transcribed (Chikhalya et al., 2009). The inhibition of this essential transactivator resulted in the disruption of the gene expression cascade resulting in a failure to produce infectious viral particles.

4.2 Late phase – DNA replication

Baculovirus early gene products are required for viral DNA replication. In AcMNPV-infected cells, genome replication occurs between 6 hpi and to 18 hpi after which time it starts to decline (Erlandson et al., 1985; Erlandson & Carstens, 1983). Six AcMNPV genes were found to be essential for DNA replication in transient DNA replication assays (Kool et al., 1994; Lu & Miller, 1995a). The genes directly involved in DNA replication (reviewed in Rohrman, 2011) encode for a homologous-region (hr) binding protein and transcriptional activator (*ie-1*) (Leisy et al., 1995; Rodems & Friesen., 1995), a single-stranded DNA binding protein (*lef-3*) (Hang et al., 1995), DNA binding helicase (*p143*) (McDougal & Guarino, 2000), a putative primase (*lef-1*), a primase-associated protein (*lef-2*) (Mikhailov & Rohrmann, 2002) and a DNA polymerase (*dnapol*) (Hang & Guarino, 1999; McDougal & Guarino, 1999). Four of these genes, *dnapol*, *p143*, *lef-1*, and *lef-2* are core genes found in all baculoviruses (Okano et al., 2006). Other genes such as *ie-2*, *lef-7*, *pe38*, and *p35*, stimulate viral DNA replication in transient assays (Chen & Thiem, 1997; Lu & Miller, 1995a; Milks et al., 2003) and were found to be differentially required in cell lines from various lepidopteran origins as well as *in vivo* (Chen & Thiem, 1997; Lu & Miller 1995b; Milks et al., 2003; Prikhod'ko et al., 1999).

BmNPV and AcMNPV genomes are highly homologous (Gomi et al., 1999) but their host ranges are very different (Gröner, 1986). *Bombyx mori* larvae and cell lines such as BmN, are fully permissive to BmNPV but their infection by AcMNPV is non-permissive (Morris & Miller, 1993). Though delayed in BmN cells, AcMNPV temporal gene expression occurs as in the permissive *Spodoptera* cells until very late times post infection (Iwanaga et al., 2004; Morris & Miller, 1993). AcMNPV DNA replication also takes place in BmN cells but the infection is arrested before BV or OBs are produced (Morris & Miller, 1993). AcMNPV-infected BmN cells also showed marked cytopathic effects (Maeda et al., 1993) which led to a drop in gene expression. Though DNA replication seemingly occurred as in the permissive Sf-21 cells, the defect in AcMNPV-BmN cells was shown to be caused by differences in the *DNA helicase* gene (*p143*) (Maeda et al., 1993). A few amino acid changes in AcMNV P143 were sufficient to overcome the defect in *B. mori* cells and larvae (Argaud et al., 1998; Croizier et al., 1994). The cytotoxicity and block in AcMNPV infection of *B. mori* cells are suggested to stem from aberrant DNA replication (reviewed in Thiem & Cheng, 2009; Rohrmann, 2011).

4.3 Late phase – Late gene expression

The final step in the replication cycle of baculoviruses is the expression of late and very late genes that mainly code for structural proteins. AcMNPV genes encoding structural proteins of nucleocapsids and BVs are transcribed at their peak during the late phase (6 - 24 hpi), while occlusion-related genes are transcribed at very late times post-infection (18 - 76 hpi) (Thiem & Miller, 1990; Wu & Miller, 1989). The increase in late viral transcription parallels the decline in host and early viral transcription (Nobiron et al., 2003). Late promoter sequences are conserved in baculoviruses with the TAAG sequence being the essential component for the recognition of late and very late promoters by viral RNA polymerase factors with cis-acting sequences dictating the differential levels and temporal expression of late and very late genes (Ooi et al., 1989). In AcMNPV, 19 genes were found to be required for optimal transcription of late (*vp39* and *p6.9*) and very late promoters (*polh* and *p10*) but not early promoters (*etl* and *pcna*) (Li et al., 1993; Lu & Miller, 1995b; Passarelli & Miller, 1993a, 1993b, 1993c; Rapp et al., 1998; Todd et al., 1995). Being required for the transcription of late and very late genes, these 19 genes have been defined as late expression factors (lefs). Since the late transcriptional events are dependent upon early transcription and DNA replication, nine of these lefs have indirect effects on late transcription by being involved in early gene transcription and DNA replication (*ie-1, ie-2, lef-1, lef-2, lef-3, p143, dnapol, p35,* and *lef-7*) (Rapp et al., 1998). Only four of these lefs (*p47, lef-4, lef-8,* and *lef-9*) have been shown to form the viral RNA polymerase that is directly responsible for the *in vitro* transcription of baculovirus late and very late promoters (Guarino et al., 1998; Rapp et al., 1998). A unique feature of baculoviruses is the hyperexpression of the very late genes, *polyhedrin* and *p10*. Increases in transcription levels of RNA polymerase occurs through the binding of very-late expression factor (VLF-1) (McLachlin & Miller, 1994; Ooi et al., 1989; Yang & Miller, 1999) to very late promoters stimulating expression of the high levels of polyhedrin required for OB formation.

Gypsy moth Ld652Y cells are semi-permissive to AcMNPV and, although all temporal classes of virus genes are transcribed and viral DNA replication is detected, translation of both viral and host proteins is arrested by about 12 hpi (reviewed in Thiem & Cheng, 2009; Guzo et al., 1992; McClintock et al., 1986; Morris & Miller, 1993). An LdMNPV gene, named the host range factor (*hrf-1*) was found to rescue the translational arrest in AcMNPV-infected Ld652Y cells (Du & Thiem, 1997a; Thiem et al., 1996). AcMNPV does not encode for a *hrf-1* homologue. Gypsy moth *in vivo* resistance was also overcome by a hrf-1-bearing, recombinant AcMNPV (Chen et al., 1998).

5. Registration of baculoviruses

Agencies responsible for pesticide product regulations were initially put in place by governments to evaluate the efficacy and non-target safety of synthetic chemical pesticides. Their extension into environmental safety assessments came in response to concerns related to the increasing number of reports of environmental damage due to pesticide toxicity and the accumulation of chemical residues such as those from DDT (Hauschild et al., 2011). Gradually, mounting public pressure and the implementation and enforcement of stricter rules and regulations led to the ban and rejection of many chemical pesticides and the need for lower-impact pesticides such as those classified as biological control agents (BCA) and microbial pest control agents (MPCA). MPCAs are those products that have a microorganism (i.e., virus, bacterium, fungus, protozoan or nematode) as the active ingredient. When submitted for registration, MPCAs were initially evaluated using existing regulatory processes that were developed for broad-spectrum chemical pesticides. Since then, a trend has emerged that now favours the development of lower-risk products such as BCAs and MPCAs and specific regulations have been developed that are better suited to the requirements of MPCAs. To facilitate the registration process for MPCAs, many countries have established departmental branches that specialize in the registration of MPCAs and other low-risk product submissions (Hauschild et al., 2011; Kabaluk et al., 2010). Despite these efforts, the MPCA registration process in many countries still requires a number of toxicity tests that might not be necessary based on the biology of the microbe that acts as the active ingredient of a MPCA. Compared to chemical pesticides, MPCAs typically target small niche markets and unnecessary registration requirements can impose burdensome costs on the biopesticide industry (Chandler et al., 2011).

The most important function of any pesticide regulatory agency is to insure that unsafe pesticides are not registered for use. Although the registration procedures, costs and processing times differ between countries, data requirements are generally similar (Hauschild et al., 2011). Typically, numerous categories of data are required for the registration of MPCAs. These categories are designed so that applicants provide 1) product identity, physical, chemical and technical properties; 2) methods of analysis, manufacturing and quality control; 3) toxicological studies and exposure data geared towards human and veterinary health and safety; 4) product residue data; 5) product fate and behaviour in the environment; 6) environmental and non-target toxicity and 7) efficacy data (Hauschild et al., 2011).

The majority of the topics included in registration packages are essential and desirable. For baculoviruses in particular, it is essential that the active ingredient has been identified as belonging to the family, *Baculoviridae* and that the species of its primary host is known (e.g. *Neodiprion abietis* gammabaculovirus) (category 1), that the method of manufacturing, analysis and quality control is appropriate and reliable (category 2) and that it has been proven to be efficacious (category 7) for the purpose intended. The regulations as applied to some of the more recently registered baculovirus-based products may, however, demonstrate the redundancy of certain data requirements involving vertebrate toxicity (categories 3 and 6) and some aspects of environmental and non-target effects (categories 4 and 5). The Pest Management Regulatory Agency (PMRA) in Canada and the United States Environmental Protection Agency (EPA) accept that certain data requirements can be met using "waivers" that provide scientific arguments based on published, peer-reviewed, scientific literature and data while, in the European Union (EU), less formal procedures allow for similar science-based evidence to replace the actual, newly-generated data (Hauschild et al., 2011). A combination of fulfilling specific data requirements and the use of waivers were used in the successful registration of the baculovirus-based product, Abietiv™, for the suppression of balsam fir sawfly populations in Canada (Lucarotti et al., 2006, 2007).

5.1 Identity of baculovirus products

MPCAs are usually registered at the strain level where the active ingredient is derived from a single host, colony or spore. While this selection method is mostly feasible for MCPAs such as bacteria and fungi, it may not be feasible or desirable for baculovirus-based products. In nature, baculoviruses consist of mixtures of different genotypes of the same species (Cory et al., 2005) and it has been shown that this diversity is naturally favored in wild type virus populations (Clavijo et al., 2011) where these genomic variants are known to impact virulence in the target organism (López-Ferber, 2003; Simón et al., 2008). The different viral genotypes may compensate for variations that occur in the larval host and/or its environment (Berling et al., 2009; Hitchman et al., 2007; Hodgson et al., 2002). For this reason, the evaluation of baculoviruses should be carried out at the species level rather than at the level of a single isolated genotype (Hauschild et al., 2011).

5.2 Human toxicity and infectivity

The scientific literature on the health and environmental safety of baculoviruses is extensive and has been well reviewed (see reviews by Black et al., 1997; Burges et al., 1980a, 1980b; Gröner, 1986; Ignoffo, 1975; OECD, 2002) and more recently by the Food and Agriculture Organization of the United Nations (FAO) (McWilliam, 2007) and the European Food Safety Authority (EFSA) (EFSA, 2009; Leuschner et al., 2010). The host range of baculoviruses is restricted to terrestrial arthropods (Barber et al., 1993; Doyle et al., 1990; Cory, 2003; Cory & Hails, 1997; Miller & Lu, 1997; Thiem & Cheng, 2009). Baculovirus products that are commercially available for biological control of insect pests have been extensively tested to determine effect on humans and other non-target animals (Hauschild et al., 2011).

Data required for assessment of human infectivity and toxicity typically involve mammalian toxicological studies of the product in laboratory test mammals (e.g. mice, rats, rabbits) *in vivo* as well as in mammalian cell cultures. Baculovirus active ingredients and end products have been ingested and inhaled by, injected (intravenous, intraperitoneal, intramuscular) into, and applied to the skin and eyes of test animals without detrimental effects that could be attributed to the baculovirus tested (Table 1) (Ashour et al., 2007; Gröner, 1986; Ignoffo, 1975; Lightner et al., 1973). Many species of baculoviruses have been tested on numerous species of animals at doses that are many times those that could be acquired in the field. For example, for the registration of Abietiv™ (NeabNPV), typical toxicity data were presented where rats had been fed single dose of 1×10^8 NeabNPV OBs (Health Canada PMRA, 2009; Lucarotti et al., 2006). All of the rats survived to the end of the observation period and showed no adverse clinical effects. At the application rate given on the Abietiv™ product label (1×10^9 OBs in 2.5 L of 20% aqueous molasses/ha), this would be the equivalent of a 70-kg man ingesting 16 L of the tank-mixed product. Thus, taking into account the label application rates and volumes at which the products are applied, the concentrations used for toxicity tests are well beyond what could be expected to be acquired in the field.

In vitro, mammalian and other vertebrate cells lines are non-permissive to baculoviruses (reviewed in Gröner, 1986; OECD, 2002). Although BV uptake has been observed, there has been no evidence that viral DNA replication, production of viral proteins or cytopathological effects have occurred. *In vivo*, the uptake of baculovirus OBs by various animals did not lead to the production of baculovirus-specific antibodies (reviewed in Gröner, 1986). Human carcinoma cell lines, HepG2 and A549, were recently challenged with AcMNPV ODVs that had been chemically extracted from OBs (Mäkelä et al., 2008). The non-permissive infection of HepG2 and A549 cells by AcMNPV ODVs was shown to be caused by the inefficient binding and internalization in the cell. The ODV-derived nucleocapsids did not reach the nucleus to release the viral genome. In addition to the lack of infection and replication in vertebrate cells, no evidence for baculovirus induced cytogenic, carcinogenic, mutogenic or teratogenic effects has ever been found (Gröner, 1986; Ignoffo, 1975; McWilliam, 2007; OECD, 2002).

Alphabaculovirus	Vertebrate Test Animals
Amscacta albistriga NPV	chicken
Autographa californica NPV	rat, guinea pig, rabbit, shrimp[a], fish[b]
Choristoneura fumiferana NPV	rat, rabbit, duck, quail, rainbow trout, white sucker
Erranis tilliara NPV	mouse
Galleria mellonella NPV	mouse
Heliothis zea NPV	mouse, rat, guinea pig, rabbit, dog, monkey, man, quail,chicken, sparrow, mallard, killifish, spotfish, rainbow trout, black bullhead, white sucker, sheepshead minnow
Lymantria dispar NPV	mouse, rat, guinea pig, rabbit, dog, blackcap chickadee,duck quail, sparrow, bluegill, brown trout
Malacosoma disstria NPV	guinea pig, rabbit, chicken
Mamestra brassicae NPV	mouse, guinea pig, pig, chicken
Orgyia pseudotsugata NPV	mouse, rat, rabbit, mule deer, duck pheasant, sparrow, chinook salmon, coho salmon, steelhead trout
Spodoptera exempta NPV	rat
Spodoptera exigua NPV	mouse, guinea pig
Spodoptera frugiperda NPV	mouse, guinea pig
Spodoptera littoralis NPV	Rat, fish[b]
Spodoptea litura NPV	chicken
Thymelicus lineola NPV	mouse, sheep, goldfish
Trichoplusia ni NPV	mouse, guinea pig, sparrow
Betabaculovirus	
Cydia pomonella GV[c]	mouse, rabbit
Estigmene acrea GV	mouse, guinea pig
Gammabaculovirus	
Neodiprion abietis NPV[d]	rat, mouse, rabbit
Neodiprion lecontei NPV	rat, hamster, rabbit, chicken, turkey, rainbow trout
Neodiprion sertifer NPV	rat, guinea pig, rabbit, duck, quail
Neodiprion swainei NPV	mouse, rat, guinea pig, rabbit, duck, quail

Table 1. List of vertebrate test animals exposed to baculoviruses from a variety of Lepidoptera and sawflies to which no adverse effects of exposure to the baculovirus could be attributed. (from Gröner, 1986; Ignoffo, 1975). a, b, c, d: Data obtained from Lightner et al. (1973), Ashour et al. (2007), and Health Canada PMRA Regulatory Notes REG2000-10 and RD2009-05, respectively.

Currently, all baculovirus-based MCPs are produced, *in vivo*, in permissive larval hosts. While baculovirus OBs are inert and non-allergenic, the larvae in which they are produced can produce dermatitis and contact urticaria where larval setae cause mechanical irritation or contain histamine or other irritating substances (Hossler, 2010a, 2010b). Although anaphylactic shock has not been reported to be caused by lepidopteran insects (Hossler, 2010a, 2010b), eye irritation studies on rabbit (Reardon et al., 2009) and limited skin eruptions have been reported from human exposure to gypsy moth larvae during heavy infestations (Tuthill et al., 1984). As a result, the Gypchek product label warns of potential eye irritation. While this is the case for some of the baculovirus products, the majority of products are not considered as sensitizers (Hauschild et al., 2011: Ignoffo, 1975).

5.3 Baculoviruses and biomedical applications

Additional evidence of the safety of baculoviruses to humans comes from their use in biomedical applications. The unique baculovirus properties, coupled with recent advances in molecular and cell biology, have broadened the scope of their application in basic and applied biomedical fields. To date, the prototype baculovirus, AcMNPV, is the most widely used baculovirus for the production of biologics for therapeutic purposes (Aucoin et al., 2010; Cox & Hollister, 2009; van Oers, 2006). This has been accomplished in part by use of baculovirus expression vector systems (BEVs) for heterologous recombinant protein production and gene transfer. The principle behind the use of BEVs is based on use of *polyhedrin* and *p10* promoters to drive the expression of foreign genes in cell culture or *in vivo* (van Oers, 2011; Summers, 2006). BEVs continue to evolve to new and robust systems. For instance, the latest vectors (flashBACultra/BacMagic3) can be semi-automated for high quality, yield, and stable recombinant protein (van Oers, 2011). Many advantages of using BEVs over other expressions systems have been reviewed in previous reports (Airenne et al., 2009, 2010; Hu, 2005). Also, there are available insect cell lines such as those derived from *S. frugiperda* (Sf9 cells) and *T. ni* (High Five™ cells), which have been extensively characterized for optimal, high quality recombinant protein production (Aucoin et al., 2010). Some of these cell lines have been adapted to grow as continous suspension cultures in serum-free media thus, allowing for high-throughput scale-up production in bioreactors (Elias et al., 2007; Feng et al., 2011). Furthermore, use of serum-free cell cultures have been recognized by regulatory agencies including the United States, Food and Drug Administration (FDA), and European Medicines Agency (EMA) as a standard approach for limiting potential adventitious agents in therapeuatic products (FDA, 2010). Many advantages of using insect cell substrate compared to embryonated eggs have been reported leading to simplified regulatory avenues for licensing baculovirus-based biologics (Cox & Hollister, 2009, Treanor et al., 2007). In addition to cell substrates, insect larvae such as *T. ni* (Chen et al., 2011), and *B. mori* (Kato et al., 2010) have been reported as potential biofactories for *in vivo* therapeutic production. For instance, *in vivo* production of antiviral agents including human interferon-γ against influenza virus H1N1 in *T. ni* larvae have been demonstrated (Chen et al., 2011; Gomez-Casado et al., 2011). Nevertheless, insect cell cultures and BEVs platforms continue to expand the applications of baculoviruses as novel tools for vaccine development, drug screening, and gene therapy (Airenne et al., 2010, 2011; Cox & Hollister, 2009, Kost et al., 2005, van Oers, 2011). The recent initiative of having a standard baculovirus reference material repository (BRM) will further boost their application and perhaps hasten the regulatory process for registering new baculovirus products (Kamen et al., 2011). This initiative was mainly proposed in order to have a proper standard that is acceptable to all researchers in academic institutions, regulatory agencies, and industries (Kamen, et al., 2011).

5.4 Baculoviruses and vaccine development

To date, there are different baculovirus-based vaccines for human and veterinary use (van Oers, 2011). Also, vaccines targeting highly pathogenic viruses that are transmitted by arthropod vectors (arboviruses) are being developed (Metz & Pijlman, 2011). Characteristics of baculovirus-based human vaccines that are currently approved or are in later phases of clinical trials are given in Table 2. The different strategies employed in the production of baculovirus-based vaccines include: (i) BEVs-based subunit vaccines; here, recombinant viral proteins or peptides are produced using BEVs in cell culture. Subunit vaccines can be efficiently produced in insect cells and have additional safety advantages over live attenuated vaccines (Madhan et al., 2010). A good example is the influenza vaccine (FluBlok), which is based on recombinant Hemagglutin (HA) proteins selected from three influenza virus strains as determined by the World Health Organization (WHO) and the Centre for Disease Control (CDC) (Airenne, 2009; Cox & Hollister, 2009). Subunit vaccines developed in BEVs have been approved based on the standards stipulated by various regulatory agencies especially clinical data on toxicology and efficacy assessment (Cox & Hollister, 2009; Cox & Hashimoto, 2011; FDA, 2009). (ii) BEVs-based virus like particles (VLPs); for example, a prophylactic, bivalent human papillomavirus vaccine for cervical cancer (Cervarix) consisting of C-terminally truncated HPV-16/18 L1 proteins is produced using BEVs in *T. ni* High- Five™ cells and assembled as VLP (Harper et al., 2006). VLPs mimic the real virus but are non-infectious due to lack of viral genome, and are safe for human use. Detailed safety data for human papillomavirus types 16 and 18 recombinant vaccine have been outlined by USA and Canada health regulatory agencies (FDA, 2009; Health Canada, 2010). (iii) Active cellular-based vaccine; a classical example and the first vaccine of this kind to be approved by FDA is Provenge® for prostate cancer. This vaccine is composed of fusion proteins consisting of a prostate cancer marker, prostatic acid phosphatase (PAP), linked to granulocyte macrophage colony stimulating factor (PAP-GM-CSF) and generated in insect cells via BEVs. The fusion protein is in turn loaded *ex vivo* in dendritic cells, the most potent antigen presenting cell (APC), leading to stimulation of cytotoxic T-cell immune response against patients cancer cells (Small et al., 2006; Vergati et al., 2010). The prostate cancer cells expressing these recombinant proteins are recognized by the patient's cell-mediated immune system. In addition to the aforementioned baculovirus strategies for vaccine development, there are other baculovirus technologies, such as baculovirus surface display technology, that are being considered for production of vaccines. Here, the desired foreign antigens are displayed on the surface of the baculovirus envelope or capsid (Mäkelä & Oker-Blom, 2006; Oker-Blom et al., 2003). More recently, a novel system based on the use of a defective baculovirus vector incapable of self assembly has been developed (Marek et al., 2010). In this approach, the baculovirus vector is engineered to produce biologics that are free from contaminating BVs and ODVs.

Vaccine	Producer	Disease	Status	Reference
Cervarix™	GlaxoSmithKlines, Rixensart, Belgium	Cervical cancer	Approved	Harper et al., 2004; 2006
Provenge	Dendreon Inc., Seattle, WA, USA	Prostate cancer	Approved	Kantoff et al., 2010 ; Small et al., 2006;
Chimigen	Virexx Medical Corp., Calgary, Canada	Hepatitis B and C	Clinical trial	
FluBlok	Protein Biosciences Corp., CT, USA	Influenza virus	Clinical trial	Cox & Holister, 2009, Cox & Hashimoto, 2011
Dyamid	Diamyd Medical AB, Stockholm, Sweden	Type-1 diabetes mellitus	Clinical trial	

Table 2. List of baculovirus-derived vaccines.

5.5 Baculovirus and mammalian gene delivery/ therapy platforms

Although baculoviruses replicate in the nucleus of specific insect hosts, mammalian cells have been shown to internalize baculoviruses, but no progeny virions are produced (Volkman & Goldsmith, 1983). Similarly, recombinant baculoviruses carrying a reporter gene under the control of human cytomegalovirus (CMV) and Rous sarcoma virus (RSV) promoters were shown to efficiently transduce mammalian cells and express foreign proteins (Boyce & Bucher 1996; Hofmann et al., 1995). These studies showed varying levels of reporter gene expression in mammalian cells of different origins. Although all cells were reported to internalize the same amount of virus, the block to expression or low expression observed in epithelial cells compared to human and rabbit hepatocytes was attributed to a specific receptor on the hepatocyte cell membranes and inhibition of endosomal maturation. Additional blocks have been linked to poor cytoplasmic transport or entry of nucleocapsid to the nucleus (reviewed in Airenne et al., 2009).

The basis of baculovirus gene delivery/ transfer in mammalian cells has been accomplished using BacMam vectors (Invitrogen Corporation, Carlesbad, CA). Unlike BEVs, which relies on baculovirus late promoters, the gene of interest in BacMam vectors is placed under the transcriptional control of mammalian active-promoters such as those of CMV, RSV, chicken beta-actin (CAG), among others (Madhan et al., 2010). Various cellular and viral promoters have been shown to affect transduction efficiency in different mammalian cells implying that promoter selection is critical to efficient use of baculovirus vectors (Kim et al., 2007; Shoji et al., 1997). Nonetheless, safety of baculovirus vectors in gene delivery is supported by extensive safety data that show lack of toxicity or pathogenicity in various mammalian species (reviewed in Airenne et al., 2009). Based on this, baculoviruses have been considered as ideal candidate for future gene therapy. Gene therapy is a novel approach for treating various forms of genetic diseases through the use of viral or non-viral shuttle vectors. This has been successfully demonstrated through *in vivo*, and *ex vivo* (human-derived tissues) studies as previously reviewed (Airenne et al., 2009, 2010; Hu, 2005). To date, viral-based vectors including those of DNA and RNA animal viruses are increasingly being tested as potential agents for gene therapy. Health Canada, like other regulatory bodies, recognizes various viral-based vectors for *in vivo* and *ex-vivo* gene therapy. Baculoviruses are included in these lists due to recent studies on their potential as tools for gene therapy and requirements for extensive preclinical studies. Baculovirus-based vectors remain promising candidates for gene delivery primarily due to the following attributes: (i) natural occurrence, (ii) host specificity (iii), well characterized genomes (Cohen et al., 2009), (iv) genetic stability due to lack of reversions or genome integration, (v) rapid and relatively low cost of production to high titers (~ 10^{12} pfu/ml) (Airenne, 2010), (vi) lack of cell substrates associated with animal serum, (vi) large transgene capacity (~100 kb), and allowance for multiple gene inserts, (viii) ability to transduce a myriad of dividing and non- dividing cells (Airenne et al., 2011; Kost & Condreay, 2002) and, (ix) lack of pre-existing immunity.

Although baculovirus gene therapy technology is relatively recent, there is a wealth of safety data in animal models (Airenne et al., 2009) and preclinical trials based on *ex-vivo* experiments (Georgopoulos et al., 2009). Their safety is augmented by early toxicity studies using intravenous, oral, intracerebral, and intramuscular inoculation of animal models, and feeding tests on voluntary humans (Gröner, 1986; Ignoffo, 1973, 1975). Similarly, techniques to assess the toxicity and transformation potential of baculovirus in mammalian cells have been developed (Hartig et al., 1989, Gonin & Gaillard, 2004). Here, quantitative PCR (qPCR), using SYBR Green or TaqMan probes is viewed as the standard tool for assessing the biodistribution of the transgene and expression of shuttle vectors (Gonin & Gaillard, 2004).

5.6 Environmental toxicity

The environmental and ecological impacts of baculovirus products are mirrored by characteristics of their pathogenesis and host range. In essence, every study that is required to assess their potential for environmental toxicity will be influenced by their limited host range and lack of infectivity to non-target animals. Data required for assessment of environmental toxicity typically involve environmental fate and environmental toxicological studies on birds, fish, plants, microorganisms, aquatic arthropods and non-target insects including beneficial insects. Baculoviruses are ubiquitous and persistent in aquatic, terrestrial and forest ecosystems (England et al., 2004; Hewsen et al., 2011; Podgwaite et al.,1979) yet, there has been no report of negative impact of baculoviruses on ecosystems other than the effect on the target host insect (Black et al., 1997; Cory, 2003; FAO, 2007; OECD, 2002). As a matter of fact, when applied in the context of pest control, the persistence and amplification of baculoviruses in the larval host population has been recognized as being an essential component of plant protection, in particular for forestry (Moreau et al. 2005; Moreau & Lucarotti 2007). Field application of baculoviruses into the environment, such as occurs when LdMNPV is applied for the control of gypsy moth, does not increase virus levels beyond those that would occur naturally (Reardon et al., 2009). Also, through water run off or direct deposit of contaminated material (insects, frass, etc.), aquatic systems are recipients of baculoviruses (Hewsen et al., 2011). None of the non-target arthropods, such as shrimps, *Daphnia* spp. and *Notonecta* spp., or fresh-water, estuarine and marine fishes that have been tested by exposure to several NPVs have shown evidence of infection, toxicity or mortality (Table 1) (Dejoux & Elouard, 1990; Lightner et al., 1973; Couch et al., 1984).

Non-target insect toxicity studies are complex and the results usually depend upon the natural host range of any given baculovirus. Fundamental studies that were aimed at determining putative host range factors generally found that a different barrier to infection occurred for every virus and non-host system examined (See sections 2 and 3). The determination of the host range for any given baculovirus has, therefore, proven to be difficult to predict. Most baculoviruses, however, are very specific to their host species or closely-related ones (reviewed in Miller & Lu, 1997; Thiem & Cheng, 2009) and cross-order infections do not occur. In addition, SNPV alphabaculoviruses, betabaculoviruses and gammabaculoviruses appear to be the most restricted in host range, while some of the MNPV alphabaculoviruses (e.g. AcMNPV and *Mamestra brassicae* NPV [MabrNPV]) can infect over to 30 species crossing over 10 families of Lepidoptera (reviewed in Miller & Lu, 1997; Thiem & Cheng, 2009). Even within the MNPVs, however, those infecting Lymantriidae hosts such as LdMNPV in the gypsy moth, *L. dispar*, appear to be truly specific to a single host (Barber et al., 1993; Cory, 2003; Cory & Myers, 2003; Glare et al., 1995). Most *in vivo* host range studies have been carried out in the laboratory and at extremely high baculovirus dosage rather than at a range of concentration that might allow for the determination of a range of lethal doses (e.g., LD_{50} – LD_{95}). This artificial system does not accurately reflect the field situation and additional caution must be given to older toxicology

– host-range studies, where the evaluation of permissiveness is based only on mortality rates. Often these studies lack confirmation of productive infections which could lead to an overestimate of the host range of a given viral isolate. Therefore, confirmation of infectivity and host range through the use of molecular techniques to identify patent infections is recommended (Cory, 2003; OECD, 2002; Thiem & Cheng, 2009). Unfortunately, the insect species that have been selected for non-target toxicity tests have often been ones that have been shown not to be susceptible to baculoviruses. Only Lepidoptera, hymenopteran sawflies and a few species of Diptera have been confirmed to host baculoviruses. There is no cross-infection of baculoviruses between these orders. Baculoviruses do not infect cockroaches, grasshoppers, aphids, neither have they been shown to infect non-phytophagous beneficial and predatory insects such as lady beetles, parasitoids and honey bees (Doyle et al., 1990; Huang et al., 1997; Ignoffo, 1975). Although not infecting parasitoids, baculoviruses can cause the premature death of the larval host and competition for resources that can affect the fitness and survival of parasitoids (Hochberg, 1991; Nakai & Kunimi, 1997). Parasitoids are often generalists and while the depletion of virally-treated insect populations will occur, the lack of non-target effects on other potential lepidopteran hosts would presumably provide alternate hosts for the parasitoids (Strazanac & Butler, 2005). In addition, some studies suggest that some parasitoids such as *Cotesia melanoscela*, *Parasetigena silvestris* and *Apanteles melanoscelus* transmit baculoviruses (e.g., LdMNPV) and contribute to the viral epizootic (Reardon & Podgwaite, 1976).

6. Summary

Viruses in the family *Baculoviridae* are host specific, infecting only one or a few closely related species of insects. They are ubiquitous in the environment and are known to be an important contributor to insect population regulation. These characteristics make them good candidates for management of crop and forest insect pests with minimal or no off-target impacts. Commercial production of baculoviruses for use as biological control agents of insect pests is carried out worldwide at different scales depending on the market. Over 50 baculovirus products have been used worldwide as microbial insecticides. Five viruses are registered for use in Canada, mostly for the control of forest insect pests. As is the case in other industrialized countries, the commercialization of baculoviruses as microbial insecticides in Canada is dependent upon the submission of a number of scientific studies that establish proof that the products are efficacious and safe. Given the extensive and long standing use of synthetic pesticides, regulatory policies are often geared toward chemical pesticides requiring extensive safety testing that could be considered to be superfluous and unwarranted given the long history of safe and efficacious use of baculoviruses. Most recently, baculovirus safety has been substantiated further by fundamental research geared towards understanding the molecular basis for the events that regulate baculovirus life cycles, pathogenesis, and host range and by the increased application of baculoviruses for pharmaceutical and therapeutic use.

The safety of baculovirus products is innately linked to the pathogenesis and host range of this family of viruses. For a productive baculovirus infection to occur, the viral replication process must successfully cross multiple environmental, temporal and organism-specific barriers. Every step in the life cycle of baculoviruses is challenged beginning with the external environment and the long periods between host availability. Once in contact with a potential host, viral particles must be released from the OBs, enter permissive cells and successfully take over the host cell transcriptional machinery to initiate the viral replication cycle. Dependence on the host-cell molecular machinery is reduced over the course of the infection as baculovirus gene expression and regulatory proteins take over. However, host-and/or tissue-specific interactions continue to play a role as the infection progresses within the infected host which will determine whether or not a patent infection will occur.

Prompted by the publication of the OECD consensus document on the "Assessment of Environmental Applications involving Baculovirus", the Regulation of Biological Control Agents (REBECA) entered baculoviruses in the positive list of "low risk" candidate microbial pest control agents (Strasser et al., 2007). In addition, baculoviruses have recently been included in the qualified presumption of safety (QPS) list authorized by the European Food Safety Authority (EFSA) (Leuschner et al., 2010) panel on Biological Hazards (BIOHAZ) (EFSA, 2009). Following a review of literature, EFSA concluded that baculoviruses are safe for animal and human consumption and are, therefore, acceptable for use in the control of insects that cause damage to plants (EFSA, 2010). Given that all published reviews unequivocally state that baculoviruses are safe and support their use as low-risk biological control agents for the control of insect pests, we propose that human and environmental toxicity tests and studies related to the residual fate of baculoviruses not be required for the registration of baculoviruses.

7. Acknowledgements

The financial support provided by the Atlantic Canada Opportunities Agency – Atlantic Innovation Fund, the Canadian Forest Service (Natural Resources Canada), the membership of SERG-International, BioAtlantech, Forest Protection Limited, and Sylvar Technologies Inc. is gratefully acknowledged. We thank Wendy Yerxa and Denise Philpott for assistance with the formatting of the references.

8. References

Adams, J.R. & McClintock, J.T. 1991. Baculoviridae, nuclear polyhedrosis viruses. Part 1: nuclear polyhedrosis viruses of insects, In: *Atlas of Invertebrate Viruses* , J.R. Adams & J.R. Bonami, (Eds.), pp. 87-180, CRC Press, ISBN 9780849368066, Boca Raton, FL, USA.

Afonso, C.L., Tulman, E.R., Lu, Z., Balinsky, C.A., Moser, B.A., Becnel, J.J., Rock, D.L., & Kutish, G.F. 2001. Genome sequence of a baculovirus pathogenic for *Culex nigripalpus*. *Journal of Virology* 75: 11157-11165.

Airenne, K.J., Mahonen, A.J., Laitinen, O.H., & Yla-Herttuala, S. 2009. Baculovirus-mediated gene transfer: an emerging universal concept. In: *Gene and Cell Therapy: Therapeutic Mechanisms and Strategies, 3rd ed.*, S.S. Templeton (Ed.), pp. 263-291. CRC Press, ISBN 084938768X, Boca Raton, FL, USA.

Airenne, K.J., Makkonen, K.E., Mahonen, A.J., & Yla-Herttuala, S. 2010. *In vivo* application and tracking of baculovirus. *Current Gene Therapy* 10: 1-8.

Airenne, K.J., Makkonen, K.E., Mahonen, A.J., & Yla-Herttuala, S. 2011. Baculoviruses mediate efficient gene expression in a wide range of vertebrate cells. In: *Viral Vectors for Gene Therapy: Methods and Protocols, Methods in Molecular Biology*, O-W.

Merten & M. Al-Rubeai (Eds.), Vol. 737, pp. 279-301. Springer Science-Business Media, ISBN 978-1617790942, LLC.

Andreadis, T.G., Becnel, J.J., & Whitem S.E. 2003. Infectivity and pathogenicity of a novel baculovirus, CuniNPV from *Culex nigripalpus* (Diptera: Culicidae) for thirteen species and four genera of mosquitoes. *Journal of Medical Entomology* 40: 512-7.

Argaud, O., Croizier, L., López-Ferber, M., & Croizier, G. 1998. Two key mutations in the host-range specificity domain of the p143 gene of *Autographa californica* nucleopolyhedrovirus are required to kill *Bombyx mori* larvae. *Journal of General Virology* 79: 931-935.

Ashour, M-B., Didair, A., Ragheb, D.A., El-Sheikh, E.-S.A., Gomaa, E.-A.A., Kamita, S.G., & Hammock, B.D. 2007. Biosafety of recombinant and wild type nucleopolyhedroviruses as bioinsecticides. *International Journal of Environmental Research and Public Health* 4: 111-125.

Aucoin, M.G., Mena, J.A., & Kamen, A.A. 2010. Bioprocessing of baculovirus vectors: a review. *Current Gene Therapy* 10: 174-186.

Barbehenn, R.V. 2001. Roles of peritrophic membranes in protecting herbivorous insects from ingested plant allelochemicals. *Archives of Insect Biochemistry and Physiology* 47: 86–99.

Barbehenn, R.V. & Martin, M.M. 1995. Peritrophic envelope permeability in herbivorous insects. *Journal of Insect Physiology* 41: 301-311.

Barber, K.N., Kaupp, W.J., & Holmes, S.B. 1993. Specificity testing of the nuclear polyhedrosis virus of the gypsy moth, *Lymantria dispar* (L.) (Lepidoptera: Lymantriidae). *The Canadian Entomologist* 125: 1055-1066.

Becnel, J.J. 2007. Current status of deltabaculoviruses, cypoviruses and chloriridoviruses pathogenic for mosquitoes. *Virologica Sinica* 22: 117-127.

Berling, M., Blachere-López, C., Soubabere, O. Lery, X., Bonhomme, A., Sauphanor, B. & López-Ferber, M. 2009. *Cydia pomonella* granulovirus genotypes overcome virus resistance in the codling moth and improve virus efficiency by selection against resistant hosts. *Applied and Environmental Microbiology* 75: 925-930.

Bézier, A., Annaheim, M., Herbinière, J., Wetterwald, C., Gyapay, G., Bernard-Samain, S., Wincker, P., Roditi, I., Heller, M., Belghazi, M., Pfister-Wilhem, R., Periquet, G., Dupuy, C., Huguet, E., Volkoff, A.-N., Lanzrein, B., & Drezen, J.-M. 2009. Polydnaviruses of braconid wasps derive from an ancestral nudivirus. *Science* 323: 926-930.

Black, B.C., Brennan, L.A., Dierks, P.M., & Gard, I.E. 1997. Commercialization of baculoviral insecticides. In: *The Baculoviruses*, L.K. Miller, (Ed.), pp. 341-387. Plenum Press, ISBN 0-306-45641-9, New York, NY, USA.

Blissard, G.W., Kogan, P.H., Wei, R., & Rohrmann, G.F. 1992. A synthetic early promoter from a baculovirus: Roles of the TATA box and conserved start site CAGT sequence in basal levels of transcription. *Journal of Virology* 190: 783–793.

Boucias, D. G. & Pendland, J. C. 1998. Baculoviruses. In: *Principles of Insect Pathology*, D.G. Boucias & J.C. Pendland, (Eds.), pp. 111–146, Kluwer Academic Publishers, ISBN 0-412-03591-X, Norwell, MA, USA.

Boyce, F.M. & Bucher, N.L.R. 1996. Baculovirus-mediated gene transfer into mammalian cells. *Proceedings of the National Academy for Sciences USA* 93: 2348-2352.

Braunagel, S.C., Elton, D.M., Ma, H., & Summers, M.D. 1996. Identification and analysis of an *Autographa californica* nuclear polyhedrosis virus structural protein of the occlusion-derived virus envelope: ODV-E56. *Journal of Virology* 217: 97–110.

Braunagel, S.C., Russell, W.K., Rosas-Acosta, G., Russell, D.H., & Summers, M.D. 2003. Determination of the protein composition of the occlusion-derived virus of *Autographa californica* nucleopolyhedrovirus. *Proceedings of the National Academy of Sciences USA* 100: 9797–9802.

Braunagel, S.C. & Summers, M.D. 2007. Molecular biology of the baculovirus occlusion-derived virus envelope. *Current Drug Targets* 8: 1084-1095.

Burand, J.P., Summers M.D., & Smith, G.E. 1980. Transfection with baculovirus DNA. *Virology* 101: 286–290.

Burges, H.D., Croizier, G., & Huber, J. 1980a. A review of safety tests on baculoviruses. *Entomophaga* 25: 329-40.

Burges, H.D., Huber, J., & Croizier, G. 1980b . Guidelines for safety tests on insect viruses. *Entomophaga* 25: 341-8.

Carbonell, L. F., Klowden, M. J., & Miller, L. K. 1985. Baculovirus-mediated expression of bacterial genes in dipteran and mammalian cells. *Journal of Virology* 56: 153-160.

Carstens, E.B., Tjia, S.T., & Doerfler, W. 1979. Infection of *Spodoptera frugiperda* cells with *Autographa californica* nuclear polyhedrosis virus. I. Synthesis of intracellular proteins after virus infection. *Virology* 99: 386-398.

Carstens, E.B., Tjia, S.T., & Doerfler, W. 1980. Infectious DNA from *Autographa californica* nuclear polyhedrosis virus. *Virology* 101: 311-314.

Chandler, D., Bailey, A.S., Tatchell, G.M., Davidson, G., Greaves, J., & Grant, W.P. 2011. The development, regulation and use of biopesticides for integrated pest management. *Philosophical Transactions of the Royal Society of London. Series B, Biological Sciences* 366 : 1987-1998.

Charlton, C.A. & Volkman, L.E. 1993. Penetration of *Autographa californica* nuclear polyhedrosis virus nucleocapsids into IPLB Sf 21 cells induces actin cable formation. *Journal of Virology* 197: 245–254.

Chen, C.J. & Thiem, S.M. 1997. Differential infectivity of two *Autographa californica* nucleopolyhedrovirus mutants on three permissive cell lines is the result of lef-7 deletion. *Journal of Virology* 227: 88–95.

Chen, C.J., Quentin, M.E., Brennan, L.A., Kukel, C., & Thiem, S.M. 1998. *Lymantria dispar* nucleopolyhedrovirus hrf-1 expands the larval host range of *Autographa californica* nucleopolyhedrovirus. *Journal of Virology* 72: 2526-2531.

Chen, C.Y., Wu, H.H., Chen, C.P., Chern, C.R., Hwang, S.M., Huang, S.F., Lo, W.H., Chen, G.H., & Hu, Y.C. 2011. Biosafety assessment of human mesenchymal stem cells engineered by hybrid baculovirus vectors. *Molecular Pharmaceutics* 2011. DOI:10.1021/mp100368d.

Chikhalya, A., Luu, D.D., Carrera, M., De La Cruz, A., Torres, M., Martinez, E.N., Chen, T.,. Stephens, K.D., & Haas-Stapleton, E.J. 2009. Pathogenesis of *Autographa californica* multiple nucleopolyhedrovirus in fifth-instar *Anticarsia gemmatalis* larvae. *Journal of General Virology* 90: 2023-2032.

Chisholm, G.E., & Henner, D.J. 1988. Multiple early transcripts and splicing of the *Autographa californica* nuclear polyhedrosis virus IE-1 gene. *Journal of Virology* 62: 3193-3200.

Clavijo, G., Williams, T., Muñoz, D., Caballero, P., & López-Ferber, M. 2011. Mixed genotype transmission bodies and virions contribute to the maintenance of diversity in an insect virus. *Proceedings of the Royal Society B: Biological Sciences* 277: 943-951.

Cohen, D.P.A., Marek, M., Davies, B.G., Vlak, J.M., & van Oers, M.M. 2009. Encylopedia of *Autographa californica* nucleopolyhedrovirus. *Virology Sinica* 24: 359-414.

Cory, J.S. 2003. Ecological impacts of virus insecticides. In: *Environmental Impacts of Microbial Insecticides: Need and Methods for Risk Assessment*, H.M.T., Hokkanen, & A.E., Hajek, (Eds.), pp. 73-91, Kluwer, ISBN 1-4020-0813-9, Dordrecht, The Netherlands.

Cory, J.S. & Hails, R.S. 1997. The ecology and biosafety of baculoviruses. *Current Opinion in Biotechnology* 8: 323-327.

Cory, J.S. & Myers, J.H. 2003. The ecology and evolution of insect baculoviruses. *Annual Review of Ecology, Evolution, and Systematics* 34: 239-272.

Cory, J.S., Green, B.M., Paul, R.K., & Hunter-Fujita, F. 2005. Genotypic and phenotypic diversity of a baculovirus population within an individual insect host. *Journal of Invertebrate Pathology* 89: 101–111.

Cory, J. S. & Hoover, K. 2006. Plant-mediated effects in insect–pathogen interactions. *Trends in Ecology and Evolution* 21: 278-286.

Couch, J.A., Martin, S.M., Thompkins, G., & Kinney, J. 1984. A simple system for the preliminary evaluation of infectivity and pathogenesis of insect virus in a nontarget estuarine shrimp. *Journal of Invertebrate Pathology* 43: 351-357.

Cox, M.M.J. & Hashimoto, J. 2011. A fast tract influenza virus vaccine produced in insect cells. *Journal of Invertebrate Pathology* 107: S31-S41.

Cox, M.M. & Hollister, J.S. 2009. FluBlok, a next generation influenza vaccine manufactured in insect cells. *Biologicals* 37: 182-189.

CPL Business Consultants. 2010. The 2010 Worldwide Biopesticides Market Summary, Volume 1, June 5, 2010 - Pub ID: BGEQ2703518. MarketResearch.com, Rockville, MD, USA.

Croizier, G., Croizier, L., Argaud, O., & Poudevigne, D. 1994. Extension of *Autographa-californica* nuclear polyhedrosisvirus host-range by interspecific replacement of a short DNA-sequence in the P143 helicase gene. *Proceedings of the National Academy of Sciences USA* 91: 48-52.

Cuddleford, V. 2006. Managing editor. Surveys gauge attitude of Canadian public. *Biocontrol Files* 5: 1-8.

Dall, D., Luque, T., & O'Reilly. D. 2001. Insect-virus relationships: sifting by informatics. *Bioessays* 23:184–193.

Dejoux, C. & Elouard, J.M. 1990. Potential impact of microbial insecticides on the freswater environment, with special reference to the WHO/UNDP/World Bank Onchocerciasis control program. In: *Safety of Microbial Insecticides*, M., Laird, L.A., Lacey, and E.W., Davidson, (Eds.), pp. 66-83, CRC Press, ISBN 978-0849347931, Boca Raton, FL, USA.

Derksen, A.C.G. & Granados,. R. 1988. Alteration of a lepidopteran peritrophic membrane by baculoviruses and enhancement of viral infectivity. *Virology* 167: 242-250.

Detvisitsakun, C., Cain, E.L., & Passarelli, A.L., 2007. The *Autographa californica* M nucleopolyhedrovirus fibroblast growth factor accelerates host mortality. *Virology* 365: 70–78.

Doyle, C.J., Hirst, M.L., Cory, J.S., & Entwistle, P.F. 1990. Risk assessment studies: detailed host range testing of wild-type cabbage moth, *Mamestra brassicae* (Lepidoptera: Noctuidae), nuclear polyhedrosis virus. *Applied and Environmental Microbiology* 56: 2704-2710.

Du, X.L. & Thiem, S.M. 1997. Characterization of host range factor 1 (hrf-1) expression in *Lymantria dispar* M nucleopolyhedrovirus- and recombinant *Autographa californica* M nucleopolyhedrovirus-infected IPLB- Ld652Y cells. *Virology* 227: 420-430.

Duffy, S.P., Young, A.M., Morin, B., Lucarotti, C.J., Koop, B.F., & Levin, D.B. 2006. Sequence analysis and organization of the *Neodiprion abietis* nucleopolyhedrovirus genome. *Journal of Virology* 80: 6952–63.

Duffy, S.P., Becker, E.M., Whittome, B.H., Lucarotti, C.J., & Levin, D.B. 2007. *In vivo* replication kinetics and transcription patterns of the nucleopolyhedrovirus (NeabNPV) of the balsam fir sawfly, *Neodiprion abietis*. *Journal of General Virology* 88: 1945-1951.

EFSA Panel on Biological Hazards (BIOHAZ). 2009. Scientific opinion on the maintenance of the list of QPS microorganisms intentionally added to food or feed (2009 update). *EFSA Journal*, 7: 1431. 92pp. www.efsa.europa.eu

Ehlers, R.-U. 2011. Regulation of biological control agents and the EU policy support action REBECA. In: *Regulation of Biological Control Agents*, R.-U. Ehlers (Ed.), pp. 3-23, Springer Dordrecht, ISBN 978-90-481-3663-6, Heidelberg, Germany.

Elam, P., Vail, P.V., & Schreiber, F. 1990. Infectivity of *Autographa californica* nuclear polyhedrosis virus extracted with digestive fluids of *Heliothis zea*, *Estigmene acrea*, and carbonate solutions. *Journal of Invertebrate Pathology* 55: 278-283.

Elias, C.B., Jardin, B., & Kamen, A. 2007. Recombinant protein production in large-scale agitated bioreactors using the baculovirus expression vector system. *Methods in Molecular Biology* 388: 225-46.

Engelhard, E.K., Kam-Morgan, L.N.W., Washburn, J.O., & Volkman, L.E. 1994. The insect tracheal system: a conduit for the systemic spread of *Autographa californica* M nuclear polyhedrosis virus. *Proceedings of the National Academy of Sciences USA* 91: 3224-3227.

Engelhard, E. K. & Volkman, L. E. 1995. Developmental resistance in fourth instar *Trichoplusia ni* orally inoculated with *Autographa californica* M nuclear polyhedrosis virus. *Virology* 209: 384–389.

England, L.S., Vincent, M.L., Trevors, J.T., & Holmes. S.B. 2004. Extraction, detection and persistence of extracellular DNA in forest litter microcosms. *Molecular and Cellular Probes* 18: 313–319.

Entwistle, P.F., Adams, P.H.W., & Evans, H.F. 1977. Epizootiology of a nuclear-polyhedrosis virus in European spruce sawfly, *Gilpinia hercyniae*: birds as dispersal agents of the virus during winter. *Journal of Invertebrate Pathology* 30: 15-19.

Erlandson, M.A. & Carstens, E.B. 1983. Mapping early transcription products of *Autographa californica* nuclear polyhedrosis virus. *Virology* 126: 398-402.

Erlandson, M.A., Gordon, J., & Carstens, E.B. 1985. Size and map locations of early transcription products on the *Autographa californica* nuclear polyhedrosis virus genome. *Virology* 142: 12-23.

Fang, M., Nie, Y., Harris, S., Erlandson, M.A., & Theilmann, D.A. 2009. *Autographa californica* multiple nucleopolyhedrovirus core gene ac96 encodes a *per os* infectivity factor (pif-4). *Journal of Virology* 83: 12569–12578.

Farris, S.M. & Schulmeister, S. 2011. Parasitoidism, not sociality, is associated with the evolution of elabrate mushroom bodies in the brains of hymenopteran insects. *Proceedings of the Royal Society B: Biological Sciences* 278: 940-951.

Faulkner, P., Kuzio, J., Williams, G. V., & Wilson, J. A. 1997. Analysis of p74, a PDV envelope protein of *Autographa californica* nucleopolyhedrovirus required for occlusion body infectivity *in vivo*. *Journal of General Virology* 78: 3091-3100.

FDA. 2009. *Review of MPL-Relevant Toxicology Information in Cervarix BLA (STN 125259)*, 1-41 pp., U.S. Department of Health and Human Services, FDA, Center for Biologics Evaluation and Research, Rockville, MD, USA.

FDA. 2010. *Guidelines for Industry. Characterization and qualification of cell substrate and other biological materials used in the production of viral vaccines for infectious disease indications*. 1-47 pp., U.S. Department of Health and Human Services, FDA, Center for Biologics Evaluation and Research, Rockville, MD, USA.

Federici, B.A. 1997. Baculovirus pathogenesis, In: *The Baculoviruses*, L.K. Miller, (Ed.), pp. 8-59, Plenum Press, ISBN 9780306456411, New York.

Federici, B.A. & Stern, V.M. 1990. Replication and occlusion of a granulosis virus in larval and adult midgut epithelium of the western grapeleaf skeletonizer, *Harrisina brillians*. *Journal of Invertebrate Pathololology* 56: 401-414.

Feng, S.Z., Jiao, P.R., Qi, W.B., Fan, H.Y., & Liao, M. 2011. Development and strategies of cell-culture technology for influenza vaccine. *Applied Microbial Biotechnology* 89: 893-902.

Friesen, P.D. 2007. Insect Viruses, In: *Fields Virology 5th ed.*, D.M. Knipe and P.M. Howley (Eds.), pp. 707-736, Lippincott Williams & Wilkins, ISBN 9780781760607, Philadelphia, PA, USA.

Fuchs, L. Y., M. S. Woods, & R. F. Weaver. 1983. Viral transcription during *Autographa californica* nuclear polyhedrosis virus infection: a novel RNA polymerase induced in infected *Spodoptera frugiperda* cells. *Journal of Virology* 48: 641-646.

Garcia-Maruniak, A., Maruniak, J.E., Zanotto, P.M., Doumbouya, A.E., Liu, J.C., Merritt, T.M., & Lanoie, J.S. 2004. Sequence analysis of the genome of the *Neodiprion sertifer* nucleopolyhedrovirus. *Journal of Virology* 78: 7036–7051.

Georgopoulos, L.J., Elgue, G., Sanchez, J., Dussupt, V., Magotti, P., Lambris, J.D., Totterman, T.H., Maitland, N.J., & Nilsson, B. 2009. Preclinical evaluation of innate immunity to baculovirus gene therapy vectors in whole human blood. *Molecular Immunology* 46: 2911-2917.

Glare, T., Newby, E., and Nelson, T. 1995. Safety testing of a nuclear polyhedrosis virus for use against gypsy moth, *Lymantria dispar*, in New Zealand. *Proceedings of the Forty-Eighth New Zealand Plant Protection Congress*, pp. 264-269, Hastings, New Zealand, August 8-10, 1995.

Gomez-Casado, E., Gomez-Sebastian, S., Núñez, M.C., Lasa-Covarrubias, R., Martínez-Pulgarín, S., & Escribano, J.M. 2011. Insect larvae biofactories as a platform for influenza vaccine production. *Protein Expression and Purification* 79: 35-43.

Gomi, S., Majima, K., & Maeda, S. 1999. Sequence analysis of the genome of *Bombyx mori* nucleopolyhedrovirus. *Journal of General Virology* 80: 1323–1337.

Gonin, P. & Gaillard C. 2004. Gene transfer vector biodistribution: pivotal safety studies in clinical gene therapy development. *Gene Therapy* 11: S98-108.

Granados, R.R. & Lawler, K.A. 1981. *In vivo* pathway of *Autographa californica* baculovirus invasion and infection. *Virology* 108: 297-308.

Gröner, A. 1986. Specificity and safety of baculoviruses. In: *The Biology of Baculoviruses Vol. II: Practical Application for Insect Control*, R.R. Granados & B.A. Federici (Eds.), pp.177-202, CRC Press, ISBN 0849359864, Boca Raton, FL, USA.

Gross, C. & Rohrmann, G.F. 1993. Analysis of the role of 50 promoter elements and 30 flanking sequences on the expression of a baculovirus polyhedron envelope protein gene. *Virology* 192: 273–281.

Gross, C.H., Russell, R.L.Q., & Rohrmann, G.F. 1994. *Orgyia pseudotsugata* baculovirus p10 and polyhedron envelope protein genes: analysis of their relative expression levels and role in polyhedron structure. *Journal of General Virology* 75: 1115–1123.

Guarino, L.A. & Summers, M.D. 1986. Functional mapping of a trans-activating gene required for expression of a baculovirus delayed-early gene. *Journal of Virology* 57: 563–571.

Guarino, L.A. & Summers, M.D. 1987. Nucleotide sequence and temporal expression of a baculovirus regulatory gene. *Journal of Virology* 61: 2091-2099.

Guarino, L.A., Xu, B., Jin, J., & Dong, W. 1998. A virus-encoded RNA polymerase purified from baculovirus-infected cells. *Journal of Virology* 72: 7985–7991.

Guyton, A.C., & Hall. J.E. 2006. *Textbook of Medical Physiology* (11 ed.). Elsevier Saunders ISBN 9780721602400, Philadelphia, PA, USA.

Guzo, D., Rathburn, H., Guthrie, K., & Dougherty E. 1992. Viral and host cellular transcription in *Autographa-californica* nuclear polyhedrosis virus-infected gypsy-moth cell-lines. *Journal of Virology* 66: 2966-2972.

Haas-Stapleton, E.J., Washburn, J.O., & Volkman, L.E. 2003. Pathogenesis of *Autographa californica* M nucleopolyhedrovirus in fifth instar *Spodoptera frugiperda*. *Journal of General Virology* 84: 2033–2040.

Haas-Stapleton, E.J., Washburn, J.O., & Volkman, L.E. 2004. P74 mediates specific binding of *Autographa californica* M nucleopolyhedrovirus occlusion-derived virus to primary cellular targets in the midgut epithelia of *Heliothis virescens* larvae. *Journal of Virology* 78: 6786–6791.

Haas-Stapleton, E.J., Washburn, J.O., & Volkman, L.E. 2005. *Spodoptera frugiperda* resistance to oral infection by *Autographa californica* multiple nucleopolyhedrovirus linked to aberrant occlusion-derived virus binding in the midgut. *Journal of General Virology* 86: 1349–1355.

Hang, X., Dong, W., & Guarino, L.A. 1995. The lef-3 gene of *Autographa californica* nuclear polyhedrosis virus encodes a single-stranded DNA-binding protein. *Journal of Virology* 69: 3924–3928.

Hang, X. & Guarino, L.A. 1999. Purification of *Autographa californica* nucleopolyhedrovirus DNA polymerase from infected insect cells. *Journal of General Virology* 80: 2519–2526.

Harper, D.M., Franco, E.L., Wheeler, C., Ferris, D.G., Jenkins, D., Schuind, A., Zahaf, T., Innis, B., Naud, P., De Carvalho, N.S., Roteli-Martins, C.M., Teixeira, J., Blatter, M.M., Korn, A.P., Quint, W., & Dubin, G. 2004. Efficacy of a bivalent L1 virus-like particle vaccine in prevention of infection with human papillomavirus types 16 and 18 in young women: a randomized controlled trial. *Lancet* 364: 1757-1765.

Harper, D.M., Franco, E.L., Wheeler, C.M., Moscicki, A.B., Romanowski, B., Roteli-Martins, C.M., Jenkins, D., Schuind, A., Costa, C.S.A., & Dubin, G. 2006. Sustained efficacy up to 4.5 years of a bivalent L1 virus-like particle vaccine against human papillomavirus type 16 and 18: Follow-up from a randomized control trial. *Lancet* 367: 1247-1255.

Hartig, P.C., Chapman, M.A., Hatch, G.G., & Kawanishi, C.Y. 1989. Insect virus assay for toxic effects and transformation potential in mammalian cells. *Applied and Environmental Microbiology* 55: 1916-1920.

Hashimoto, Y., Corsaro, B.G., & Granados, R.R. 1991. Location and nucleotide sequence of the gene encoding the viral enhancing factor of the *Trichoplusia ni* granulosis virus. *Journal of General Virology* 72: 2645-2651.

Hauschild, R., Speiser, B., & Tamm, L. 2011. Regulation according to EU Directive 91/414: Data requirements and procedure compared with regulation practice in other OECD countries. pp. 25-77. In: *Regulation of Biological Control Agents*, Ehlers, R.-U. (Editor). Springer Science+Business Media B.V. ISBN 978-90-481-3663-6, Dordrecht Heildelberg London New York.

Hawtin, R.E., Zarkowska, T., Arnold, K., Thomas, C.J., Gooday, G.W., King, L.A., Kuzio, J.A., & Possee, R.D. 1997. Liquefaction of *Autographa californica* nucleopolyhedrovirus-infected insects is dependent on the integrity of virus-encoded chitinase and cathepsin genes. *Journal of Virology* 238: 243-253.

Hayakawa, T., Ko, R., Okano, K., Seong, S., Goto, C., & Maeda, S. 1999. Sequence analysis of the *Xiestia c-nigrum* granulovirus genome. *Journal of Virology* 262: 277-297.

Health Canada. 2010. Summary Basis of Decision (SBD) CERVARIX®. Human papillomavirus Types 16 and 18 Recombinant AS04 adjuvanted vaccine GlaxoSmithKline Inc. Submission Control Number: 127987. Health Products and Food Branch (October 26, 2010), pp. 1-17.

Health Canada - Pest Management Regulatory Agency. 2009. *Neodiprion abietis* Nucleopolyhedrovirus Newfoundland Strain. Registration Decision RD2009-05. http://www.hc-sc.gc.ca

Health Canada - Pest Management Regulatory Agency. 2000. Virosoft CP4 *Cydia pomonella* granulosis virus. Regulatory Note. REG2000-10. http://www.hc-sc.gc.ca

Hefferon, K.L., Oomens, A.G., Monsma, S.A., Finnerty, C.M., & Blissard, G.W. 1999. Host cell receptor binding by baculovirus GP64 and kinetics of virion entry. *Journal of Virology* 258: 455-468.

Heimpel, A.M. 1955. The pH in the gut and blood of the larch sawfly, *Pristiphora erichsonii* (Htg.), and other insects with reference to the pathogenicity of *Bacillus cereus* Fr. and Fr.1. *Canadian Journal of Zoology* 33: 99-106.

Herniou, E.A., Olszewski, J.A., O'Reilly, D.R., & Cory, J.S. 2004. Ancient coevolution of baculoviruses and their insect hosts. *Journal of Virology* 78: 3244-3251.

Hess, R.T. & Falcon, L.A. 1987. Temporal events in the invasion of the codling moth, *Cydia pomonella* , by a granulosis virus: an electron microscope study. *Journal of Invertebrate Pathology* 50: 85-105.

Hewson, I., Brown, J.M., Gitlin, S.A., & Doud. D.F. 2011. Nucleopolyhedrovirus detection and distribution in terrestrial, freshwater, and marine habitats of Appledore Island, Gulf of Maine. *Microbial Ecology* 62: 48-57.

Hitchman, R.B., Hodgson, D.J., King, L.A., Hail, R.S., Cory, J.S., & Possee, R.D. 2007. Host mediated selection of pathogen genotypes as a mechanism for the maintenance of baculovirus diversity in the field. *Journal of Invertebrate Pathology* 94: 153–162.

Hochberg, M.E. 1991. Intra-host interactions between a braconid endoparasitoid, *Apanteles glomeratus*, and a baculovirus for larvae of *Pieris brassicae. Journal of Animal Ecology* 60: 51-63.

Hodgson, D.J., Vanbergen, A.J., Hartley, S.E., Hails, R.S., & Cory, J.S. 2002. Differential selection of baculovirus genotypes mediated by different species of host food plant. *Ecology Letters* 5: 512–518.

Hofmann, C., Sandig, V., Jennings, G., Rudolph, M., Schlag, P., & Strauss, M. 1995. Efficient gene transfer into human hepatocytes by baculovirus vectors. *Proceedings of the National Academy of Sciences USA* 92: 10099-10103.

Hom, L.G., Ohkawa, T., Trudeau, D., & Volkman, L.E. 2002. *Autographa californica* M nucleopolyhedrovirus ProV-CATH is activated during infected cell death. *Journal of Virology* 296: 212–218.

Hoopes, R. R., Jr. & G. F. Rohrmann. 1991. *In vitro* transcription of baculovirus immediate early genes: accurate mRNA initiation by nuclear extracts from both insect and human cells. *Proceedings of the National Academy of Sciences USA* 88: 4513-4517.

Hoover, K., Grove, M., Gardner, M., Hughes, D.P., McNeil, J. & Slavicek, J. 2011. A gene for an extended phenotype. *Science* 333: 1401.

Hoover, K., Humphries, M.A., Gendron, A.R., & Slavicek, J.M. 2010. Impact of viral *enhancin* genes on potency of *Lymantria dispar* multiple nucleopolyhedrovirus in *L. dispar* following disruption of the peritrophic matrix. *Journal of Invertebrate Pathology* 104: 150-152.

Horton, H. M. & J. P. Burand. 1993. Saturable attachment sites for polyhedron-derived baculovirus on insect cells and evidence for entry via direct membrane fusion. *Journal of Virology* 67: 1860–1868.

Hossler, E.W. 2010a Caterpillars and moths: Part I. Dermatologic manifestations of encounters with Lepidoptera. *Journal of the American Academy of Dermatology* 62: 1-10.

Hossler, EW. 2010b. Caterpillars and moths: Part II. Dermatologic manifestations of encounters with Lepidoptera. *Journal of the American Academy of Dermatology* 62: 13-28.

Hu, Y.C. 2005. Baculovirus as a highly efficient expression vector in insect and mammalian cells. *Acta Pharmacologica Sinica* 26: 405-16.

Huang, X.-P., Davis, T.R., Hughes, P., & Wood. A. 1997. Potential replication of recombinant baculoviruses in nontarget insect species: reporter gene products as indicators of infection. *Journal of Invertebrate Pathology* 69: 234-245.

Huh, N.E. & Weaver R.F. 1990. Identifying the RNA polymerases that synthesize specific transcripts of the *Autographa califomica* nuclear polyhedrosis virus. *Journal of General Virology* 71:195-201.

Ignoffo, C. M. 1975. Evaluation of *in vivo* specificity of insect viruses. In: *Baculoviruses for Insect Pest Control*, M. Summers, R. Engler, L. A. Falcon and P. V. Vail (Eds.), pp. 52-57, American Society for Microbiology, ISBN 0914826077, Washington, DC, USA..

Iwanaga, M., Takaya, K., Katsuma, S., Ote, M., Tanaka, S., Kamita, S.G., Kang, W., Shimada T., & Kobayashi, M. 2004. Expression profiling of baculovirus genes in permissive and nonpermissive cell lines. *Biochemical and Biophysical Research Communications* 323: 599-614.

Jahn, E. 1967. Über eine Massenvermehrung der stahlblauen Kieferngespinstblattwespe, *Acantholyda erythrocephela* Chr., im Steinfeld, Niederösterreich, in den Jahren 1964-1967. *Anzeiger für Schädlingskunde* 39: 145-152.

Jehle, J.A, Blissard, G.W., Bonning, B.C., Cory, J.S., Herniou, E.A., Rohrmann, G.F, Theilmann, D.A., Thiem, S.M., & Vlak, J.M. 2006. On the classification and nomenclature of baculoviruses: A proposal for revision. *Archives of Virology* 151: 1257-1266.

Johnston, K.A., Lee, M.J., Brough, C., Hilder, V.A., Gatehouse, A.M.R., & Gatehouse, J.A. 1995. Protease activities in the larval midgut of *Heliothis virescens*: evidence for trypsin and chymotrypsin- like enzymes. *Insect Biochemistry and Molecular Biology* 25: 375-383.

Kabaluk, J.T., Svircev, A.M., Goettel, M.S., & Woo, S.G. (Eds.). 2010. *The Use and Regulation of Microbial Pesticides in Representative Jurisdictions Worldwide*. IOBC Global. 99pp. Available online through www.IOBC-Global.org.

Kamen, A.A., Aucoin, M.G., Merten, O.W., Alves, P., Hashimoto, Y., Airenne, K., Hu, Y.C., Mezzina, M., & van Oers, M.M. 2011. An initiative to manufacture and characterize baculovirus reference material. *Journal of Invertebrate Pathology* 107: 113-117.

Kamita, S.G., Nagasaka, K., Chua, J.W., Shimada, T., Mita, K., Kobayashi, M., Maeda, S., & Hammock, B.D. 2005. A baculovirus-encoded protein tyrosine phosphatase gene induces enhanced locomotory activity in a lepidopteran host. *Proceedings of the National Academy of Science USA* 102: 2584-9.

Kantoff, P.W., Higano, C.S., Shore, N.D., Berger, E.R., Small, E.J., Penson, D.F., Redfern, C.H., Ferrari, A.C., Dreicer, R., Sims, R.B., Frohlich, M.W., & Schelhammer, P.F. 2010. Sipuleucel-T immunotherapy for castration-resistantant prostate cancer, *New England Journal of Medicine* 363: 411-422.

Kato, T., Kajikawa, M., Maeneka, K., & Park, E.Y. 2010. Silkworm expression system as a platform technology in life science. *Applied Microbiology and Biotechnology* 85: 459-470.

Katou, Y., Ikeda, M., & Kobayashi M. 2006. Abortive replication of *Bombyx mori* nucleopolyhedrovirus in Sf9 and High Five cells: defective nuclear transport of the virions. *Virology* 347: 455-465.

Katsuma, S., Horie, S., Daimon, T., Iwanaga, M., Shimada, T., 2006. *In vivo* and *in vitro* analyses of a *Bombyx mori* nucleopolyhedrovirus mutant lacking functional *vfgf*. *Virology* 355: 62-70.

Keddie, B.A., Aponte, G.W., & Volkman, L.E. 1989. The pathway of infection of *Autographa californica* nuclear polyhedrosis virus in an insect host. *Science* 243: 1728-1730.

Kikhno, I., Gutierrez, S., Croizier, L., Croizier, G., & Ferber, M. L. 2002. Characterization of *pif*, a gene required for the *per os* infectivity of *Spodoptera littoralis* nucleopolyhedrovirus. *Journal of General Virology* 83: 3013-3022.

Kim, Y.K., Jiang, H.L., Je, Y.H., Cho, M.H., Cho, C.S. 2007. Modification of baculovirus for gene therapy. In: *Communicating Current Research and Educational Topics and Trends in Applied Microbiology Vol. II.* A. Mendez-Vilas (Ed.), pp. 875-884, Formatex Research Center, ISBN-13 978-84-611-9423-0, Badajoz, Spain.

Kirkpatrick, B.A., Washburn, J.O., Engelhard, E.K., & Volkman, L.E. 1994 Primary infection of insect tracheae by *Autographa californica* M nuclear polyhedrosis virus.*Virology* 203: 184-186.

Kool, M., Ahrens, C.H., Goldbach, R.W., Rohrmann G.F., & Vlak J.M. 1994. The pathway of infection of *Autographa californica* nuclear polyhedrosis virus in an insect host. *Proceedings of the National Academy of Sciences USA* 91: 11212–11216.

Kost, T.A. & Condreay, J.P. 2002. Recombinant baculoviruses as mammalian cell gene-delivery vectors. *Trends in Biotechnology* 20: 173-180.

Kost, T.A., Condreay, J.P., & Jarvis, D.L. 2005. Baculovirus as versatile vectors for protein expression in insect and mammalian cells. *Nature Biotechnology* 23: 567-575.

Lacey, L.A., Vail, P.V., & Hoffmann, D.F. 2002. Comparative activity of baculoviruses against the codling moth, *Cydia pomonella*, and three other tortricid pests of tree fruit. *Journal of Invertebrate Pathology* 80: 64 -68.

Lanier, L.M. & Volkman, L.E. 1998. Actin binding and nucleation by *Autographa california* M nucleopolyhedrovirus. *Journal of Virology* 243: 167–177.

Lapointe, R., Popham, H.J.R., Straschil, U., Goulding, D., O'Reilly, D.R., & Olszewski, J.A. 2004. Characterization of two *Autographa californica* nucleopolyhedrovirus proteins, Ac145 and Ac150, which affect oral infectivity in a host-dependent manner. *Journal of Virology* 78: 6439-6448.

Lautenschlager, R.A., Podgwaite, J.D., & Watson, D.E. 1980. Natural occurrence of the nucleopolyhedrosis virus of the gypsy moth, *Lymantria dispar* [Lep.: Lymantriidae] in wild birds and mammals. *BioControl* 25: 261-267.

Lauzon, H.A.M., Lucarotti, C.J., Krell, P.J., Feng, Q., Retnakaran, A., & Arif, B.M. 2004. Sequence and organization of the *Neodiprion lecontei* nucleopolyhedrovirus genome. *Journal of Virology* 78: 7023-7035.

Lauzon, H.A.M., Garcia-Maruniak, A., Zanotto, P.M.de A., Clemente, J.C., Herniou, E.A., Lucarotti, C.J., Arif, B.M., & Maruniak, J.E. 2006. Genomic comparison of *Neodiprion sertifer* and *Neodiprion lecontei* nucleopolyhedroviruses and identification of potential hymenopteran baculovirus-specific open reading frames. *Journal of General Virology* 87: 1477–1489.

Lehane, M.J. 1997. Peritrophic matrix: structure and function. *Annual Review of Entomology* 42: 525–550.

Leisy, D.J., Rasmussen, C., Kim, H.T., & Rohrmann, G.F. 1995. The *Autographa californica* nuclear polyhedrosis virus homologous region 1a: identical sequences are essential for DNA replication activity and transcriptional enhancer function. *Virology* 208: 742-52.

Lepore, L.S., Roelvink, P.R., & Granados, R.R. 1996. Enhancin, the granulosis virus protein that facilitates nucleopolyhedrovirus (NPV) infections, is a metalloprotease. *Journal of Invertebrate Pathology* 68: 131–140.

Leuschner, R.G.K., Robinson, T.P., Hugas, M., Sandro Cocconcelli, P., Richard-Forget, F., Klein, G., Licht, T.R., Nguyen-The, C., Querol, A., Richardson, M., Suarez, J.E., Thrane, U. Vlak, J.M., & von Wright. A. 2010. Qualified presumption of safety

(QPS): a generic risk assessment approach for biological agents notified to the European Food Safety Authority (EFSA). *Trends in Food Science & Technology* 21: 425-435.

Li, Y., Passarelli, A.L., Miller, L.K. 1993. Identification, sequence, and transcriptional mapping of *lef-3*, a baculovirus gene involved in late and very late gene expression. *Journal of Virology* 67: 5260-8.

Liang, Z. Zhang, X., Yin, X., Sumei Cao, S., & Xu, F. 2011. Genomic sequencing and analysis of *Clostera anachoreta* granulovirus. *Archives of Virology* 156: 1185–1198.

Lightner, D.V., Procter, R.R., Sparks, A.A., Adams, J.R., & Heimpel, A.M. 1973. Testing of penaid shrimps for susceptibility to an insect nuclear polyhedrosis virus. *Environmental Entomology* 2: 611-614.

Long, G., Pan, X., Kormelink, R., & Vlak, J.M. 2006. Functional entry of baculovirus into insect and mammalian cells is dependent on clathrin-mediated endocytosis. *Journal of Virology* 80: 8830–8833.

Lòpez-Ferber, M., Simòn, O., Williams, T., & Caballero, P. 2003. Defective or effective? Mutualistic interactions between virus genotypes. *Proceedings of the Royal Society B: Biological Sciences* 270: 2249–2255.

Lu., A. & Carstens, E.B. 1991. Nucleotide sequence of a gene essential for viral DNA replication in the baculovirus *Autographa californica* nuclear polyhedrosis virus. *Virology* 181: 336-347.

Lu, A. & Carstens, E.B. 1992. Transcription analysis of the EcoRI-D region of the baculovirus *Autographa californica* nuclear polyhedrosis virus identifies an early 4-kilobase RNA encoding the essential *p143* gene. *Journal of Virology* 66: 655-663.

Lu, A. & Miller, L.K. 1995a. The roles of eighteen baculovirus late expression factor genes in transcription and DNA replication. *Journal of Virology* 69: 975–982.

Lu, A. & Miller, L.K. 1995b. Differential requirements for baculovirus late expression factor genes in two cell lines. *Journal of Virology* 69: 6265–6272.

Lucarotti, C.J., Kettela, E.G., & Mudryj, G. 2006. The registration of Abietiv™: A biological control product based on *Neodiprion abietis* nucleopolyhedrovirus for use against its natural host, the balsam fir sawfly. SERG International Report, 47 pp.

Lucarotti, C. J., Moreau, G., & Kettela, E. G. 2007. Abietiv™ – a viral biopesticide for control of the balsam fir sawfly. In: *Biological Control: A Global Pespective*, C. Vincent, M. Goettel, G. Lazarovits (Eds.), pp. 353-361, CABI Publishing, ISBN-10: 184593265X, Wallingford, UK.

Madhan, S., Prabakaran, M., & Kwang, J. 2010. Baculovirus as vaccine vectors. *Current Gene Therapy* 1: 201-213.

Maeda, S., Kamita, S.G., & Kondo, A. 1993. Host-range expansion of *Autographa californica* nuclear polyhedrosis virus (NPV) following recombination of a 0.6- kilobasepair DNA fragment originating from *Bombyx mori* NPV. *Journal of Virology* 67: 6234-6238.

Mäkelä, A.R. & Oker-Blom, C. 2006. Baculovirus display: a multifunctional technology for gene delivery and eukaryotic library development. *Advances in Virus Research* 68: 91-112.

Mäkelä, A.R., Tuusa, J.E., Volkman, L.E., & Oker-Blom, C. 2008. Occlusion-derived baculovirus: interaction with human cells and evaluation of the envelope protein P74 as a surface display platform. *Journal of Biotechnology* 135: 145-156.

Marek, M., van Oers, M.M., Devaraj, F.F., & Vlak, J.M. 2010. Engineering of baculovirus vector for the manufacture of virion-free biopharmaceuticals. *Biotechnology and Bioengineering* 108: 1056-1067.

Marrone, P.G. 2007. Barriers to adoption of biological control agents and biological pesticides. *CAB Reviews: Perspectives in Agriculture, Veterinary Science, Nutrition and Natural Resources* 2007 2, No. 051. doi: 10.1079/PAVSNNR20072051 ISSN 1749-8848.

Martignoni, M.E., & Iwai, P.J. 1981. A catalogue of viral diseases of insects, mites and ticks, In: *Microbial Control of Pests and Plant Diseases*, H.D. Burgess, (Ed.), pp. 897-911, Academic Press, ISBN 9780121433604, London, UK.

McClintock, J. T., Dougherty, E. M., & Weiner, R. M. 1986. Semipermissive replication of a nuclear polyhedrosis virus of *Autographa californica* in a gypsy-moth cell-line. *Journal of Virology* 57: 197-204.

McConnell, E.L., Basit, A.W., & Murdan. S. 2008. Measurements of rat and mouse gastrointestinal pH, fluid and lymphoid tissue, and implications for *in-vivo* experiments. *Journal of Pharmacy and Pharmacology* 60: 63–70.

McDougal, V.V. & Guarino, L.A. 1999. *Autographa californica* nuclear polyhedrosis virus DNA polymerase: measurements of processivity and strand displacement. *Journal of Virology* 73: 4908–4918.

McDougal, V.V. & Guarino, L.A. 2000. The *Autographa californica* nuclear polyhedrosis virus p143 gene encodes a DNA helicase. *Journal of Virology* 74: 5273–5279.

McLachlin, J.R. & Miller, L.K. 1994. Identification and characterization of *vlf-1*, a baculovirus gene involved in very late gene expression. *Journal of Virology* 68: 7746–7756.

McWilliam, A. 2007. Environmental impact of baculoviruses. FAO. R7299_FTR_anx3. http://www.fao.org/docs/eims/upload/agrotech/2003/R7299_FTR_anx3.pdf.

Means, J.C. & Passarelli, A.L. 2010. Viral fibroblast growth factor, matrix metalloproteases, and caspases are associated with enhancing infection by baculoviruses. *Proceedings of the National Academy of Science USA* 107: 9825–9830.

Metz, S.W. & Pijlman, G.P. 2011. Arbovirus vaccines: opportunities for the baculovirus-insect cell expression system. *Journal of Invertebrate Pathology* 107: S16-S30.

Miele, S.A.B., Garavaglia, M.J., Belaich, M.N., & Ghiringhelli, P.D. 2011. Baculovirus: molecular insights on their diversity and conservation. *International Journal of Evolutionary Biology* 2011: article ID 379424, 15 pp. doi:10.4061/2011/379424.

Mikhailov, V.S. & Rohrmann, G.F. 2002. The baculovirus replication factor LEF-1 is a DNA primase. *Journal of Virology* 76: 2287–2297.

Milks, M.L., Washburn, J.O., Willis, L.G., Volkman, L.E., & Theilmann, D.A. 2003. Deletion of *pe38* attenuates AcMNPV genome replication, budded virus production, and virulence in *Heliothis virescens*. *Virology* 310: 224-34.

Miller, L.K. & Lu, A. 1997. The molecular basis of baculovirus host range. In: *The Baculoviruses*, L.K. Miller (Ed.), pp. 217-235, Plenum Press, ISBN 0-306-45641-9, New York, NY, USA.

Monsma, S.A., Oomens, A.G.P., & Blissard G.W. 1996. The gp64 envelope fusion protein is an essential baculovirus protein required for cell-to-cell transmission of infection. *Journal of Virology* 70: 4607–4616.

Moreau, G. & Lucarotti, C.J. 2007. A brief review of the past use of baculoviruses for the management of eruptive forest defoliators and recent developments on a sawfly virus in Canada. *The Forestry Chronicle* 83: 105–112.

Moreau, G., Lucarotti, C.J., Kettela, E.G., Thurston, G.S., Holmes, S., Weaver, C., Levin, D.B., & Morin, B. 2005. Aerial application of nucleopolyhedrovirus induces decline in increasing and peaking populations of *Neodiprion abietis*. *Biological Control* 33: 65–73.

Morris, T. D. & Miller, L, K. 1993. Characterization of productive and nonproductive AcMNPV infection in selected insect-cell lines. *Virology* 197: 339-348.

Moscardi, F., Lobo de Souza, M., Batista de Castro, M.Ee., Moscardi, L.M., & Szewczyk, B. 2011. Baculovirus pesticides – present state and future perspectives, In: *Microbes and Microbial Technology*, I. Ahmad, F. Ahmad & P. Pichtel, (Eds.), pp. 415-445, Springer, ISBN 978-1-4419-7930-8, New York, NY, USA.

Moser, B.A., Becnel, J.J., White, S.E., Afonso, C., Kutish, G., Shanker, S., & Almira, E. 2001. Morphological and molecular evidence that *Culex nigripalpus* baculovirus is an unusual member of the family Baculoviridae. *Journal of General Virology* 82: 283–297.

Nakai M. & Kunimi Y. 1997. Granulosis virus infection of the smaller tea tortrix (Lepidoptera: Tortricidae): effect on the development of the endoparasitoid, *Ascogaster reticulatus* (Hymenoptera: Braconidae). *Biological Control* 8: 74-80.

Nobiron, I., O'Reilly, D.R., & Olszewski, J.A. 2003. *Autographa californica* nucleopolyhedrovirus infection of *Spodoptera frugiperda* cells: a global analysis of host gene regulation during infection, using a differential display approach. *Journal of General Virology* 84: 3029-3039.

OECD (Organization for Economic Co-operation and Development). 2002. *Consensus Document on Information Used in the Assessment of Environmental Applications Involving Baculovirus*. OECD Environment, Health and Safety Publications, Series on Harmonization of Regulatory Oversight in Biotechnology Number 20. Paris, France. 79 pp. http://www.rebeca- net.de/downloads/report/deliverable %2012.pdf.

Ohkawa, T., Washburn, J.O., Sitapara, R., Sid, E., & Volkman, L.E. 2005. Specific binding of *Autographa californica* M nucleopolyhedrovirus occlusion-derived virus to midgut cells of *Heliothis virescens* larvae is mediated by products of pif genes Ac119 and Ac022 but not Ac115. *Journal of Virology* 79: 15258–15264.

Ohkawa, T., Volkman, L.E., & Welch. M.D. 2010. Actin-based motility drives baculovirus transit to the nucleus and cell surface. *Journal of Cell Biology* 190: 187–195.

Okano, K., Vanarsdall, A.L., Mikhailov, V.S., & Rohrmann, G.F. 2006. Conserved molecular systems of the Baculoviridae. *Journal of Virology* 344: 77–87.

Oker-Blom, C., Airenne, K.J., & Grabherr, R. 2003. Baculovirus display strategies: emerging tools for Eukaryotic libraries and gene delivery. *Briefings in Functional Genomics and Proteomics* 2: 224-253.

Ooi, B.G., Rankin, C., & Miller, L.K. 1989. Downstream sequences augment transcription from the essential initiation site of a baculovirus polyhedrin gene. *Journal of Molecular Biology* 210: 721–736.

Oomens, A.G. & Blissard, G.W. 1999. Requirement for GP64 to drive efficient budding of *Autographa californica* multicapsid nucleopolyhedrovirus. *Journal of Virology* 254: 297–314.

Passarelli, A.L. 2011. Barriers to success: how baculovirus establish efficient systemic infections. *Journal of Virology* 411: 383-392.

Passarelli, A.L. & Miller, L.K. 1993a. Identification and characterization of *lef-1*, a baculovirus gene involved in late and very late gene expression. *Journal of Virology* 67: 3481-3488.

Passarelli, A.L. & Miller, L.K. 1993b Three baculovirus genes involved in late and very late gene expression: *ie-1, ie-n*, and *lef-2. Journal of Virology* 67: 2149-2158.

Passarelli, A.L. & Miller, L.K. 1993c. Identification of genes encoding late expression factors located between 56.0 and 65.4 map units of the *Autographa californica* nuclear polyhedrosis virus genome. *Virology* 197: 704-714.

Pearson, M.N., Groten, C. & Rohrmann, G.F. 2000. Identification of the *Lymantria dispar* nucleopolyhedrovirus envelope fusion protein provides evidence for a phylogenetic division of the Baculoviridae. *Journal of Virology* 74: 6126– 6131.

Peng, K., van Lent, J.W., Vlak, J.M., Hu, Z., & van Oers, M.M. 2011. *In situ* cleavage of the baculovirus occlusion derived virus receptor binding protein P74 in the peroral infectivity complex. *Journal of Virology* 85: 10710-10718.

Peng, K., van Oers, M.M., Hu, Z., van Lent, J.W., Vlak, J.M. 2010. Baculovirus *per os* infectivity factors form a complex on the surface of occlusion-derived virus. Journal of *Virology* 84: 9497-504.

Perera, O.P., Valles, S.M., Green, T.B., White, S., Strong, C.A., & Becnel, J.J. 2006. Molecular analysis of an occlusion body protein from *Culex nigripalpus* nucleopolyhedrovirus (CuniNPV). *Journal of Invertebrate Pathology* 91: 35-42.

Perera, O., Green, T.B., Stevens, S.M., White, J.S., & Becnel, J.J. 2007. Proteins associated with *Culex nigripalpus* nucleopolyhedrovirus occluded virions. *Journal of Virology* 81: 4585–4590.

Peters, W. 1992. *Peritrophic Membranes*, Springer, ISBN 3540536353, New York, NY, USA.

Pijlman, G. P., Pruijssers, A. J., & Vlak, J. M. 2003. Identification of pif-2, a third conserved baculovirus gene required for per os infection of insects. *Journal of General Virology* 84: 2041–2049.

Plymale, R., Grove, M.J., Cox-Foster, D., Ostiguy, N., & Hoover. K. 2008. Plant-mediated alteration of the peritrophic matrix and baculovirus infection in lepidopteran larvae. *Journal of Insect Physiology* 54: 737–749.

Podgwaite, J., Shields, K., Zerillo, R., & Bruen, R. 1979. Environmental persistence of the nucleopolyhedrosisvirus of the gypsy moth. *Environmental Entomology* 8: 523-536.

Possee, R.D. & Rohrmann, G.F. 1997. Baculovirus genome organization and evolution. *The Baculoviruses*, L.K. Miller, (Ed.), pp. 109-140, ISBN 0-306-45641-9, Plenum Press, NewYork, NY, USA.

Prikhod'ko, E.A., Lu, A., Wilson, J.A., & Miller, L.K. 1999. *In vivo* and *in vitro* analysis of baculovirus ie-2 mutants. *Journal of Virology* 73: 2460-2468

Pullen, S.S., & Friesen, P.D. 1995. The CAGT motif functions as an initiator element during early transcription of the baculovirus transregulator ie-1. *Journal of Virology* 69: 3575–3583.

Quinlan, R, & Gill, A. 2006. *The World Market for Microbial Biopesticides. Overview Volume.* CPL Business Consultants, October 2006. 26 pp.

Rapp, J.C., Wilson, J.A., & Miller, L.K. 1998. Nineteen baculovirus open reading frames, including LEF-12, support late gene expression. *Journal of Virology* 72: 10197–10206.

Ravensberg, W.J. 2011. *Progress in Biological Control 10. A Roadmap to the Successful Development and Commercialization of Microbial Pest Control Products for the Control of Arthropods.* ISBN 978-94-007-0436-7, Springer Science + Business Media B.V.

Raymond, B., Hartley, S. E., Cory, J. S., & Hails, R. S. 2005. The role of food plant and pathogen-induced behaviour in the persistence of a nucleopolyhedrovirus. *Journal of Invertebrate Pathology* 88: 49–57.

Reardon, R., & Podgwaite, J. 1976. Disease-parasitoid relationships in natural populations of *Lymantria dispar* in the Northeastern United States. *Entomophaga* 21: 333-341.

Reardon, R.C., Podgwaite, J.D., & Zerillo, R. 2009. Gypchek - bioinsecticide for the gypsy moth. USDA Forest Health Technology Enterprise Team Report, FHTET-2009-01.

Rodems, S.M. & Friesen, P.D. 1995. Transcriptional enhancer activity of *hr5* requires dual-palindrome half sites that mediate binding of a dimeric form of the baculovirus transregulator IE1. *Journal of Virology* 69: 5368–5375.

Rohrbach, D.H. & Timpl, R. 1993. *Molecular and Cellular Aspects of Basement Membranes.* ISBN 0125931654, Academic Press, New York, NY, USA.

Rohrmann, G.F. 2011. *Baculovirus Molecular Biology, 2nd edition,* National Center for Biotechnology Information (US), Bethesda, MD, USA. http://www.ncbi.nlm.nih.gov/books/NBK1764/.

Royama, T. 1992. *Analytical Population Dynamics.* Population and Community Biology Series 10. Chapman and Hall, London, UK.

Russell, R.L.Q., & Rohrmann, G.F. 1990. A baculovirus polyhedron envelop protein: immunogold localization in infected cells and mature polyhedral. *Virology* 174: 177-184.

Sarauer, B.L., Gillott, C., & Hegedus, D. 2003. Characterization of an intestinal mucin from the peritrophic matrix of the diamondback moth, *Plutella xylostella. Insect Molecular Biology* 12: 333-343.

Savard, J., Tautz, D., Richards, S., Weinstock, G.M., Gibbs, R.A., Werren, J.H., Tettelin, H., & Lercher, M.J. 2006. Phylogenomic analysis reveals bees and wasps (Hymenoptera) at the base of the radiation of Holometabolous insects. *Genome Research* 16: 1334-1338.

Shoji, I., Aizaki, H., Tani, H., Ishii, K., Chiba, T., Saito, I., Miyamura, T., & Matsuura, Y. 1997 Efficient gene transfer into various mammalian cells, including non-hepatic cells, by baculovirus vectors. *Journal of General Virology* 78: 2657-2664

Simón, O., Williams, T., López-Ferber, M., Taulemesse J-M., & Caballero, P. 2008. Population genetic structure determines speed of kill and occlusion body production in *Spodoptera frugiperda* multiple nucleopolyhedrovirus. *Biological Control* 44: 321–330.

Slack, J. & Arif, B.M. 2007. The baculoviruses occlusion-derived virus: virion structure and function. *Advances in Virus Research* 69: 99-165.

Slack, J.M., Lawrence, S.D., Krell, P.J., & Arif, B.M. 2008. Trypsin cleavage of the baculovirus occlusion derived virus attachment protein P74 is prerequisite in *per os* infection. *Journal of General Virology* 89: 2388-2397.

Slavicek, J. M. & Popham, H.J.R., 2005. The *Lymantria dispar* nucleopolyhedrovirus enhancins are components of occlusion-derived virus. *Journal of Virology* 79: 10578–10588.

Small, E.J., Schellhammer, P.J., Higano, C.J., Redfern, C.H., Nemunaitis, J.J., Valone, F.H., Verjee,. S.S., Jones, L.A., & Hershberg, R.M. 2006. Placebo-controlled phase II trial of immunogenic therapy with Stipuleucel-T (APC8015) in patients with metastatic, asymptomatic hormone refractory prostate cancer. *Journal of Clinical Oncology* 24: 3089-3094.

Sparks, W.O., Harrison, R. L., & Bonning, B. C. 2011. *Autographa californica* multiple nucleopolyhedrovirus ODV-E56 is a *per os* infectivity factor, but is not essential for binding and fusion of occlusion-derived virus to the host midgut. *Journal of Virology* 409: 69-76.

Stewart, T.M., Huijskens, I., Willis, L.G., & Theilmann, D.A. 2005. The *Autographa californica* multiple nucleopolyhedrovirus *ie0-ie1* gene complex is essential for wild-type virus replication, but either IE0 or IE1 can support virus growth. *Journal of Virology* 79: 4619-4629.

Strasser, H. Strauch, O., Ehlers, R-U., & Hauschild, R. 2007. Positive list of "low risk" candidates. REBECA, Deliverable 12. Project no. SSPE-CT-2005-022709. 37 pp.

Strazanac, J. & Butler, L. (Eds.) 2005. Long-term evaluation of the effects of Btk, Gypchek, and *Entomophaga maimaiga* on nontarget organisms in mixed broadleaf-pine forests in the Central Appalachians. USDA, Forest Health Technology Enterprise Team Report, FHTET-2004-14, 81 pp.

Summers, M.D. 2006. Milestones leading to the genetic engineering of baculoviruses as expression vector systems and viral pesticides. *Advances in Virus Research* 68: 3-73.

Szewczyk, B., Rabalski, L., Krol, E., Sihler, W., & Lobo de Souza, M. 2009. Baculovirus biopesticides - a safe alternative to chemical protection of plants. *Journal of Biopesticides* 2: 209-216.

Tani, H., Nishijima, M., Ushijima, H., Miyamura, T., & Matsuura Y. 2001. Characterization of cell-surface determinants important for baculovirus infection. *Journal of Virology* 279: 343-353.

Tellam, R.L., Wijffels, G., & Willadsen, P. 1999. Peritrophic matrix proteins. *Insect Biochemistry and Molecular Biology* 29: 87-101.

Terra, W.R., Ferreira, C., & Baker, J.E. 1996. Compartmentalization of digestion. In. *Biology of the Insect Midgut*, M.J. Lehane and P.F. Billingsley, (Eds.), ISBN 0 412 61670 X, pp. 206-235, Chapman and Hall, London, UK.

Thiem, S. & Cheng, X-H. 2009. Baculovirus host-range. *Virologica Sinica* 24: 436-457.

Thiem, S. M., Du, X. L., Quentin, M. E., & Bernier, M.M. 1996. Identification of a baculovirus gene that promotes *Autographa californica* nuclear polyhedrosis virus replication in a nonpermissive insect cell line. *Journal of Virology* 70: 2221-2229.

Thiem, S.M. & Miller, L.K. 1990. Differential gene expression mediated by late, very late and hybrid baculovirus promoters. *Gene* 91: 87-94.

Todd J.W., Passarelli, A.L., & Miller L.K. 1995. Eighteen baculovirus genes, including *lef-11, p35, 39K*, and *p47*, support late gene expression. *Journal of Virology* 69: 968-974.

Treanor, J.J., Schiff, G.M., Hayden, F.G., Brady, R.C., Hay, C.M., Meyer, A.L., Wiltse, J.H., Laing, H., Gilbert, A., & Cox, M. 2007. Safety and immunogenicity of a baculovirus expressed hemagglutinin influenza vaccine. *Journal of American Medical Association* 297: 1577-1582.

Tuthill, R.W., Canada, A.T., Wilcock, K., Etkind, P.H., O'Dell, T.M., & Shama, S.K. 1984. An epidemiologic study of gypsy moth rash. *American Journal of Public Health* 74: 799-803.

Uwo, M.F., Ui-Tei, K., Park, P., & Takeda, M. 2002. Replacement of midgut epithelium in greater wax moth, *Galleria mellonella*, during larval-pupal moult. *Cell and Tissue Research* 308: 319-331.

van Oers, M.M. 2006. Vaccines for viral and parasitic diseases produced with baculovirus vectors. *Advances in Virus Research* 68: 193-253

van Oers, M.M. 2011. Opportunities and challenges for the baculovirus expression system. *Journal of Invertebrate Pathology* 107: 3-15.

Vasconcelos, S.D., Williams, T., Hails, R.S., & Cory, J.S. 1996. Prey selection and baculovirus dissemination by carabid predators of Lepidoptera. *Entomological Entomology* 21: 98-104.

Vergati, M., Intrivici, C., Huen, N.Y, Schlom, J., & Tsang, K.W. 2010. Strategies for cancer vaccine development. *Journal of Biomedicine and Biotechnology* 2010: 1-13 pp. doi:10.1155/2010/596432.

Volkman, L.E. & Summers, M.D. 1977. *Autographa californica* nuclear polyhedrosis virus: comparative infectivity of the occluded, alkali-liberated, and nonoccluded forms. *Journal of Invertebrate Pathology* 30: 102–103.

Volkman, L.E. & Goldsmith, P.A. 1983. *In vitro* survey of *Autographa californica* nuclear polyhedrosis virus interaction with nontarget vertebrate host cells. *Applied and Environmental Microbiology* 45: 1085-1093.

Wang, L., Salem, T.Z., Lynn, D.E., & Cheng, X.-W., 2008. Slow cell infection, inefficient primary infection and inability to replicate in the fat body determine the host range of *Thysanoplusia orichalcea* nucleopolyhedrovirus. *Journal of General Virology* 89: 1402–1410.

Wang, P. & Granados, R.R. 1997a. An intestinal mucin is the target substrate for a baculovirus enhancin. *Proceedings of the National Academy of Sciences USA* 94: 6977–6982.

Wang, P. & Granados, R.R. 1997b. An intestinal mucin is the target substrate for a baculovirus enhancin. *Proceedings of the National Academy of Sciences USA* 94: 6977–6982.

Wang, P. & Granados, R.R. 2000. Calcofluor disrupts the midgut defense system in insects. *Insect Biochemistry and Molecular Biology* 30: 135–143.

Wang, P., Hammer, D.A. & Granados, R.R. 1994. Interaction of *Trichoplusia ni* granulosis virus-encoded enhancin with the midgut epithelium and peritrophic membrane of four lepidopteran insects. *Journal of General Virology* 75: 1961–1967.

Wang P., Li G., & Granados R.R. 2004. Identification of two new peritrophic membrane proteins from larval *Trichoplusia ni*: structural characteristics and their functions in the protease rich insect gut. *Insect Biochemistry and Molecular Biology*. 34: 215-227.

Wang, R. R., Deng, F., Hou, D. H., Zhao, Y., Guo, L., Wang, H. L., & Hu, Z. H. 2010. Proteomics of the *Autographa californica* nucleopolyhedrovirus budded virions. *Journal of Virology* 84: 7233–7242.

Wang, X-F., Zhang,, B-Q. Xu, H-J., Cui, Y-J., Xu, Y-P., Zhang, M-J., Han, Y. S., Lee, Y.S., Bao, Y-Y., & Zhang, C-X. 2011. ODV-associated proteins of the *Pieris rapae* granulovirus. *Journal of Proteome Research* 10: 2817–2827.

Wang, Y. & Jehle, J.A. 2009. Nudiviruses and other large, double-stranded circular DNA viruses of invertebrates: new insights on an old topic. *Journal of Invertebrate Pathology* 101:187-193.

Wang, Y., Bininda-Emonds, O.R.P., Oers, M.M., Vlak, J.M., & Jehle, J.A. 2011. The genome of *Oryctes rhinoceros* nudivirus provides novel insight into the evolution of nuclear arthropod-specific large circular double-stranded DNA viruses. *Virus Genes* 42: 444-456.

Washburn, J.O., Kirkpatrick, B.A., & Volkman, L.E. 1995. Comparative pathogenesis of *Autographa californica* M nuclear polyhedrosis virus in larvae of *Trichoplusia ni* and *Heliothis virescens*. *Virology* 209: 561-568.

Washburn, J.O., Lyons, E.H., Haas-Stapleton, E.J., & Volkman, L.E. 1999. Multiple nucleocapsid packaging of *Autographa californica* nucleopolyhedrovirus accelerates the onset of systemic infection in *Trichoplusia ni*. *Journal of Virology* 73: 411-416.

Washburn, J.O., Chan, E.Y., Volkman, L.E., Aumiller, J.J., & Jarvis D.L. 2003a. Early synthesis of budded virus envelope fusion protein GP64 enhances *Autographa californica* multicapsid nucleopolyhedrovirus virulence in orally infected *Heliothis virescens*. *Journal of Virology* 77: 280-290.

Washburn, J.O., Trudeau, D., Wong, J.F., & Volkman, L.E. 2003b. Early pathogenesis of *Autographa californica* multiple nucleopolyhedrovirus and *Helicoverpa zea* single nucleopolyhedrovirus in *Heliothis virescens*: a comparison of the 'M' and 'S' strategies for establishing fatal infection. *Journal of General Virology* 84: 343-351.

Westenberg, M., Uijtdewilligen, P., & Vlak. J.M. 2007. Baculovirus envelope fusion proteins F and GP64 exploit distinct receptors to gain entry into cultured insect cells. *Journal of General Virology* 88: 3302-3306.

Wickham, T.J., Shuler, M.L., Hammer, D.A., Granados, R.R., & Wood, H.A. 1992. Equilibrium and kinetic analysis of *Autographa californica* nuclear polyhedrosis virus attachment to different insect cell lines. *Journal of General Virology* 73: 3185-3194.

Winstanley D. & Crook N.E. 1993. Replication of *Cydia pomonella* granulosis virus in cell cultures. *Journal General Virology* 74: 1599-1609.

Witt, D. J. 1984. Photoreactivation and ultraviolet-enhanced reactivation of ultraviolet irradiated nuclear polyhedrosis virus by insect cells. *Archives of Virology* 79: 95-107.

Wu, J.G. & Miller, L.K. 1989. Sequence, transcription and translation of a late gene of the *Autographa californica* nuclear polyhedrosis virus encoding a 34.8K polypeptide. *Journal of General Virology* 70: 2449-2459.

Xiang, X., Chen, L., Guo, A., Yu, S., Yang, R., & Wu, X. 2011. The *Bombyx mori* nucleopolyhedrovirus (BmNPV) ODV-E56 envelope protein is also a *per os* infectivity factor. *Virus Research* 155: 69-75.

Yang S. & Miller L.K. 1999. Activation of baculovirus very late promoters by interaction with very late factor. *Journal of Virology* 73: 3404-3409.

Yao, L., Zhou, W., Xu, H., Zheng, Y., & Qi, Y. 2004. The *Heliothis armigera* single nucleocapsid nucleopolyhedrovirus envelope protein P74 is required for infection of the host midgut. *Virus Research* 104: 111-121.

Zhang J.H., Washburn J.O., Jarvis D.L., & Volkman L.E. 2004. *Autographa californica* M nucleopolyhedrovirus early GP64 synthesis mitigates developmental resistance in orally infected noctuid hosts. *Journal of General Virology* 85: 833-842.

Zhang, J-H., Ohkawa, Washburn, T.J.O., & Volkman, L.E. 2005. Effects of Ac150 on virulence and pathogenesis of *Autographa californica* multiple nucleopolyhedrovirus in noctuid hosts. *Journal of General Virology* 86: 1619–1627.

Part 2

Integrated Pest Management – Future Challenges

9

Evaluating Surface Seals in Soil Columns to Mitigate Methyl Isothiocyanate Volatilization

Shad D. Nelson[1,2], Catherine R. Simpson[2],
Husein A. Ajwa[3] and Clinton F. Williams[4]
[1]Texas A&M University-Kingsville, Kingsville
[2]Texas A&M University -Kingsville Citrus Center, Weslaco
[3]University of California-Davis, Salinas
[4]US Department of Agriculture-Agricultural Research Service, Maricopa
USA

1. Introduction

The banning of methyl bromide (MeBr) as a pre-plant soil fumigant due to its implication as an ozone depleting substance, has led to increased interest in finding alternative soil fumigants to replace MeBr (United States Environmental Protection Agency [USEPA], 2009). One of the promising alternatives for certain crops is methyl isothiocyanate (MITC). Several MITC generating compounds, such as metam sodium®, metam potassium®, and dazomet® are being used to control a wide variety of fungal pathogens, weeds, and nematodes in soils. The physiochemical characteristics of MITC are significantly different than that of MeBr, such as that its effectiveness in regards to dissipation and movement in the soil is altered by multiple factors, such as soil type, texture, and soil moisture content. The largest challenge to soil fumigation is the prevention of fumigant loss to the atmosphere and especially to the nearby communities and homes adjacent to farm land. Rapid off-gassing or non-target release of the fumigant to the atmosphere can lead to poor pesticide performance and ineffective pest control. To combat this problem that is common to all soil fumigants currently on the market, various methods have been employed to reduce chemical off-gassing. A few of these methods are tarping the soil surface immediately following chemical application with high density polyethylene plastic, incorporation of organic matter to the soil surface to absorb the fumigant, or altering chemical formulations. Another method of reducing fumigant loss can be applying a surface water application as a means of sealing the soil surface to prevent chemical volatilization. On-farm field scale studies have been performed to evaluate all of these methods to better evaluate the potential for reducing fumigant loss to the atmosphere. However, field-scale studies are expensive to perform, and experimental error is challenging to control and replicate due to diurnal temperature fluctuations, varying soil physical properties, and air current differences. Thus, the volatilization loss in one study will not represent the typical fumigant loss from site to site. A more controlled laboratory environment is needed to more adequately predict fumigant loss under specific conditions. Laboratory-scale columns can be used to study soil fumigant release from soils under a wide array of conditions and under controlled circumstances. The aim of this study was to evaluate the amount of water applied to the surface of a specific soil type to reduce MITC volatilization in soil columns. Furthermore, evaluating the impact of various soil physical properties have on MITC loss is important, such as varied soil type, soil bulk density, organic matter additions and various MITC generating compound formulas have on MITC loss and mitigation. In short, the results of these studies will summarize the effectiveness of the use of soil columns to adequately assess MITC loss at the laboratory scale as a tool to predict chemical fate prior to the expense of large-scale on-farm studies.

1.1 History of fumigants

Soil fumigants are commonly used in high-value horticultural crop production to control soil originating pests such as plant-parasitic nematodes, soil-borne pathogens, insects and weeds. The intrinsic volatility of a fumigant is essential for a chemical to disperse laterally and vertically throughout the soil profile in order to control soil-borne diseases. Fumigants are typically applied via shank/chisel injection directly into the soil. After being applied, the fumigants quickly change into a gaseous phase whereby it is dispersed within the soil and results in pest control. Many compounds are classified as soil fumigants, with various rates of efficacy, with MeBr considered the most effective broad-spectrum pest control fumigant due to its high efficacy level. MeBr was one of the most widely used soil fumigants until, under the Montreal Protocol; it was officially phased out in 2005 as an ozone depleting compound (USEPA, 2009). MeBr is still used in developing countries, but must be phased out by 2015 (United States Government Printing Office [USGPO], 2005).

In effort to meet the challenge to find a suitable replacement for MeBr that has similar efficacy capabilities for crop protection a concentrated effort of research and funding has occurred. Although these studies on alternative fumigants to MeBr have been occurring for approximately two decades, there is still no fumigant replacement as effective in almost all soil types like MeBr. Currently there are still several instances where MeBr can be used; such as critical use exemptions (CUE), quarantine and pre-shipment (QPS), and emergency exemption (EE). However, these uses are highly restricted and subjected to strict regulation. In recent years, there has been a movement to find alternatives to MeBr that are as effective but less harmful to the atmosphere and environment. While this has proved a formidable challenge to scientists, there are several fumigants used in agriculture today that are effective under specific soil and cropping conditions. Table 1 shows the five most used soil fumigants in the United States.

1.2 Methyl bromide

MeBr has been the most effective soil fumigant for most soil borne pathogens and pests since it was introduced as a pesticide in 1932. Due to its harmful effects on the atmosphere as an ozone depletor, MeBr production has been phased out in most developed and developing countries in accordance with the Montreal Protocol (USEPA, 2009). MeBr can still be used under critical use and emergency exemptions but its use is strictly regulated by state and governmental agencies.

MeBr is a volatile gas at room temperature and 1 atm pressure and can be produced commercially or by plants and algae (National Pesticide Information Center [NPIC], 2000).

MeBr is a odorless, gaseous chemical above 4°C that is highly toxic to humans and vertebrate animals that can result in death under acute exposure. Thus, commercial formulations of MeBr include a certain percent of chloropicrin (tear-gas) added to act as a warning agent to indicate presence of MeBr to prevent overexposure. MeBr is applied under pressure as a liquid using shank injection into the soil, usually in conjunction with covering the soil with plastic tarps to suppress and prevent volatilization loss of the gas to the atmosphere (Papiernik et al., 2001; Wang et al., 1997). The gas then diffuses through soil pores and cracks and allows for control of soil borne pests and pathogens.

Rank	Fumigant	Formulations	Application	Amount Used per Year
1	Metam sodium/ Metam potassium	Liquid, soluble concentrate	Shank injection, chemigation	51-55 million lbs/ 1-2 million lbs (2002)
2	Methyl bromide	Pressurized gas	Shank injection, hot gas	14.76 million lbs (2007)*
3	Chloropicrin	Liquid, pressurized gas, pressurized liquid, emulsifiable concentrate	Shank injection, drip irrigation	10 million lbs (2007)
4	1,3- Dichloropropene	Liquid	Soil injection, deep drip irrigation	40,420 lbs (1998 estimate)
5	Dazomet	Granule, pellet, liquid, water soluble solids	Spreader	15,000 lbs (2003)

*Critical use exemption and emergency exemption usage
(USEPA, 2009).

Table 1. Top five most commonly used soil fumigants in the United States.

1.3 Methyl bromide alternatives

While no fumigant has proven as effective as MeBr for the control of soil-borne pests and pathogens, the reasons why the four most widely used fumigant alternatives are currently in use today are discussed below.

1.3.1 Metam sodium and metam potassium

Metam sodium (MS) is among the most widely used soil fumigant available for use (USEPA, 2008b; Sullivan et al., 2004). MS and metam potassium (MK) are broad-spectrum soil fumigants and are also used in sewers, drains, and ponds to control weeds and roots (USEPA, 2008b). MS is a sodium salt formulation of methyldithiocarbamate which breaks down into the active ingredient methyl isothiocyanate (MITC) when injected into the soil. MK is a potassium salt of N-methyldithiocarbamate and breaks down into MITC similarly to MS. MITC is a volatile gas used for soil borne pest control, it is mobile and water soluble. While it has minimal effects on impacting ozone, it does have potential as a groundwater contaminant (El Hadiri et al., 2003). Because of its relative ease in water solubility it can be used in chemigation applications and it leaves no residue on food crops (Noling and Becker, 1994). MS and MK are applied via shank injection and chemigation into the soil as a liquid.

1.3.2 Chloropicrin

Chloropicrin (trichloronitromethane) is a common fumigant used to control fungi, insects and nematodes. It is used as a pre-plant soil fumigant, warning agent and in wood treatment (USEPA, 2008a). It is a volatile gas that does not have a significant impact on ozone depletion. However it does have the potential to be a groundwater contaminant. Chloropicrin is also commonly mixed with another fumigant to increase the fumigants effectiveness (Shaw & Larson, 1999). It is shank injected into the soil or can be applied via chemigation.

1.3.3 1,3–Dichloropropene

1,3-dichloropropene (1,3-D) is a volatile gas used for the control of nematodes, fungi, insects and weeds (USEPA, 1998). 1,3-D is commonly applied as a pre-plant soil fumigant for many crops. It is considered by many to be one of the more important soil fumigant replacements for MeBr (Noling & Becker, 1994). It is typically shank injected into the soil, after which a soil sealing method is required to prevent off-gassing. 1,3-D is mobile and persistent and has the potential for groundwater contamination (USEPA, 1998). It has been estimated that 1,3-D emission loss to the atmosphere can range from 30 to 60% of the total amount applied to the soil (Gan et al., 1998a, 1998b; Gan et al., 2000b)

1.3.4 Dazomet

Dazomet is another MITC generating compound used in pathogen control. It is a broad spectrum soil fumigant used in controlling weeds, nematodes and fungi. It also has applications as a material preservative, as a biocide, and in wood treatment. It is most commonly sold and is applied in a granular form through spreaders.

1.4 Preventing emmissions of soil fumigants

Common methods used to reduce fumigant emission loss (off-gassing) to the atmosphere include using polyethylene (PE) tarps, other improved plastic barrier films, use of clear PE films for soil solarization (Chase et al., 1998; Gamliel et al., 1997; Nelson et al., 2000), soil amendment additions, drip application (Ajwa et al., 2002; Schneider et al., 1995), and surface water sealing. The on-farm fumigant emission reduction practice most readily used is the covering of the soil with PE plastic films. Emission of MeBr can still be extensive regardless of PE film use, therefore, improved formulations of high density polyethylene (HDFE) films or 'virtually impermeable films (VIF) that have lower permeability to MeBr have been investigated and used at the farm level (Wang et al., 1997). Many of these films are of high cost and limit their use in commercial production for crops that do not supply a high economic return to the grower. Various chemical additions have also been used in film formulations that may further suppress the volatilization loss of fumigants through PE, HDFE and VIFs.

Clear PE films have been used in locations such as Florida to suppress noxious weeds, such as purple and yellow nutsedge, and nematode populations. This practice of using clear plastic films can create a natural greenhouse effect and heating the upper soil rooting depth to temperatures that kill soil-borne pests, nematodes, or burns the foliage of weeds, but the pest control efficacy of this practice is limited and unpredictable making it an unreliable cultural practice for most growers (Chase et al., 1998). Incorporation of organic matter or fertilizer amendments into the soil surface in concert with PE film use have also been employed to lower fumigant emissions. Enhanced degradation of the fumigant 1,3-D have been observed after soil incorporation of organic matter (Dungan et al., 2001) and ammonium thiosulfate by chemical reactions with 1,3-D (Wang et al., 2001; Gan et al., 2000a).

Drip fumigation integrates the use of soil fumigant chemical application within drip irrigation lines. To achieve success, drip fumigation requires that the fumigant is diluted in water below its solubility or carried in conjunction with an emulsifier and dispersed throughout the rooting depth of crops by water through the dripline. In crops and soils where drip irrigation lines are utilized, drip fumigation has the potential to use lower fumigant rates than shank injection (Ajwa et al., 2002; Gan et al., 1998b), while reducing the amount of labor needed to apply the fumigant where drip lines are pre-installed (Schneider et al., 1995).

Another form of soil surface sealing is the application of water to act as a barrier to soil fumigants volatilization from the soil surface (Gan et al., 1998a, 1998b). Soil surface sealing with water application is used to change the chemical exposure within the soil being fumigated. Additional water can prolong the amount of time that MITC remains exposed to soil-borne pathogens, extending the efficacy of the chemical. There have been many studies that have shown reduced fumigant volatilization from the soil surface after irrigation water has been applied immediately following fumigant application. Results have been promising for lowering fumigant off-gassing whether the water was applied in a single event or in an intermittent method following soil fumigant application. The use of water seals is impractical for many of the highly volatile, low water soluble fumigants, such as MeBr and chloropicrin. These compounds will typically escape too quickly from the soil surface as they rapidly convert from the liquid to gaseous phase after application. Water seals are generally applied via overhead sprinkler systems, which do not apply water fast enough to prevent the gaseous fumigant's release into the atmosphere. Therefore, surface water seals typically work best for soil fumigants that have greater water solubility and will stay in solution longer before transformation into its volatile form, like MS and other MITC generating compounds (Simpson et al., 2010).

2. Field and laboratory methods

When dealing with volatile chemicals such as soil fumigants, both laboratory and field scale experiments are needed to estimate and measure off-gassing in a wide variety of conditions and situations. While field scale studies are of the utmost importance, they are labor intensive, and require more time and expense in order to test experimental variations. Laboratory, bench-scale experiments can be an inexpensive, fast way to test theories and experimental methods before performing larger scale field studies (Gan et al., 2000b).

2.1 Field methods

Most fumigants are used in conjunction with tarps to seal the surface of the soil and prevent off-gassing of the chemical, thus allowing more time for the pest control properties of the fumigant to occur. For on-farm field scale water seal investigations it typically requires shank injection of soil fumigants into the soil followed by irrigation of the soil surface to create the surface water seal. A challenge for growers to implement this into practice is the fact that they must set out standing pipe in the field equipped with sprinkler heads and risers prior to soil fumigation. The conversion of the chemical into a gaseous phase generally occurs too quickly not to have this done in advance, furthermore human fumigant exposure becomes a high risk if working in the field after application. Irrigation lines in the field can restrict blanket soil fumigant applications throughout the entire site, as pipe may limit where tractors can drive. Despite these challenges, surface water seals have been accomplished at the on-farm level with promising results for fumigant suppression (Sullivan et al., 2004). A limiting factor that makes field-scale studies challenging, is that they are typically good for that site only, and seldom reflect the potential fumigant loss for other locations that have different soil types and physical characteristics. Soils are highly variable systems, and small changes in organic matter content, soil water content, temperature, bulk density, and the fraction of sand, silt and clay will alter fumigant behavior (Dungan et al., 2001).

2.2 Laboratory methods

The use of stationary, bench-scale soil columns has been shown to a reliable means of estimating the emission potential of soil fumigants under many different soil conditions and soil types (Gan et al., 2000b). Artificial soil profile conditions under a controlled environment can be created and manipulated to more quickly assess fumigant behavior under restricted conditions. In many ways, these conditions can provide data that is less costly and cumbersome than field-scale conditions, and yet give appropriate estimates of fumigant loss comparable to that observed from field trials.

The following describes the experimental conditionals and results of one such soil column study aimed at determining the proper amount of water needed to best suppress MITC release from a sandy loam soil after MS application.

2.2.1 Experimental setup

To simulate a soil profile in laboratory scale studies, stainless steel soil columns were constructed. The soil columns constructed were 60 cm high with a 10 cm I.D. as shown in Fig.1a. Gas sampling ports were installed and spaced 10 cm apart located at soil depths of 15, 25, 35, 45, and 55 cm down the length of the soil columns. All gas sampling ports were sealed with Swagelock® fittings and septa to create an air tight environment to prevent gas leaking. A sandy clay loam soil (fine-loamy, mixed, hyperthermic Typic Ochraqualfs), used to pack the soil columns, and was collected from an area not previously exposed to soil fumigants. Soil was air dried and sieved to 2.0 mm, then brought to 8% moisture with distilled water. Each column was packed to a bulk density of 1.5 g cm^{-3}. A headspace sampling chamber was attached to the top of the soil column in order to collect gas samples and to apply a uniform water seal through a microjet spray sprinkler attached to the inside of the chamber. The upper chamber was sealed to the lower column using aluminum air-conditioning duct construction tape to preserve an airtight chamber. To promote airflow through the chamber, two holes were drilled on opposite sides of the headspace chamber, one with access to outside airflow and the other attached to a vacuum source. Charcoal tubes were connected to the ends of each port to act as filters to collect any volatile MITC that was released during the study. The vacuum airflow rate was maintained at 150 +/- 10 mL min^{-1} from the 1mmHg vacuum source.

Fig. 1a. Soil columns with charcoal filters.   Fig. 1b. MS injection at 15 cm soil depth.

MS was applied to the soil columns via simulated soil drip fumigation by injecting the fumigant in the center of the soil through a side port located 10 cm below the soil surface (Fig. 1b). The MS was applied at a rate of 420 g L^{-1} EC (Vapam® 42; Amvac Chemical Corp., Los Angeles, CA) with 112 mL of distilled water, thus MS was diluted in water sufficient to simulate a 1.3 cm chemigation event. The equivalent amount of MITC applied to each column was 121.2 mg. Additional water application through the microjet spray sprinkler located inside the top of the soil column cap to simulate water seals of 0, 1.3, 2.5 and 3.8 cm applied to the soil surface and this was performed immediately following the injection of MS in order to prevent chemical off-gassing. Each treatment was replicated in triplicate for statistical analysis.

2.2.2 Chemical analysis

Analysis of MITC can be done in many ways. Gas chromatography (GC) with flame ionization detector (FID) was used in this research, but other detectors such as electron capture detectors and nitrogen phosphorus detectors can be used in MITC analysis for greater sensitivity.

After MS was applied to each column, air samples were taken at predetermined times from the side ports along each column. MITC concentrations within the soil air space were determined by filling a gas-tight syringe with 250 μL of air and injecting it into the GC-FID. The charcoal filters attached to the columns were sealed and replaced every 4 to 8 hours

(Fig. 2a) to ensure that no MITC was escaping undetected. These filters were then frozen until analyzed. To determine the amount of MITC volatilized from the soil surface, each glass charcoal filter tube was broken and the charcoal dispensed into 10-mL headspace sampling vials (Fig. 2b).

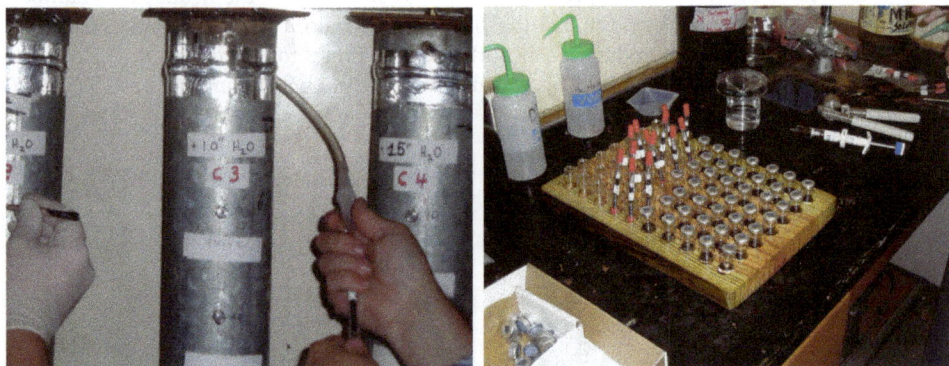

Fig. 2a. Charcoal filters replaced periodically.   Fig. 2b. Charcoal filter extracted into vials.

Afterwards, 5 mL of organic solvent (methanol) was used to extract the MITC off the charcoal, the vials were immediately cap sealed, then shaken (Fig. 3a) overnight in the dark, as it was determined in a preliminary trial that 12 h was sufficient time to extract over 99% of all MITC from the charcoal. Charcoal was placed on the counter for 2 h to allow it settle to the bottom of the vial. 1-mL of the solvent supernatant was then extracted and transferred to 2-mL GC vials (Fig. 3b), and a GC syringe was used to extract the solvent from small GC vials (Fig. 3c) followed by injection into the GC for MITC analysis by FID (Fig. 3d).

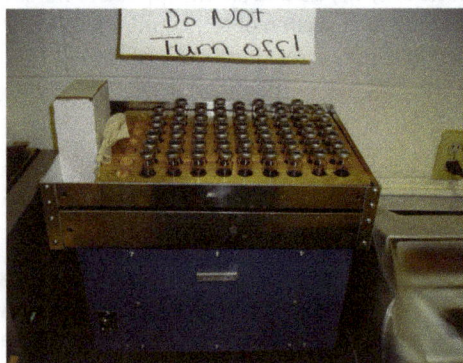

Fig. 3a. Vials shaken to extract MITC.   Fig. 3b. Transfer of solvent to small GC vial.

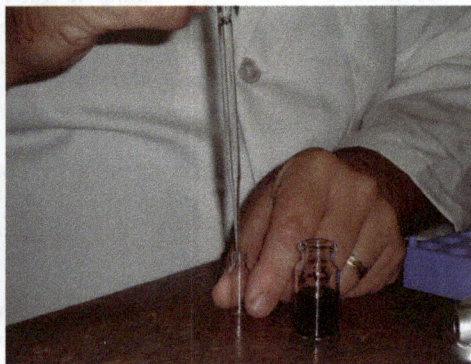

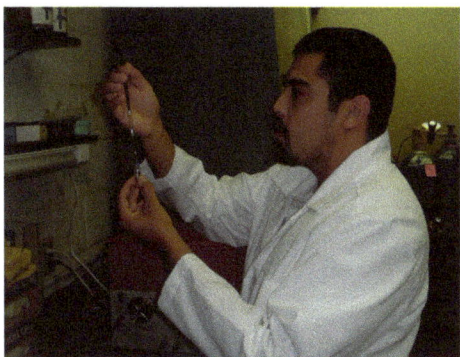

Fig. 3c. Syringe extraction of solvent.

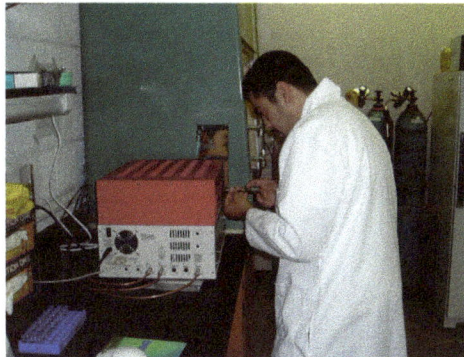

Fig. 3d. Injection of solvent into GC for analysis.

3. Results of water seal column study

The movement of MS within soil systems can be described in regards to its partitioning from the liquid phase into the gaseous phase after transformation to MITC. For analytical simplicity, only MITC was within the gaseous phase was analyzed during this study, although MITC does partition in water as well. The amount of MITC volatilized was monitored over time after MS chemical injection in two parts: 1) the soil-air movement of MITC within the soil column profile, and 2) the flux of MITC evolved from the soil surface.

3.1 Soil air movement of MITC

The distribution of MITC within the soil-air space within the soil profile was measured at periodic times, but only data from 0.3, 1, 2, 3 and 5 days after treatment (DAT) are displayed here for simplicity (Fig. 4a-d). As expected, the soil columns that did not receive additional water to the soil surface (0-cm water seal) had rapid release of the fumigant after application (Fig. 4a) because of a lack of a barrier film of water to restrict MITC volatilization. This is evident by the bulk of MITC located at the 20 cm soil depth within hours after application (0.3 DAT). Although the MS was applied at the 10-cm injection port, the bulk of the chemical moved down the soil profile as apparent by the bulk MITC concentration located at the 20 cm soil depth 0.3 DAT. This was due to the total initial amount of water applied with the diluted MS solution and a lower fumigant amount near the 10 cm soil depth of the column. The highest level of MITC was observed 1.0 DAT at the 10 cm soil depth, indicating that the majority of the fumigant was moving upward throughout the soil column. Thereafter the amount of MITC within the soil-air phase progressively decreased each DAT (Fig. 4a).

Similar to the 0-cm water seal treatment, the 1.3-cm water seal treatment had MITC distributed in a like manner, with the highest amount of MITC observed at the 10 cm sampling depth 1.0 DAT (Fig. 4b). However, the concentration level of MITC observed within the soil profile was higher than that of the 0-cm water seal treatment at sampling times after chemical application. This indicates that although the water seal amount was low (1.3 cm) it is sufficient to restrict and delay the volatilization loss of MITC. This is apparent by MITC levels 2 to 3 times greater within the soil-air phase 2.0 and 3.0 DAT at the upper 30 cm soil sampling depths when compared to the no water seal treatment (Fig. 4b).

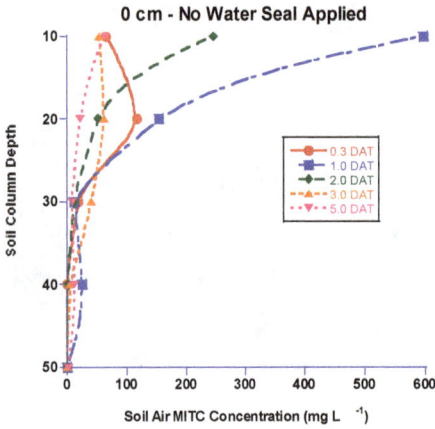

Fig. 4a.

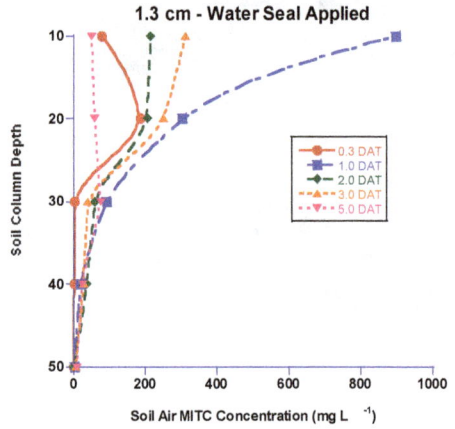

Fig. 4b.

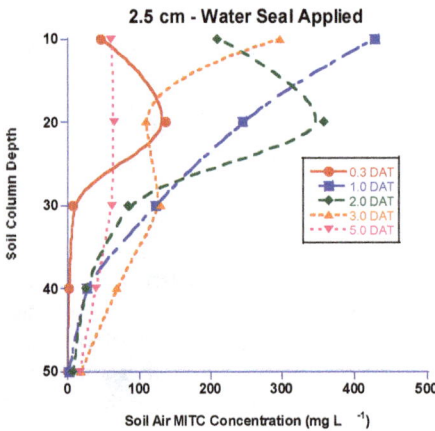

Fig. 4c.

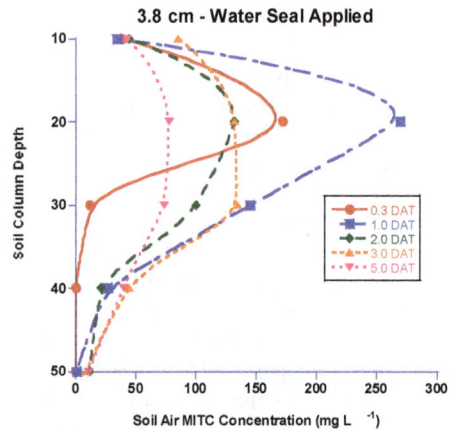

Fig. 4d.

Fig. 4. MITC distribution in soil airspace throughout the soil profile of columns over time (DAT=days after treatment, or after injection of metam sodium at 10-cm column depth); data shown represent the mean of three replications for each water seal treatment [0-cm (4a), 1.3-cm (4b), 2.5-cm (4c), and 3.8-cm (4d) water seal application depth].

The real impact of the water seal treatment at suppressing MITC volatilization was observed in the 2.5-cm water seal treatment (Fig. 4c). This is especially apparent when looking at the level of MITC over time at the 10 cm soil depth. The concentration of MITC at the 10 cm soil depth was lower 1.0 DAT for the 2.5-cm than the 0-cm and 1.3-cm water seal treatments (Fig. 4a-c), suggesting a restriction in the volatilization loss of MITC through the soil surface. Furthermore, the bulk amount of MITC resided at the 20 cm soil depth for a longer period of time after chemical application when compared to the lower water seal treatments, allowing the MITC to distribute vertically throughout the soil column with MITC concentrations observed at the 50 cm by 5.0 DAT (Fig. 4c).

Application of a 3.8-cm water seal resulted in the longest retention of MITC within the soil profile, along with the greatest suppression of MITC from the soil surface as evident by low MITC soil-air phase levels at the 10 cm soil depth up to 5.0 DAT (Fig. 4d). The extra water applied to the soil surface in the 3.8-cm treatment moved the MS further down the soil profile resulting in high MITC concentrations at both the 20 and 30 cm soil depths 1.0 to 5.0 DAT. The higher water amount within the soil profile was confirmed at the end of the study as soil moisture levels were higher at the 25 cm soil depth of the 3.8-cm than the 2.5-cm water seal treatments (data not shown).

3.2 Soil surface flux of MITC

The highest amount of MITC volatilized through the soil surface was observed from soil columns with no (0-cm) water seal applied after MS application (Fig. 5). The greatest MITC flux was observed within the initial 36 h after chemical application and decreased over time thereafter. A similar trend was observed for the 1.3-cm water seal treatment, but the amount of MITC evolved was substantially less than that from the 0-cm treatment. The lowest amount of MITC flux observed occurred from both the 2.5-cm and 3.8-cm water seal treatments, with the 2.5-cm treatment releasing slightly more MITC by 120 h after MS application.

Mean MITC Volatilization

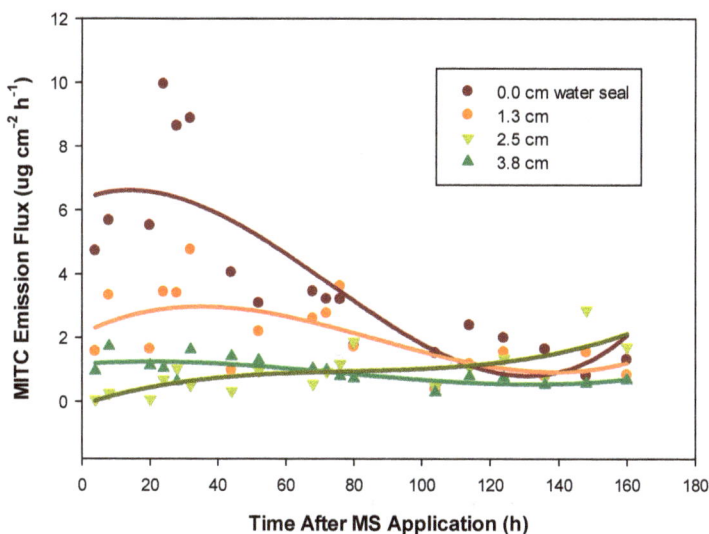

Fig. 5. The mount of MITC volatilized and captured on charcoal filters over time. Data represents mean of three replicates per water seal treatment.

In order to determine to total amount of MITC volatilized from the surface of the soil columns, cumulative MITC levels were calculated and plotted (Fig. 6). In this respect it is easily apparent that the highest MITC emissions occurred from soil columns without a water seal treatment. But more importantly, for soil columns that received additional surface water irrigation, total MITC volatilization decreased with increasing water seal depth (Fig. 6a). The total mean MITC volatilization loss from the 0-, 1.3-, 2.5- and 3.8-cm water seal treatments was respectively 24, 14, 9 and 6% of the total initial MITC applied (Fig. 6b). The highest variability in MITC loss was observed in the low to no water seal treatments, suggesting that neither of these treatments would be acceptable for suppressing MITC fumigant loss from soils. Whereas a low amount of variability (small error bars) was observed for the higher water seal treatments, with no statistical difference in total MITC loss (Fig. 6b).

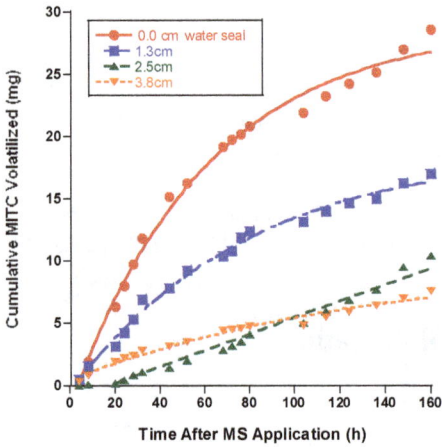

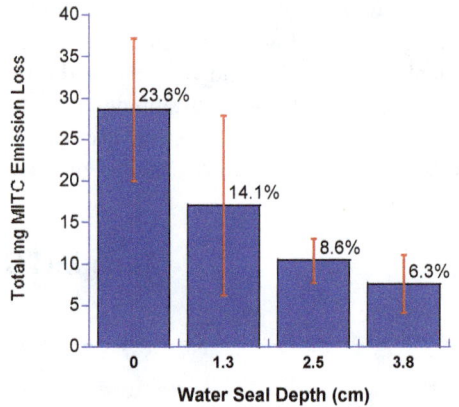

Fig. 6a. Mean MITC emitted from soil columns.   Fig. 6b. MITC release ± std error mean.

Fig. 6. Total cumulative MITC evolved from soil columns as captured on charcoal filters.

4. Conclusion

These findings illustrate how effective bench-scale soil column studies are at assessing the volatilization potential of MS after varying surface water seal treatments. Keeping in mind that this study represents specific and restricted conditions, it does provide a good estimate of the proper water seal depth needed for a sandy clay loam soil type. Although a 3.8-cm water seal led to the least amount of fumigant loss, it is recommended that a 2.5 -cm water seal be applied in the field for similar soil types. This is suggested due to the fact that applying large amounts of water can significantly alter chemical behavior by further diluting the MS to a level below the critical threshold for MITC to be effective for pest control. Furthermore, in areas where water tables are high, adding too much water via supplemental overhead irrigation may lead to groundwater contamination and result in other environmental concerns. The 2.5-cm water seal application suppressed MITC volatilization to level statistically equivalent to that of the 3.8-cm treatments and therefore, it is a good practice to reduce fumigant emissions to the atmosphere while minimizing excessive chemical movement beyond the crop rooting depth. On-farm field investigations will be needed to back up these laboratory scale findings to provide confirmation that the suppressive loss of MITC is ultimately achievable.

5. Acknowledgement

The authors would like to thank the financial support of the US Department of Agriculture-CSREES Methyl Bromide Transition Program, grant award # 2002-51102-1922, and USDA-CSREES Hispanic Serving Institution program, grant award # 2004-38422-14608, and the dedicated work by student research aides like Mr. LeeRoy Rock (shown in Fig. 3c, 3d with permission) for the long hours required for data collection.

6. References

Ajwa, H. A.; Trout, T.; Mueller, J.; Wilhelm, S.; Nelson, S. D.; Soppe, R. & Shatley, D. (2002). Application of alternative fumigants through drip irrigation systems, *Phytopathology*, Vol.92, pp. 1349-1355.

Chase, C.A.; Gilreath, J.P.; Locascio, S.J.; Sinclair, T. R.; & Shilling, D. G. (1998). Light effects on rhizome morphogenesis in nutsedges (Cyperus spp.): implications for control by soil solarization, *Weed Science*, Vol. 46, No. 5, pp. 575-580.

Dungan, R. S.; Gan, J., & Yates S. R. (2001). Effect of temperature, organic amendment rate and moisture content on the degradation of 1,3dichloropropene in soil, *Pest Management Science*, Vol.57, pp. 1107-1113.

El Hadiri N.; Ammati, M.; Chgoura, M. & Mounir, K. (2003). Behavior of 1,3-dichloropropene and methyl isothiocyanate in undistubed soil columns. *Chemosphere*, Vol. 52, pp. 893-899.

Gamliel, A.; Grinstein, A. & Katan, J. (1997). Improved technologies to reduce emission of methyl bromide from fumigated soil, *Phytoparasitica*, Vol. 25, pp. 21S-30S.

Gan, J.; Yates, S. R.; Crowley, D. & Becker, J. O. (1998a). Acceleration of 1,3dichloropropene degradation by organic amendments and potential application for emissions reduction, *Journal of Environmental Quality*, Vol.27, pp. 408-414.

Gan, J.; Yates, S. R. Wang, D. & Ernst, F. F. 1998b. Effect of application methods on 1,3-dichloropropene volatilization from soil under controlled conditions, *Journal of Enviromental. Quality*, Vol.27, pp. 432-438.

Gan, J.; Becker, J. O.; Ernst, F. F.; Hutchinson, C.; Knuteson, J. A. & Yates, S. R. (2000a). Surface application of ammonium thiosulfate fertilizer to reduce volatilitzation of 1,3-dichloropropene from soil, *Pest Management Science*, Vol.56, pp. 264-270.

Gan, J.; Hutchinson, C.; Ernst, F. F.; Becker, J. O. & Yates, S. R. (2000b). Column system for concurrent assessment of emission potential and pest control of soil fumigants, *Journal of Environmental Quality*, Vol.29, pp. 657-661.

Gan, J.; Yates, S. R.; Knuteson, J. A. & Becker, J. O. (2000c). Transformation of 1,3-dichloropropene in soil by thiosulfate fertilizers, *Journal of Envionmental Quality*, Vol.29, pp. 1476-1481.

National Pesticide Information Center. (2000). Methyl bromide, *National Pesticide Telecommunications Network Technical Fact Sheet*. 26.06.2011. Available from http://npic.orst.edu/factsheets/MBtech.pdf

Nelson, S.D.; Allen, L.H. Jr.; Gan, J.; Riegel, C.; Dickson, D.W. & Locascio, S. J. (2000) Can virtually impermeable films reduce the amount of fumigant required for pest-pathogen management in high value crops? *Proceedings of Soil and Crop Science Society of Florida*, Vol. 59 pp. 85-89.

Noling, J. W. & Becker, J. O. (1994). The challenge of research and extension to define and implement alternatives to methyl bromide, *Journal of Nematology Supplemental,* Vol.26, pp. 573-586.

Papiernik, S. K.; Yates, S. R. & Gan, J. (2001). An approach for estimating the permeablility of agricultural films, *Environmental Science and Technology*, Vol.35, pp. 1240-1246.

Schneider, R. C.; Green, R. E.; Wolt, J. D.; Loh, R. K. H.; Schmitt, D. P. & Sipes, B. S. (1995). 1,3-dichloropropene distribution in soil when applied by drip irrigation of injection in pineapple culture. *Pesticide Science*, Vol.43, pp. 97-105.

Shaw, D.V. & Larson, K. D. (1999). A meta-analysis of strawberry yield response to preplant soil fumigation with combinations of methyl bromide-chloropicrin systems. *HortScience*, Vol 34, No. 5, pp. 839-845.

Simpson, C.R.; Nelson, S.D.; Stratmann, J.E. & Ajwa, H.A. (2010). Surface water seal application to minimize volatilization loss of methyl isothiocyanate from soil columns, *Pest Management Science*, Vol. 66, pp. 686-692.

Sullivan, D. A.; Holdsworth, M. T. & Hlinka, D. J. (2004). Control of off-gassing rates of methyl isothiocyanate from the application of metam-sodium by chemigation and shank injection, *Atmospheric Environment*, Vol.38, pp. 2457-2470.

United States Government Printing Office. (2005). Final rule on continued production of methyl bromide for export to developing countries, In: *Federal Register*, Vol. 70, No. 249, pp. 1-7.

U.S. Environmental Protection Agency. (1998). RED fact sheet: 1,3-dichloropropene, In: *US Environmental Protection Agency Office of Pesticide Programs*. 26.06.2011. Available from http://www.epa.gov/pesticides/reregistration/REDs/factsheets/0328fact.pdf

U.S. Environmental Protection Agency. (2008a). RED fact sheet: chloropicrin, In: *US Environmental Protection Agency Office of Pesticide Programs*. 26.06.2011. Available from http://www.epa.gov/pesticides/reregistration/REDs/factsheets/chloropicrin-fs.pdf

U.S. Environmental Protection Agency. (2008b). RED fact sheet: Methyldithiocarbamate salts-metam sodium/potassium and MITC, In: *US Environmental Protection Agency Office of Pesticide Programs*. 26.06.2011. Available from http://www.epa.gov/oppsrrd1/REDs/factsheets/metamsodium-fs.pdf

U.S. Environmental Protection Agency. (2009). The phaseout of methyl bromide: the phaseout schedule for methyl bromide, In: *US Environmental Protection Agency, Ozone Depletion Rules and Regulations section*. 25.06.2011. Available from http://www.epa.gov/ozone/mbr

Wang, D.; Yates, S. R.; Ernst, F. F.; Gan, J. & Jury, W. A. (1997). Reducing methyl bromide emission with a high barrier plastic film and reduced dosage, *Environmental Science and Technology* 31:3686-3691.

Wang, Q.; Gan, J.; Papiernik, S. K. & Yates, S. R. (2001). Isomeric effects on thiosulfate transformation and detoxification of 1,3-dichloropropene, *Journal of Environmental Toxicology and Chemistry*, Vol.20, pp. 960-964.

10

Wheat Midges and Thrips Information System: Decision Making in Central Germany

Nabil El-Wakeil[1,2,*], Nawal Gaafar[1,2],
Mostafa El-Wakeil[3] and Christa Volkmar[2]

[1]*Pests and Plant Protection Department, National Research Centre, Dokki, Cairo,*
[2]*Institute of Agric. and Nutritional Sciences, Martin-Luther-University Halle-Wittenberg,*
[3]*Kuwaiti Ministry of Interior – Information Systems Directorate, Dajij, Farwania,*
[1]*Egypt*
[2]*Germany*
[3]*Kuwait*

1. Introduction

Wheat Midges and Thrips Expert System (WMTES) is constructive computer software, giving the users a recommendation based on pheromone and water traps catches as well as infestation levels. These results were collected from our field experiments which conducted in three locations in central Germany (Halle, Silstedt and Salzmünde) during three years 2007 to 2009. Computer programs can help in information recovery and decision support when dealing with pest problems. These decision support tools can provide farmers with easy, rapid access to accurate information that can help them to obtain the threshold to make adequate management decisions. Plans for future field testing and expert system implementation are also discussed. Using such as expert system for controlling wheat ear insects can be successfully applied to the solution of daily problems in plant protection programs for wheat producers. Finally, the obtained results would give a good guide for decision making which proved an efficient method of integrated plant protection for wheat ear insects as well as other insects in another crop.

An expert system is a computer program, which mimics behaviour of an expert in a particular area of knowledge. Expert systems (ES) have been developed and applied in many agriculture fields i.e. diagnose insects and diseases of various crops. Farmers across the world face problems like soil erosion, increasing cost of chemical pesticides, weather damage recovery, the need to spray, mixing and application, yield loses and pest resistance. On the other hand researchers in the field of agriculture are constantly working in Pakistan on new management strategies to promote farm success (Khan et al., 2008). Pest management is a highly challenging problem. Globally, annual losses from pests and diseases had increased year after year (Sharma, 2001).

The development of an agricultural expert system requires the combined efforts of specialists from many fields of agriculture, and must be developed with the cooperation of

---

*Correspondence Author

the farmers and extension officers who will use them (Chakraborty & Chakrabarti, 2008; ESICM, 1994). Expert Systems (ES) can be used by decision makers for predictions, such as on the needs for water, fertilizers and pesticides for a particular crop in the region given the area cultivated with such a crop. This generated information is important for different users: the traders, the exporters, the importers of these materials (Rafea et al., 1993; Rafea & Shaalan, 1996). Edrees et al. (2003) performed an expert system (NEPER) for wheat production dealing with all agricultural practices. This system are verified, validated, and, tested in the wheat fields in Egypt. There are some expert systems which are used in management systems, for example for aphids in Germany as reported by Freier et al. (1996); Gonzalez-Andujar et al. (1993); Gosselke et al. (2001) and in UK as recorded by Knight et al. (1992); Mann et al. (1986). ES is dealing also with development method for insects forecasting (Jörg et al., 2007) and diseases (Räder et al., 2007) on plants to optimize control. Up to date more than 20 met-data -based forecasting models have been developed and introduced into agricultural practice in Germany (Kleinhenz & Roßberg, 2000; Kleinhenz & Zeuner, 2007; Tiedemann & Kleinhenz, 2008).

Wheat ear insects are perceived as being of major importance. As a result, international surveillance schemes have been established, aimed at providing advance warning of pest outbreaks that will allow public and private sector agencies, including farmers for performing agricultural extension services, to make appropriate preparations for insect control (Sivakami & Karthikeyan, 2009). In Europe, wheat midge and thrips are two of the most important groups of insect pests (El-Wakeil et al., 2010; Gaafar, 2010; Gaafar & Volkmar, 2010; Gaafar et al., 2009, 2011 a,b; Volkmar et al., 2008, 2009) some species cause damage directly, through feeding, and indirectly from the fungi infestation. This work is aimed at providing the decision support tools for farmers with rapid access to accurate information that can help them to obtain the threshold to make adequate control decisions.

2. Model verification study (methodology)

Model verification was done at three sites; two research fields in Halle and Silstedt and one large scale field in Salzmünde, which were selected for detailed study in 2007, 2008 and 2009. The sites were chosen to cover a range of soil types and locations representative of the infested area of central Germany, and to be cover by meteorological stations.

S. mosellana males were monitored using pheromone traps and ear samples taken to assess the ultimate level of midge larvae infestation in all sites and in 2 growth stages; flowering (GS 65) and milky (GS 73) based on Tottman (1987) scales. White water traps were used to sample the migrated midge larvae to soil. For all of these sites the highest catch of male midges in pheromone traps was recorded. A correlation analysis was used to investigate the relationship between midge catches and the ultimate level of grain damaged. Levels of wheat midge infestation were relatively correlated with low/ high throughout the monitoring methods to use in the expert system.

The observations of variability in trap catch, and how it related to subsequent infestations, were very relevant when deciding how best to use the traps for wheat midge risk assessment and were used to develop a decision support model. With this in mind it has been kept as simple and user-friendly as possibly being based on a stepwise decision tree involving yes/no answers to questions (Fig. 1).

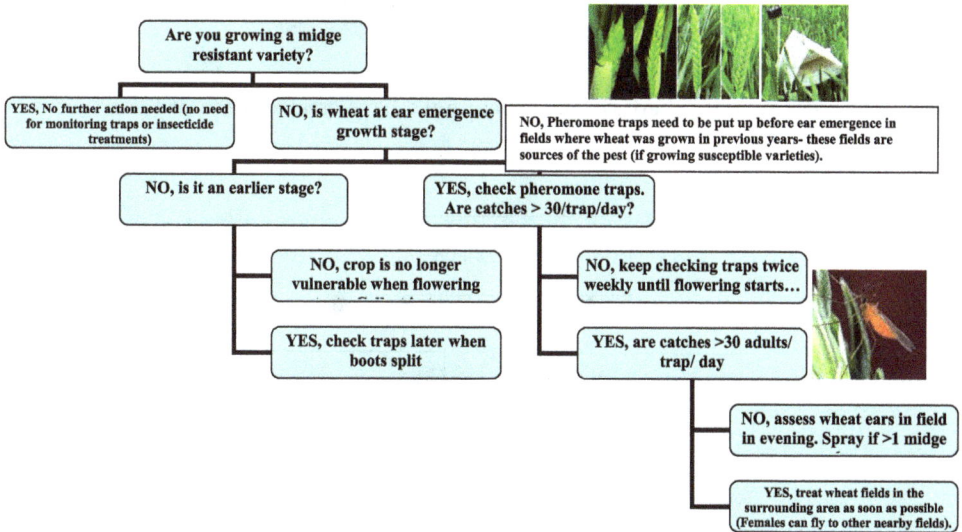

Fig. 1. Wheat midge decision support model (modified after Ellis et al., 2009).

3. Expert system development life cycle and structure

3.1 Development life cycle

The first step is creating the knowledge base and the first task in the creation of knowledge base is knowledge acquisition. Knowledge acquisition is considered as one of the most important phases in the expert system development life cycle. Knowledge acquisition is to obtain facts and rules from the domain expert so that the system can draw expert level conclusions (Gonzalez-Diaz et al., 2009). Some commonly used approaches of knowledge acquisition are interviews, observations, taking experts through case studies and rule induction by machines. Knowledge acquisition is crucial for the success of an expert system and regarded as a bottleneck in the development of an expert system (Saini et al., 2002). After the knowledge acquisition is done, the process of representing that knowledge begins. There are many approaches used for knowledge representation, for example rules, logic expressions and semantic networks. In rule-based expert systems Rules are made on the basis of the hierarchy and these rules lead to proper treatment that the user has to use.

The domain must be compact and well organized. The quality of knowledge highly influences the quality of expert system (Suo & Shi, 2008). The first step in the development of any expert system is problem identification. The problem here is a diagnostic problem aimed to identify ailments in the wheat using symptoms of insect pests. The problems occur frequently and the consequences on farmer's financial status are enormous. The demand for help is increasing rapidly. Diagnosis or diagnostic problem solving is the process of understanding what is wrong in a particular situation. Thus gathering of information and then interpreting the gathered information for determining what is wrong are of central importance in diagnostic problem solving (Lucus, 1997).

3.2 System structure

Figure (2) shows typical expert system structure we have created. Each of these blocks is explained below.

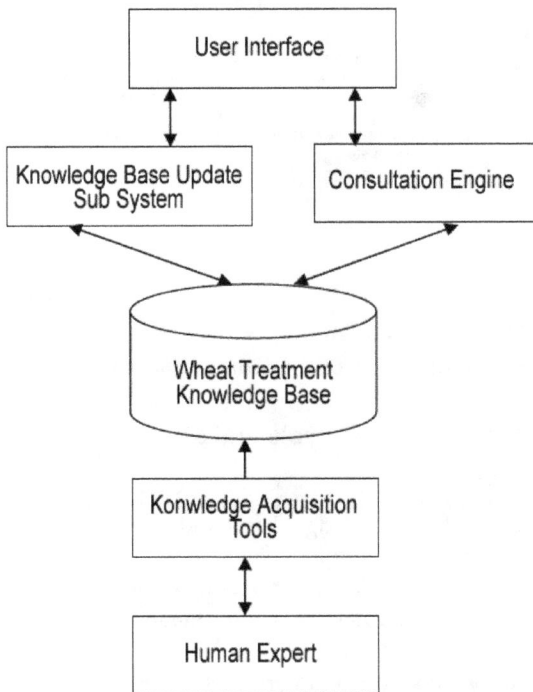

Fig. 2. Expert system structure of wheat ear insects (modified after Khan et al., 2008).

3.3 User interface

This is the interface the end user will use to interact with the system by providing parameters to it and having recommendations and consultation results out of it.

Knowledge base update sub system

As we know the utmost drawback of using expert system is that it has fixed knowledge base. If this base is not updated periodically, the results of consultation by time will be out of date. Thus, we developed this sub system to have the ability to update and enhance the knowledge base at any time easily and smoothly.

Consultation engine

Consultation Engine is the communication channel between the end user and the system; this is where user submits his consultation. Engine has 2 operation modes one is system wizard and the other is manual entry.

Wheat treatment knowledge base

This is the heart and the core of the system where it holds all the knowledge that we process to give the right decision to the user.

Knowledge acquisition tools

This is the ways we acquire knowledge from different sources and save it in the knowledge base.

Human expert

Everything in the end must return back to humans without the help of human expert we can not by any means have computerized expert system.

3.4 System user interface

Our system has 3 main modules:

1. System main data entry module
2. Knowledge base update module
3. Consultation engine module

3.4.1 System main data entry module: System main screen (fig. 3a)

Main Data Menu: (Fig. 3B)

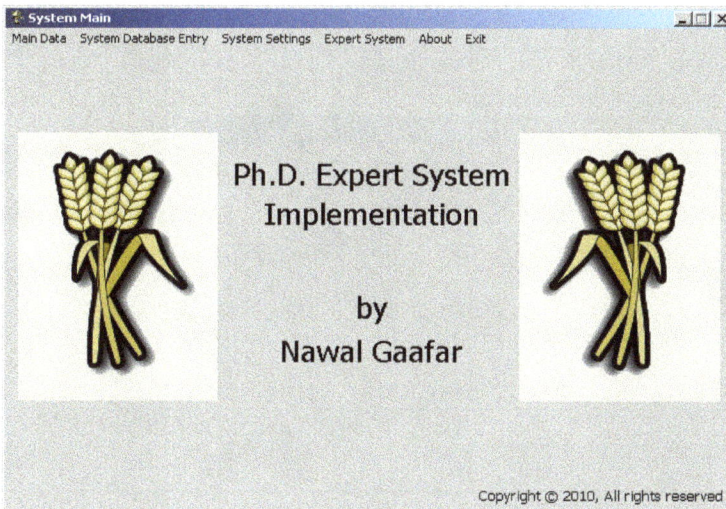

Fig. 3A. System Main Data Entry Module.

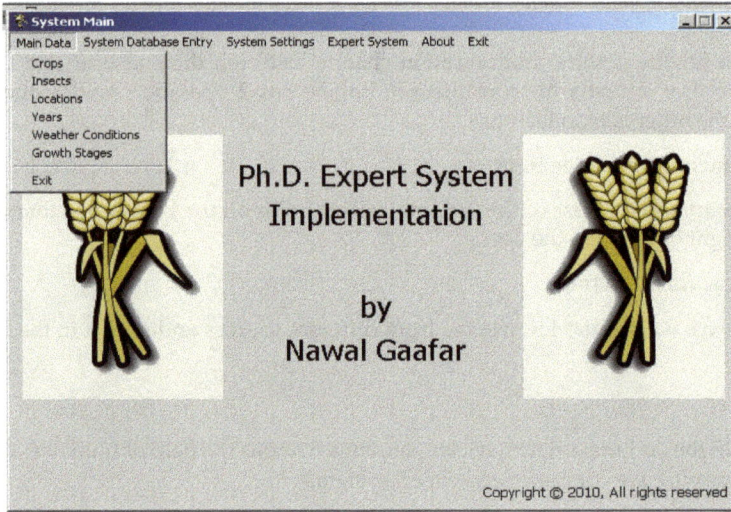

Fig. 3B. Main Data Menu.

Crops Window

Here we can add, update and delete any kind of crops that we are dealing with now or may be in need to deal with in the future (Fig. 4A).

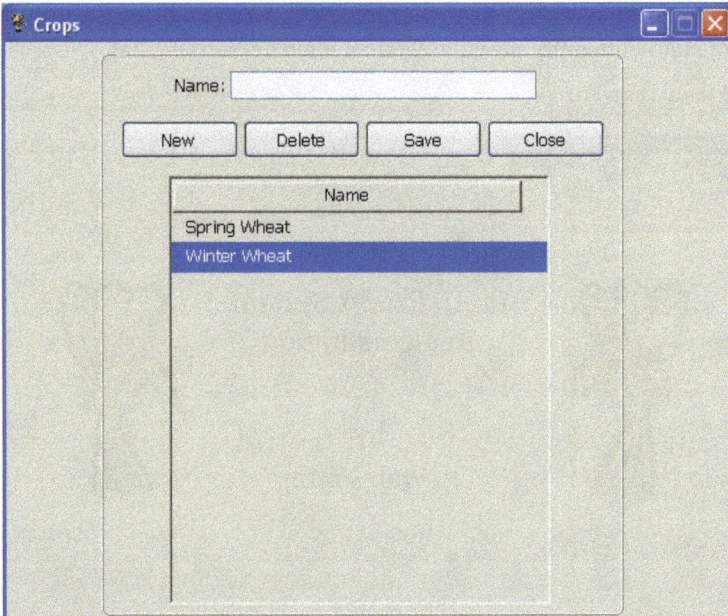

Fig. 4A. Crops window.

Insects Window

From this window we can add, update and delete any kinds of insects that we are dealing with now or may be in need to deal with in the future(Fig. 4B)

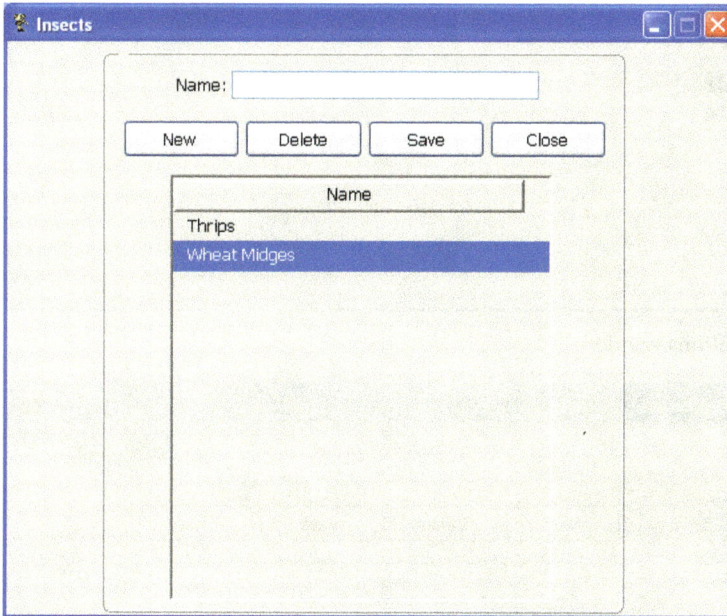

Fig. 4B. Insects window.

Locations Window

From this window we can add, update and delete any study locations that we are using now or may use in the future (Fig. 5A).

Weather Conditions Window

From this window we can add, update and delete any weather conditions that may be affected either on crop or insects (Fig. 5B).

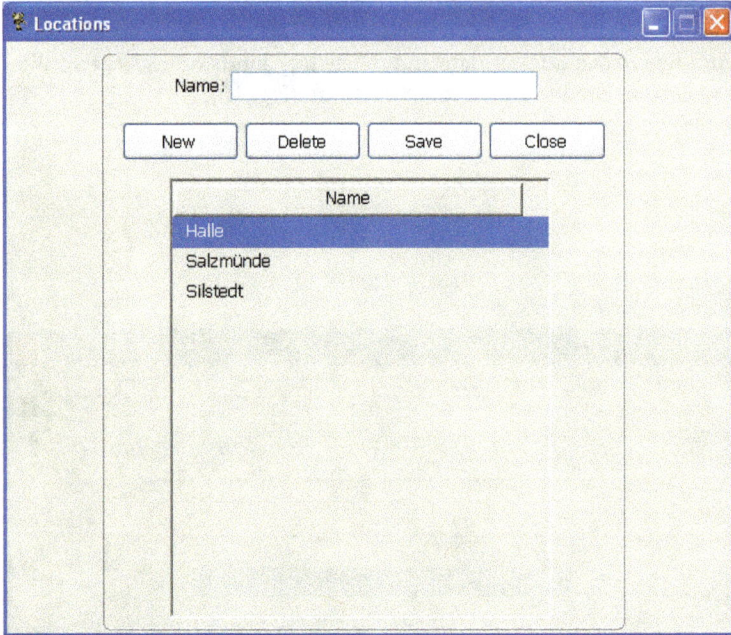

Fig. 5A. Locations window.

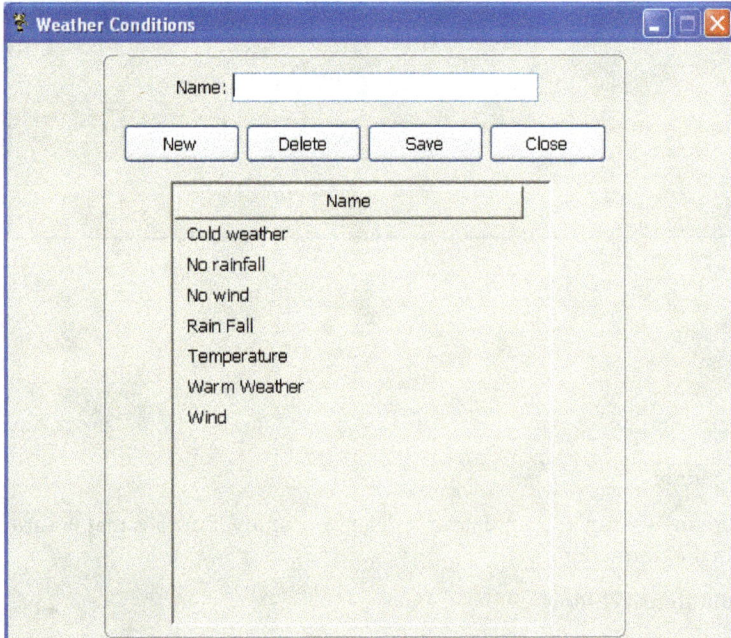

Fig. 5B. Weather conditions window.

Growth stages window

From this window we can add, update and delete any growth stages that we are interesting to study the population dynamic of insects (Fig. 6).

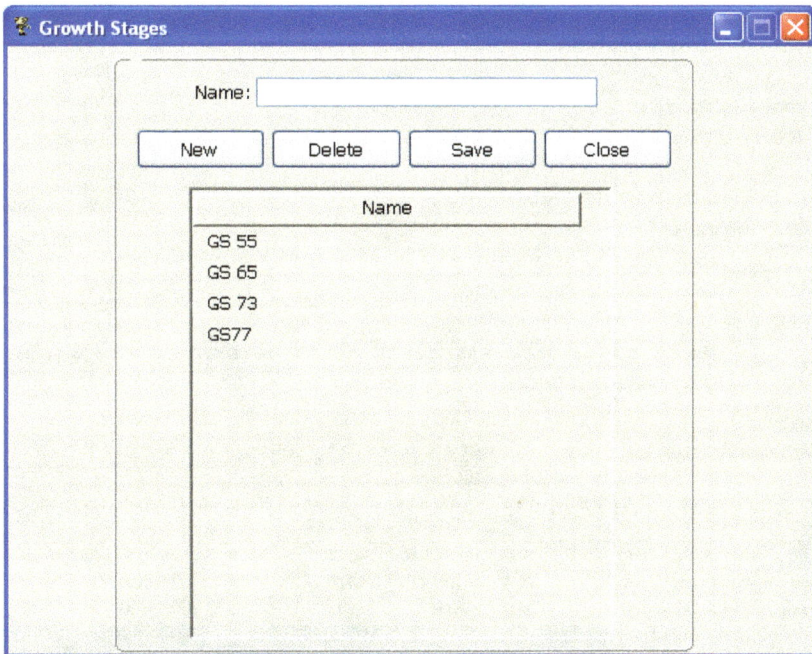

Fig. 6. Growth stages window.

3.4.2 Knowledge base update module

In this window we will be able to modify and update the knowledge base we have to be concurrent with the latest researches and results we got from different data acquisition techniques (Fig. 7).

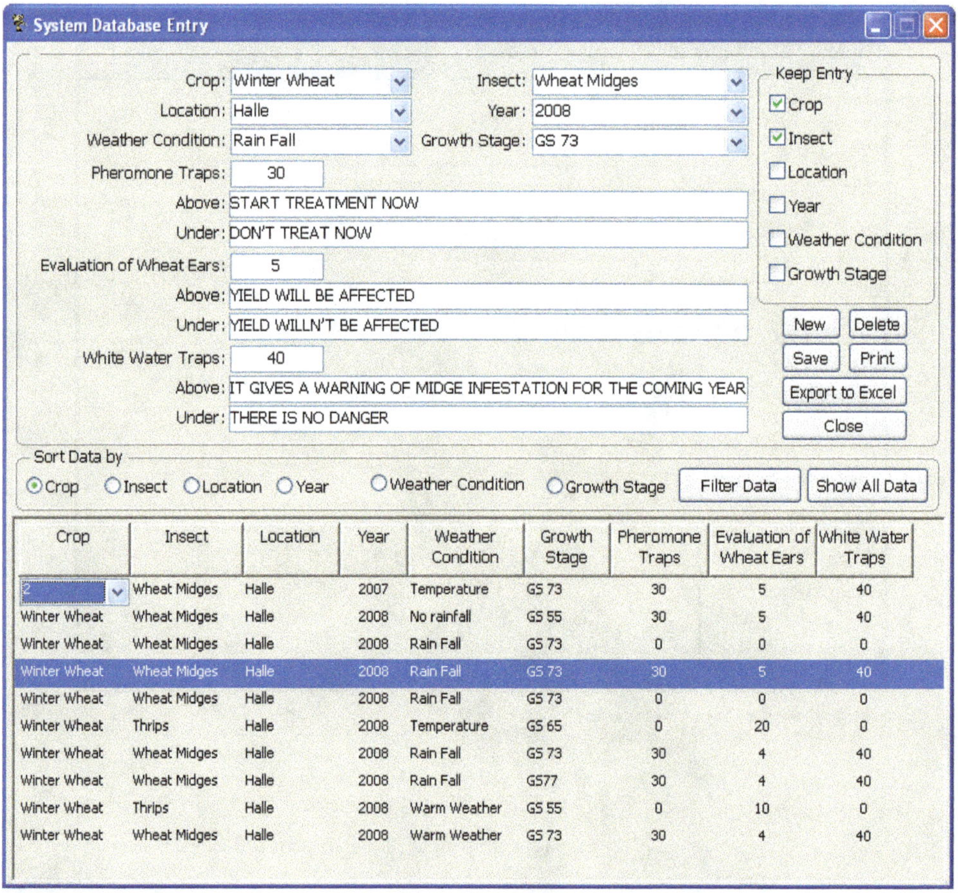

Fig. 7. Knowledge Base Update Module.

3.4.3 Consultation engine

System settings

In this window we can change the consultation engine setting by choosing the defaults of the system parameters and even saving it permanently in the system (if you click on save) or just change it for the current consultation session (if you click on apply) (Fig. 8).

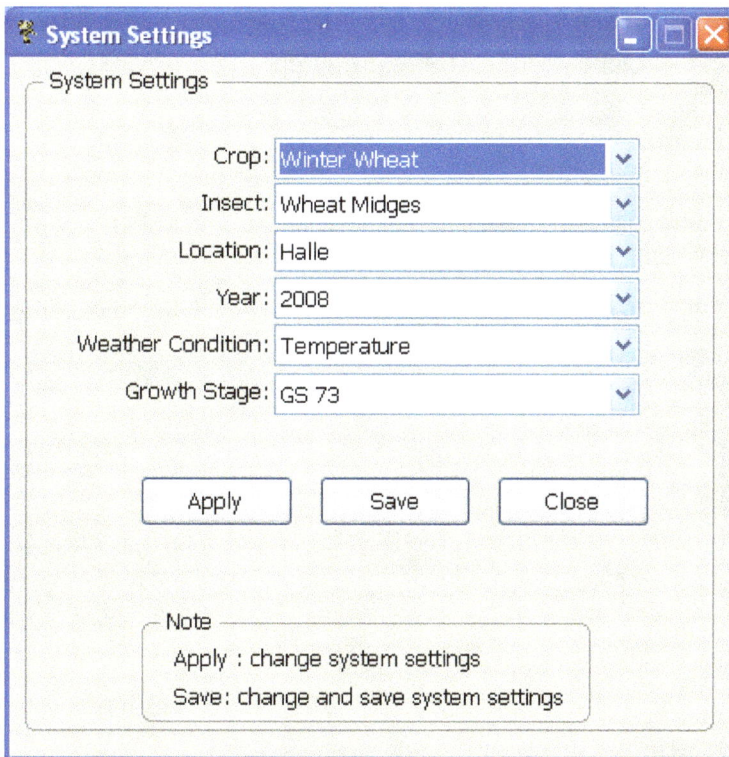

Fig. 8. System Settings.

• Consultation Engine Menu
This menu contains menu items for System Wizard, Pheromone Traps, Evaluation of Wheat Ears and Water Traps

• System Wizard
In system wizard the system will keep asking questions to select insect species, year, location, growth stage and weather conditions for getting answers from user till it has all the required information to give right decision for the user (Fig. 9).

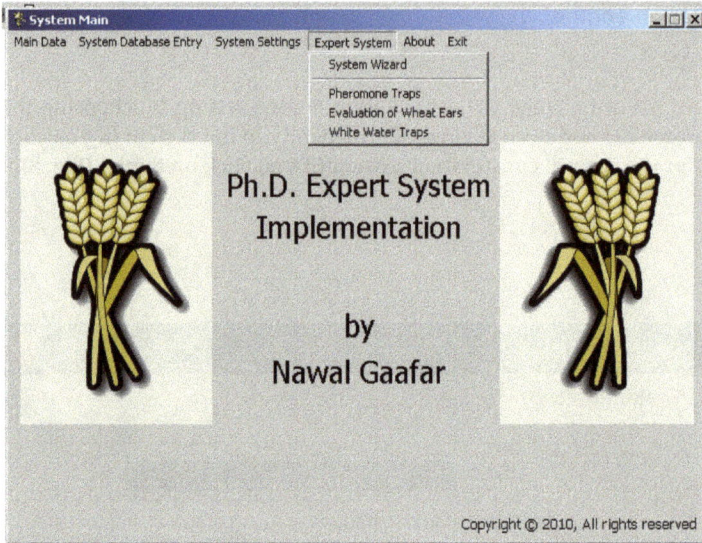

Fig. 9. System Wizard.

Summary window after gathering all the required information from the user (Fig. 10A)

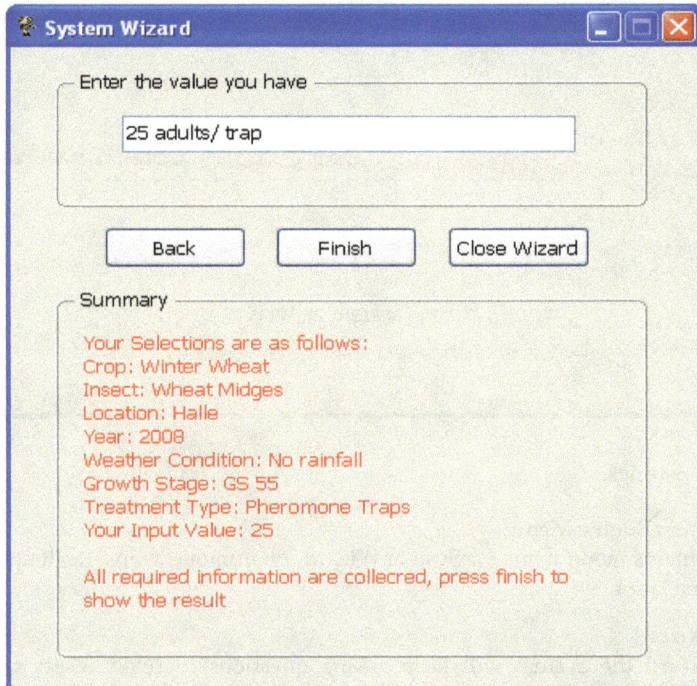

Fig. 10A. All the required information.

This is an example for the consultation result out of the system (Fig. 10B).

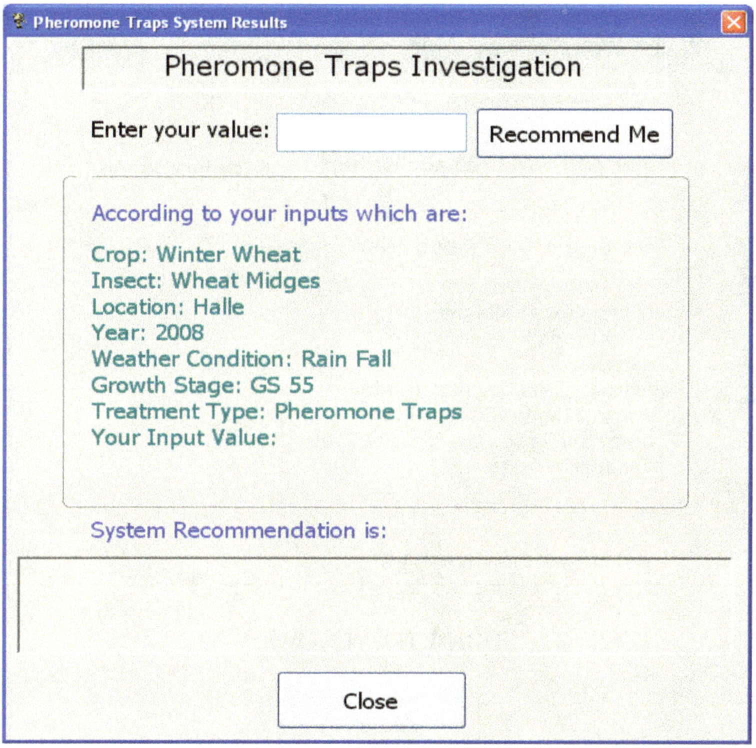

Fig. 10B. The consultation result out of the system.

Pheromone Traps: In this window system will use the defaults assigned in system settings in the consultation where user will only submit the OWBM value and click on recommend me and the system will process the value and give recommendation to user (Fig. 11 A&B)

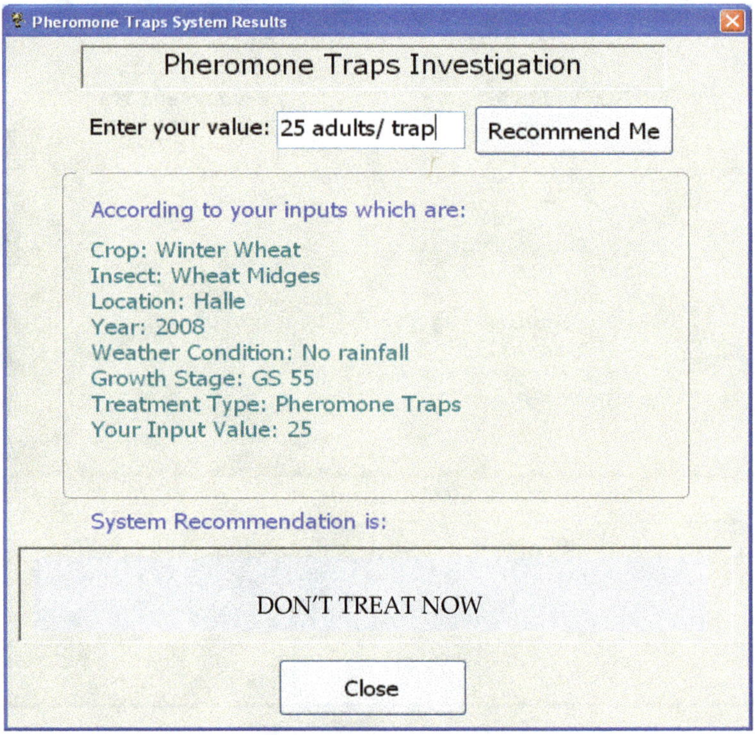

Fig. 11A. Recommendation without treatment.

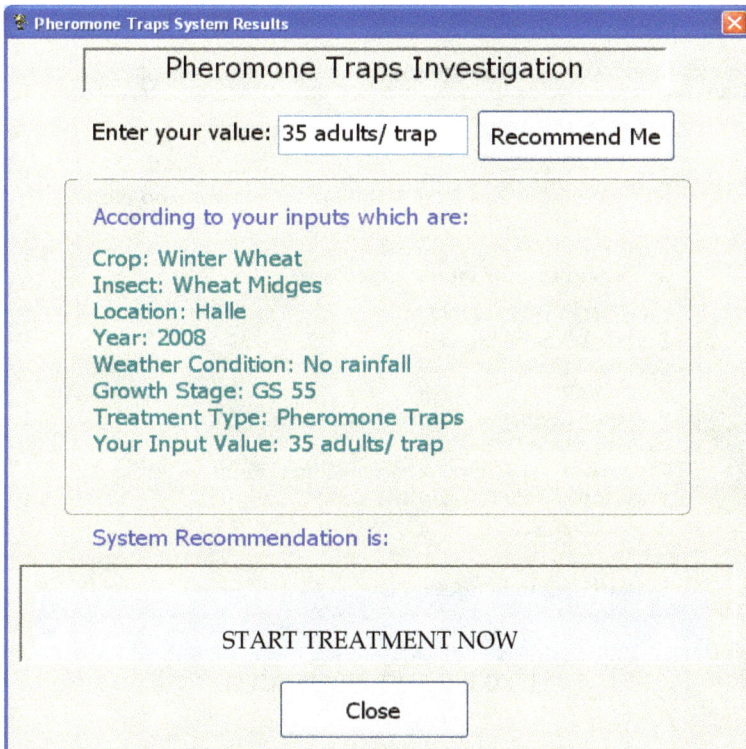

Fig. 11B. Recommendation with treatment.

Evaluation of wheat ears (wheat midges)

Here, this system will use the defaults assigned in system settings in the consultation where the user will only submit midge larvae value and click on recommend me and the system will process the value and give recommendation (Fig. 12 A&B).

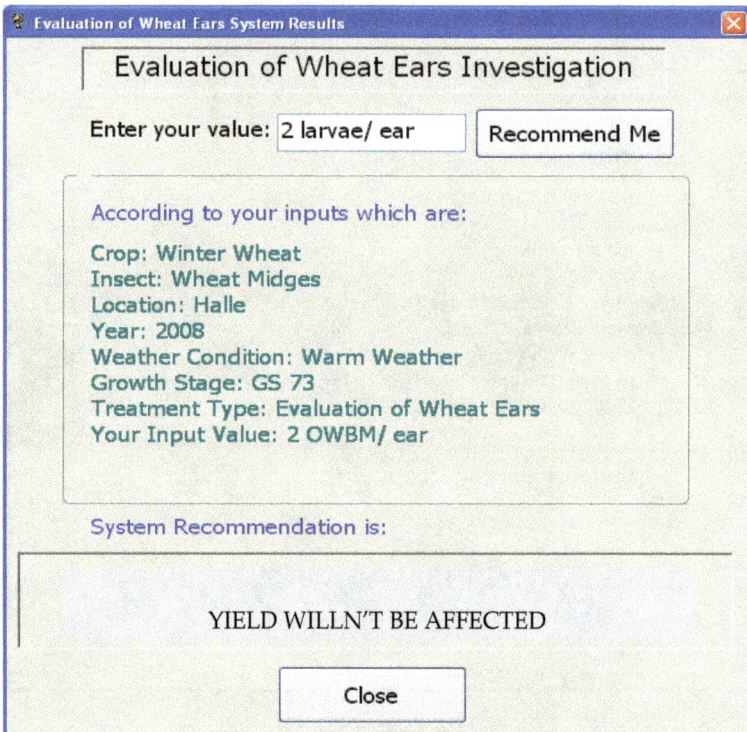

Fig. 12A. Expectation without yield losses.

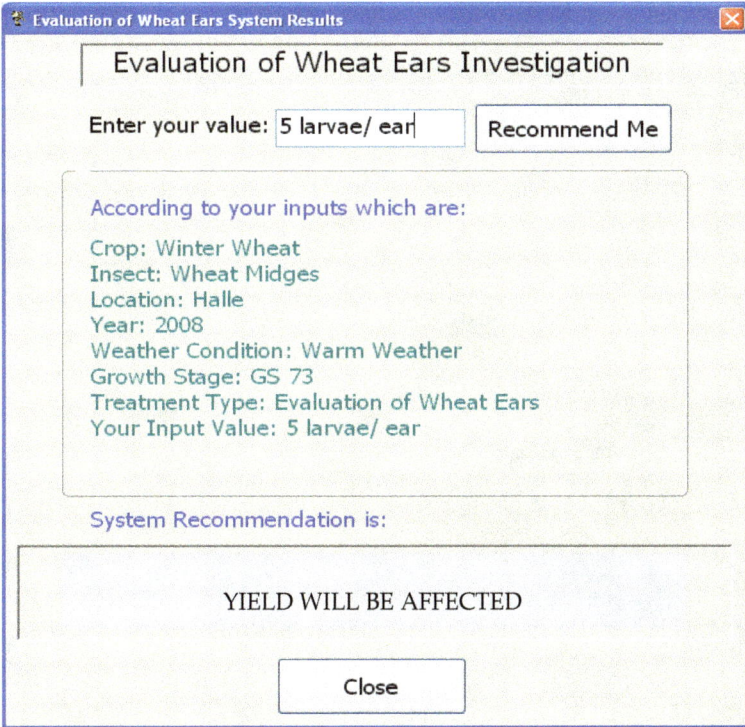

Fig. 12B. Expectation with yield losses.

Evaluation of wheat ears (thrips)

This system will use the defaults assigned in system settings in the consultation where the user will only submit thrips value and click on recommend me and the system will process the value and give recommendation to the user (Fig. 13 A&B).

Fig. 13A. Recommendation without treatment.

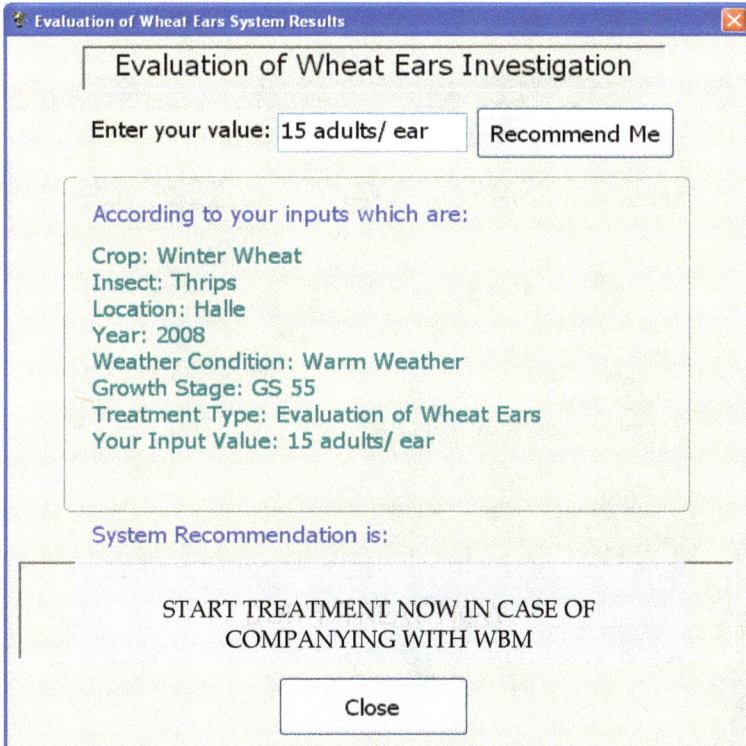

Fig. 13B. Recommendation with treatment.

White water traps

In this window, the system will use the defaults assigned in system settings in the consultation where the user will only submit the midge larvae value and click on recommend me and the system will process the value and give recommendation (Fig. 14 A&B).

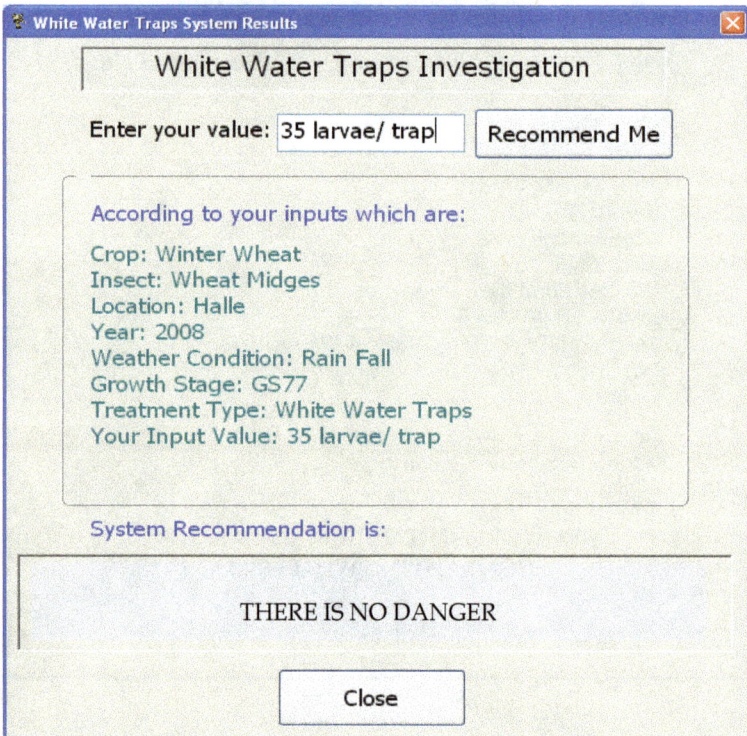

Fig. 14A. Recommendation without danger.

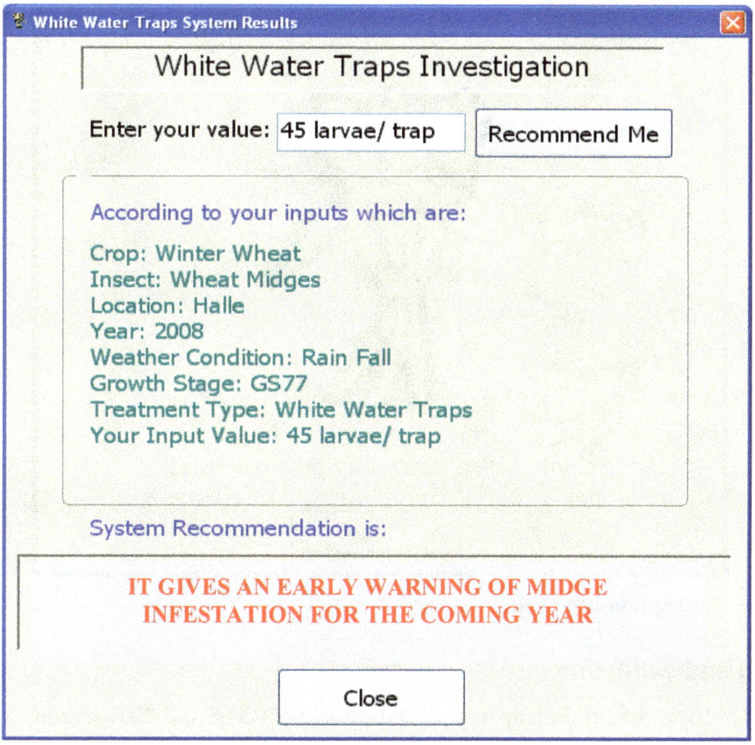

Fig. 14B. Recommendation with warning.

About

This is the about and copyright of the system (Fig. 15)

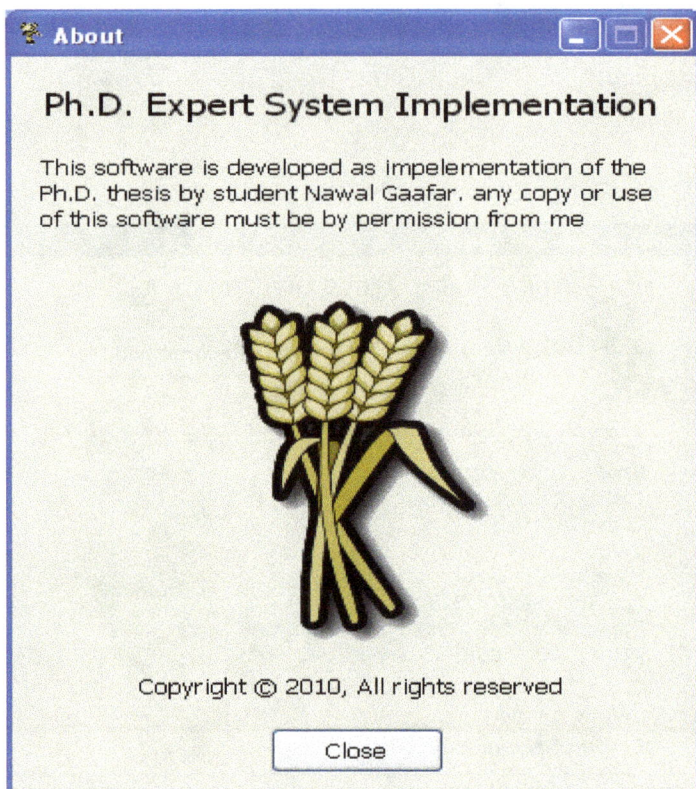

Fig. 15. The copyright of this system.

4. Testing and validation

WMTES field trials (based on crop samples) during 2007, 2008 and 2009 seasons, were done for the samples received from three different regions. As it is unwise to implement any ES from day one after completion of its development, parallel consultation from the IPM experts was found useful in improvement and validation of ES results. Also, real world ES need testing and validation in the real world environment i.e. field testing. Comparison of results produced from system as well as suggested by IPM experts has been used to improve the quality of inferences and consultation. Feedback forms have been used for preparation of validation case sheets.

5. Discussion

Pheromone traps monitor only male midge whereas it is female WBM that lay eggs from which the damaging stage of the pest emerges. Therefore, it is not possible to set a simple trap catch threshold above which economic damage occurs and below which it does not. A decision support model that can be used by farmers was developed using a stepwise decision tree involving yes/no answers to questions (Fig. 1). When growing a susceptible wheat variety pheromone traps need to be put out before ear emergence in fields where wheat was grown in previous years and provide a source of the pest. These traps should be monitored daily or at least twice weekly during the susceptible growth stage. When trap catches exceed threshold (30 midges/ day) crop inspections provide additional information to help decide whether to treat a field. It is advantageous that the pheromone traps are so sensitive and catch as many WBM as possible because they provide an early warning of midge flight during ear heading time and the suitable weather conditions, thus avoiding situations in which insecticide sprays are applied too late when they are needed. The threshold of midge larvae infestation is 3-4 larvae/ ear, where as in water traps, used to sample the migrated midge larvae is 30-40 larvae/ trap. Similar results were recorded by Ellis et al. (2009). The later should be also monitored carefully after the heavy rain, particularly at late milky stage; it gives an early caution of midge infestation for the coming year, especially intend cultivating wheat after wheat.

Midge infestation was higher in 2008 than in 2007 and 2009. The low levels of midge infestation hindered the verification of the decision flow tree as it was not possible to examine the impact of midge catches on infested ears. There can be some confidence in the proposed threshold of greater than 30 midges/ trap/ day to indicate a need to inspect crops for the pest as reported by Ellis et al. (2009).

The main objectives of WMTES have been met in allowing better data provision at all levels. However, the system can easily be changed and updated to meet new demands. For example, the current system automatically produces bulletins. This function could be developed further to produce automatically customized pest reports to include the interpretation of current pheromone or water traps data and to provide pertinent advice to growers and/or advisors. These could include comparisons between ear insect numbers in previous years and changes in the numbers between dates for specific regions, or a more forecast for weather conditions on the population dynamic of ear insect in any year. A knowledge base developed for a specific region in a problem can be tuned to make it more appropriate for other regions. Our results are consistent with Edrees et al. (2003) and Khan et al. (2008).

The current expert system has improved understanding of the WBM problem. Improving risk prediction: the decision flow chart proposes thresholds to help predict the need for insecticide treatment. The verification study suggested that this is a good basis for risk management. However, thresholds are based on data from a limited number of sites and years; further work is required to confirm the initial findings and improve the precision with which it is possible to predict the risk of pest attack. Risk of damage is also primarily dependent upon the coincidence among midge activity, the susceptible stage and weather conditions. Being able to predict the likely timing of the susceptible stage in relation to midge emergence would be a significant development, and help to limit unnecessary spray as stated by Freier et al. (1996).

The expert system will be used in training new experts. It will allow less experienced users to examine the reasoning process of an expert, to improve their understanding of how one takes control decision and to learn how to approach different situations to take the adequate decision. Agricultural extension services require more effective ways of handling, communicating, and using information. The program for the control of wheat ear insects is an example of one way that expert system technology can be successfully applied to daily problem solution in plant protection as recommended by Räder et al. (2007) and Suo & Shi (2008).

Much of the power and flexibility of expert systems are due to the fact that the knowledge base is separated from the inference mechanism (Rauscher, 1990; Waterman, 1985). The knowledge base can be modified without interfering with the operation of the system or the performance of other rules. In this way the program is now being enriched with rules for other insects, crops and sites, as well as by using of the user's feedback to improve the system. It is as important to develop the system as to maintain it. This task will be carried out for the extension services. WMTES is designed to enhance the quality and availability of knowledge required by decision makers in wheat insect management. It depends on a knowledge base that contains all the knowledge required to give useful, accurate and adequate consultations to wheat farmers.

6. References

Chakraborty, P. & Chakrabarti D.K. (2008). A brief survey of computerized expert systems for crop protection being used in India. *Progress in Natural Science* 18: 469-473.

Edrees, S.A., Rafea, A., Fathy, I. & Yahia, M. (2003). NEPER: a multiple strategy wheat expert system. *Computer & Electron in Agriculture* 40: 27-43.

ESICM (1994). Expert Systems for Improved Crop Management (FAO): A study of the needs assessment for expert systems in the agriculture sector in Egypt: *Technical Report No. TR-88-024-33.*

Ellis, S.A., Bruce, T.J.A., Smart, L.E., Martin, J.L., Snape, J. & Self, M. (2009). Integrated management strategies for varieties tolerant and susceptible to wheat midge. *HGCA Project Report number 451 May 2009*, 148 pages.

Freier, B., Triltsch, H. & Roßberg, D. (1996). GTLAUS — a model of wheat–cereal aphid–predator interaction and its use in complex agroecological studies. *Zeitschrift Pflanzenkrankh. Pflanzenschutz* 103: 543–554.

El-Wakeil, N., Gaafar, N. & Volkmar C. (2010). Susceptibility of spring wheat to infestation with wheat midges and thrips. *Journal of Plant Diseases and Protection* 117: 261–267.

Gaafar, N. (2010). Wheat midges and thrips information system: Monitoring and decision making in central Germany. PhD Dissertation Martin- Luther- University Halle-Wittenberg, 109 pages.

Gaafar, N. & Volkmar, C. (2010). Evaluation of wheat ear insects in large scale field in central Germany. *Agricultural Sciences* 1: 68-75.

Gaafar, N., Cöster, H. & Volkmar, C. (2009). Evaluation of ear infestation by Thrips and wheat blossom midges in winter wheat cultivars. In: Feldmann F., Alford D., Furk C., (eds.). *Proceedings 3rd International Symposium on Plant Protection & Plant Health*, pp. 349-359.

Gaafar, N., El-Wakeil, N. & Volkmar, C. (2011a). Assessment of wheat ear insects in winter wheat varieties in central Germany. *Journal of Pest Science* 84: 49-59.

Gaafar, N., Volkmar, C., Cöster, H. & Spilke J (2011b). Susceptibility of winter wheat cultivars to wheat ear insects in central Germany. *Gesunde Pflanzen* 62: 107-115.

Gonzalez-Andujar, J.L., Garcia De Ceca, J.L. & Fereres, A., (1993). Cereal aphids expert system" Identification and decision making. *Computer & Electron in Agriculture* 8: 293-300.

Gonzalez-Diaz, L., Martínez-Jimenez, P., Bastida, F. & Gonzalez-Andujar, J.L. (2009). Expert system for integrated plant protection in pepper (*Capsicum annuun* L.). *Expert System with Application* 36: 8975-8979.

Gosselke, U., Triltsch, H., Rossberg, D. & Freier, B. (2001). GETLAUS01 – the latest version of a model for simulating aphid population dynamics in dependence on antagonists in wheat. *Ecological Modelling* 145: 143–157.

Jörg, E., Racca, P., Preiß, U., Buttutini, A., et al. (2007). Control of Colorado potato beetle with the SIMLEP decision support system. *OEPP/EPPO* 37: 353-358.

Khan, F.S., Razzaq, S., Irfan, K., Maqbool, F., Farid, A., Illahi, I. &, Ul Amin, T. (2008). Dr. Wheat: a web-based expert system for diagnosis of diseases and pests in Pakistani wheat. World Congress on Engineering, London, July 2 - 4, 2008. *Proceedings of World Congress on Engineering* I: 549-554.

Kleinhenz, B. & Rossberg, D. (2000). Structure and development of decision-support systems and their use by the State Plant Protection Services in Germany. *OEPP/EPPO Bulletin* 30: 93-97.

Kleinhenz, B. & Zeuner, T. (2007). Introduction of GIS in decision support systems for plant protection. In Alord D., Feldman F., Hasler J., Tiedemann A. *Proceedings 2rd International Symposium on Plant Protection & Plant Health in Europe, Humboldt Universität, 2007, 82: 24-25.*

Knight, J. D., Tatchell, G. M., Norton, G. A. & Harrington, R. (1992). FLYPAST: an information management system for the Rothamsted Aphid Database to aid pest control research and advice. *Crop Protection* 11: 419-426.

Lucus, P. (1997). Symbolic diagnosis and its formalization, Knowle. *Engineering Review* 12: 109–146.

Mann, B.P., Wratten, S.D., Watt, A.D. (1986). A computer-based advisory system for cereal aphid control. *Computer Electron Agriculture* 1: 263-270.

Räder, T., Racca, P., Jörg, E. & Hau, B. (2007). PUCREC/PUCTRI - a decision support system for the control of leaf rust of winter wheat and winter rye. *OEPP/EPPO* 37: 378-382.

Rafea, A. & Shaalan, K. (1996). Using expert systems as a training tool in the agriculture sector in Egypt. *Expert Systems with Applications* 11: 343-349.

Rafea, A., El-Dessouki, A., Hassan, H. & Mohamed, S. (1993). Development and implementation of a knowledge acquisition methodology for crop management expert systems. *Computer & Electron in Agriculture* 8: 129-146.

Rauscher, H.M. (1990). Practical expert system development in Prolog. *Artificial Intelligence in Natural Resource Management* 4: 51-55.

Saini, H.S., Kamal, R. & Sharma, A.N. (2002). Web Based Fuzzy Expert System for Integrated Pest Management in Soybean. *International Journal Information Technology* 8: 54-74.

Sharma, M.C. (2001). Integrated pest management in developing countries: Special Reference to India. *IPM Mitr*, pp. 37-44.

Sivakami, S. & Karthikeyan, C. (2009). Evaluating the effectiveness of expert system for performing agricultural extension services. *Expert System with Application* 36: 9634-9636.

Suo, X. & Shi, N. (2008). Web-based expert system of wheat and corn growth management. *IFIP International Federation for Information Processing* 258: 111-119.

Tiedemann, A. & Kleinhenz, B. (2008). Prognose contra Praxis? *DLG- Mitteilungen* 3: 54-58.

Tottman, D.R. (1987). The decimal code for the growth stages of cereals, with illustrations. *Annual of Applied Biology* 110: 441-454.

Volkmar C., Werner C. & Matthes P. (2008). On the occurrence and crop damage of wheat blossom midges *Contarinia tritici* (Kby.) and *Sitodiplosis mosellana*. *Mitteilungen Deutschen Gesellschaft für Allgemeine und Angewandte Entomologie* 16: 305-308.

Volkmar C., Schröder A., Gaafar N., Cöster, H. & Spilke, J. (2009). Evaluierungsstudie zur Befallssituation von Thripsen in einem Winterweizensortiment. *Mitteilungen Deutschen Gesellschaft für Allgemeine und Angewandte Entomologie* 17: 227-230.

Waterman, D.A. (1985). A Guide to Expert Systems. *Addison-Wesley, Reading, MA,* 441 pp.

11

Transgenesis, Paratransgenesis and Transmission Blocking Vaccines to Prevent Insect-Borne Diseases

Marcelo Ramalho-Ortigão and Iliano Vieira Coutinho-Abreu
Kansas State University,
Department of Entomology
USA

1. Introduction

Insect-borne diseases are responsible for severely affecting human life around the world, causing significant morbidity and mortality. Malaria alone is responsible for 1-2 million deaths annually, and approximately 300 million are at risk of becoming infected. Insect-borne diseases also are responsible for an estimated 50% of all neglected tropical diseases (NTDs), which affect over 1 billion people – one sixth of the world population – and include such diseases as leishmaniasis, filariasis, Chagas' disease, African trypanosomiases, onchocerciasis, schistosomiasis, trachoma and others. Such NTDs cause severe morbidity and are frequently referred to as "poverty causing diseases". The lack of effective vaccines or drugs for many insect-borne diseases makes control mainly dependent on insecticides. However, the appearance of insecticide resistance requires the development of new strategies to reduce pathogen transmission in the field [1]. Among the research themes with potential to generate new tools to control vector borne diseases, major efforts have been carried out to establish transgenesis, paratransgenesis, and transmission-blocking vaccines (TBVs) as new weapons to reduce vector competence.

While vector competence encompasses the intrinsic genetic factors that define the ability of a vector to transmit a pathogen (and it is a component of vectorial capacity), vectorial capacity is a measurement of the efficiency of vector-borne disease transmission (i.e., total number of infective bites delivery to a single host in one day), and influenced by vector density and longevity [2]. Regarding vector competence, several molecular techniques, such as quantitative trait loci (QTL) mapping, and gene knock-down, can be used to identify the intrinsic genetic factors (i.e., molecules expressed by the vector) involved with the ability of vectors to transmit pathogens. Molecules involved in vector competence can be directly targeted by antibodies (as in the case of TBVs), or overexpressed in transgenic insects or paratransgenic symbionts in order to reduce pathogen development and transmission.

In the late 1990's, the establishment of stable lines of genetically modified mosquitoes opened new avenues for studying molecules with potential to reduce vector competence [3]. Transgenic mosquitoes expressing dsRNAs (i.e., to induce RNAi pathways) targeting RNAs associated with mosquito immune-related proteins [4], or overexpressing microbial peptides [3, 5], or expressing a truncated transcription factor to generate a dominant-negative phenotype [6] were generated in order to investigate the role of these molecules in vector competence. Understanding such mechanisms is considered a pre-requisite for the development of molecular strategies to control vector-borne diseases.

For a heterologous protein (exogenous protein introduced into a disease vector) to be used to reduce vector competence, a gene drive mechanism is required to spread the gene encoding the protein throughout the targeted insect vector population. A gene drive system is spread within a population by increasing its frequency to ratios greater than those expected by traditional Mendelian rules. Thus, the combination of a given transgene (expressing a heterologous protein) with a gene drive system also can increase the frequency of the transgene in a population. Gene drive systems currently known include transposons, homing endonuclease, engineered under-dominance, meiotic drive, endosymbionts, and Medea element [7, 8]. Yet, only transposons are currently available to be used in the genetic transformation of insect disease vectors [8], and Wolbachia endosymbionts are thought to be a feasible way to spread paratransgenic symbionts in natural populations of these insect vectors [9].

The use of transposons to generate stably transformed insect germ lines, i.e., with exogenous DNA inserted into the genome and capable of being transferred into following generations (as depicted in Figure 1A), is well established [10] for a couple of insect vector species. Different species of mosquitoes, representing *Aedes* [11-14], *Anopheles* [15-24], and *Culex* [25] genera, have been genetically altered or transformed (Table 1), and, in some cases, the transformed mosquitoes expressed proteins targeting pathogen development [3, 20, 21, 26, 27]. Here, the common goal is to transform insect vectors with gene(s) whose protein(s) impair(s) pathogen development. As indicated above, genes that reduce pathogen development are to be associated with a gene drive system that increases the frequency of the transgenic vector when they are released into their natural habitats.

Paratransgenesis is an alternate approach to reduce vector competence via the genetic manipulating of symbionts commonly found in insect disease vectors (Figure 1B). The main characteristics of paratransgenesis are the simplicity with which symbionts are transformed (through viral or bacterial genetic transformation), the feasibility of the transformed symbiont to be spread across a population (maternally or via coprophagy), and the reduced fitness cost associated with the transformation of symbionts [28]. Symbionts currently targeted for paratransgenesis include bacteria that inhabit triatomine hindguts [28-30] and tsetse fly tissues [31], and densoviruses infecting *An. gambiae* and *Ae. aegypti* mosquitoes [32, 33]. To date, insect vector symbionts have been genetically modified to express antimicrobial peptides [28], single chain antibodies [29, 30, 34], and dsRNAs [35-38]. In all three approaches the expressed molecules proved harmful to the pathogens transmitted by each vector.

Transmission-blocking vaccines (TBVs) are intended to prevent the transmission of pathogens from infected to uninfected hosts (Figure 1C) by a disease vector. Such vaccines do not protect an individual from infection but rather can reduce transmission. TBVs target molecule(s) that are expressed on the surface of parasites during their developmental phase

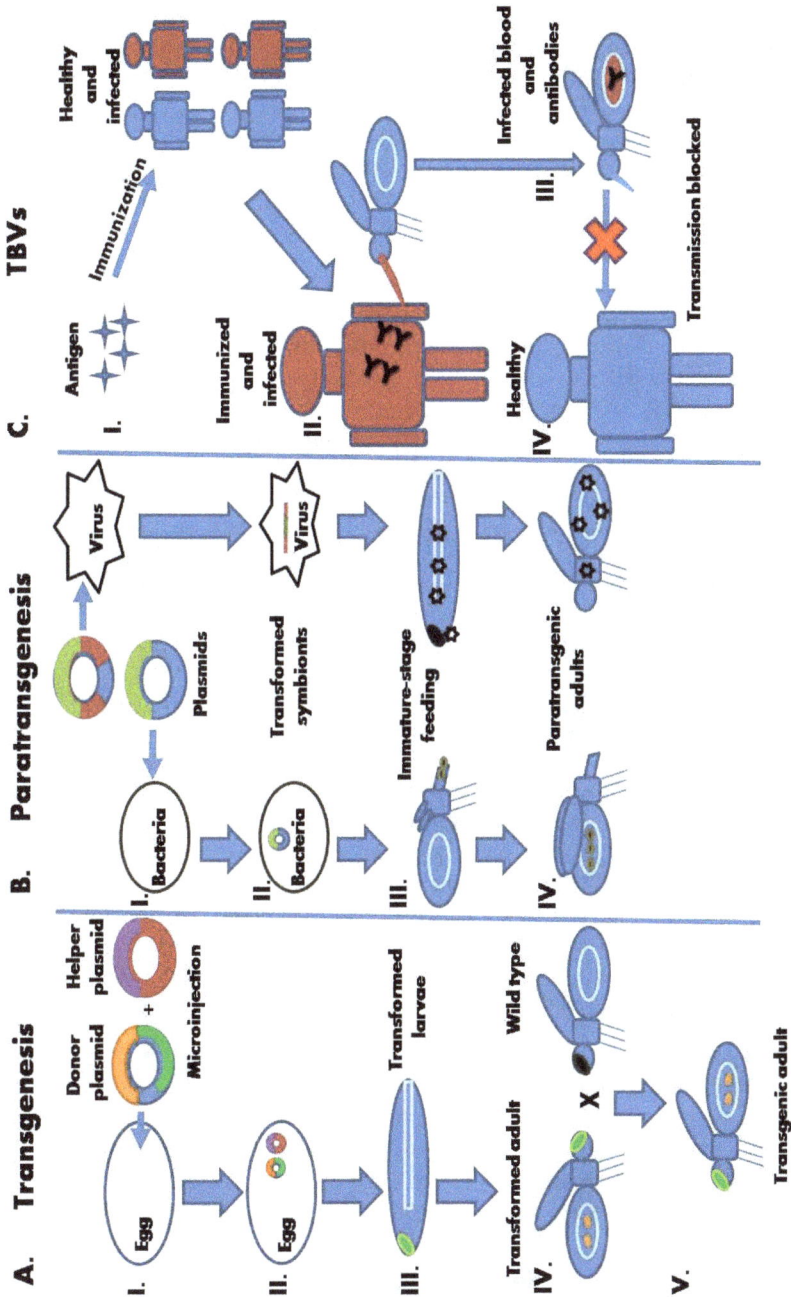

Fig. 1. Transgenesis, paratransgenesis, and transmission blocking vaccines (TBVs) (modified from (Coutinho-Abreu & Ramalho-Ortigao 2010; Coutinho-Abreu et al. 2010)). (**A**)

Transgenesis; the general technique for production of transgenic insects by germ line transformation is shown. (I) Insect eggs are microinjected with a donor plasmid [expressing a transgene (orange) and a reporter gene (green)], and with a helper plasmid, [expressing a transposase (purple)]. (II) After inoculation into the eggs, plasmids are taken up by some (or all) of the germ line cells. (III) transgenic and non-transgenic larvae can be separated by the expression of the reporter gene (green eye phenotype), controlled by an eye specific promoter. (IV) Transgenic insects are crossed with wild type to confirm that the transposon-carrying transgene was inserted into the chromosome. (V) Transgenic adult insect expressing the transgene (e.g., orange proteins in the insect midgut) is shown. (B) Paratransgenesis; the general technique used to obtain insect transformation via their symbionts is shown. Two insect orders are represented: Hemiptera/reduviids (left panel) and Diptera/mosquitoes (right panel). (I) Bacteria or viruses symbionts can be genetically modified to express a gene blocking parasite development in vectors' tissues. (II) Symbiotic bacteria are transformed with plasmids (blue) expressing a gene (green) to inhibit parasite development in insect gut. Alternatively, viral genome (red) is inserted into a plasmid (blue) and manipulated to express a transgene (green). Viral particles can be generated by expression of such a plasmid in insect cells. (III) The transformed symbionts are acquired by insect hosts through larvae or nymph feeding, or through thoracic injection. (IV) Once insects acquire the transformed symbionts, these microorganisms can express proteins to inhibit pathogen development. (C) Transmission-blocking vaccines (TBVs). TBV is a strategy to prevent transmission of a pathogen by the bite of an infected vector. Frequently, TBVs rely on generating antibodies against vector molecules that are involved in pathogen development. (I) Healthy (blue) and infected (red) individuals are immunized with a TBV antigen; (II) Insect-vectors take an infected blood meal containing TBV antigen-specific antibodies; (III) Specific antibodies produced against the antigen inhibit pathogens development within the insect vector, (IV) preventing transmission to uninfected host(s).

within the insect vector (Table 2) [39-48], or that are expressed on the surface of vector tissues with which pathogens may be required to interact during their development within the vector (Table 3) [48-52]. The potential application of proteins (antigens) expressed on the surface of the malaria parasites *Plasmodium falciparum* and *Plasmodium vivax* as TBV has been tested [39-48]. Two of these proteins (Pfs25 and Pvs25) were deemed safe following phase-1 human trials [53, 54]. Specific antibodies against molecules expressed in midgut tissues of *An. gambiae* and *Phlebotomus papatasi* are also capable of reducing parasite loads in these vectors, pointing to their potential as TBVs candidates [48-52].

In the following pages, details of each of these three approaches are provided. Technical aspects of the strategy utilized as well as results from studies *in vitro* and *in vivo*, or depending on the case animal and human tests, as well as semi-field or field release of modified insects are also indicated. Due to the massive amount of information that has been generated in recent years regarding some of the topics (e.g., TBVs), we intend to limit our analyses to some specific points we find critical for the readers information.

2. Transgenesis in insect disease vectors

Generally, the goal for vector transgenesis is the interruption of pathogen transmission through introduction of exogenous DNA fragment (i.e., gene) into the genome of a disease vector, followed by expression of the gene to inhibit pathogen development within the vector. Various mosquito species, vectors of different parasites and viruses, have been transformed (Table 1). Some of the transformed mosquitoes were shown capable of blocking pathogen development via tissue-specific expression of molecules that impair pathogen attachment to the midgut (Ito et al. 2002), or activate biochemical pathways detrimental to pathogen survival (Franz et al. 2006). However, vector transgenesis is a complex approach, highlighted by the fact that insect germ line transformation technique is only successfully performed by a handful of laboratories. Various issues related with transgenic vectors, including stability of the transgene in the genome, and fitness of the transformed insect in the field, need to be fully resolved prior to the successful application of transgenic insects fat body and hemocoel against insect-borne diseases.

A few aspects of development of transgenic vectors, how interference or blockage of parasite is achieved, and what may lie ahead for vector transgenesis are discussed below.

2.1 Germ line transformation

Stable genetic transformation of insects is accomplished by inserting an exogenous gene into the insect genome via the inoculation of plasmids containing the transgene (donor plasmids) into insect eggs (Figure 1). Donor plasmids are constructed to carry an engineered transposon element lacking the gene that encodes the transposase, the enzyme that mediates transposon activity by a cut and paste mechanism. *Hermes, Mariner (Mos1), minus*, and *piggyBac*, for the most part, have been the transposable elements of choice. Transposases are then supplied *in trans* (expressed in a separate plasmid) by the co-inoculation of a transposase-encoding helper plasmid along with the donor plasmid. Transposase expression is usually driven by a heat shock protein promoter that is activated upon raising the temperature, as it is frequently performed with injected mosquito eggs (Wimmer 2003).

Transformed offspring can be identified by a specific phenotype alteration, mediated by the expression of a reporter protein encoded by the donor vector, such as eye color (Wimmer 2003). However, this strategy is restricted to insect species displaying polymorphic eye phenotypes, such as *Ae. aegypti*. Alternatively, phenotypic markers such as firefly luciferase, or green (EGFP), red (dsRED) and cyan fluorescent (CFP) proteins can be used as transformation markers (Moreira et al. 2000; Kokoza et al. 2001; Nolan et al. 2002; Perera et al. 2002; Nirmala et al. 2006).

2.2 Tissue-specific transgene expression

Within the insect vector, a pathogen may interact with specific tissues, such as the midgut (as is the case for *Trypanosoma cruzi* and *Leishmania sp.*), or midgut, hemocoel and salivary glands (as is the case for *Plasmodium sp*). Frequently, molecules that can block pathogen development within its vector can be expressed in a tissue-specific manner to increase effectiveness. Tissue-specific expression of a transgene is accomplished by the use of a tissue-specific promoter (Table 1). Most promoters used in vector transgenesis drive protein expression specifically within the midgut, the hemocoel, or the salivary glands, as these are sites where pathogens are commonly found within an infected vector (Kokoza et al. 2000; Moreira et al. 2000; Abraham et al. 2005; Lombardo et al. 2005; Yoshida & Watanabe 2006; Chen et al. 2007b; Rodrigues et al. 2008).

In mosquitoes, promoters for carboxypeptidase and peritrophin have been widely used to drive midgut-specific expression of several transgenes. In *Ae. aegypti*, a vitellogenin promoter driving expression of transgenes in the fat body was used to express innate immune defense-related genes (Kokoza et al. 2000; Kokoza et al. 2001; Shin et al. 2003a; Bian et al. 2005), as well as dsRNA targeting *REL1* transcripts (Bian et al. 2005). The same promoter was used to express CFP in *An. stephensi* (Nirmala et al. 2006). The robustness of the vitellogenin promoter was confirmed by its capacity to function following multiple gonotrophic cycles in transgenic *An. stephensi* (Chen et al. 2007b).

Salivary gland-specific promoters also have been used in germ-line transformation of mosquitoes. *D7* and *apyrase* promoters from *An. gambiae* and *antiplatelet* from *An. stephensi* were used to transform *An. stephensi* (Lombardo et al. 2005; Yoshida & Watanabe 2006). *Ae. aegypti* also was transformed and successfully expressed luciferase within salivary glands using *maltase-like 1* and *apyrase* promoters (Coates et al. 1999).

2.3 Transgenes targeting *Plasmodium* development

In spite of the many mosquito species successfully transformed, only a handful has been transformed with molecules that impair pathogen development (Ito et al. 2002; Moreira et al. 2002; Kim et al. 2004; Abraham et al. 2005; Franz et al. 2006; Rodrigues et al. 2006). A list of genetically modified mosquitoes obtained to date, including the transposon and reporter genes used, tissue-expression specificity and target pathogens, among others is seen on Table 1.

In one example of a transgene targeting *Plasmodium*, expression of phospholipase-2 (PLA2) in *An. stephensi* (Moreira et al. 2002) led to an 87% reduction of *P. berghei* oocyst intensity compared to non-PLA2-expressing controls. When a peritrophin promoter was used to drive the expression of PLA2 in *An. stephensi*, inhibition of *P. berghei* oocyst intensity ranged from 74% to 94% (Abraham et al. 2005). Nevertheless, expression of PLA2 in the mosquito midgut did not exert a direct effect on the parasite, but rather led to structural damage of the midgut epithelium (Moreira et al. 2002; Abraham et al. 2005).

Synthetic peptides that block *Plasmodium* development in mosquitoes also have been identified and tested. SM1, identified using a phage display library (Ghosh et al. 2001), blocked *P. berghei* invasion of *An. stephensi* midgut and salivary glands. In *An. stephensi* transformed with *piggyBac* expressing a four tandem repeat of SM1 under a carboxypeptidase promoter, *P. berghei* intensity was inhibited by 81.6%. Interestingly, these transgenic mosquitoes even with sporozoites in their salivary glands were unable to transmit *P. berghei* to mice (Ito et al. 2002).

Another molecule tested was the C-type lectin (CELIII) from sea cucumber. When expressed in *An. stephensi*, CELIII is was shown to be cytotoxic to *P. berghei* ookinetes, reducing prevalence and intensity by 84% and 90%, respectively (Yoshida et al. 2007). CELIII expression was driven by a midgut-specific *An. gambiae*-derived carboxypeptidase promoter.

Besides PLA2, SM1, and CELIII, mosquitoes also were transformed with antimicrobial peptides (Kokoza et al. 2000; Kim et al. 2004; Bian et al. 2005). *An. gambiae* transformed with a *cecropin A* (driven by a carboxypeptidase promoter from *Ae. aegypti*) inhibited *P. berghei* intensity by 61% on average when compared with non-transformed controls (Kim et al. 2004). Recently, *An. stephensi* were transformed with the *piggyBac* transposon expressing single-chain antibodies (scFvs) targeting *P. falciparum* proteins. Some of these scFv genes have been attached to a *Cecropin A* gene so as to improve the effectiveness of the antibody against *P. falciparum*. In fact, the scFv-Cecropin A construct (m2A10) targeting the *P. falciparum* circunsporozoite protein reduced *P. falciparum* intensity by 97%, prevalence by 86%, and sporozoites load by 84% (Isaacs et al. 2011).

2.4 Transgenes inducing gene silencing and targeting viral transmission

Stable transformation of mosquitoes with two inverted repeats of the same gene to induce assembly of double-strand RNAs (dsRNAs) and activation of the RNAi pathway has also been obtained (Brown et al. 2003; Bian et al. 2005). This strategy takes advantage of the RNAi mechanism to block expression of insect molecules associated with vectorial competence (Bian et al. 2005), or it can directly target viral replication within insect tissues (Franz et al. 2006). In spite of the fact that both approaches are technically feasible, only the latter has led to substantial reduction in the development of any human pathogen (Franz et al. 2006).

Gene silencing via dsRNA was first demonstrated with *An. stephensi* expressing sense and anti-sense RNAs targeting EGFP (Brown et al. 2003). EGFP dsRNA-expressing mosquitoes were crossed with a transgenic mosquito line that expresses EGFP. The double transgenic offspring displayed lower level of EGFP expression than the parental line expressing it, indicating the effect of the RNAi machinery reducing the expression of the EGFP transgene (Brown et al. 2003). *Ae. aegypti* expressing dsRNA targeting *REL1*, a gene involved in innate immune response, also inhibited expression of *REL1* via RNAi (Bian et al. 2005).

As indicated, the activation of the RNAi pathway is intended to affect the replication of infecting RNA viruses transmitted by mosquitoes. *Ae. aegypti* expressing DEN2 sense and antisense RNAs reduced viral load by fivefold, confirming the effectiveness of the RNAi in controlling virus replication in disease vectors (Franz et al. 2006).

2.5 Future of vector transgenesis

Despite advances in the development of stable lines of genetically modified disease vectors (Moreira et al. 2000; Ito et al. 2002; Lobo et al. 2002; Perera et al. 2002), many challenges exist to the application of transgenesis to control vector-borne diseases outside the laboratory. Beyond issues dealing with social and environmental impact(s) that are inherent to the potential use of genetically modified organisms (not the scope of this chapter), of significance also is the fact that, for example, none of the most import human malaria vectors (i.e. *An. gambiae s.l.* and *An. funestus*) has been successfully transformed and displayed reduced vector competence. Moreover, only one strain of transgenic mosquitoes blocked the development of *P. falciparum*, the principal human malaria parasite (Yoshida et al. 2007). The only exceptions of transgenic insect lines robustly impairing development of a human pathogen in its human vector are the transgenic strain of *Ae. aegypti* capable of inhibiting DEN virus development (Franz et al. 2006; Mathur et al. 2010).

Fitness of transgenic mosquitoes in natural habitats is also an important issue. Laboratory tests demonstrated that in four lines of transgenic *An. stephensi* the frequency of transgenic individuals declined over time (Catteruccia et al. 2003). Although SM1-transgenic hemizygous *An. stephensi*, carrying a single transgene copy in the genome, exhibited higher fitness than wild type when fed on infected mice (Marrelli et al. 2007), transgenic homozygous *An. stephensi* (harboring two transgene copies), possibly advantageous for field releases, displayed lower fitness than non-transformed mosquitoes (Li et al. 2008). Transgenic lines of *Ae. aegypti* expressing either EGFP or a transposase also displayed lower fitness than wild type (Irvin et al. 2004). Such fitness load issue may be overcome by taking advantage of a site-specific recombination strategy, as shown for *An. stephensi* transgenic lines containing phi C31 attP 'docking' sites and expressing ECFP. (Amenya et al. 2011).

Although fitness in natural habitats is one of the main constraints of transgenic disease vectors, mathematical models suggest that a highly efficient transposon can spread through natural populations if it affects fitness by less than 50% (Hickey 1982; Ribeiro & Kidwell 1994). Nevertheless, pathogen refractoriness needs to be at or very close to 100% to substantially decrease disease prevalence in high endemic areas (Boete & Koella 2002). Future studies mimicking field conditions likely will uncover the importance of fitness to the establishment of transgenic mosquitoes in natural habitats.

Problems also are associated with transposons as genetic drive systems for transgenes. Transposons can remobilize in somatic tissues possibly causing damage in other regions of the genome (Atkinson 2004). Interestingly, none of the transposable elements (*Hermes, Mos1, minos,* and *piggyBac*) appears to remobilize in *Ae. aegypti* germ line, possibly reflecting a resistance mechanism, since the same elements can remobilize in *Drosophila* germ line tissues (O'Brochta et al. 2003). In addition, a mechanism to drive transposase expression and restrict gene drive system activity to germ-line tissues has been created using the regulatory sequence of *nanos*, a gene involved in early embryonic development (Adelman et al. 2007).

Further issues regarding the design and potential field release of transgenic disease vectors include: i) non-canonical transposition reactions, such as transgene insertion by a mechanism other than cut-and-paste, resulting in integration of donor plasmid fragments into the insect genome, as observed in transposition events accomplished by the transposons *Hermes, Mos1,* and *piggyBac* in *Ae. aegypti* (O'Brochta et al. 2003); ii) transgene size influencing transposon activity, as shown for *Mariner* (Lampe et al. 1998) and iii) inhibition of transgene expression after some generations due to unknown mechanisms, as observed with *Ae. aegypti* expressing a anti-Dengue virus dsRNA (Franz et al. 2009). Another issue is the possibility of horizontal transfer of the transgene between mosquito sibling species, as proposed for the introgression of the *P*-element between *Drosophila* lines (Engels 1997). Horizontal transfer also can be virus mediated, such as the case of *piggyBac*, initially identified in a *Tricoplusia ni* virus (Fraser et al. 1983). Technology to prevent the potential horizontal transposon transfer by viruses and to inhibit transposition activity mediated by endogenous transposases still needs to be developed.

Other gene-drive mechanisms have been developed to assist with problems associated with transposon elements in insect germ-line transformation (Sinkins & Gould 2006). Recently, a driving mechanism known as *Medea* (maternal-effect dominant embryonic arrest) was shown capable of driving population replacement in *Drosophila* without an apparent fitness cost (Chen et al. 2007a). This gene drive system consists of a DNA segment encoding a protein lethal to insects and an antidote that neutralize the lethal protein. A heterozygous female (*Medea*/+) expresses the toxin within all oocytes, killing all the +/+ offspring as they do not express the antidote to neutralize the maternal toxin. *Medea* can be designed to restrict transgene activity to the host species through the utilization of siRNAs-encoding genes as toxin genes (Chen et al. 2007a). Although *Medea* has been postulated to function in *An. gambiae* population replacement, it has yet to be developed for mosquitoes (Marshall & Taylor 2009). Transgenic insects also can be developed to express female dominant-lethal genes to reduce the number of females in an insect population (Thomas et al. 2000; Horn & Wimmer 2003). RIDL, or release of insects carrying a dominant lethal, was originally designed to overcome issues associated with SIT (sterile insect technique). Although SIT was successfully applied against the screwworm fly *Cochliomyia homonivorax* (Krafsur et al. 1987), the fruit fly *Ceratitis capitata* (Robinson et al. 1999), and the tsetse fly *G. austeni* (Vreysen et al. 2000), drawbacks such as reduced sterile male fitness and sterile female contamination, were detected (Thomas et al. 2000). RIDL consists of release of transgenic male insects expressing the female dominant-lethal genes, causing a reduction on the numbers of females in the following generations (Robinson et al. 1999). Robust transgenic vectors approaches could also be used with RIDL (Thomas et al. 2000), and a transgenic mosquito sexing-system has already been developed (Catteruccia et al. 2005). In fact, an *Ae. aegypti* RIDL line has been successful developed and tested in natural settings (Fu et al. ; James 2011). Taking advantage of the *Actin* gender-specific alternative splicing, female *Ae. aegypti* displays a flightless phenotype, reducing potential mating and consequently mosquito densities (Fu et al. 2010). Despite the substantial achievement, its release in field setting has been the subject of much criticism (James 2011).

In our view, hurdles to the establishment of an efficient transgenic vector approach include the lack of a transgene(s) that effectively reduce pathogen load, and the inefficiency of transposons as gene-drive mechanism(s). Further studies to identify traits associated with vector competence will likely pinpoint candidate genes that, when targeted, may effectively block pathogen development and transmission. The availability of gene drive mechanisms to overcome issues associated with the use of transposons, such as remobilization, fitness load, and the potential to introgress to closely related species, also is of interest. *Medea* was suggested as a gene drive that could overcome such issues, but it is yet to be developed for insect vectors. As for RIDL, the current absence of a gene drive mechanism also prevents its application against insect vectors.

The recent establishment of a binary Gal4/UAS system in *Ae. aegypti* (Kokoza & Raikhel) may also speed the establishment of other transgenic mosquito lines, as this system represents an invaluable tool for refinement of genetic tools in mosquitoes, and possibly for the discovery of new molecular targets for control of vector-borne diseases via transgenesis.

Clearly, much work remains before genetically modified insect vectors can be systematically released into natural habitats. When realized, transgenesis may provide a significant tool in the fight against vector-borne diseases.

Insect species	Transposon	Promoter	Reporter gene	Transgene	Targeted pathogen	Reference
Ae. aegypti	Hermes	D. melanogaster Cinnabar	D. melanogaster Cinnabar	Cinnabar	-	(Jasinskiene et al. 1998)
	Mos1	D. melanogaster Cinnabar	D. melanogaster Cinnabar	Cinnabar	-	(Coates et al. 1998)
	Hermes	Ae. aegypti maltase-like 1 / apyrase	Firefly luciferase	Firefly luciferase	-	(Coates et al. 1999)
	Hermes	D. melanogaster Actin 5C	EGFP	EGFP	-	(Pinkerton et al. 2000)
	Hermes	Ae. aegypti vitellogenin	D. melanogaster Cinnabar	Defensin A	Micrococcus luteus	(Kokoza et al. 2000)
	Hermes	Ae. aegypti carboxypeptidase	Firefly luciferase	Firefly luciferase	-	(Moreira et al. 2000)
	Mos1	An. gambiae carboxypeptidase	Firefly luciferase	Firefly luciferase	-	(Moreira et al. 2000)
	piggyBac	Ae. aegypti vitellogenin	EGFP	Defensin A	Enterobacter cloacae/ P. gallinaceum	(Kokoza et al. 2001; Shin et al. 2003b)
	Minos	D. melanogaster Actin 5C	dsRED	dsRED	-	(Nolan et al. 2002)
	piggyBac	D. melanogaster Actin 5C	dsRED	dsRED	-	(Nolan et al. 2002)
	piggyBac	D. melanogaster Cinnabar	D. melanogaster Cinnabar	Cinnabar	-	(Lobo et al. 2002)
	Mos1	D. melanogaster Cinnabar/hsp70	D. melanogaster Cinnabar/dsRED	Cinnabar/dsRED	-	(Wilson et al. 2003)
	piggyBac	D. pseudoobscura hsp82	EGFP	Mos1 transposase	-	(Wilson et al. 2003)
	piggyBac	Ae. aegypti vitellogenin	EGFP	ΔREL1[a]	-	(Shin et al. 2003a)
	piggyBac	Ae. aegypti vitellogenin	EGFP	Cecropin A	Enterobacter cloacae	(Shin et al. 2003a)
	piggyBac	Ae. aegypti vitellogenin	EGFP	ΔREL1-A[b]	-	(Bian et al. 2005)
	piggyBac	Ae. aegypti vitellogenin	EGFP	dsRNA against REL1 transcripts	-	(Bian et al. 2005)

Insect species	Transposon	Promoter	Reporter gene	Transgene	Targeted pathogen	Reference
Ae. aegypti	*Mos1*	*An. gambiae* carboxypeptidase	EGFP	dsRNA against prM of DENV-2	Dengue-2 virus	(Franz et al. 2006; Franz et al. 2009)
	Mos1	*Ae. aegypti* nos, *Ae. aegypti* D7	dsRED	*Mos1* transposase	-	(Adelman et al. 2007)
	Mos1		EGFP	*Gal4*	-	
	Mos1	*Ae. aegypti* 30K	EGFP	*Mtp*	Dengue-2 virus	(Mathur et al. 2010)
Ae. fluviatilis	*piggyBac*	*An. gambiae* peritrophin	EGFP	*mPLA2*	*P. gallinaceum*	(Rodrigues et al. 2008)
An. albimanus	*piggyBac*	*D. melanogaster* polyubiquitin	EGFP	*EGFP*	-	(Perera et al. 2002)
An. gambiae	*piggyBac*	*Baculovirus* hr5-ie1	EGFP	*EGFP*	-	(Grossman et al. 2001)
	piggyBac	*Ae. aegypti* carboxypeptidase	EGFP	*Cecropin A*	*P. berghei*	(Kim et al. 2004)
An. stephensi	*minos*	*D. melanogaster* actin 5C	EGFP	*EGFP*	-	(Catteruccia et al. 2000)
	piggyBac	*D. melanogaster* actin 5C	dsRED	*dsRED*	-	(Nolan et al. 2002)
	piggyBac	*An. gambiae* carboxypeptidase	EGFP	*PLA2*	*P. berghei*	(Moreira et al. 2002)
	piggyBac	*An. gambiae* carboxypeptidase	EGFP	$[SM1]_4$	*P. berghei*	(Ito et al. 2002)
	minos	*D. melanogaster* actin 5C	dsRED	dsRNA against EGFP	-	(Brown et al. 2003)
	minos	*An. gambiae* apyrase *An. gambiae* D7r4	EGFP	*LacZ*	-	(Lombardo et al. 2005)
	piggyBac	*An. gambiae* peritrophin	EGFP	*PLA2*	*P. berghei*	(Abraham et al. 2005)

Insect species	Transposon	Promoter	Reporter gene	Transgene	Targeted pathogen	Reference
An. stephensi	piggyBac	D. melanogaster β2-tubulin	dsRED	EGFP	-	(Catteruccia et al. 2005)
	piggyBac	An. stephensi vitellogenin	EGFP	CFP	-	(Nirmala et al. 2006)
	minos	An. gambiae carboxypeptidase	EGFP	CELIII	P. berghei	(Yoshida et al. 2007)
	minos	An. stephensi antiplatelet	EGFP	dsRED	P. falciparum	(Yoshida & Watanabe 2006)
	piggyBac	An. gambiae vitellogenin	dsRED	EGFP	-	(Chen et al. 2007b)
	piggyBac	D. melanogaster hsp70	dsRED	minos transposase EGFP	-	(Scali et al. 2007)
	minos	D. melanogaster actin 5C	EGFP		-	
	piggyBac + attP docking site	3XP3	ECFP	-	-	(Amenya et al. 2010)
Cx. quinquefasciatus	Hermes	D. melanogaster actin 5C	EGFP	EGFP	-	(Allen et al. 2001)

(a) *Drosophila* Relish-related gene lacking the transactivator domain.
(b) REL1-A lacking a C-terminal domain.

Table 1. Germ line transformed mosquitoes (Modified from (Coutinho-Abreu et al. 2010)).

3. Paratransgenesis to reduce vector competence

Paratransgenesis usually refers to the use of genetically modified symbiotic organisms expressing molecules that can block pathogen development or transmission by vectors. Bacteria symbionts of blood sucking bugs (Durvasula et al. 1997; Durvasula et al. 1999; Durvasula et al. 2008), tsetse flies (Cheng & Aksoy 1999), and mosquitoes (Favia et al. 2007; Jin et al. 2009), and symbiotic viruses of *An. gambiae* (Ren et al. 2008) and *Ae. aegypti* (Carlson et al. 1995; Ward et al. 2001; Carlson et al. 2006), have been used (Figure 2). Recently, genetically modify entomopathogenic fungi strains have induced high levels of P. falciparum mortality in An. gambiae (Fang et al. 2011). Current data indicate that symbionts expressing molecules targeting pathogen development have the potential to reduce transmission in endemic regions, and appear unrelated to any fitness load (Durvasula et al. 1997; Cheng & Aksoy 1999). As with transgenesis, spread of transformed symbionts also would benefit from the availability of a gene drive system to replace non-transformed symbionts present in natural vector populations.

3.1 Transformation of reduviids, tsetse, and mosquitoes with bacterial symbionts

Paratransgenesis in disease vectors was demonstrated through the expression of cecropin A by *Rhodococcus rhodnii* within the midgut of the kissing bug (reduviid) *Rhodnius prolixus* (Durvasula et al. 1997). A 99% reduction in the intensity of *Trypanosoma cruzi* infection in the hindgut of *R. prolixus* was observed without interfering with insect fitness. Additionally, transformed symbionts were shown to be horizontally transmitted to *R. prolixus* carrying non-transformed symbionts via reduviid coprophagic habits (Durvasula et al. 1997). Subsequently, functionally active antibody fragments also were successfully expressed in the guts of *R. prolixus* (Durvasula et al. 1999) and *Triatoma infestans* (Durvasula et al. 2008) utilizing symbionts. Transformed symbionts were stably maintained within the gut of the insects without need for antibiotic selection (Durvasula et al. 1997; Durvasula et al. 1999; Durvasula et al. 2008).

Paratransgenesis seems to be a promising strategy to reduce African trypanosomes transmission by tsetse flies. Genetically transformed *Sodalis*, a symbiont of tsetse flies commonly found in the midgut and hemolymph of *Glossina m. morsitans*, *Glossina p. palpalis*, *Glossina austeni*, and *Glossina brevipalpis*, and the salivary glands of *Gl. p. palpalis*, is transmitted vertically via the female milk glands (Cheng & Aksoy 1999; Weiss et al. 2006; Aksoy et al. 2008). In addition, when *Sodalis* originally isolated from of *Gl. m. morsitans* and *Gl. fuscipes* was transformed with GFP, the *recSodalis* obtained colonized septic non-native tsetse host species at a density similar to a native colonization and without reducing host fitness (Weiss et al. 2006).

Symbiotic bacteria also have been isolated from *An. stephensi* (Favia et al. 2007; Riehle et al. 2007). One such symbiont, *Asaia sp.*, was successfully transformed with plasmids expressing GFP (Damiani et al. 2008) or with dsRED gene cassette inserted into bacterium genome (Damiani et al. 2008). *Asaia* was found in mosquito tissues, such as midgut and salivary gland, which are sites for pathogen development, as well as in male and female reproductive tracts, supporting bacteria transovarial and venereal transmission (Riehle et al. 2007; Damiani et al. 2008). Additionally, larval stages can acquire such bacteria strain from the environment (Riehle et al. 2007).

3.2 Transformation of viral symbionts

Symbiotic densovirus also can be genetically manipulated to express molecules to reduce vector competence. Densoviruses are linear single-stranded DNA viruses with the genome packaged in a non-enveloped particle. These viruses are suitable vectors for expression of foreign genes in mosquitoes because they are highly specific, environmentally stable, kill mosquito larvae in a dose-dependent manner, decrease lifespan of surviving adults, and are transmitted vertically (Carlson et al. 1995; Carlson et al. 2006). In *Ae. aegypti*, densoviruses can spread to fat body, muscles, and nerves (Ward et al. 2001) following infection through the anal papillae. Densoviruses infecting *Ae. aegypti* (AeDNV) and *An. gambiae* (AgDNV) were isolated and modified to express GFP (Ward et al. 2001; Ren et al. 2008). The green phenotype obtained by the expression of GFP in recombinant AgDNV-infecting *An. gambiae* was observed in 20% of F2 and F3 generations, suggesting that transformed densoviruses may be used to express molecules targeting pathogen development in mosquitoes (Ren et al. 2008).

3.3 Transformation of entomopathogenic fungi

Several studies demonstrated that entomopathogenic fungi are capable of reducing mosquito life span as well as vector competence (Carlson et al. 2006; Thomas & Read 2007; Cook et al. 2008; Read et al. 2009). Blanford et al. (2005) observed a high mortality rate (55-80%) in mosquitoes 7-14 days following infection with the fungus *Beauveria bassiana*, which was suggestive of the ability of entomopathogenic fungi to drastically reduce pathogen transmission in endemic areas. However, fungus-mediated killing is a slower process compared to chemical approaches, and critics have suggested that the use of entomopathogenis fungi alone is incapable of controlling mosquitoes in malaria-endemic areas.

More recently, genetically modified entomopathogenic fungus *Metarhizium anisopliae* expressing molecules that affect the development of *P. falciparum* in *An. gambiae* were generated (Fang et al. 2011). Specifically, *M. anisopliae* expressing the SM1 peptide ([SM]$_8$), the anti-microbial peptide scorpin, and a single chain antibody targeting a plasmodium surface protein were shown to reduce sporozoite load by 90% without affecting mosquito fitness. Moreover, co-expression of a [SM]$_8$-scorpin fusion protein along with scorpin led to nearly elimination of sporozoite infection in the salivary glands of the infected mosquitoes (Fang et al, 2011). The results using the transgenic *Metarhizium* suggest that this could be a powerful approach to control malaria transmission.

3.4 Future of vector paratransgenesis

Although a low number of bacteria symbionts have been transformed to date, a potential advantage of this approach over transgenesis is lack of fitness load (Durvasula et al. 1997; Weiss et al. 2006). Also, the alternate use of genetically modified symbiotic viruses (instead of bacterial symbionts) may provide additional tools against pathogen development. Densoviruses efficiently expresses heterologous proteins in *An. gambiae* and *Ae. aegypti* and are transmitted vertically (Carlson et al. 2006; Ren et al. 2008). Viral symbionts can be engineered to express single chain antibodies (scFv), blocking pathogen development (de Lara Capurro et al. 2000). Recombinant Sindbis expressing transcripts from an infecting virus genome can reduce viral load (of the infecting virus) in mosquitoes (Powers et al. 1996; Travanty et al. 2004). The results from Sindbis expressing transcripts from LaCrosse (LAC), dengue (DEN), or yellow fever (YF) viruses indicated a substantial interference with the replication of these viruses in *Aedes triseriatus* (LAC) and in *Ae. aegypti* (DEN and YF) (Olson et al. 1996; Powers et al. 1996; Higgs et al. 1998; Adelman et al. 2001). Such viral replication inhibition is accomplished by the mosquito RNAi machinery (Cirimotich et al. 2009).

Despite early successes with transformation of insect vector symbionts, it is not known if transformed symbionts can replace non-transformed in natural insect populations, and potentially affect pathogen development and transmission in natural habitats. Symbionts seem to have no fitness load on insect hosts and are capable of being transmitted vertically (via trans-ovarian transmission) or laterally (due to feeding habits). Thus, a strong gene drive system can potentiate the effectiveness of paratransgenesis. *Wolbachia* endosymbionts have been proposed as such gene drive system (Aksoy et al. 2008).

Wolbachia are intracellular, maternally inherited bacteria that manipulate reproduction of insects via cytoplasmatic incompatibility (CI) (Sinkins & Gould 2006). Due to the effects of CI, a *Wolbachia*-uninfected female will not breed with infected males successfully, reducing the frequency of uninfected individuals and increasing the frequency of *Wolbachia*-infected insects in a population (Sinkins & Gould 2006). Thus, other maternally inherited transformed symbionts would be spread within an insect population in association with *Wolbachia* (Aksoy et al. 2008), increasing the frequency of the transformed symbiont. This mechanism has been observed in *Ae. aegypti*, *Aedes albopictus*, and *Culex quinquefasciatus* (Sinkins & Gould 2006), representing a potential manner to spread transformed symbionts, such as densovirus, in natural populations of mosquitoes.

A life-shortening strain of *Wolbachia* (*w*MelPop) identified in *D. melanogaster* was recently introduced into *Ae. aegypti* (McMeniman et al. 2009) and *An. gambiae* (Jin et al. 2009). Beyond promoting the spread of the transformed symbiont across the mosquito population (i.e., acting as a gene drive mechanism), this strain of *Wolbachia* was also thought to reduce the time frame available (i.e., mosquito life span) for pathogen development within the mosquito (known as the extrinsic incubation period or EIP) (McMeniman et al. 2009). Thus, the application of *w*MelPop to eliminate disease vectors may lead to a reduction of the pathogen developmental time (or EIP) within the vector (Read & Thomas 2009). Counter to this argument is the possibility that, *Wolbachia*, as well as densovirus and entomopathogenic fungi potentially target older mosquitoes over younger ones, and are considered evolution-proof mosquitocidal biocontrols agents (Read et al. 2009). For *Anopheles*, due to loss of fecundity per gonotrophic cycle in natural conditions (from 20% to 40% per gonotrophic cycle, (Killeen et al. 2000)), a selective pressure on pathogen developmental time already exists, especially in the case of *Plasmodium*-infected mosquitoes (Read et al. 2009). Thus, such selective pressure from the addition of *Wolbachia* is likely not strong enough to shorten the parasite life cycle within the vector.

wMelPop infection reduces the feeding ability of old mosquitoes [16] and activates the mosquito immune system (particularly antimicrobial peptides), leading to reduction of filarial worms [17], dengue virus, and plasmodium parasite load (Moreira et al. 2009), including *P. falciparum* (Hughes et al. 2011).

As wMelPop and entomopathogenic fungi are capable of reducing vector competence (linear parameter) and vector survivorship (exponential parameter), these two effects combined may significantly reduce vectorial capacity and human malaria burden in endemic areas.

4. Transmission-blocking vaccines (TBVs)

Transmission-blocking vaccines (or TBVs) aim at interfering and/or blocking pathogen development within the vector, halting transmission to non-infected vertebrate host (depicted in Figure 1C). TBVs usually rely on immunization of vertebrate hosts (either infected or uninfected) with molecules derived from the pathogen or the vector in order to reduce pathogen transmission from infected to uninfected hosts. Such molecules (i.e., antigens) may be inoculated into the vertebrate host as purified proteins inducing the host immune system to produce specific antibodies (Singh & O'Hagan 1999). Alternatively, antibodies can be raised by inoculating the host with recombinant DNA plasmids containing the gene encoding such molecules (Lobo et al. 1999; Coban et al. 2004; Kongkasuriyachai et al. 2004; LeBlanc et al. 2008). The expression and secretion of the specific protein into host tissues induce the immune system to produce antibodies against such proteins (Abdulhaqq & Weiner 2008). To boost the immune response of the vertebrate, antigens are usually inoculated in conjunction with adjuvants. The mechanisms by which adjuvants improve the immune response are still poorly understood (Singh & O'Hagan 1999; Aguilar & Rodriguez 2007). The specific antibodies produced against pathogen and/or vector antigens will interfere with the development of the pathogen within the vector following a blood meal on a vaccinated and infected individual. Other insect-based vaccines, such as sialome-based vaccines (Valenzuela 2004; Oliveira et al. 2009) and insecticidal vaccines (Foy et al. 2003) that are, in some cases, dependent on cell-mediated immune-response in order to prevent vertebrate host infection and reduce insect lifespan, respectively, are not discussed in details this chapter. These can be found in (Willadsen 2004; Billingsley et al. 2006; Titus et al. 2006; Billingsley et al. 2008; Dinglasan & Jacobs-Lorena 2008; Oliveira et al. 2009). Insecticidal vaccines, due to their potential to reduce vectorial capacity exponentially (Billingsley et al. 2006), are briefly discussed.

For a molecule to be an effective TBV candidate it has to induce high antibody titers in order to block completely pathogen development within the insect (Kubler-Kielb et al. 2007). Additionally, the antigen/adjuvant combination has to be safe enough to the vertebrate host so as to avoid side effects after immunization (Saul et al. 2007; Wu et al. 2008). Ideally, a TBV candidate antigen will display low levels of polymorphisms (in field isolates) so that a unique antigen may be used to produce a TBV capable of recognizing all the field variants of that specific antigen (Kocken et al. 1995; Drakeley et al. 1996; Duffy & Kaslow 1997; Sattabongkot et al. 2003). Alternatively, an effective TBV may need to combine different antigens because the combined action of the antibodies against such antigens may produce a more efficient transmission-blocking result (Duffy & Kaslow 1997; Gozar et al. 1998; Kongkasuriyachai et al. 2004).

4.1 Parasite antigen-based TBVs

Most of the studies on TBVs to date were conducted using antibodies targeting antigens expressed on the surface of sexual stage of malaria parasites (Figure 1C; Table 2). *P. falciparum* proteins Pfs25, Pfs28, Pfs48/45, and Pfs230, and their orthologs in *Plasmodium vivax*, have been tested in transmission-blocking assays (Quakyi et al. 1987; Kaslow et al. 1988; Duffy & Kaslow 1997; Hisaeda et al. 2000; Sattabongkot et al. 2003; Malkin et al. 2005; Outchkourov et al. 2008).

4.1.1 *P. falciparum*-derived TBV candidate – Pfs25

Pfs25 is a 25kDa protein expressed on the surface of zygote and ookinete stages of *P. falciparum* and consists of four tandem epidermal growth factor (EGF) domains (Kaslow et al. 1988). The TBV potential of Pfs25 was demonstrated using the Vaccinia virus as delivery systems of this antigen to mammalian hosts (Kaslow et al. 1991), or using recombinant Pfs25 expressed in yeast (Barr et al. 1991; Kaslow et al. 1991).

4.1.2 *P. falciparum*-derived TBV candidate – Pfs28

Pfs28 is a 28kDa *P. falciparum* conserved protein expressed on the surface of retorts, a transitional stage between zygote and ookinete. This antigen also was tested in transmission blocking activity assays. Antibodies produced by the injection of yeast-expressed Pfs28 (yPfs28), in the presence of alum, significantly reduced the infectivity of *An. freeborni* mosquitoes with *P. falciparum*. Lower infectivity was exhibited when vaccination was carried out with yPfs28 and yPfs25 antigens injected together (Duffy & Kaslow 1997).

Transmission blocking activity against *P. falciparum* was further improved when Pfs25 and Pfs28 were expressed as a unique chimeric protein in yeast, the 25-28c recombinant protein. Vaccination with the 25-28c recombinant protein led to complete arrest of oocyst development earlier, using a lower dose and for a greater amount of time, than vaccination with either Pfs25 or Pfs28 alone or a combination of both (Gozar et al. 1998).

4.1.3 *P. falciparum*-derived TBV candidate – Pfs48/45

Another TBV candidate to control spread of *P. falciparum* is Pfs48/45. The *Pfs48/45* gene encodes a unique protein that migrates as a double band under non-reducing conditions (Milek et al. 2000). This protein is expressed on *P. falciparum* gametocyte and gamete surfaces and has a central role in male gamete fertility (van Dijk et al. 2001). Immunization of mice with this recombinant protein led to production of antibody titers that were capable of reducing *P. falciparum* oocyst intensity in *An. stephensi* by at least 88% in 11 out of 12 assays (Outchkourov et al. 2008).

Furthermore, with regards to the application of Pfs48/45 as a potential TBV against malaria, the variability of Pfs48/45 from culture and field isolates from many countries was analyzed (Kocken et al. 1995; Drakeley et al. 1996). The results obtained indicated low levels of polymorphism in the overall gene among either *in vitro* cultures or field isolates (Kocken et al. 1995; Drakeley et al. 1996).

4.1.4 *P. falciparum*-derived TBV candidate – Pfs230

Another *P. falciparum* protein tested in TBV assays was Pfs230, a 230kDa protein expressed on the surface of gametocytes. Although antibodies against Pfs230 blocked the development of *P. falciparum* in the midguts of *An. freeborni*, the transmission blocking activity of anti-Pfs230 monoclonal antibodies was completely lost when complement was inactivated. Thus, the blocking activity of anti-Pfs-230 antibodies was detected only when complement proteins were present (Quakyi et al. 1987).

4.1.5 *P. vivax*-derived TBV candidates – Pvs25 and Pvs28

P. vivax sexual stage surface proteins, orthologs of *P. falciparum* TBV candidates, also have been isolated and tested in transmission blocking experiments. Pvs25, a Pfs25 ortholog, is expressed on the surfaces of the insect-stages, zygotes and mature ookinetes, whereas Pvs28, a Pfs28 ortholog, is mainly expressed on retorts and mature ookinetes (Hisaeda et al. 2000).

Transmission blocking experiments using antibodies against either Pvs25 or Pvs28 were tested (Hisaeda et al. 2000). Four species of mosquitoes were artificially fed on a mixture of *P. vivax*-infected chimpanzee blood in the presence of antibodies (raised in mice co-injected with alum). *P. vivax* ookinete development was completely blocked by the anti-serum against Pvs25 (Hisaeda et al. 2000). Vaccination against Pvs25 and Pvs28 also presented efficient transmission blocking activity against *P. vivax* isolated from human patients, despite polymorphism in these proteins (Sattabongkot et al. 2003).

Transmission blocking activity of Pvs25 has been evaluated in phase 1 human trials. The results from the study revealed significant interference in *P. vivax* development within mosquito midgut caused by the human anti-Pvs25 sera. Additionally, long lasting antibody titers were elevated, and no reactogenicity (side effects) was observed (Malkin et al. 2005). Nevertheless, higher antibody titers are necessary for successful control of *P. falciparum* transmission by mosquitoes in endemic areas (Malkin et al. 2005). Unfortunately, a second phase 1 trial, using Pvs25 as a potential TBV using Montanide ISA 51 as an adjuvant was halted due to induced local and systemic reactions in the vacinees (Wu et al. 2008).

4.2 Other pathogen molecule-based TBV candidates

In regard to proteins expressed on the surface of parasites (other than *Plasmodium*) transmitted to humans by insect vectors (Table 2), only a limited number has been tested as potential TBVs (Tonui et al. 2001a; Saraiva et al. 2006).

In *Leishmania major*, the two most abundant surface antigens, LPG and gp63, were tested as transmission blocking vaccines. *Phlebotomus dubosqci* sand flies were partially fed on mice immunized with purified native LPG, recombinant gp63 (rgp63) expressed in bacteria, crude *L. major* lysate (WPA), or a cocktail of LPG and rgp63. The sand flies were subsequently fed on *L. major*-infected mice. The results indicated that serum against WPA and the two protein-cocktail exhibited greater *L. major* blocking activity than sera against either LPG or gp63 (Tonui et al. 2001a). However, blocking of *L. major* development was due to damage of the midgut epithelial layer, probably caused by immune-active substances present in the blood of the pre-vaccinated mice (Tonui et al. 2001b).

Interestingly, a commercially available treatment for canine visceral leishmaniasis (Leishmune®) was recently shown to function as a TBV in sand flies (Saraiva et al. 2006). Leishmune® (FML-vaccine) is a protective vaccine made of *L. donovani* fucose-manose ligand and the adjuvant saponin, which was successfully tested in a phase III vaccine trial (da Silva et al. 2000). Although the surface molecule (FML) was isolated from *L. donovani*, Leishmune® exhibited transmission blocking activity in the New World sand fly *Lutzomyia longipalpis* when infected with *Leishmania infantum chagasi* (Saraiva et al. 2006). Antibodies produced in dogs following Leishmune® injection reduced *Lu. longipalpis* infectivity by 79.3% and parasite load by 74.3% even after 12 months of immunization (Saraiva et al. 2006).

4.3 Vector-based TBVs

Proteins expressed within insect vector tissues and that may interact with pathogens also have been tested as TBV candidates (Table 3). Vector-based TBV candidates include (structural) proteins that are expressed by the insect midgut (Lal et al. 2001), midgut enzymes that play a role in blood digestion (Lavazec et al. 2007), and parasite receptors expressed by the epithelial cells lining the midgut (Kamhawi et al. 2004; Dinglasan et al. 2007)

In mosquitoes, polyclonal antibodies against *An. gambiae* midgut proteins nearly completely reduced the intensity of *P. falciparum* oocysts (98%) and sporozoites (96%) within *An. stephensi* tissues. Also, *An. gambiae*-derived anti-midgut monoclonal antibodies inhibited development of *P. falciparum* and *P. vivax* in different *Anopheles* species (Lal et al. 2001). Additionally, these antibodies also can be used to reduce insect vector densities (vector-blocking vaccines) because they reduce vector survivorship and fecundity (Lal et al. 2001). Antibodies against carboxypeptidase cpbAg1 from *An. gambiae* reduced *P. falciparum* infectivity by more than 92% seven days after an infectious artificial blood feeding (Lavazec et al. 2007). In addition to the effect on the number of oocysts per infected mosquito, anti-cpbAg1 strongly reduced mosquito progeny (Lavazec et al. 2007). Antibodies to a midgut aminopeptidase (AgAPN1), which is one of the *P. falciparum* receptors in the *An. gambiae* midgut, were used to reduce *P. falciparum* oocyst intensity in *An. gambiae* and *An. stephensi* by 73% and 67%, respectively (Dinglasan et al. 2007).

Another molecule expressed on the surface of midgut cells that may serve as receptor for parasite attachment has also been assessed as a TBV candidate (Kamhawi et al. 2004). The galectin-like PpGalec characterized from the midgut of the sand fly *P. papatasi* is a receptor for *L. major* lipophosphoglycan (LPG). *P. papatasi* artificially fed on blood mixed sera from PpGalec-immunized mice displayed a reduction of 86% of *L. major* midgut infection. Moreover, no infectious metacyclic forms were detected from the flies fed on anti-PpGalec sera (Kamhawi et al. 2004).

4.4 Future of TBVs

In addition to identifying TBV candidates that are effective and may span different insect vector species, studies on TBV development must include antigenic variability present in field isolates (Kocken et al. 1995; Drakeley et al. 1996; Duffy & Kaslow 1997; Sattabongkot et al. 2003), immunogenicity of such antigens (Kubler-Kielb et al. 2007), reactogenicity caused by adjuvants (Saul et al. 2007; Wu et al. 2008), non-specific responses (Quakyi et al. 1987; Tonui et al. 2001a), and improper folding of antigens (Kaslow et al. 1994; Milek et al. 1998a; Milek et al. 1998b; Milek et al. 2000). Natural antigenic boosting is another important issue that must be dealt with (Arevalo-Herrera et al. 2005).

Antigens expressed on the surface of insect-stage parasites have been postulated as TBV candidates because they seem not to be under the selective pressure mediated by the vertebrate immune system.

Another interesting aspect of TBVs is the possibility of natural boosting of the immune response of animals infected with a pathogen (i.e., pre-immunized) (Milek et al. 1998a; Arevalo-Herrera et al. 2005). Hence, candidate TBV proteins expressed on the surface of both insect-stage and blood-stage pathogens may induce activation of the immune response in infected hosts vaccinated with the same antigens (Arevalo-Herrera et al. 2005). However, this approach may not be suitable to every TBV, such as Pvs25 which displays low expression in blood-stage *P. vivax* (Arevalo-Herrera et al. 2005), and has yet to be demonstrated for the *Plasmodium* TBV-antigen candidates that are expressed during gametocytogenesis, for example, Pfs230 (Quakyi et al. 1987) and Pfs48/45 (Milek et al. 1998a). Proper folding of the TBV candidate protein following expression via recombinant techniques also may affect the efficacy of the vaccinating antigen. Thus, the system of choice for recombinant expression can significantly affect the outcome of the TBV candidate.

In regards to vector-based TBV candidate molecules, the number of TBV antigens available is limited and must to be increased to target other vector species. In addition to assessing a TBV candidate molecule that prevents pathogen development within insect vector tissues, an effect on the vector survivorship is also one of the main objectives.

Reduction of vector survival is thought to interfere exponentially with vectorial capacity (Black IV & Moore 2004; Billingsley et al. 2006; Billingsley et al. 2008), as the time available for pathogen development within the vector is significantly shortened. Despite several studies showing that insect feeding on blood of animals immunized with insect tissue homogenates exhibit reduced survivorship, most of these studies suffered from high experimental variability (Billingsley et al. 2006). However, one study has shown that immunization with a unique insect molecule (mucin) can induce an immune response capable of killing insect vectors via a cell-mediated response (Foy et al. 2003). Thus, an ideal TBV antigen should reduce parasite development, reducing vector competence (a linear parameter in the vectorial capacity equation), as well as vector survivorship (the exponential parameter). These two effects associated can lead to thorough reduction of vectorial capacity and disease burden in endemic areas.

TBV could also be able to reduce survivorship of different species of insect vectors, via immunization with conserved antigens, as proposed by (Canales et al. 2009), providing protection to pathogens transmitted by different vectors. However, significant cross-species effects have yet to be demonstrated.

TBV antigen (antigen origin)	Antigen production	Antigen name	Vaccinated animal (adjuvant)	Sera dilution (concentration)	Targeted pathogen	Insect vector	Mean oocyst number	Infectivity %	Reference
Pfs25 (P. falciparum)	Virus (Vaccinia virus)	rPfs25^	Mouse (Ribi)	1:2 (200µg/ml) 1:4 (100µg/ml) 1:8 (50µg/ml) 1:16 (25µg/ml)	P. falciparum	An. freeborni	0.1 1 1.4 3.5	1% 11% 16% 39%	(Kaslow et al. 1991)
	Yeast (S. cerevisiae)	Pfs25-B	Mouse (FCA) (MTP*)	1:2 1:2	P. falciparum	An. freeborni	0 0	<1% 0%	(Barr et al. 1991)
			Monkey@ (MTP*)	1:2			0-2.6#	0-10%	
	Yeast (S. cerevisiae)	Pfs25-B	Mouse (Alum) Monkey@ (Alum)	Neat Neat	P. falciparum	An. freeborni	0 0-3.8& 0-12.8$ 19.9-28.2$$	0% - -	(Kaslow et al. 1994)
	DNA vaccination	VR1020/25	Mouse	1:5 1:10	P. falciparum	An. stephensi	0.17-0.26 0.19-0.39	3.4-3.8% 2.6-4.3%	(Lobo et al. 1999)
	DNA vaccination/ Yeast (S. cerevisiae)	VR1020/25 and Pfs25	Monkey< (Montanide ISA 720)	1:2 1:4 1:8	P. falciparum	An. stephensi An. gambiae?	0.9-3.1 2.4-6.4 3.2-10.4	5-17% 14-38% 19-62%	(Coban et al. 2004)
	Yeast (P. pastoris)	Pfs25	Mouse (cholera toxin)	1:2 1:8 1:32	P. falciparum	An. dirus	0 0 10	0% 0% 21.3%	(Arakawa et al. 2005)
	Yeast (P. pastoris)	Pfs25	Human (Montanide ISA 51)	-	P. falciparum	An. stephensi	-	<10%	(Wu et al. 2008)
	DNA vaccination (in vivo electroporation)	Pfs25	Mouse	1:2> 1:4> 1:8> 1:16>	P. falciparum	An. gambiae	1.0 3.4 9.5 51.9	2.5% 9% 24% 100%	(LeBlanc et al. 2008)

TBV antigen (antigen origin)	Antigen production	Antigen name	Vaccinated animal (adjuvant)	Sera dilution (concentration)	Targeted pathogen	Insect vector	Mean oocyst number	Infectivity %	Reference
Pfs28 (P. falciparum)	Yeast (S. cerevisiae)	yPfs28	- (FCA+Ribi) (Alum)	Neat##	P. falciparum	An. freeborni	0-0.33& 0.21&	0-8% 3%	(Duffy & Kaslow 1997)
		yPfs25+yPfs28+	- (FCA+Ribi)	1:40			0.047-0.16	7.6-9%	
Pfs25-Pfs28 (P. falciparum)	Yeast (S. cerevisiae)	25-28C	Mouse (Alum)	Neat##	P. falciparum	An. freeborni	0	0%	(Gozar et al. 1998)
Pfs45/48 (P. falciparum)	Bacteria (refolded in vitro)	Pfs45/48-10C	Mouse (FCA)	1:2 (10µg/ml)	P. falciparum	An. stephensi	0.45	0.06%	(Outchkourov et al. 2007)
	Bacteria (w/chaperonins)	Pfs45/48-10C	Mouse (FCA)	1:2 (10µg/ml)	P. falciparum	An. stephensi	0-5.1	0-12%	(Outchkourov et al. 2008)
Pfs230 (P. falciparum)	Purification	Pfs230^	Rabbit (FCA)	- (100µg/ml)	P. falciparum	An. freeborni	0.2-4	0.3-5.8%	(Quakyi et al. 1987)
Pvs25 (P. vivax)	Yeast (S. cerevisiae)	Pvs25	Mouse (Alum)	1:2	P. vivax	An. stephensi	0	0%	(Hisaeda et al. 2000)
	Yeast (-)	Pvs25	Mouse (Alum) Rabbit (Alum)	1:2 1:8 1:2 1:8	P. vivax	An. dirus	0.18 1.26 4.25 4.06	- - - -	(Sattabongkot et al. 2003)
	DNA vaccination	DV25 DV28 DV25+DV28	Mouse	1:8 1:10 1:10 1:10 1:8 1:10 1:10	P. vivax	An. freeborni An. freeborni An. gambiae An. freeborni An. gambiae	3.17 0.4 0.4 1.5 0.8 0.04 0.8	86% 87% 87% 86% 93% 99% 93%	(Kongkasuriyachai et al. 2004)
	Yeast (S. cerevisiae)	Pvs25	Monkey[1] (Montanide ISA 720)	1:4	P. vivax	An. albimanus	0.0-0.04	-	(Arevalo-Herrera et al. 2005)
	Yeast (S. cerevisiae)	Pvs25H	Human (Alhydrogel®)	Neat## 1:1	P. vivax	An dirus	- -	- -	(Malkin et al. 2005)

TBV antigen (antigen origin)	Antigen production	Antigen name	Vaccinated animal	Sera dilution (concentration)	Targeted pathogen	Insect vector	Mean oocyst number	Infectivity %	Reference
Pvs25 (P. vivax)	Yeast (S. cerevisiae)	Pvs25	Human (Montanide ISA 51)	-	-	-	-	-	(Wu et al. 2008)
	Yeast (S. cerevisiae)	Pvs25	Monkey< (Montanide ISA 720) (Alum)	1:2 1:8 1:32 1:2 1:8 1:32	P. vivax	An. freeborni	0+ 1.41 1.63 1.71 1.3 2.05	0 1.6 32.8 2.9 10.6 34.2	(Collins et al. 2006)
Pvs28 (P. vivax)	Yeast (S. cerevisiae)	Pvs28	Mouse (Alum)	1:2	P. vivax	An. freeborni	0.91	0.7%	(Hisaeda et al. 2000)
	Yeast (S. cerevisiae)	Pvs28	Mouse (Alum) Rabbit (Alum)	1:2 1:8 1:2 1:8	P. vivax	An. dirus	0.11 1.31 10.73 4.79	- - - -	(Sattabongkot et al. 2003)
WPA (L. major)	Purified	Whole cell lysate	Mouse	-	L. major	P. duboscqi	-	25%	(Tonui et al. 2001a)
rgp63 (L. major)	Bacteria	gp63	Mouse	-	L. major	P. duboscqi	-	40%	(Tonui et al. 2001a)
LPG (L. major)	Purified	LPG	Mouse	-	L. major	P. duboscqi	-	43.3%	(Tonui et al. 2001a)
LPG+rgp63 (L. major)	Purified + Bacteria	LPG+rgp63	Mouse	-	L. major	P. duboscqi	-	37.5%	(Tonui et al. 2001a)
Leishmune® (L. donovani)	Purification	FML	Dog (Saponin)	1:1	L. major	L. longipalpis	-	30.6%	(Saraiva et al. 2006)

(^) Monoclonal antibodies were used in transmission blocking assays; (') Freund's complete adjuvant; (*) Muramyl tripeptide; (®) Aotus trivirgatus; (#) Oocysts present in midguts of mosquitoes fed on sera from monkeys immunized 22 weeks before challenge. No oocysts were present in mosquitoes that fed on sera from animals immunized 12 weeks before challenge; (-) Undetermined; (&) Seven days after 3rd immunization; ($) Sixty one days after 3rd immunization; ($$) Eighty nine days after 3rd immunization; (<) Macaca mulatta; (?) Similar results using An. stephensi; (>)Immunization with 20μg of plasmid; (+) Sera diluted 1:40; (++) TBV assayed 204 days after immunization; (¹) Aotus lemurinus griseimembr; (!!) Sera were not previously diluted prior to mixing with equal or greater amount of blood for insect feeding.

Table 2. **Transmission blocking vaccines based on pathogens molecules** (Modified from (Coutinho-Abreu & Ramalho-Ortigao 2010)).

5. Conclusion

The technology for generating transgenic and paratransgenic insects to fight vector borne diseases is well established in several laboratories, and the use of such strategies is upon us. However, many studies still need to be performed in order to improve both the design and the efficiency of transgenic insects in preventing disease transmission. In addition, many aspects related to potential environmental impacts of the release of transgenic insects in nature need to be clarified. On the other hand, TBVs may also emerge as a feasible approach against several vector borne diseases, including leishmaniasis and malaria (Malkin et al. 2005; Saraiva et al. 2006). This assumption is supported by at least two recent cases. The first, a fuccose-mannose ligand-based TBV was previously tested during a phase III trial (da Silva et al. 2000; Saraiva et al. 2006) and became a commercialized drug in Brazil (Palatnik-de-Sousa et al. 2008) against canine visceral leishmaniasis. Another is based on the *P. vivax* Pvs25 antigen use as a TBV, based on which was approved during a phase I trial (Malkin et al. 2005; Saraiva et al. 2006).

However, regardless of the approach to be developed, it is clear to many investigators that new technologies to be combined with existing approaches against vector-borne disease are necessary to reduce the burden of such diseases.

6. References

Abdulhaqq, S. A. and D. B. Weiner (2008). "DNA vaccines: developing new strategies to enhance immune responses." *Immunol Res* 42(1-3): 219-32.

Abraham, E. G., M. Donnelly-Doman, et al. (2005). "Driving midgut-specific expression and secretion of a foreign protein in transgenic mosquitoes with AgAper1 regulatory elements." *Insect Mol Biol* 14(3): 271-9.

Adelman, Z. N., C. D. Blair, et al. (2001). "Sindbis virus-induced silencing of dengue viruses in mosquitoes." *Insect Mol Biol* 10(3): 265-73.

Adelman, Z. N., N. Jasinskiene, et al. (2007). "nanos gene control DNA mediates developmentally regulated transposition in the yellow fever mosquito Aedes aegypti." *Proc Natl Acad Sci U S A* 104(24): 9970-5.

Aguilar, J. C. and E. G. Rodriguez (2007). "Vaccine adjuvants revisited." *Vaccine* 25(19): 3752-3762.

Aksoy, S., B. Weiss, et al. (2008). "Paratransgenesis applied for control of tsetse transmitted sleeping sickness." *Adv Exp Med Biol* 627: 35-48.

Allen, M. L., D. A. O'Brochta, et al. (2001). "Stable, germ-line transformation of Culex quinquefasciatus (Diptera: Culicidae)." *J Med Entomol* 38(5): 701-10.

Amenya, D. A., M. Bonizzoni, et al. (2010). "Comparative fitness assessment of Anopheles stephensi transgenic lines receptive to site-specific integration." *Insect Mol Biol* 19(2): 263-9.

Amenya, D. A., M. Bonizzoni, et al. (2011). "Comparative fitness assessment of Anopheles stephensi transgenic lines receptive to site-specific integration." *Insect Mol Biol* 19(2): 263-9.

Arakawa, T., A. Komesu, et al. (2005). "Nasal immunization with a malaria transmission-blocking vaccine candidate, Pfs25, induces complete protective immunity in mice against field isolates of Plasmodium falciparum." *Infect Immun* 73(11): 7375-80.

Arevalo-Herrera, M., Y. Solarte, et al. (2005). "Induction of transmission-blocking immunity in Aotus monkeys by vaccination with a Plasmodium vivax clinical grade PVS25 recombinant protein." *Am J Trop Med Hyg* 73(5 Suppl): 32-7.

Atkinson, P. W. (2004). Trangenic Mosquitoes and DNA Research Safeguards. *Biology of Disease Vectors*. W. C. Marquardt. Burlington, Elservier Academic Press: 785.

Barr, P. J., K. M. Green, et al. (1991). "Recombinant Pfs25 protein of Plasmodium falciparum elicits malaria transmission-blocking immunity in experimental animals." *J Exp Med* 174(5): 1203-8.

Bian, G., S. W. Shin, et al. (2005). "Transgenic alteration of Toll immune pathway in the female mosquito Aedes aegypti." *Proc Natl Acad Sci U S A* 102(38): 13568-73.

Billingsley, P. F., J. Baird, et al. (2006). "Immune interactions between mosquitoes and their hosts." *Parasite Immunol* 28(4): 143-53.

Billingsley, P. F., B. Foy, et al. (2008). "Mosquitocidal vaccines: a neglected addition to malaria and dengue control strategies." *Trends Parasitol* 24(9): 396-400.

Black IV, W. C. and C. G. Moore (2004). Population biology as a tool to study vector-borne diseases. *Biology of disease vectors*. W. C. Marquardt. Burlington, Elservier Academic Press: 785.

Blanford, S., B. H. Chan, et al. (2005). "Fungal pathogen reduces potential for malaria transmission." *Science* 308(5728): 1638-41.

Boete, C. and J. C. Koella (2002). "A theoretical approach to predicting the success of genetic manipulation of malaria mosquitoes in malaria control." *Malar J* 1: 3.

Brown, A. E., L. Bugeon, et al. (2003). "Stable and heritable gene silencing in the malaria vector Anopheles stephensi." *Nucleic Acids Res* 31(15): e85.

Canales, M., V. Naranjo, et al. (2009). "Conservation and immunogenicity of the mosquito ortholog of the tick-protective antigen, subolesin." *Parasitol Res*.

Carlson, J., K. Olson, et al. (1995). "Molecular genetic manipulation of mosquito vectors." *Annu Rev Entomol* 40: 359-88.

Carlson, J., E. Suchman, et al. (2006). "Densoviruses for control and genetic manipulation of mosquitoes." *Adv Virus Res* 68: 361-92.

Catteruccia, F., J. P. Benton, et al. (2005). "An Anopheles transgenic sexing strain for vector control." *Nat Biotechnol* 23(11): 1414-7.

Catteruccia, F., H. C. Godfray, et al. (2003). "Impact of genetic manipulation on the fitness of Anopheles stephensi mosquitoes." *Science* 299(5610): 1225-7.

Catteruccia, F., T. Nolan, et al. (2000). "Stable germline transformation of the malaria mosquito Anopheles stephensi." *Nature* 405(6789): 959-62.

Chen, C. H., H. Huang, et al. (2007a). "A synthetic maternal-effect selfish genetic element drives population replacement in Drosophila." *Science* 316(5824): 597-600.

Chen, X. G., O. Marinotti, et al. (2007b). "The Anopheles gambiae vitellogenin gene (VGT2) promoter directs persistent accumulation of a reporter gene product in transgenic Anopheles stephensi following multiple bloodmeals." *Am J Trop Med Hyg* 76(6): 1118-24.

Cheng, Q. and S. Aksoy (1999). "Tissue tropism, transmission and expression of foreign genes in vivo in midgut symbionts of tsetse flies." *Insect Mol Biol* 8(1): 125-32.

Cirimotich, C. M., J. C. Scott, et al. (2009). "Suppression of RNA interference increases alphavirus replication and virus-associated mortality in Aedes aegypti mosquitoes." *BMC Microbiol* 9: 49.

Coates, C. J., N. Jasinskiene, et al. (1998). "Mariner transposition and transformation of the yellow fever mosquito, Aedes aegypti." *Proc Natl Acad Sci U S A* 95(7): 3748-51.

Coates, C. J., N. Jasinskiene, et al. (1999). "Promoter-directed expression of recombinant firefly luciferase in the salivary glands of Hermes-transformed Aedes aegypti." *Gene* 226(2): 317-25.

Coban, C., M. T. Philipp, et al. (2004). "Induction of Plasmodium falciparum transmission-blocking antibodies in nonhuman primates by a combination of DNA and protein immunizations." *Infect Immun* 72(1): 253-9.

Collins, W. E., J. W. Barnwell, et al. (2006). "Assessment of transmission-blocking activity of candidate Pvs25 vaccine using gametocytes from chimpanzees." *Am J Trop Med Hyg* 74(2): 215-21.

Cook, P. E., C. J. McMeniman, et al. (2008). "Modifying insect population age structure to control vector-borne disease." *Adv Exp Med Biol* 627: 126-40.

Coutinho-Abreu, I. V. and M. Ramalho-Ortigao (2010). "Transmission blocking vaccines to control insect-borne diseases: a review." *Mem Inst Oswaldo Cruz* 105(1): 1-12.

Coutinho-Abreu, I. V., K. Y. Zhu, et al. (2010). "Transgenesis and paratransgenesis to control insect-borne diseases: current status and future challenges." *Parasitol Int* 59(1): 1-8.

da Silva, V. O., G. P. Borja-Cabrera, et al. (2000). "A phase III trial of efficacy of the FML-vaccine against canine kala-azar in an endemic area of Brazil (Sao Goncalo do Amaranto, RN)." *Vaccine* 19(9-10): 1082-92.

Damiani, C., I. Ricci, et al. (2008). "Paternal transmission of symbiotic bacteria in malaria vectors." *Curr Biol* 18(23): R1087-8.

de Lara Capurro, M., J. Coleman, et al. (2000). "Virus-expressed, recombinant single-chain antibody blocks sporozoite infection of salivary glands in Plasmodium gallinaceum-infected Aedes aegypti." *Am J Trop Med Hyg* 62(4): 427-33.

Dinglasan, R. R. and M. Jacobs-Lorena (2008). "Flipping the paradigm on malaria transmission-blocking vaccines." *Trends Parasitol* 24(8): 364-70.

Dinglasan, R. R., D. E. Kalume, et al. (2007). "Disruption of Plasmodium falciparum development by antibodies against a conserved mosquito midgut antigen." *Proc Natl Acad Sci U S A* 104(33): 13461-6.

Drakeley, C. J., M. T. Duraisingh, et al. (1996). "Geographical distribution of a variant epitope of Pfs48/45, a Plasmodium falciparum transmission-blocking vaccine candidate." *Mol Biochem Parasitol* 81(2): 253-7.

Duffy, P. E. and D. C. Kaslow (1997). "A novel malaria protein, Pfs28, and Pfs25 are genetically linked and synergistic as falciparum malaria transmission-blocking vaccines." *Infect Immun* 65(3): 1109-13.

Durvasula, R. V., A. Gumbs, et al. (1997). "Prevention of insect-borne disease: an approach using transgenic symbiotic bacteria." *Proc Natl Acad Sci U S A* 94(7): 3274-8.

Durvasula, R. V., A. Gumbs, et al. (1999). "Expression of a functional antibody fragment in the gut of Rhodnius prolixus via transgenic bacterial symbiont Rhodococcus rhodnii." *Med Vet Entomol* 13(2): 115-9.

Durvasula, R. V., R. K. Sundaram, et al. (2008). "Genetic transformation of a Corynebacterial symbiont from the Chagas disease vector Triatoma infestans." *Exp Parasitol* 119(1): 94-8.

Engels, W. R. (1997). "Invasions of P elements." *Genetics* 145(1): 11-5.

Fang, W., Vega-Rodriguez, J., et al. (2011). "Development of trangenic fungi that kill human malaria parasites in mosquitoes". *Science* 331: 1074-7.

Favia, G., I. Ricci, et al. (2007). "Bacteria of the genus Asaia stably associate with Anopheles stephensi, an Asian malarial mosquito vector." *Proc Natl Acad Sci U S A* 104(21): 9047-51.

Foy, B. D., T. Magalhaes, et al. (2003). "Induction of mosquitocidal activity in mice immunized with Anopheles gambiae midgut cDNA." *Infect Immun* 71(4): 2032-40.

Franz, A. W., I. Sanchez-Vargas, et al. (2006). "Engineering RNA interference-based resistance to dengue virus type 2 in genetically modified Aedes aegypti." *Proc Natl Acad Sci U S A* 103(11): 4198-203.

Franz, A. W., I. Sanchez-Vargas, et al. (2009). "Stability and loss of a virus resistance phenotype over time in transgenic mosquitoes harbouring an antiviral effector gene." *Insect Mol Biol* 18(5): 661-72.

Fraser, M. J., G. E. Smith, et al. (1983). "Acquisition of Host Cell DNA Sequences by Baculoviruses: Relationship Between Host DNA Insertions and FP Mutants of Autographa californica and Galleria mellonella Nuclear Polyhedrosis Viruses." *J Virol* 47(2): 287-300.

Fu, G., R. S. Lees, et al. "Female-specific flightless phenotype for mosquito control." *Proc Natl Acad Sci U S A* 107(10): 4550-4.

Fu, G., R. S. Lees, et al. (2010). "Female-specific flightless phenotype for mosquito control." *Proc Natl Acad Sci U S A* 107(10): 4550-4.

Garrett-Jones, C. (1964). "Prognosis for interruption of the malaria transmission through assessment of the mosquito vectorial capacity." *Nature* 204(4964): 1173-1175.

Garrett-Jones, C. and G. R. Shidrawi (1969). "Malaria vectorial capacity of a population of Anopheles gambiae." *Bull. Wld. Hlth. Org.* 40: 531-545.

Ghosh, A. K., P. E. Ribolla, et al. (2001). "Targeting Plasmodium ligands on mosquito salivary glands and midgut with a phage display peptide library." *Proc Natl Acad Sci U S A* 98(23): 13278-81.

Gozar, M. M., V. L. Price, et al. (1998). "Saccharomyces cerevisiae-secreted fusion proteins Pfs25 and Pfs28 elicit potent Plasmodium falciparum transmission-blocking antibodies in mice." *Infect Immun* 66(1): 59-64.

Grossman, G. L., C. S. Rafferty, et al. (2001). "Germline transformation of the malaria vector, Anopheles gambiae, with the piggyBac transposable element." *Insect Mol Biol* 10(6): 597-604.

Hickey, D. A. (1982). "Selfish DNA: a sexually-transmitted nuclear parasite." *Genetics* 101(3-4): 519-31.

Higgs, S., J. O. Rayner, et al. (1998). "Engineered resistance in Aedes aegypti to a West African and a South American strain of yellow fever virus." *Am J Trop Med Hyg* 58(5): 663-70.

Hisaeda, H., A. W. Stowers, et al. (2000). "Antibodies to malaria vaccine candidates Pvs25 and Pvs28 completely block the ability of Plasmodium vivax to infect mosquitoes." *Infect Immun* 68(12): 6618-23.

Horn, C. and E. A. Wimmer (2003). "A transgene-based, embryo-specific lethality system for insect pest management." *Nat Biotechnol* 21(1): 64-70.

Hughes, G. L., R. Koga, et al. (2011). "Wolbachia infections are virulent and inhibit the human malaria parasite Plasmodium falciparum in anopheles gambiae." *PLoS Pathog* 7(5): e1002043.

Irvin, N., M. S. Hoddle, et al. (2004). "Assessing fitness costs for transgenic Aedes aegypti expressing the GFP marker and transposase genes." *Proc Natl Acad Sci U S A* 101(3): 891-6.

Isaacs, A. T., F. Li, et al. (2011). "Engineered resistance to Plasmodium falciparum development in transgenic Anopheles stephensi." *PLoS Pathog* 7(4): e1002017.

Ito, J., A. Ghosh, et al. (2002). "Transgenic anopheline mosquitoes impaired in transmission of a malaria parasite." *Nature* 417(6887): 452-5.

James, A. A. (2011). "Genetics-based field studies prioritize safety." *Science* 331(6016): 398.

Jasinskiene, N., C. J. Coates, et al. (1998). "Stable transformation of the yellow fever mosquito, Aedes aegypti, with the Hermes element from the housefly." *Proc Natl Acad Sci U S A* 95(7): 3743-7.

Jin, C., X. Ren, et al. (2009). "The virulent Wolbachia strain wMelPop efficiently establishes somatic infections in the malaria vector Anopheles gambiae." *Appl Environ Microbiol* 75(10): 3373-6.

Kamhawi, S., M. Ramalho-Ortigao, et al. (2004). "A role for insect galectins in parasite survival." *Cell* 119(3): 329-41.

Kaslow, D. C., I. C. Bathurst, et al. (1994). "Saccharomyces cerevisiae recombinant Pfs25 adsorbed to alum elicits antibodies that block transmission of Plasmodium falciparum." *Infect Immun* 62(12): 5576-80.

Kaslow, D. C., S. N. Isaacs, et al. (1991). "Induction of Plasmodium falciparum transmission-blocking antibodies by recombinant vaccinia virus." *Science* 252(5010): 1310-3.

Kaslow, D. C., I. A. Quakyi, et al. (1988). "A vaccine candidate from the sexual stage of human malaria that contains EGF-like domains." *Nature* 333(6168): 74-6.

Killeen, G. F., F. E. McKenzie, et al. (2000). "A simplified model for predicting malaria entomologic inoculation rates based on entomologic and parasitologic parameters relevant to control." *Am J Trop Med Hyg* 62(5): 535-44.

Kim, W., H. Koo, et al. (2004). "Ectopic expression of a cecropin transgene in the human malaria vector mosquito Anopheles gambiae (Diptera: Culicidae): effects on susceptibility to Plasmodium." *J Med Entomol* 41(3): 447-55.

Kocken, C. H., R. L. Milek, et al. (1995). "Minimal variation in the transmission-blocking vaccine candidate Pfs48/45 of the human malaria parasite Plasmodium falciparum." *Mol Biochem Parasitol* 69(1): 115-8.

Kokoza, V., A. Ahmed, et al. (2000). "Engineering blood meal-activated systemic immunity in the yellow fever mosquito, Aedes aegypti." *Proc Natl Acad Sci U S A* 97(16): 9144-9.

Kokoza, V., A. Ahmed, et al. (2001). "Efficient transformation of the yellow fever mosquito Aedes aegypti using the piggyBac transposable element vector pBac[3xP3-EGFP afm]." *Insect Biochem Mol Biol* 31(12): 1137-43.

Kokoza, V. A. and A. S. Raikhel (2011)"Targeted gene expression in the transgenic Aedes aegypti using the binary Gal4-UAS system." *Insect Biochem Mol Biol* 41(8): 637-44.

Kongkasuriyachai, D., L. Bartels-Andrews, et al. (2004). "Potent immunogenicity of DNA vaccines encoding Plasmodium vivax transmission-blocking vaccine candidates

Pvs25 and Pvs28-evaluation of homologous and heterologous antigen-delivery prime-boost strategy." *Vaccine* 22(23-24): 3205-13.

Krafsur, E. S., C. J. Whitten, et al. (1987). "Screwworm eradication in North and Central America." *Parasitol Today* 3(5): 131-7.

Kubler-Kielb, J., F. Majadly, et al. (2007). "Long-lasting and transmission-blocking activity of antibodies to Plasmodium falciparum elicited in mice by protein conjugates of Pfs25." *Proc Natl Acad Sci U S A* 104(1): 293-8.

Lal, A. A., P. S. Patterson, et al. (2001). "Anti-mosquito midgut antibodies block development of Plasmodium falciparum and Plasmodium vivax in multiple species of Anopheles mosquitoes and reduce vector fecundity and survivorship." *Proc Natl Acad Sci U S A* 98(9): 5228-33.

Lampe, D. J., T. E. Grant, et al. (1998). "Factors affecting transposition of the Himar1 mariner transposon in vitro." *Genetics* 149(1): 179-87.

Lavazec, C., C. Boudin, et al. (2007). "Carboxypeptidases B of Anopheles gambiae as targets for a Plasmodium falciparum transmission-blocking vaccine." *Infect Immun* 75(4): 1635-42.

LeBlanc, R., Y. Vasquez, et al. (2008). "Markedly enhanced immunogenicity of a Pfs25 DNA-based malaria transmission-blocking vaccine by in vivo electroporation." *Vaccine* 26(2): 185-92.

Li, C., M. T. Marrelli, et al. (2008). "Fitness of transgenic Anopheles stephensi mosquitoes expressing the SM1 peptide under the control of a vitellogenin promoter." *J Hered* 99(3): 275-82.

Lobo, C. A., R. Dhar, et al. (1999). "Immunization of mice with DNA-based Pfs25 elicits potent malaria transmission-blocking antibodies." *Infect Immun* 67(4): 1688-93.

Lobo, N. F., A. Hua-Van, et al. (2002). "Germ line transformation of the yellow fever mosquito, Aedes aegypti, mediated by transpositional insertion of a piggyBac vector." *Insect Mol Biol* 11(2): 133-9.

Lombardo, F., T. Nolan, et al. (2005). "An Anopheles gambiae salivary gland promoter analysis in Drosophila melanogaster and Anopheles stephensi." *Insect Mol Biol* 14(2): 207-16.

Malkin, E. M., A. P. Durbin, et al. (2005). "Phase 1 vaccine trial of Pvs25H: a transmission blocking vaccine for Plasmodium vivax malaria." *Vaccine* 23(24): 3131-8.

Marrelli, M. T., C. Li, et al. (2007). "Transgenic malaria-resistant mosquitoes have a fitness advantage when feeding on Plasmodium-infected blood." *Proc Natl Acad Sci U S A* 104(13): 5580-3.

Marshall, J. M. and C. E. Taylor (2009). "Malaria control with transgenic mosquitoes." *PLoS Med* 6(2): e20.

Mathur, G., I. Sanchez-Vargas, et al. (2010). "Transgene-mediated suppression of dengue viruses in the salivary glands of the yellow fever mosquito, Aedes aegypti." *Insect Mol Biol* 19(6): 753-63.

McMeniman, C. J., R. V. Lane, et al. (2009). "Stable introduction of a life-shortening Wolbachia infection into the mosquito Aedes aegypti." *Science* 323(5910): 141-4.

Milek, R. L., A. A. DeVries, et al. (1998a). "Plasmodium falciparum: heterologous synthesis of the transmission-blocking vaccine candidate Pfs48/45 in recombinant vaccinia virus-infected cells." *Exp Parasitol* 90(2): 165-74.

Milek, R. L., W. F. Roeffen, et al. (1998b). "Immunological properties of recombinant proteins of the transmission blocking vaccine candidate, Pfs48/45, of the human malaria parasite Plasmodium falciparum produced in Escherichia coli." *Parasite Immunol* 20(8): 377-85.

Milek, R. L., H. G. Stunnenberg, et al. (2000). "Assembly and expression of a synthetic gene encoding the antigen Pfs48/45 of the human malaria parasite Plasmodium falciparum in yeast." *Vaccine* 18(14): 1402-11.

Moreira, L. A., M. J. Edwards, et al. (2000). "Robust gut-specific gene expression in transgenic Aedes aegypti mosquitoes." *Proc Natl Acad Sci U S A* 97(20): 10895-8.

Moreira, L. A., J. Ito, et al. (2002). "Bee venom phospholipase inhibits malaria parasite development in transgenic mosquitoes." *J Biol Chem* 277(43): 40839-43.

Moreira, L. A., I. Iturbe-Ormaetxe, et al. (2009). "A Wolbachia symbiont in Aedes aegypti limits infection with dengue, Chikungunya, and Plasmodium." *Cell* 139(7): 1268-78.

Nirmala, X., O. Marinotti, et al. (2006). "Functional characterization of the promoter of the vitellogenin gene, AsVg1, of the malaria vector, Anopheles stephensi." *Insect Biochem Mol Biol* 36(9): 694-700.

Nolan, T., T. M. Bower, et al. (2002). "piggyBac-mediated germline transformation of the malaria mosquito Anopheles stephensi using the red fluorescent protein dsRED as a selectable marker." *J Biol Chem* 277(11): 8759-62.

O'Brochta, D. A., N. Sethuraman, et al. (2003). "Gene vector and transposable element behavior in mosquitoes." *J Exp Biol* 206(Pt 21): 3823-34.

Oliveira, F., R. C. Jochim, et al. (2009). "Sand flies, Leishmania, and transcriptome-borne solutions." *Parasitol Int* 58(1): 1-5.

Olson, K. E., S. Higgs, et al. (1996). "Genetically engineered resistance to dengue-2 virus transmission in mosquitoes." *Science* 272(5263): 884-6.

Outchkourov, N., A. Vermunt, et al. (2007). "Epitope analysis of the malaria surface antigen pfs48/45 identifies a subdomain that elicits transmission blocking antibodies." *J Biol Chem* 282(23): 17148-56.

Outchkourov, N. S., W. Roeffen, et al. (2008). "Correctly folded Pfs48/45 protein of Plasmodium falciparum elicits malaria transmission-blocking immunity in mice." *Proc Natl Acad Sci U S A* 105(11): 4301-5.

Palatnik-de-Sousa, C. B., F. Barbosa Ade, et al. (2008). "FML vaccine against canine visceral leishmaniasis: from second-generation to synthetic vaccine." *Expert Rev Vaccines* 7(6): 833-51.

Perera, O. P., I. R. Harrell, et al. (2002). "Germ-line transformation of the South American malaria vector, Anopheles albimanus, with a piggyBac/EGFP transposon vector is routine and highly efficient." *Insect Mol Biol* 11(4): 291-7.

Pinkerton, A. C., K. Michel, et al. (2000). "Green fluorescent protein as a genetic marker in transgenic Aedes aegypti." *Insect Mol Biol* 9(1): 1-10.

Powers, A. M., K. I. Kamrud, et al. (1996). "Molecularly engineered resistance to California serogroup virus replication in mosquito cells and mosquitoes." *Proc Natl Acad Sci U S A* 93(9): 4187-91.

Quakyi, I. A., R. Carter, et al. (1987). "The 230-kDa gamete surface protein of Plasmodium falciparum is also a target for transmission-blocking antibodies." *J Immunol* 139(12): 4213-7.

Read, A. F., P. A. Lynch, et al. (2009). "How to make evolution-proof insecticides for malaria control." *PLoS Biol* 7(4): e1000058.

Read, A. F. and M. B. Thomas (2009). "Microbiology. Mosquitoes cut short." *Science* 323(5910): 51-2.

Ren, X., E. Hoiczyk, et al. (2008). "Viral paratransgenesis in the malaria vector Anopheles gambiae." *PLoS Pathog* 4(8): e1000135.

Ribeiro, J. M. and M. G. Kidwell (1994). "Transposable elements as population drive mechanisms: specification of critical parameter values." *J Med Entomol* 31(1): 10-6.

Riehle, M. A., C. K. Moreira, et al. (2007). "Using bacteria to express and display anti-Plasmodium molecules in the mosquito midgut." *Int J Parasitol* 37(6): 595-603.

Robinson, A. S., G. Franz, et al. (1999). "Genetic sexing strains in the medfly, *Ceratitis capitata*: Development, mass rearing and field application." *Trends in Entomology* 2: 81-104.

Rodrigues, F. G., S. B. Oliveira, et al. (2006). "Germline transformation of Aedes fluviatilis (Diptera:Culicidae) with the piggyBac transposable element." *Mem Inst Oswaldo Cruz* 101(7): 755-7.

Rodrigues, F. G., M. N. Santos, et al. (2008). "Expression of a mutated phospholipase A2 in transgenic Aedes fluviatilis mosquitoes impacts Plasmodium gallinaceum development." *Insect Mol Biol* 17(2): 175-83.

Saraiva, E. M., A. de Figueiredo Barbosa, et al. (2006). "The FML-vaccine (Leishmune) against canine visceral leishmaniasis: a transmission blocking vaccine." *Vaccine* 24(13): 2423-31.

Sattabongkot, J., T. Tsuboi, et al. (2003). "Blocking of transmission to mosquitoes by antibody to Plasmodium vivax malaria vaccine candidates Pvs25 and Pvs28 despite antigenic polymorphism in field isolates." *Am J Trop Med Hyg* 69(5): 536-41.

Saul, A., M. Hensmann, et al. (2007). "Immunogenicity in rhesus of the Plasmodium vivax mosquito stage antigen Pvs25H with Alhydrogel and Montanide ISA 720." *Parasite Immunol* 29(10): 525-33.

Scali, C., T. Nolan, et al. (2007). "Post-integration behavior of a Minos transposon in the malaria mosquito Anopheles stephensi." *Mol Genet Genomics* 278(5): 575-84.

Shin, S. W., V. Kokoza, et al. (2003a). "Relish-mediated immune deficiency in the transgenic mosquito Aedes aegypti." *Proc Natl Acad Sci U S A* 100(5): 2616-21.

Shin, S. W., V. A. Kokoza, et al. (2003b). "Transgenesis and reverse genetics of mosquito innate immunity." *J Exp Biol* 206(Pt 21): 3835-43.

Singh, M. and D. O'Hagan (1999). "Advances in vaccine adjuvants." *Nat Biotechnol* 17(11): 1075-81.

Sinkins, S. P. and F. Gould (2006). "Gene drive systems for insect disease vectors." *Nat Rev Genet* 7(6): 427-35.

Thomas, D. D., C. A. Donnelly, et al. (2000). "Insect population control using a dominant, repressible, lethal genetic system." *Science* 287(5462): 2474-6.

Thomas, M. B. and A. F. Read (2007). "Can fungal biopesticides control malaria?" *Nat Rev Microbiol* 5(5): 377-83.

Titus, R. G., J. V. Bishop, et al. (2006). "The immunomodulatory factors of arthropod saliva and the potential for these factors to serve as vaccine targets to prevent pathogen transmission." *Parasite Immunol* 28(4): 131-41.

Tonui, W. K., P. A. Mbati, et al. (2001a). "Transmission blocking vaccine studies in leishmaniasis: I. Lipophosphoglycan is a promising transmission blocking vaccine molecule against cutaneous leishmaniasis." *East Afr Med J* 78(2): 84-9.

Tonui, W. K., P. A. Mbati, et al. (2001b). "Transmission blocking vaccine studies in leishmaniasis: II. Effect of immunisation using Leishmania major derived 63 kilodalton glycoprotein, lipophosphoglycan and whole parasite antigens on the course of L. major infection in BALB/c mice." *East Afr Med J* 78(2): 90-2.

Travanty, E. A., Z. N. Adelman, et al. (2004). "Using RNA interference to develop dengue virus resistance in genetically modified Aedes aegypti." *Insect Biochem Mol Biol* 34(7): 607-13.

Valenzuela, J. G. (2004). Blood-feeding arthropod salivary glands and saliva. *Biology of Disease Vectors*. W. C. Marquardt, Black IV, W. C., Freier, J. E., Hagedorn, H. H., Hemingway, J., Higgs, S., James, A. A., Kondratieff, B., Moore, C. G. Burlington, Elsevier Academic Press: 785.

van Dijk, M. R., C. J. Janse, et al. (2001). "A central role for P48/45 in malaria parasite male gamete fertility." *Cell* 104(1): 153-64.

Vreysen, M. J., K. M. Saleh, et al. (2000). "Glossina austeni (Diptera: Glossinidae) eradicated on the island of Unguja, Zanzibar, using the sterile insect technique." *J Econ Entomol* 93(1): 123-35.

Ward, T. W., M. S. Jenkins, et al. (2001). "Aedes aegypti transducing densovirus pathogenesis and expression in Aedes aegypti and Anopheles gambiae larvae." *Insect Mol Biol* 10(5): 397-405.

Weiss, B. L., R. Mouchotte, et al. (2006). "Interspecific transfer of bacterial endosymbionts between tsetse fly species: infection establishment and effect on host fitness." *Appl Environ Microbiol* 72(11): 7013-21.

Willadsen, P. (2004). "Anti-tick vaccines." *Parasitology* 129 Suppl: S367-87.

Wilson, R., J. Orsetti, et al. (2003). "Post-integration behavior of a Mos1 mariner gene vector in Aedes aegypti." *Insect Biochem Mol Biol* 33(9): 853-63.

Wimmer, E. A. (2003). "Innovations: applications of insect transgenesis." *Nat Rev Genet* 4(3): 225-32.

Wu, Y., R. D. Ellis, et al. (2008). "Phase 1 trial of malaria transmission blocking vaccine candidates Pfs25 and Pvs25 formulated with montanide ISA 51." *PLoS ONE* 3(7): e2636.

Yoshida, S., Y. Shimada, et al. (2007). "Hemolytic C-type lectin CEL-III from sea cucumber expressed in transgenic mosquitoes impairs malaria parasite development." *PLoS Pathog* 3(12): e192.

Yoshida, S. and H. Watanabe (2006). "Robust salivary gland-specific transgene expression in Anopheles stephensi mosquito." *Insect Mol Biol* 15(4): 403-10.

12

Role of GPI-Anchored Membrane Receptors in the Mode of Action of *Bacillus thuringiensis* Cry Toxins

Fernando Zúñiga-Navarrete, Alejandra Bravo,
Mario Soberón and Isabel Gómez
Instituto de Biotecnología-UNAM. Depto.
Microbiología Molecular.Cuernavaca, Morelos
Mexico

1. Introduction

Insect pests are the major cause of damage to commercially important agricultural crops. Chemical pesticides have long-term detrimental effects, leading to irreversible damage to the environment and elimination of natural predators. Also, several hundred insect species have developed resistance to one or more chemical insecticides. There is, therefore, a need for environmentally safe pest control to maintain sustainability of the environment. *Bacillus thuringiensis* (Bt) emerged as a valuable biological alternative in pest control, because of its advantages of specific toxicity against target insects, lack of polluting residues and safety to non-target organisms such as humans, other vertebrates and plants, and is completely biodegradable. Bt has been used as a biopesticide in agriculture, forestry and mosquito control and accounts for 95% of the 1% market share of biopesticides in the total pesticide market. However, insect resistance against Bt has been reported in many cases. Insects develop resistance to insecticides through mechanisms that reduce the binding of toxins to gut receptors (de Maagd et al., 2001).

Bacillus thuringiensis Cry toxins have been widely used in the control of insect pests either as spray products or expressed in transgenic crops. These proteins are pore-forming toxins with a complex mechanism of action that involves the sequential interaction with several toxin-receptors. Cry toxins are specific against susceptible larvae and although they are often highly effective, some insect pests are not affected by them or show low susceptibility. In addition, the development of resistance threatens their effectiveness, so strategies to cope with all these problems are necessary. In this chapter we will discuss and compare the different proteins that are involved in the mechanism of action of Cry toxins with special emphasis on GPI-receptors: Aminopeptidases and alkaline phosphatases. We will discuss how the mechanism of toxin-receptor interaction has an important role to design new strategies to improve insecticidal activity of Cry toxins. In addition we will discuss other insect gut proteins that have recently been shown to bind Cry toxins and that may be involved in Cry toxin action.

2. Cry toxins

Bacillus thuringiensis (Bt) is an endospore-forming bacterium characterized by the presence of a protein crystal within the cytoplasm of the sporulating cell. Individual Cry toxin has a defined spectrum of insecticidal activity, usually restricted to a few species in one particular order of Lepidoptera (butterflies and moths), Diptera (flies and mosquitoes), Coleoptera (beetles and weevils), Hymenoptera (wasps and bees), and also to nematodes, respectively (Rajamohan et al., 1998). A few toxins have an activity spectrum that spans two or three insect orders. For example, Cry1Ba is most notably active against the larvae of moths, flies, and beetles. The combination of toxins in a given strain, therefore, defines the activity spectrum of that strain. Cry proteins are defined as: a parasporal inclusion protein from Bt that exhibits toxic effects to a target organism, or any protein that has obvious sequence similarity to a known Cry protein (Schnepf et al., 1998).

To date, the tertiary structures of seven different Cry proteins, Cry1Aa, Cry2Aa, Cry3Aa, Cry3Bb, Cry4Aa, Cry4Ba and Cry8Ea have been determined by X-ray crystallography (Li et al., 1991; Grochulski et al., 1995; Galitsky et al., 2001; Morse et al., 2001; Boonserm et al., 2005; Boonserm et al., 2006; Guo et al., 2009). All these structures display a high degree of similarity with a three-domain organization (Figure 1), suggesting a similar mode of action of the Cry protein family even though they show very low amino acid sequence similarity. Cry toxins are classified by their primary amino acid sequence and more than 500 different *cry* gene sequences have been classified into 67 groups (Cry1–Cry67). They are globular molecules composed of three structural domains connected by single linkers. Domain I, a seven α-helix bundle, is implicated in membrane insertion, toxin oligomerization and pore formation. Domain II is a beta-prism of three anti-parallel β-sheets packed around a hydrophobic core with exposed loop regions that are involved in receptor recognition, and domain III, is a β-sandwich of two anti-parallel β-sheets. Both domain II and III are implicated in insect specificity by mediating specific interactions with different insect gut proteins (Bravo et al., 2007).

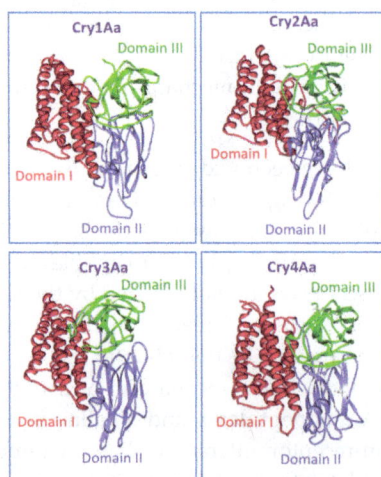

Fig. 1. Crystal structure of Cry1Aa (PDB code, 1CIY); Cry2Aa (PDB code, 1I5P); Cry3A (PDB code, 1DLC) and Cry4Aa (PDB code, 2C9K). Figures were generated using PyMol program.

2.1 Structure-function relationship

Domain I was immediately recognized as being equipped for pore formation, since shares structural similarities with other PFT like colicin Ia and N and diphtheria toxin, supporting the role of this domain in pore-formation. Isolated domain I fragments have been demonstrated to partition into model membranes to form pores (de Maagd et al., 2003); conversely, the domains II–III segment expressed without domain I was shown to bind to midgut membrane (Flores et al., 1997). Domain II was suspected to determine the specificity, because it represents the most divergent part of the toxin sequence, and exchanging domain II, or domains II and III, between closely related toxins has resulted in active hybrids showing altered specificity. Site-directed mutagenesis in its three hypervariable apical loops have identified residues involved in specific binding to membranes of several lepidopteran and coleopteran insects (Rajamohan et al., 1998; Schnepf et al., 1998). The specific binding consisted of reversible and irreversible steps, but only the irreversible binding is correlated with toxicity. More recent domain-exchange studies have found that toxicity of the hybrids to the insect host followed the movement of domain III, which would point to domain III as being responsible for the prerequisite step in toxicity, namely receptor binding. Site-directed mutagenesis in domain III located a small number of residues affecting membrane binding affinity and toxicity (de Maagd et al., 2001). However, direct observation of a toxin with a bound specificity determinant from the insect receptor is still needed to identify the receptor-binding site.

2.2 Molecular mode of action

The mode of action of Cry toxins has been characterized principally in lepidopteran insects and involves several steps and interactions with different receptors that depend on the oligomeric state of the toxin in a ping-pong binding mechanism. Cry1A toxins are produced as crystal inclusion bodies, which need to be ingested by the susceptible larvae to be toxic. These crystals are dissolved in the alkaline and reducing environment of the larval midgut, releasing soluble protoxins of 130 kDa. The inactive protoxins are then cleaved by midgut proteases yielding 60 kDa monomeric toxins (Soberon et al., 2009).

In figure 1, the activated monomeric toxins bind to highly abundant low affinity receptors, glycosylphosphatidylinositol (GPI)-anchored proteins, such as aminopeptidase N (APN) and alkaline phosphatase (ALP), localizing the toxin in the brush border microvilli. Specifically, loop 3 of domain II and β-16 of domain III are involved in this first interaction (Pacheco et al., 2009; Arenas et al., 2010). After this, the toxin binds to low abundant cadherin receptor, in a high affinity and complex interaction involving participation of loop 2, loop 3, and loop α-8 of domain II the toxin. Binding with cadherin facilitates additional protease cleavage of the N-terminal end of the toxin eliminating helix α-1 of domain I (Gomez et al., 2002). This cleavage induces assembly of an oligomeric form of the toxin. The conformational changes in toxin oligomers results in 100-fold increased binding affinity to APN and ALP receptors, through loop 2 (Bravo et al., 2004; Arenas et al., 2010). After the oligomers bind to these receptors they insert into membrane microdomains, creating pores in the apical membrane of midgut cells causing osmotic shock, bursting off the midgut cells and finally ending with the death of the insect (Soberon et al., 2009).

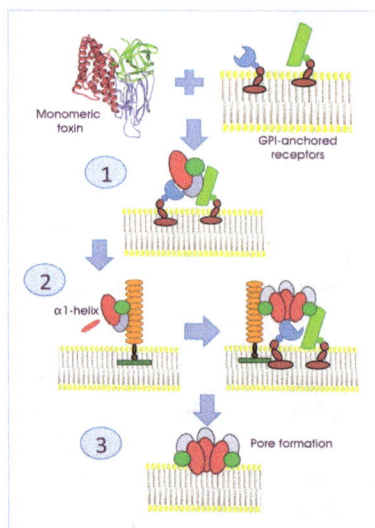

Fig. 2. Model of the mode of action of Cry toxins. 1) First interaction between monomeric toxin with low affinity-high abundant GPI-anchored receptors. 2) Second interaction occurs between monomeric toxin with cadherin receptor to induce oligomerization of the toxin. 3) The interaction of oligomeric toxin with GPI-anchored receptors result in pore formation.

2.3 Specifity=Binding receptor?

Although a given toxin has quite a narrow host range *in vivo*, *in vitro* many are non-specific and activated forms are even capable of forming pores in simple lipid bilayers. However, excessively high concentrations are required for lipid bilayer pore formation and the resulting pores have properties different than those formed in the presence of insect midgut membrane components. Many proteins have been identified in the midgut membrane that are capable of binding to Bt toxins but the ability to bind does not always correlate with susceptibility. For example, the pink bollworm (*Pectinophora gossypiella*) is susceptible to Cry1Ab and Cry1Ac toxins, both of which bind to a common site. A strain of pink bollworm (AZP-R) resistant to these two toxins could no longer bind Cry1Ab but, surprisingly, Cry1Ac binding was unaffected (Gonzalez-Cabrera et al., 2003). One explanation for such a result is that not all toxin binding is productive; that is, not all toxin binding results in pore-formation. Nevertheless, for several insect midgut proteins have been shown to have functional relevance to the toxic mechanism. The first is aminopeptidase N, whose expression in a *Drosophila* cell line resulted in an acquired susceptibility to toxin (Gill & Ellar 2002) and whose silencing by RNAi in a susceptible insect resulted in a reduced sensitivity to toxin (Rajagopal et al., 2002). Another membrane protein, cadherin, demonstrated a high toxin affinity, and mutations in cadherin genes result in a loss of sensitivity (Gahan et al., 2001; Morin et al., 2003).

3. Cry toxins receptors

As mentioned previously, Cry toxins are highly selective and kill only a limited number of insect species. This selectivity is mainly due to the interaction of Cry toxins with larval proteins located in the midgut epithelium cells. The crucial role of this receptor binding for toxicity is emphasized by the observation that insects selected for resistance to a Cry toxin often have no or reduced binding capacity for that toxin(Ferre & Van Rie 2002). A major research effort has taken place in the identification of insect proteins that bind Cry toxins and mediate toxicity. Among these, two major types of receptors have been identified: transmembrane proteins, such as cadherins, and proteins anchored to the membrane such as the GPI-anchored proteins that have been proposed to be involved in the action of Cry toxins (Gomez et al., 2007).

3.1 What kind of molecules are receptors for Cry toxins?

After it was demonstrated that specific high-affinity toxin binding sites are present in the midgut of susceptible insects, efforts to identify and clone this molecules has been intensified. Several putative Cry toxin receptors have since been reported, of which the best characterized are the aminopeptidase N (APN) receptors and the cadherin-like receptors identified in lepidopterans. In nematodes, glycolipids are believed to be an important class of Cry toxin receptors. Other putative receptors include alkaline phosphatases (ALPs), a 270-kDa glycoconjugate, and a 252-kDa protein (Pigott & Ellar 2007).

Protein	Insect order	Species
Cadherin	Lepidoptera	*Manduca sexta, Heliothis virescens, Ostrinia nubilalis, Helicoverpa armigera, Bombix mori, Pectinophora gossypiella*
	Diptera	*Anopheles gambiae, Aedes aegypti*
	Coleoptera	*Tenebrio molitor Diabrotica virgifera virgifera*
BTR-270	Lepidoptera	*Lymantria dispar*
Chlorophyllide-binding protein (252 protein)	Lepidoptera	*Bombix mori*
ADAM 10 metalloprotease	Coleoptera	*Leptinotarsa decemlineata*

Table 1. Receptors of Cry toxins (non-GPI anchored to the membrane) described in different insects.

3.1.1 Cadherins

Cadherin proteins represent a large family of glycoproteins that are responsible for intercellular contacts. These proteins are composed of an ectodomain formed by 11 to 12 cadherin repeats (CR), a transmembrane domain and an intracellular domain (Bel & Escriche 2006). In the case of CADR of the lepidopteran *Manduca sexta*, it was shown that this protein is located in the microvilli of midgut cells (Chen et al., 2005). Cry1A toxins binds to cadherin proteins of at least six lepidopteran species, *Manduca sexta, Bombyx mori, Heliothis virescens, Helicoverpa armigera, Pectinophora gossypiella* and *Ostrinia nubilalis* (Pigott & Ellar 2007).

The role of cadherin-like proteins as Cry1A toxin receptor was supported by the selection of a *Heliothis virescens* Cry1Ac-resistant line YHD2 that has a retrotransposon insertion mutation in the cadherin-like gene (Gahan et al., 2001). In addition, the characterization of CADR alleles in field-derived and in laboratory-selected Cry toxin-resistant strains of the cotton pest *Pectinophora gossypiella* (pink bollworm) revealed three mutated CADR alleles that were associated with Cry toxin resistance (Morin et al., 2003).

The interaction of Cry1A toxins with the CADR receptor is a complex process. Three regions in CADR proteins have been shown to interact with three domain II loop regions. Cry1Ab loop 2 interacts with CADR residues [865]NITIHITDTNN[875] located in repeat 7, whereas loops α-8 and 2 interact with CADR residues [1331]IPLPASILTVTV[1342] located in repeat 11. A third Cry1A binding region was located in CADR in the repeat 12. In the case of *H. virescens* cadherin, this binding epitope was narrowed to [1423]GVLTLNFQ[1431] and demonstrated that it interacts with Cry1Ac domain II loop 3 (Gomez et al., 2007).

A cadherin protein from *T. molitor* was identified as a Cry3Aa binding protein, and it was shown to facilitate Cry3Aa oligomer formation. Moreover, silencing of the cadherin gene by feeding dsRNA showed that the silenced beetles were resistant to Cry3Aa indicating an active role of cadherin on Cry3Aa toxicity (Fabrick et al., 2009). A cadherin protein was also identified as a Cry3Aa receptor in *Diabrotica virgifera virgifera*. A fragment of this cadherin protein containing the membrane proximal cadherin repeats 8-10 bound Cry3Aa and Cry3Bb toxins with high affinity (Kd of 12 and 1.4 nM, respectively) and enhanced Cry3Aa and Cry3Bb toxicity to different coleopteran insects (Park et al., 2009).

As in lepidopteran insects, cadherin proteins have been identified in *Ae. aegypti* and *An. gambiae* showing binding to Cry11Aa and Cry4Ba respectively (Hua et al., 2008; Chen et al., 2009). In *Ae. aegypti*, cadherin also serves a receptor of Cry11Ba toxin that was isolated from the Bt var *jegathesan* strain but showed lower affinity to Cry4Ba protein (Gill et al., 2011). An *An. gambiae* cadherin fragment containing the Cry4Ba binding site enhanced the toxicity of Cry4Ba in both *An. gambiae* and *Ae. aegypti* larvae suggesting its active role as a receptor of Cry4Ba in these mosquitoes species. In the case of *Ae. aegypti* cadherin, it was shown that an anti-cadherin antibody competed binding of Cry11Aa to *Ae. aegypti* BBMV. In both *Ae. aegypti* and *An. gambiae*, cadherin is located in the microvilli of the caeca and in the microvilli of the posterior gut cells, that are the same sites where Cry11Aa and Cry4Ba bind (Hua et al., 2008; Chen et al., 2009; Park et al., 2009)

3.2 Other molecules

In lepidopteran insects, another proteins and molecules different from cadherin have been identified as a 270 kDa glycoconjugate and a 250 kDa protein named P252 (Pigott & Ellar 2007).

The 270 kDa glycoconjugate was identified as Cry1Ac binding protein in *L. dispar* (Valaitis et al., 2001). Recently *B. mori* P252 that binds Cry1Ac was identified as a choraphyllide-binding protein (Pandian et al., 2008). In addition, glycolipids were proposed to act as Cry toxin receptors in lepidopteran insects as was demonstrated for the nematode *Caenorhabditis elegans* (Griffitts et al., 2005). In the case of *L. decemlineata*, an ADAM 3 metalloprotease was identified as Cry3Aa receptor. Binding of Cry3Aa to ADAM-3 through domain II loop 1 enhanced Cry3Aa pore-formation activity suggesting that this binding interaction is important for Cry3Aa toxicity (Ochoa-Campuzano et al., 2007). The only GPI-anchored protein identified in coleopteran insects as a putative Cry receptor was an ALP from *A. grandis* that bound Cry1B toxin (Martins et al., 2010).

3.2.1 GPI-anchored receptors: APN, ALP and glucosidase

Glycosylphosphatidylinositol-anchored proteins (GPI proteins) are eukaryotic exoplasmic membrane proteins that play very diverse biological functions including hydrolytic enzyme activity, transmembrane signaling, intracellular sorting and cell adhesion interaction. Besides these biological functions, GPI proteins such as alkaline phosphatases (AP-GPI), 5-nucleotidase, dipeptidase, aminopeptidase P, have been used primarily as markers of plasma membranes during their purification procedure. Later, most of GPI proteins were found to be resistant to Triton X-100 solubilization at low temperature in kidney brush border membranes, whereas transmembrane enzymes were solubilized (Nosjean et al., 1997). Similar observations were done on brush border membranes from insects where APN and ALP are both GPI-anchored proteins; these proteins are proposed to be selectively included in lipid rafts that are conceived as spatially differentiated liquid-ordered microdomains in cell membranes. The APN and ALP in *M. sexta* and *H. virescens*, in contrast to the CADR receptors, were shown to be located in lipid rafts. The interaction of pore-forming toxins with lipid rafts could result in additional cellular events, including toxin internalization, signal transduction and cellular response (Zhuang et al., 2002; Bravo et al., 2004). In table 2 are presented the GPI-ancored proteins identified as Cry toxins receptors.

The first Cry1A toxin-binding protein that was described was an APN protein in the lepidopteran *M. sexta*. This protein was glycosilated and anchored to the membrane by a GPI anchor. Since then, other GPI-anchored APNs have been recognized as Cry toxin receptors in different lepidopteran species such as *H. virescens*, *Spodoptera litura*, *H. armigera*, *B. mori*, *Lymantria dispar*, *Plutella xylostella*, and in the dipteran *An. Quadrimaculatus*, and *A. aegypti*. Phylogenetic analyses suggest that in lepidopteran insects there are at least five different APN families and at least three of them have been shown to bind Cry1 toxins in different insect species (Gomez et al., 2007; Pigott & Ellar 2007).

The APN has been implicated in toxin insertion, since cleavage of APN by phosphatidyl-inositol specific phospholipase C treatment which cleaves out the GPI anchored proteins substantially decreased the levels of Cry1Ab incorporation into insoluble lipid raft membranes (Bravo et al., 2004) and drastically reduced the pore formation activity of the toxin assayed in BBMV from *Trichoplusia ni* (Lorence et al. 1997). In addition, the incorporation of APN into the lipid bilayer enhanced Cry1Aa pore formation activity (Schwartz et al., 1997). The sugar GalNAc in the APN receptor is an important epitope in the interaction with Cry1Ac toxin (Burton et al., 1999; de Maagd et al., 1999). In the case of the lepidopteran *Lymantria dispar*, it was proposed that the monomeric Cry1Ac toxin interacts with APN following a sequential binding model. In this model, APN is first recognized by domain III of Cry1Ac through the GalNAc moiety, followed by a protein-protein contact of the domain II loop region of Cry1Ac. The first contact is fast and reversible, and mutations close to a domain III cavity affect this initial binding, while mutations in domain II affect the rate constants of the second interaction step which is slower and tighter (Jenkins et al., 2000). Li et al. reported that the binding of GalNAc to monomeric Cry1Ac correlates with an increase of temperature factors in the pore-forming domain I. However, there was no indication of a clear conformational change in the monomeric-Cry1Ac toxin (Li et al., 2001). In contrast, the fluorescence spectroscopy studies of Cry1Ac in its oligomeric state showed that GalNAc induces a conformational change in domain III of the oligomeric structure of Cry1Ac in the vicinity of the sugar pocket. The interaction of Cry1A-oligomer with GalNAc enhanced membrane insertion of the soluble pre-pore oligomeric structure, supporting the model that interaction of the Cry1A pre-pore with GPI-anchored receptors facilitates membrane insertion and pore-formation. The APN-oligomer interaction may be especially critical when low toxin protein concentrations reach the midgut epithelium, conditions that may occur in vivo in the larvae gut where the Cry toxins are exposed to high concentration of proteases (Pardo-Lopez et al., 2006).

A resistant *S. exigua* population that is resistant to Cry1Ca toxin was shown to lack the RNA transcript of APN-1, suggesting that this APN is involved in Cry1C toxicity to this insect species (Herrero et al., 2005). Finally, in the case of *S. litura*, silencing an APN with dsRNA resulted in a lower susceptibility to Cry1Ca toxin, also indicating a role of APN in Cry1C toxicity in this insect species (Herrero et al., 2005).

Two APN isoforms (AaeAPN1 and AaeAPN2) were identified in *Ae. aegypti* by Cry11Aa pull down experiments. Protein fragments from both APN isoforms were produced in *E. coli* and shown to inhibit binding of Cry11Aa to BBMV, suggesting their active role in Cry11Aa binding to insect membranes (Chen et al., 2009). In the case of *An. gambiae* and *An. quadrimaculatus* larvae, two APN's were also identified as Cry11Ba binding proteins. Interestingly, Cry11Ba binds both *An. quadrimaculatus* and *An. gambiae* APN molecules with a very high binding affinity of 0.56 nM and 6.4 nM respectively (Abdullah et al., 2006; Zhang et al., 2008). These results suggest that APN may have a more important role in the toxicity of Cry11Ba in these two Anopheline species. In fact, it was recently shown that certain *A. gambiae* APN protein fragments enhanced Cry11Ba toxicity as has been shown for cadherin protein fragments (Zhang et al., 2008).

In regard to the APN binding epitopes involved in Cry toxin interaction, a region of 63 residues (I135-P198) involved in Cry1Aa binding was identified in *B. mori* APN1 (Nakanishi et al., 2002). The domain III of Cry toxins is involved in the APN–Cry interaction, as shown by the interchange of domain III between Cry1Ac and Cry1Ab toxins (de Maagd et al., 1999). Domain III residues of Cry1Aa [508]STRVN[513] and [582]VFTLSAHV[589] were shown to be involved in binding I135-P198 APN fragment (Atsumi et al., 2005).

ALPs are found in all animals and as is expected are mainly localized in microvilli of columnar cells and of insect midgut epithelium cells (Eguchi 1995). ALPs can be divided into two groups: soluble (s-ALP) and membrane-bound (m-ALP) (Eguchi et al., 1990; Itoh et al., 1991). In insects, both ALPs are found in larval midgut epithelium cells; however, they are expressed in different cell types. The s-ALP is found exclusively in the cavity of goblet cells and in the apical region of the midgut; whereas, m-ALP is localized in the brush border membrane of columnar cells and particularly restricted to the middle and posterior midgut. Moreover, s- and m-ALPs show distinct differences in enzymatic activity (such as optimal pH) and also the structure of sugar side chain, suggesting that they perform different functions *in vivo* (Eguchi et al., 1990).

GPI-anchored Cry toxin-binding receptors [a]				
ORDER	SPECIES	PROTEIN	BINDING TOXIN	IDENTIFICATION METHOD
Lepidoptera	*Manduca sexta*	Class 1 APN	Cry1Ac[11], Cry1Aa, Cry1Ab[14]	Affinity chromatography [11], Chromatography purification, ligand blot, SPR [14]
		Class 2 APN	Cry1Ab5 [5]	Affinity chromatography, ligand blot [5]
		ALP	Cry1Ac [15]	PLPC treatment, 2D-SDS-PAGE-Mass spectrometry [15]
	Bombix mori	Class 1 APN	Cry1Aa [23], Cry1Aa, Cry1Ab [16]	Ion exchange chromatography purification [23], Heterologous expression, toxin overlay assay [16]
		Class 3 APN, Class 4 APN	Cry1Aa, Cry1Ab [16]	Heterologous expression, toxin overlay assay [16]
	Helicoverpa armigera	Class 1 APN	Cry1Aa, Cry1Ab, Cry1Ac [18]	Heterologous expression, toxin overlay assay [18]
		Class 3 APN	Cry1Ac [18]	Heterologous expression, ligand blot [18]
		ALP	Cry1Ac [21]	Anion exchange chromatography [21]
	Heliothis virescens	Class 1 APN	Cry1Aa, Cry1Ab, Cry1Ac [13], Cry1Fa [3]	Affinity chromatography, SPR [13], Affinity chromatography, ligand blot [3]
		Class 3 APN	Cry1Ac [9]	Affinity chromatography, ligand blot [9]
		Class 4 APN	Cry1Fa [3]	Affinity chromatography, SPR [3]
		ALP	Cry1Ac [10]	PLPC treatment, western blot [10]
	Lymantria dispar	Class 1 APN	Cry1Ac [22]	Immunolocalization [22]
		Class 3 APN	Cry1Ac [8]	Chromatographic purification [8]
	Plutella xylostella	Class 1 APN, Class 2 APN, Class 3 APN, Class 5 APN	Cry1Aa, Cry1Ab [16]	Heterologous expression, toxin overlay assay [16]
	Spodoptera litura	Class 3 APN, Class 5 APN	Cry1C [2]	*In silico* identification, heterologous expression, toxin overlay assay [2]
	Epiphyas postvittana	Class 3 APN	Cry1Aa, Cry1Ba [19]	Ion exchange chromatography purification [19]
Diptera	*Aedes aegypti*	AaeAPN1, AaeAPN2	Cry11Aa [4]	Pull-down, heterologous expression, ligand blot [4]
		APN	Cry4Ba [4]	2D SDS-PAGE, ligand blot [4]
		APN2778, APN2783, APN5808	Cry4Ba [20]	*In silico* identification, RNA interference [20]
		ALP	Cry4Ba [4], Cry11Aa [7]	2D SDS-PAGE, ligand blot [4], PLPC treatment, ligand blot [7]
	Anopheles albimanus	α-amilase	Cry4Ba, Cry11Aa [8]	PLPC treatment, ligand blot [8]
	Anopheles quadrimaculatus	APN	Cry11Aa [1]	Anion-exchange chromatography, enzymatic activity, SPR [1]
	Anopheles gambiae	APN	Cry11Ba [24]	*In silico* identification, heterologous expression, PLPC treatment, anion- exchange chromatography [24]
Coleoptera	*Anthonomus grandis*	ALP	Cry8Ka5 [17]	2D-SDS-PAGE, ligand blot, mass spectrometry [17]

(a) Compiled from: (1) Abdulah et. al., 2006; (2) Agrawal et. al., 2002; (3) Banks et. al., 2001; (4) Bayyareddy et. al., 2009; (5) Chen et. al., 2009; (6) Denolf et. al., 1997; (7) Fernández et. al., 2006; (8) Fernández-Luna et. al., 2010; (9) Garner et. al., 1999; (10) Gill et. al., 1995; (11) Jurat-Fuentes & Adang 2004; (12) Knight et. al., 1994; (13) Luo et. al., 1997; (14) Masson et. al., 1995; (15) McNall & Adang 2003; (16) Nakanishi et. al., 2002; (17) Nakasu et. al., 2010; (18) Rajagopal et. al., 2003; (19) Simpson & Newcomb 2000; (20) Saengwiman et. al., 2011; (21) Santosh & Singh P, 2011; (22) Valaitis et. al., 1997; (23) Yaoi et. al., 1997; (24) Zhang et. al., 2008.

Table 2. Description of GPI-anchored receptors founded in insects that bind to Cry toxins.

A GPI-anchored ALP that binds Cry toxins has been described in the lepidopterans *M. sexta*, *H. virescens* and *H. armigera*; in the dipteran *Ae. aegypti*, and in the coleopteran *Anthonomus grandis* (McNall et al., 2003; Jurat-Fuentes et al., 2004; Martins et al., 2010; Ning et al., 2010). In the case of *M. sexta*, ALP protein that bind Cry1A toxins was shown to be located in the microvilli of epithelial cells that is the site of action of Cry1A toxins (Chen et al., 2005). Also, an *H. virescens* laboratory selected population, YHD2, contained a retrotransposon insertion in the cadherin gene (Gahan et al., 2001). However, the mutation in the cadherin gene only accounted for 40-80% of the resistance phenotype. Additional mutations were responsible for the rest of the resistant phenotype in YHD2. These additional mutations were shown to affect GPI-ALP production, indicating that ALP is likely a functional receptor of Cry1Ac toxin in *H. virescens* (Jurat-Fuentes & Adang 2004).

Blocking the interaction of Cry toxins with GPI-anchored receptors has been useful in some cases to show the role of these proteins in Cry insecticidal activity. In the case of *M. sexta*, a scFv-phage that bound Cry1Ab toxin through β16-β22 of domain III blocked binding of Cry1Ab with APN but not with Bt-R1 and inhibited the toxicity of Cry1Ab in bioassays (Gomez et al., 2006). Nevertheless, in *B. mori* detached midgut cells, an anti-APN antibody did not affect toxicity of Cry1Aa in contrast to an anticadherin antibody that inhibited toxicity, suggesting either that this APN may not be involved in toxicity or that other additional GPI-anchored proteins or lipids could substitute APN function (Hara et al., 2003; Ibiza-Palacios et al., 2008). In *Ae. aegypti*, a peptide-phage that bound the 65 kDa ALP competed binding of the Cry11Aa to BBMVof mosquito and inhibited Cry11Aa toxicity in bioassays, suggesting that GPI-anchored ALP is a functional receptor of Cry11Aa (Fernandez et al., 2006).

Recently, ligand blot assays indicated that a 70 kDa GPI-anchored protein present in midgut brush border membrane vesicles of *A. albimanus* interacts with Cry4Ba and Cry11Aa toxins. This protein was identified as an a-amylase by mass spectrometry and enzymatic activity assays (Fernandez-Luna et al., 2010).

The fact that similar Cry binding proteins are involved in the mechanism of action of Cry toxins in both lepidopteran and dipteran insects suggests the Cry toxins have a conserved mode of action. However, the precise role of the Cry toxins receptors identified in mosquitoes in the mode of action of Cry toxins still remains to be determined. As in lepidopteran insects cadherin binding might facilitate oligomer formation while binding of Cry oligomer to GPI-anchored ALP or APN receptors might be necessary to facilitate membrane insertion. Nevertheless, the high binding affinity of Anopheline APN's to Cry11Ba is substantially different from what has been reported in lepidopteran insects and further studies on the differential role of APN and cadherin in monomer/oligomer binding in mosquitoes are necessary to determine the precise role of toxin binding to these receptor molecules.

3.3 Other insect molecules that bind Cry toxins identified using proteomic approaches

Proteomic approaches based on two-dimensional (2D) gel electrophoresis and mass spectrometry have been used to discover novel Bt toxin binding proteins and elucidate changes in midgut proteins associated with Bt resistance. Using this approach ALP was identified as a Cry1Ac binding protein in brush border of *Manduca sexta* and *Heliothis virescens* (McNall & Adang 2003; Krishnamoorthy et al., 2007). This identification was validated in *H. virescens* when ALP was demonstrated as a functional receptor molecule and loss of the enzyme correlated with Bt resistance to Cry1Ac (Jurat-Fuentes & Adang 2004). Candas et al. used differential-in-gel electrophoretic (DIGE) analysis to compare Bt susceptible and resistant larvae of *Plodia interpunctella*. These authors detected increased levels of midgut enzymes associated with oxidative metabolism and altered migration of an F_1F_0-ATPase in resistant larvae when compared to susceptible proteins on 2D gels. Candas et al. also detected reduced levels of alkaline chymotrypsin in the resistant *P. interpunctella* larvae associated with reduced capacity for protoxin activation (Candas et al., 2003).

Additional Cry1Ac binding proteins in lepidopteran brush border preparations detected by 2DE ligand blots approach includes actin, aminopeptidase, vacuolar-ATPase subunit A and a desmocollin-like protein (McNall & Adang 2003; Krishnamoorthy et al., 2007). A proteomics-based approach using differential-in-gel electrophoretic (DIGE) analysis quantified altered levels of specific proteins in Bt susceptible and resistant larvae of *Plodia interpunctella* (Candas et al., 2003). Those authors detected changes in the levels of APN, V-ATPase and an F_1F_0-ATPase in resistant larvae.

The analysis of Cry4Ba binding proteins by mass spectrometry in *Ae. aegypti* BBMV, revealed two lipid rafts associated proteins, flotillin and prohibitin, as well as cytoplasmic actin, besides ALP and APN, thus suggesting that additional proteins as well as intracellular proteins may have an active role in the mode of action of Cry toxins in mosquitoes (Bayyareddy et al., 2009).

Protein	Insect	Reference
ABC transporter	*Heliothis virescens* (Lepidoptera)	Gahan *et. al.*, 2010
	Helicoverpa armigera (Lepidoptera)	Chen *et. al.*, 2010
V-ATPase	*Heliothis virescens* (Lepidoptera) (A subunit)	Krishnamoorthy *et. al.*, 2007
	Anthonomus grandis (Coleoptera) (A subunit)	Nakasu, 2010
	Helicoverpa armigera (Lepidoptera) (B subunit)	Chen *et. al.*, 2010
	Aedes aegypti (Diptera) (E subunit)	Bayyareddy *et. al.*, 2009
Heat Shock Proteins	*Helicoverpa armigera* (Lepidoptera)	Chen *et. al.*, 2010
	Anthonomus grandis (Coleoptera)	Nakasu, 2010
Actin	*Manduca sexta* (Lepidoptera)	McNall *et al.*, 2003
	Heliothis virescens (Lepidoptera)	Krishnamoorthy, *et al.*, 2007
	Helicoverpa armigera (Lepidoptera)	Chen *et. al.*, 2010
	Aedes aegypti (Diptera)	Bayyareddy *et al.*, 2009
ATP sintase	*Aedes aegypti* (Diptera)	Bayyareddy, *et al.*, 2009
Flotilin	*Aedes aegypti* (Diptera)	Bayyareddy *et al.*, 2009
Prohibitin	*Aedes aegypti* (Diptera)	Bayyareddy *et al.*, 2009

Table 3. Receptors of Cry toxins described in different insects using proteomic analysis.

3.4 How these receptors are involved in resistance to Cry toxins

Resistance to Cry toxins can be developed by mutations in the insect pests that affect any of the steps of the mode of action of Cry toxins. The most common mechanism of toxin resistance in insect pests until now is the reduction in toxin binding to midgut cells, that in different insect species include mutations in Cry toxin receptors (Gahan et al., 2001; Ferre & Van Rie 2002; Morin, Biggs et al. 2003; Zhang, Hua et al. 2008). In fact, the most frequently phenotype of insect resistance, denoted as "Mode 1 of Resistance", is characterized by the reduction of one Cry1A toxin binding, cross resistance of Cry1Aa, Cry1Ab and Cry1Ac and lack of resistance to Cry1C. In several lepidopteran insects, the mode 1 of resistance is linked to mutations in the cadherin gene (Gahan et al., 2001; Morin et al., 2003; Xu et al., 2005). In field conditions three lepidopteran insect pests have evolved resistance to formulated Bt products, *Plodia interpunctella*, *Plutella xylostella* and *T. ni* [McGaughey, 1985; Tabashnik, et al, 1994; (Janmaat & Myers, 2003). In recent years, at least four cases of resistance to Bt crops have been documented, *H. zea* to Bt-cotton expressing Cry1Ac in United States, *S. frugiperda* to Bt-corn expressing Cry1F in Puerto Rico, *Busseola fusca* to Bt-corn expressing Cry1Ab in South Africa and *P. gossypiella* to Bt-cotton expressing Cry1Ac in India (Gill et al., 2011).

Recently, a resistant allele of a *H. virescens* resistant population was identified as a mutation in a gene coding for an ABC transporter molecule. This mutation affected binding of Cry1A toxin to brush border membrane vesicles indicating that this ABC transporter molecule is a novel Cry1A toxin receptor probably involved in the later stages of oligomer membrane insertion (Gahan et al., 2010). Also it was reported the correlation between reduced ALP protein, activity, and mALP expression levels in strains of three species in the Noctuidae Family with diverse resistance phenotypes against Cry toxins (*Heliothis virescens*, *Helicoverpa armigera* and *Spodoptera frugiperda*) (Jurat-Fuentes et al., 2011). Finally, in a recent work it was showed the role of aminopeptidase the resistance to the Bt toxin Cry1Ac in the cabbage looper, *Trichoplusia ni*, evolved in greenhouses, is associated with differential alteration of two midgut aminopeptidases N, APN1 and APN6, conferred by a trans-regulatory mechanism (Tiewsiri & Wang, 2011).

4. Conclusion

The mode of action of Cry toxins is a multi-step process that involves the interaction with several receptor molecules leading to membrane insertion and cells lysis. The characterization of the mode of action of Cry toxins in susceptible organisms will be important to fully understand the mode of action of this family of proteins. In the case of GPI-anchored receptors, APN and ALP have been identified in different insect species as Cry toxin binding molecules and in several insects they have been shown to be important for toxin action. In the case of *M. sexta*, both APN and ALP have been shown to have a similar role since Cry1Ab toxin show similar binding affinities to both molecules depending on the oligomer state of the toxin. Thus it seems that APN and ALP have redundant roles in the action of Cry toxins. Nevertheless, it was shown that ALP could have a predominant role in toxicity since it was preferentially express in young instar larvae that are more sensitive to the toxin in contrast to APN that was expressed later in larval development. Nevertheless, it could still be possible that in certain insect species both APN and ALP act as a complex receptor or that APN or ALP could be differentially important in toxicity depending in the insect species.

5. Acknowledgments

This research was supported by CONACyT 83135 and PAPIIT/UNAM IN209011-3.

IG wants to special thankful to L'ORÉAL-UNESCO-AMC fellowship to support this work.

6. References

Abdullah, M. A., A. P. Valaitis, et al. (2006). "Identification of a *Bacillus thuringiensis* Cry11Ba toxin-binding aminopeptidase from the mosquito, *Anopheles quadrimaculatus*." *BMC Biochem* 7: 16.

Atsumi, S., E. Mizuno, et al. (2005). "Location of the *Bombyx mori* aminopeptidase N type 1 binding site on *Bacillus thuringiensis* Cry1Aa toxin." *Appl Environ Microbiol* 71(7): 3966-77.

Bayyareddy, K., T. M. Andacht, et al. (2009). "Proteomic identification of *Bacillus thuringiensis subsp. israelensis* toxin Cry4Ba binding proteins in midgut membranes from *Aedes (Stegomyia) aegypti Linnaeus* (Diptera, Culicidae) larvae." *Insect Biochem Mol Biol* 39(4): 279-86.

Bel, Y. and B. Escriche (2006). "Common genomic structure for the Lepidoptera cadherin-like genes." *Gene* 381: 71-80.

Boonserm, P., P. Davis, et al. (2005). "Crystal structure of the mosquito-larvicidal toxin Cry4Ba and its biological implications." *J Mol Biol* 348(2): 363-82.

Boonserm, P., M. Mo, et al. (2006). "Structure of the functional form of the mosquito larvicidal Cry4Aa toxin from *Bacillus thuringiensis* at a 2.8-angstrom resolution." *J Bacteriol* 188(9): 3391-401.

Bravo, A., S. S. Gill, et al. (2007). "Mode of action of *Bacillus thuringiensis* Cry and Cyt toxins and their potential for insect control." *Toxicon* 49(4): 423-35.

Bravo, A., I. Gomez, et al. (2004). "Oligomerization triggers binding of a *Bacillus thuringiensis* Cry1Ab pore-forming toxin to aminopeptidase N receptor leading to insertion into membrane microdomains." *Biochim Biophys Acta* 1667(1): 38-46.

Burton, S. L., D. J. Ellar, et al. (1999). "N-acetylgalactosamine on the putative insect receptor aminopeptidase N is recognised by a site on the domain III lectin-like fold of a *Bacillus thuringiensis* insecticidal toxin." *J Mol Biol* 287(5): 1011-22.

Candas, M., O. Loseva, et al. (2003). "Insect resistance to *Bacillus thuringiensis*: alterations in the indianmeal moth larval gut proteome." *Mol Cell Proteomics* 2(1): 19-28.

Chen, J., K. G. Aimanova, et al. (2009). "*Aedes aegypti* cadherin serves as a putative receptor of the Cry11Aa toxin from *Bacillus thuringiensis* subsp. israelensis." *Biochem J* 424(2): 191-200.

Chen, J., K. G. Aimanova, et al. (2009). "Identification and characterization of *Aedes aegypti* aminopeptidase N as a putative receptor of *Bacillus thuringiensis* Cry11A toxin." *Insect Biochem Mol Biol* 39(10): 688-96.

Chen, J., M. R. Brown, et al. (2005). "Comparison of the localization of *Bacillus thuringiensis* Cry1A delta-endotoxins and their binding proteins in larval midgut of tobacco hornworm, *Manduca sexta*." *Cell Tissue Res* 321(1): 123-9.

de Maagd, R. A., P. L. Bakker, et al. (1999). "Domain III of the *Bacillus thuringiensis* delta-endotoxin Cry1Ac is involved in binding to *Manduca sexta* brush border membranes and to its purified aminopeptidase N." *Mol Microbiol* 31(2): 463-71.

de Maagd, R. A., A. Bravo, et al. (2003). "Structure, diversity, and evolution of protein toxins from spore-forming entomopathogenic bacteria." *Annu Rev Genet* 37: 409-33.

de Maagd, R. A., A. Bravo, et al. (2001). "How *Bacillus thuringiensis* has evolved specific toxins to colonize the insect world." *Trends Genet* 17(4): 193-9.

Eguchi, M. (1995). "Alkaline phosphatase isozymes in insects and comparison with mammalian enzyme." *Comp Biochem Physiol B Biochem Mol Biol* 111(2): 151-62.

Eguchi, M., M. Azuma, et al. (1990). "Genetically defined membrane-bound and soluble alkaline phosphatases of the silkworm: their discrete localization and properties." *Prog Clin Biol Res* 344: 267-87.

Fabrick, J., C. Oppert, et al. (2009). "A novel *Tenebrio molitor* cadherin is a functional receptor for Bacillus thuringiensis Cry3Aa toxin." *J Biol Chem* 284(27): 18401-10.

Fernandez, L. E., K. G. Aimanova, et al. (2006). "A GPI-anchored alkaline phosphatase is a functional midgut receptor of Cry11Aa toxin in *Aedes aegypti* larvae." *Biochem J* 394(Pt 1): 77-84.

Ferre, J. and J. Van Rie (2002). "Biochemistry and genetics of insect resistance to *Bacillus thuringiensis*." *Annu Rev Entomol* 47: 501-33.

Flores, H., X. Soberon, et al. (1997). "Isolated domain II and III from the *Bacillus thuringiensis* Cry1Ab delta-endotoxin binds to lepidopteran midgut membranes." *FEBS Lett* 414(2): 313-8.

Gahan, L. J., F. Gould, et al. (2001). "Identification of a gene associated with Bt resistance in *Heliothis virescens*." *Science* 293(5531): 857-60.

Gahan, L. J., Y. Pauchet, et al. (2010). "An ABC Transporter Mutation Is Correlated with Insect Resistance to *Bacillus thuringiensis* Cry1Ac Toxin." *Plos Genetics* 6(12).

Galitsky, N., V. Cody, et al. (2001). "Structure of the insecticidal bacterial delta-endotoxin Cry3Bb1 of *Bacillus thuringiensis*." *Acta Crystallogr D Biol Crystallogr* 57(Pt 8): 1101-9.

Gill, M. and D. Ellar (2002). "Transgenic Drosophila reveals a functional in vivo receptor for the *Bacillus thuringiensis* toxin Cry1Ac1." *Insect Mol Biol* 11(6): 619-25.

Gill, S. S., S. Likitvivatanavong, et al. (2011). "Multiple Receptors as Targets of Cry Toxins in Mosquitoes." *Journal of Agricultural and Food Chemistry* 59(7): 2829-2838.

Gomez, I., I. Arenas, et al. (2006). "Specific epitopes of domains II and III of Bacillus thuringiensis Cry1Ab toxin involved in the sequential interaction with cadherin and aminopeptidase-N receptors in Manduca sexta." *J Biol Chem* 281(45): 34032-9.

Gomez, I., I. Arenas, et al. (2010). "Role of Alkaline Phosphatase from *Manduca sexta* in the Mechanism of Action of *Bacillus thuringiensis* Cry1Ab Toxin." *Journal of Biological Chemistry* 285(17): 12497-12503.

Gomez, I., L. Pardo-Lopez, et al. (2007). "Role of receptor interaction in the mode of action of insecticidal Cry and Cyt toxins produced by *Bacillus thuringiensis*." *Peptides* 28(1): 169-73.

Gomez, I., J. Sanchez, et al. (2002). "Cadherin-like receptor binding facilitates proteolytic cleavage of helix alpha-1 in domain I and oligomer pre-pore formation of *Bacillus thuringiensis* Cry1Ab toxin." *FEBS Lett* 513(2-3): 242-6.

Gonzalez-Cabrera, J., B. Escriche, et al. (2003). "Binding of *Bacillus thuringiensis* toxins in resistant and susceptible strains of pink bollworm (*Pectinophora gossypiella*)." *Insect Biochem Mol Biol* 33(9): 929-35.

Griffitts, J. S., S. M. Haslam, et al. (2005). "Glycolipids as receptors for *Bacillus thuringiensis* crystal toxin." *Science* 307(5711): 922-5.

Grochulski, P., L. Masson, et al. (1995). "*Bacillus thuringiensis* CryIA(a) insecticidal toxin: crystal structure and channel formation." *J Mol Biol* 254(3): 447-64.

Guo, S., S. Ye, et al. (2009). "Crystal structure of *Bacillus thuringiensis* Cry8Ea1: An insecticidal toxin toxic to underground pests, the larvae of *Holotrichia parallela*." *J Struct Biol* 168(2): 259-66.

Hara, H., S. Atsumi, et al. (2003). "A cadherin-like protein functions as a receptor for *Bacillus thuringiensis* Cry1Aa and Cry1Ac toxins on midgut epithelial cells of *Bombyx mori* larvae." *FEBS Lett* 538(1-3): 29-34.

Herrero, S., T. Gechev, et al. (2005). "Bacillus thuringiensis Cry1Ca-resistant *Spodoptera exigua* lacks expression of one of four Aminopeptidase N genes." *BMC Genomics* 6: 96.

Hua, G., R. Zhang, et al. (2008). "*Anopheles gambiae* cadherin AgCad1 binds the Cry4Ba toxin of *Bacillus thuringiensis* israelensis and a fragment of AgCad1 synergizes toxicity." *Biochemistry* 47(18): 5101-10.

Ibiza-Palacios, M. S., J. Ferre, et al. (2008). "Selective inhibition of binding of *Bacillus thuringiensis* Cry1Ab toxin to cadherin-like and aminopeptidase proteins in brush-border membranes and dissociated epithelial cells from *Bombyx mori*." *Biochem J* 409(1): 215-21.

Itoh, M., S. Takeda, et al. (1991). "Cloning and sequence analysis of membrane-bound alkaline phosphatase cDNA of the silkworm, *Bombyx mori*." *Biochim Biophys Acta* 1129(1): 135-8.

Jenkins, J. L., M. K. Lee, et al. (2000). "Bivalent sequential binding model of a *Bacillus thuringiensis* toxin to gypsy moth aminopeptidase N receptor." *J Biol Chem* 275(19): 14423-31.

Jurat-Fuentes, J. L. and M. J. Adang (2004). "Characterization of a Cry1Ac-receptor alkaline phosphatase in susceptible and resistant *Heliothis virescens* larvae." *Eur J Biochem* 271(15): 3127-35.

Krishnamoorthy, M., J. L. Jurat-Fuentes, et al. (2007). "Identification of novel Cry1Ac binding proteins in midgut membranes from *Heliothis virescens* using proteomic analyses." *Insect Biochem Mol Biol* 37(3): 189-201.

Li, J., D. J. Derbyshire, et al. (2001). "Structural implications for the transformation of the *Bacillus thuringiensis* delta-endotoxins from water-soluble to membrane-inserted forms." *Biochem Soc Trans* 29(Pt 4): 571-7.

Lorence, A., A. Darszon, et al. (1997). "Aminopeptidase dependent pore formation of *Bacillus thuringiensis* Cry1Ac toxin on *Trichoplusia ni* membranes." *FEBS Lett* 414(2): 303-7.

McNall, R. J. and M. J. Adang (2003). "Identification of novel *Bacillus thuringiensis* Cry1Ac binding proteins in *Manduca sexta* midgut through proteomic analysis." *Insect Biochem Mol Biol* 33(10): 999-1010.

Miranda-Rios, J., M. T. Fernandez-Luna, et al. (2010). "An alpha-amylase is a novel receptor for *Bacillus thuringiensis* ssp israelensis Cry4Ba and Cry11Aa toxins in the malaria vector mosquito *Anopheles albimanus* (Diptera: Culicidae)." *Environmental Microbiology* 12(3): 746-757.

Morin, S., R. W. Biggs, et al. (2003). "Three cadherin alleles associated with resistance to *Bacillus thuringiensis* in pink bollworm." *Proc Natl Acad Sci U S A* 100(9): 5004-9.

Morse, R. J., T. Yamamoto, et al. (2001). "Structure of Cry2Aa suggests an unexpected receptor binding epitope." *Structure* 9(5): 409-17.

Nakanishi, K., K. Yaoi, et al. (2002). "Aminopeptidase N isoforms from the midgut of *Bombyx mori* and *Plutella xylostella* -- their classification and the factors that determine their binding specificity to *Bacillus thuringiensis* Cry1A toxin." *FEBS Lett* 519(1-3): 215-20.

Nosjean, O., A. Briolay, et al. (1997). "Mammalian GPI proteins: sorting, membrane residence and functions." *Biochim Biophys Acta* 1331(2): 153-86.

Ochoa-Campuzano, C., M. D. Real, et al. (2007). "An ADAM metalloprotease is a Cry3Aa *Bacillus thuringiensis* toxin receptor." *Biochem Biophys Res Commun.* 362(2):437-42.

Pacheco, S., I. Gomez, et al. (2009). "Domain II loop 3 of *Bacillus thuringiensis* Cry1Ab toxin is involved in a "ping pong" binding mechanism with *Manduca sexta* aminopeptidase-N and cadherin receptors." *J Biol Chem* 284(47): 32750-7.

Pandian, G. N., T. Ishikawa, et al. (2008). "*Bombyx mori* midgut membrane protein P252, which binds to *Bacillus thuringiensis* Cry1A, is a chlorophyllide-binding protein, and the resulting complex has antimicrobial activity." *Appl Environ Microbiol* 74(5): 1324-31.

Pardo-Lopez, L., I. Gomez, et al. (2006). "Structural changes of the Cry1Ac oligomeric pre-pore from *Bacillus thuringiensis* induced by N-acetylgalactosamine facilitates toxin membrane insertion." *Biochemistry* 45(34): 10329-36.

Park, Y., M. A. Abdullah, et al. (2009). "Enhancement of *Bacillus thuringiensis* Cry3Aa and Cry3Bb toxicities to coleopteran larvae by a toxin-binding fragment of an insect cadherin." *Appl Environ Microbiol* 75(10): 3086-92.

Pigott, C. R. and D. J. Ellar (2007). "Role of receptors in *Bacillus thuringiensis* crystal toxin activity." *Microbiol Mol Biol Rev* 71(2): 255-81.

Rajagopal, R., S. Sivakumar, et al. (2002). "Silencing of midgut aminopeptidase N of *Spodoptera litura* by double-stranded RNA establishes its role as *Bacillus thuringiensis* toxin receptor." *J Biol Chem* 277(49): 46849-51.

Rajamohan, F., M. K. Lee, et al. (1998). "*Bacillus thuringiensis* insecticidal proteins: molecular mode of action." *Prog Nucleic Acid Res Mol Biol* 60: 1-27.

Schnepf, E., N. Crickmore, et al. (1998). "*Bacillus thuringiensis* and its pesticidal crystal proteins." *Microbiol Mol Biol Rev* 62(3): 775-806.

Schwartz, J. L., Y. J. Lu, et al. (1997). "Ion channels formed in planar lipid bilayers by *Bacillus thuringiensis* toxins in the presence of *Manduca sexta* midgut receptors." *FEBS Lett* 412(2): 270-6.

Soberon, M., S. S. Gill, et al. (2009). "Signaling versus punching hole: How do *Bacillus thuringiensis* toxins kill insect midgut cells?" *Cell Mol Life Sci* 66(8): 1337-49.

Valaitis, A. P., J. L. Jenkins, et al. (2001). "Isolation and partial characterization of gypsy moth BTR-270, an anionic brush border membrane glycoconjugate that binds *Bacillus thuringiensis* Cry1A toxins with high affinity." *Arch Insect Biochem Physiol* 46(4): 186-200.

Xu, X., L. Yu, et al. (2005). "Disruption of a cadherin gene associated with resistance to Cry1Ac {delta}-endotoxin of *Bacillus thuringiensis* in *Helicoverpa armigera*." *Appl Environ Microbiol* 71(2): 948-54.

Zhang, R., G. Hua, et al. (2008). "A 106-kDa aminopeptidase is a putative receptor for *Bacillus thuringiensis* Cry11Ba toxin in the mosquito *Anopheles gambiae*." *Biochemistry* 47(43): 11263-72.

Zhuang, M., D. I. Oltean, et al. (2002). "*Heliothis virescens* and *Manduca sexta* lipid rafts are involved in Cry1A toxin binding to the midgut epithelium and subsequent pore formation." *J Biol Chem* 277(16): 13863-72.

13

Essential Oils of Umbelliferae (Apiaceae) Family Taxa as Emerging Potent Agents for Mosquito Control

Epameinondas Evergetis[1], Antonios Michaelakis[2]
and Serkos A. Haroutounian[1,*]
[1]*Chemistry Laboratory, Agricultural University of Athens, Athens*
[2]*Department of Entomology, Benaki Phytopathological Institute, Athens*
Greece

1. Introduction

Warm-humid areas around the globe constitute the cradle of humanity, providing their inhabitants the most favorable environments for living and agricultural production. In this "Garden of Eden", which spreads within the globe's temperate and tropical zones, is also thriving an annoying but dangerous daemon, the mosquito. This little devil constitutes the main vector of malaria and human encephalitis, both infectious diseases that account as major threats of public health (Becker et al., 2003). Recently, these threats have been spread to a broader geographical area, as a consequence of their vectors (*Aedes* sp., *Anopheles* sp. and *Culex* sp.) introduction into metropolitan areas of northern hemisphere, such as Chicago (Tedesco et al., 2010), New York (Peterson et al., 2006) and Paris (Delaunay et al., 2009). Since mosquito breeding habitats in both urban and rural areas are man-made (Imbahale et al., 2010), there are several restrictions limiting the efforts towards the development of an integrated vector management system. Todate, the history of evolutions of malaria vector interventions is directly connected with the mosquito control tools development, concerning either environmental modifications/manipulations or their chemical and/or biological control (Kilama, 2009).

In respect the chemical control, a significant milestone was the DichloroDiphenyl-Trichloroethane (DDT) synthesis by Zeidler in 1874. The DDT success was followed by the fast introduction of numerous chlorinated hydrocarbons, which were used in massive amounts for the control of mosquito-borne diseases (Ray, 2010). Despite their efficiency, the use of organochlorines had severe environmental impacts which were publicly (and dramatically) addressed by Carson (1962) in Silent Spring, initiating the development of insecticide resistant mosquito populations. These undesirable characteristics, in combination with concerns on public health risks, derived from the organochlorine residues detected in humans and animals, led to their ban in early 70's. Thus, they were replaced by less persistent chemicals, such as organophosphates, pyrethroids and avermectin derivates,

*Corresponding Author

substitutes that also display the major disadvantage of resistance development (Alves et al., 2010; Daaboub et al., 2008; Lima et al., 2011).

Recent research trend on mosquito chemical control mainly focuses on currently used compounds, aiming to enhance their potency and circumvent the problems connected with their application. In this respect, the so far developed pyrazole derivatives are quite efficient exhibiting however adverse environmental effects (Stevens et al., 2011), while the corresponding pyrroles display the desirable efficiency (Raghavendra et al., 2011) but adequate research on their environmental side effects is still underway. Amides, such as methazolamide and acetazolamide, were also evaluated as potent mosquito larvicides but were found to display significant bioaccumulation properties (Del Pilar Corena et al., 2006) which discourage their broad use. Finally, various novel pyrimido-quinolione molecules have been developed and assessed as highly toxic for other organisms (Rajanarendar et al., 2010). Ray (2010) recognized that the insecticide treated nets, in connection with the long lasting insecticidal nets, have resurrected the chemical control of malaria's mosquito vector. This may be rationalized considering that their targeted application resolves the problems connected with the environmental impacts of chemical control agents since limit their expansion, availability and environmental penetration.

Despite the numerous efforts and progress achieved, the efficacy of insecticidal nets in malaria prevention still constitutes a hot issue, since depends strongly upon a plethora of additional factors (Killeen & Smith, 2007). In particular, despite efforts (Pennetier et al., 2010) to overcome the recognized for longtime resistance development issues of insecticidal nets, todate these problems have not been resolved (Yadouleton et al., 2011). An additional drawback derives by the combined impact of herbicide application that promotes the cross-resistance to mosquito populations (Boyer et al., 2006; Riaz et al., 2009).

The corresponding biological control has dictated the development of novel-alternative mosquito control tools, including the sterile males technique (Patersson et al., 1968), the genetically modified mosquitoes (Gu et al., 2011; Lavialle-Defaix et al., 2011), the entomopathogenic fungi (Van Breukelen et al., 2011; Kanzok & Jacobs-Lorena, 2006) and bacteria. Among the tools developed the bacterial pathogens application is considered as the most prominent intervention, displaying species selective insecticidal ability (Hayes et al., 2011) which is considered as an efficient means for mosquito control without harmful impacts for the environment (Caquet et al., 2011). Major thresholds limiting the wider application of this technique are related with the induced pathogen introduction among the natural mosquito populations (Hancock et al., 2011) and the threats connected with bioaccumulation and resistance development (Tilquin et al., 2008). In general, the biological control tools are still under development, presenting todate a low degree of maturity for large-scale interventions.

Temephos was considered as one of the most potent-safe insecticides. Its recent exclusion from Annex I of the Directive 98/8/EC resulted in the discontinuation of its application in mosquito control programs by the European, emerging the development and use of new-safer insecticides. Thus, relative research directed towards the discovery-development of novel molecules, capable to control the mosquito populations without exhibiting the disadvantages of synthetic pesticides. In this respect, the plant originated natural compounds constitute a large deposit of such molecules, inherently allowing the retrieval of various commercially successful molecules like pyrethrins. Todate, the search for novel, potent and safer pesticides from this deposit has already provided several candidates, either as pure compounds and/or their extracts. Specifically, various organic acids such as lactic and orthophosporic acids (Chakraborty et al., 2010), alkaloids (Talontsi et al., 2011) and plant proteins (Chowdhury et al., 2008) have been identified as efficient mosquito control agents. Furthermore, several plants were used as the maternal material to produce bio-products which were applied against mosquitoes with hopeful results (Shaalan et al., 2005; Sukumar et al., 1991). On the other hand, the plant derived Essential Oils (EOs) constitute a special category of natural products that exhibit the major advantage -for the mosquito control endeavor- of exhibiting an insect oriented mode of action with low penetrability to the ecosystems that does not affect larger animals. In addition, the natural diversity of their constituents addresses effectively the problem of resistance development (Isman, 2000).

2. Literature review

2.1 Umbelliferae (Apiaceae) family: A source of potent natural agrochemicals

Many EOs originated from diverse plant families have been considered and studied as potential sources of natural agrochemicals. In this respect, previous research results on Umbeliferae (Apiaceae) family plant materials revealed the significant acaricidal activities of butylidenepthalides isolated from *Angelica acutiloba* Kitagawa var. *sugiyame* Hikino (Kwon & Ahn, 2002) and the similar activity of the EO of *Foeniculum vulgare,* attributed to the presence of *p*-anisaldehyde and (+,−)-fenchone in the EO (Lee, 2004). These EOs were practically inactive in fumigant toxicity tests against *Lycoriella mali* though they are known to contain the active monoterpenes *a*-pinene and *β*-pinene (Choi et al., 2006), which are common constituents of many Umbelliferae EOs. Methanolic extracts of *Angelica dahurica, Cnidium officinale,* and *Foeniculum vulgare* were also tested against the Coleoptera *Lasioderma sericorne, Sitophilus oryzae* and *Callosobruchus chinensis* exhibiting a moderate activity only the second extract (Kim et al., 2003a; Kim et al., 2003b). Other EOs of this family screened as inactive against coleoptera were originated from the species *Anethum graveolens* L., *Apium graveolens* Houtt., *Coriandrum sativum* L., *Cuminum cyminum* L. and *Petroselinum sativum* L. (Regnault-Roger & Hamraoui, 1994; Papachristos & Stamopoulos, 2002). On the contrary, the EOs of *Pimpinella anisum* L. and *Cuminum cyminum* L. displayed excellent ovicidal and insecticidal activities against the *Tribolium confusum* du Val and the *Ephestia kuehniella* Zeller (Tunc et al., 2000). In addition, the aqueous extract of *Pimpinella anisum* exhibited good repellent effect against the adults of sweet potato whitefly *Bemisia tabaci* (Ateyyat et al., 2007).

These rather controversial results are not connected with the impressive activities that Umbelliferae EOs were found to exhibit against the Diptera, with the EO of *Ammi visnaga* displaying -among 19 EOs- the most potent ovicidal activity against *Mayetiola destructor* (Lamiri et al., 2001). In addition, tests against *Droshophila melanogaster* of furanocoumarins and pthalides isolated from *Angelica acutiloba* Kitagawa var. *sugiyame* Hikino revealed the hypothesis that the insecticidal properties of the plant extracts are connected with the acetylcholinesterase inhibition (Miyazawa et al., 2004). Finally, alkylpthalides originated from *Cnidium officinale* Makino were tested as extremely effective against *Droshophila melanogaster* (Tsukamoto et al., 2005).

2.2 Umbelliferae (Apiaceae) family: A strong focal point for mosquito control

Table 1 summarizes the test results against various mosquito species reported for all extracts and EOs derived from plants belonging to the Umbeliferae family. Same table also contains the test results of fourteen EOs, which appear herein for the first time. Results indicate that the organic phase of the *Cryoptaenia canadensis* extract is the most active against fourth instars of *Culex pipiens*, leading to the isolation –from the extract- of the acetylated very toxic (LC$_{50}$ values lower than 10 mgl^{-1}) molecules of falcarinol and falcarindiol (Eckenbach et al., 1999). The larvicidal properties of hexane soluble fraction of *Apium graveolens* seeds -a plant with pleasant aroma- and three isolated compounds (sedanolide, senkyunolide-N, senkyunolide-J) against *Aedes aegypti* mosquitoes highlighted sedanolide as very active (100% mortality at 50 mgl^{-1}, Momin & Nair, 2001). As a consequence, a gel containing 5% of the *Apium graveolens* hexane extract was developed, providing full protection to volunteers from mosquito bites for two hours (Tuetun et al., 2009), while the ethanolic formulations from the same plant also provided protection against *Aedes aegypti*. Another formulation containing the aforementioned hexane extract and 5% vanillin showed strong repellent activities against different mosquito species (Tuetun et al., 2005, see also **Table 1** for details). The crude seed extract had no adverse effects on human volunteers skins when tested for several anti-mosquito properties (Choochote et al., 2004). This plant's EO exhibits potent larvicidal activity against two laboratory-reared mosquito species, the malaria vector *Anopheles dirus* and the vector of dengue *Aedes aegypti* (Pitasawat et al., 2007).

Species	Part used	Mosquito species	Bioactivity	Reference
Ammi visnaga	Seeds	*Culex quinquefasciatus*	Larvicidal	Pavela, 2008
Anethum graveolens	(not mentioned)	*Anopheles stephensi, Aedes aegypti, Culex quinquefasciatus*	Larvicidal	Amer and Mehlhorn, 2006
Anethum graveolens	Leaves, twigs	*Aedes aegypti*	Larvicidal, effects on growth and development	Promsiri et al., 2006
Angelica archangelica	Fruits	*Culex quinquefasciatus*	Larvicidal	Pavela, 2009
Angelica sylvestris	Aerial parts	*Culex pipiens*	Larvicidal	(present study)
Apium graveolens	Seeds	*Aedes aegypti*	Larvicidal	Momin & Nair, 2001
Apium graveolens	Seeds	*Aedes aegypti*	Larvicidal, adulticidal, repellent	Choochote, 2004
Apium graveolens	Seeds	*Aedes aegypti, Aedes gardnerii, Aedes lineatopennis, Anopheles barbirostris, Armigeres subalbatus, Culex tritaeniorhynchus, Culex gelicus, Culex vishnui* group, *Mansonia uniformis*	Repellent	Tuetun et al., 2005

Species	Part used	Mosquito species	Bioactivity	Reference
Apium graveolens	Seeds	*Aedes aegypti, Anopheles ditrus*	Larvicidal	Pitasawat et al., 2007
Apium graveolens	Seeds	*Aedes, Anopheles, Armigeres, Culex, Mansonia*	Repellent	Tuetun et al., 2009
Athamanta densa	Aerial parts	*Culex pipiens*	Larvicidal	(present study)
Bupleurum fruticosum	Aerial parts	*Culex pipiens*	Larvicidal	Evergetis et al., 2009
Carum carvi	Fruits	*Aedes aegypti, Culex quinquefasciatus*	Larvicidal	Lee, 2006
Carum carvi	Seeds	*Aedes aegypti, Anopheles ditrus*	Larvicidal	Pitasawat et al., 2007
Carum ptroselinum	(not mentioned)	*Culex pipiens*	Larvicidal	Khater and Shalaby, 2008
Chaerophyllum heldreichii	Aerial parts	*Culex pipiens*	Larvicidal	(present study)
Conium divaricatum	Aerial parts	*Culex pipiens*	Larvicidal	(present study)
Conopodium capillifolium	Aerial parts	*Culex pipiens*	Larvicidal	Evergetis et al., 2009
Coriander sativum	Seeds	*Ochlerotatus caspius*	Larvicidal	Knio et al., 2008
Cryptotaenia canadensis	Fresh foliage, root, fruits	*Culex pipiens*	Larvicidal	Eckenbach et al., 1999
Daucus carota	Roots	*Aedes aegypti, Culex quinquefasciatus*	Larvicidal	Lee, 2006
Eleoselinum asclepium	Aerial parts	*Culex pipiens*	Larvicidal	Evergetis et al., 2009
Ferula assa-foetida	Stems	*Culex quinquefasciatus*	Larvicidal	Pavela, 2009
Ferula galbaniflua	(not mentioned)	*Anopheles stephensi, Aedes aegypti, Culex quinquefasciatus*	Larvicidal	Amer and Mehlhorn, 2006
Ferula lancerottensis	Stems	*Culex quinquefasciatus*	Larvicidal	Pavela, 2008
Ferulago nodosa	Aerial parts	*Culex pipiens*	Larvicidal	(present study)
Foeniculum vulgare	Fruits	*Aedes aegypti*	Repellent	Kim et al., 2002
Foeniculum vulgare	(not mentioned)	*Aedes aegypti*	Larvicidal	Orozco & Lentz, 2005
Foeniculum vulgare	Flowers	*Culex pipiens*	Larvicidal, repellent	Trabousli et al., 2005
Foeniculum vulgare	Fruits	*Aedes aegypti, Anopheles ditrus*	Larvicidal	Pitasawat et al., 2007
Foeniculum vulgare	Stems, inflorencences, leaves	*Culex pipiens*	Larvicidal	Manolakou et al., 2009

Species	Part used	Mosquito species	Bioactivity	Reference
Foeniculum vulgare	Leaves	*Aedes albopictus*	Larvicidal	Conti et al., 2010
Heracleum sphondylium	Aerial parts	*Culex pipiens*	Larvicidal	Evergetis et al., 2009
Imperatoria ostruthium	Roots	*Culex quinquefasciatus*	Larvicidal	Pavela, 2009
Laserpitium pseudomeum	Aerial parts	*Culex pipiens*	Larvicidal	(present study)
Oenanthe pimpinelloides	Aerial parts	*Culex pipiens*	Larvicidal	Evergetis et al., 2009
Petroselimum crispum	Seeds	*Ochlerotatus caspius*	Larvicidal	Knio et al., 2008
Peucedanum neumayeri	Aerial parts	*Culex pipiens*	Larvicidal	(present study)
Peucedanum officinale	Aerial parts	*Culex pipiens*	Larvicidal	(present study)
Pimpinella anisum	Seeds	*Anopheles stephensi, Aedes aegypti, Culex quinquefasciatus*	Larvicidal, adulticidal, ovicidal, oviposition-deterrent, repellent	Prajapati et al., 2005
Pimpinella anisum	Seeds	*Ochlerotatus caspius*	Larvicidal	Knio et al., 2008
Pimpinella peregrina	Aerial parts	*Culex pipiens*	Larvicidal	(present study)
Pimpinella rigidula	Aerial parts	*Culex pipiens*	Larvicidal	(present study)
Pimpinella tragium ssp tragium	Aerial parts	*Culex pipiens*	Larvicidal	(present study)
Scaligeria cretica	Aerial parts	*Culex pipiens*	Larvicidal	(present study)
Seseli montanum	Aerial parts	*Culex pipiens*	Larvicidal	Evergetis et al.,2009
Seseli pallasii	Stems	*Culex quinquefasciatus*	Larvicidal	Pavela, 2009
Seseli parnassicum	Aerial parts	*Culex pipiens*	Larvicidal	(present study)
Seseli tortuosum	Stems	*Culex quinquefasciatus*	Larvicidal	Pavela, 2008
Smyrnium rotundifolium	Aerial parts	*Culex pipiens*	Larvicidal	(present study)
Thamnosciadium junceum	Aerial parts	*Culex pipiens*	Larvicidal	(present study)
Trachyspermum ammi	Seeds	*Anopheles stephensi*	Larvicidal, oviposition-deterrent, vapor toxicity, repellent	Pandey et al., 2009

Table 1. Reported phytochemicals derived from plants belonging to Apiaceae family against various mosquito species.

Another EO found to possess potent larvicidal, oviposition-deterrent, vapor toxicity and repellent activities against *Aedes aegypti* was isolated from ajowan (*Tachyspermum ammi*, Pandey et al. 2009). *Anethum graveolens* extract exhibited larval toxicity with LC$_{50}$ values from 27 to 20 mgl^{-1} (for 24 and 48 hours exposures respectively), while on growth survival and prolongation tests of the various instar larvae of *Aedes aegypti*, the second instar larvae was determined as the more susceptible. The lowest concentration of crude extracts of *Anethum graveolens* used (caused more than 50% larval mortality) was not toxic to guppy fish (*Poecilia reticulata*) at concentrations of 12.5 mgl^{-1} (Promsiri et al., 2006).

Among all EOs tested for mosquito control, the most potent was derived from *Foeniculum vulgare*, which caused the highest mortality against *Aedes albopictus* (Conti et al., 2010) and moderate against *Anopheles dirus* and *Aedes aegypti* (Pitasawat et al., 2007). Main component of this EO is methyl chavicol (more than 43%), while its methanolic extract (*trans*-anethole chemotype) was moderately active against *Aedes aegypti*, the yellow fever mosquito (Orozco & Lentz, 2005). The hexane fraction from its fruit-derived parts showed 99% repellency against *Aedes aegypti*, while the other fractions (chloroform, ethyl acetate and water: 37, 37 and 17% respectively) were practically inactive (Kim et al., 2002). Repellency and toxicity were also studied against *Culex pipiens* (Trabousli et al., 2005), indicating that the EO of *Foeniculum vulgare* was the most effective, while the repellency assays revealed protection time for almost one hour when applied at concentration of 3%.

Pimpinella anisum L. EO proved to possess equally potent larvicidal and ovicidal activities against *Anopheles stephensi*, *Aedes aegypti*, *Culex quinquefasciatus* and only larvicidal against *Ochlerotatus caspius* (Prajapati et al., 2005; Knio et al., 2008). Similar larvicidal activity results were also observed when the EOs of *Coriander sativum* and *Petroselinum crispum* were tested against *Ochlerotatus caspius* (Knio et al., 2008). The larvicidal tests of EOs of genus *Carum* were performed for *Carum carvi* against *Anopheles dirus* and *Aedes aegypti* and for *Carum petroselinum* against *Culex pipiens* (Lee, 2006; Pitasawat et al., 2007; Khater & Shalaby, 2008). The results were directly similar to those of the EO of *Daucus carota* (against *Anopheles dirus* and *Aedes aegypti*) proving their inability to cause 100% mortality at the lowest concentration (Lee, 2006).

Among the methanolic extracts of 118 Euroasiatic plants, tested for their larvicidal effects against *Culex quinquefasciatus*, the species *Ammi visnaga* and *Seseli pallasii* were determined as two of the most toxic materials tested, with LC$_{50}$ values lower than 10 mgl^{-1} (Pavela, 2008, 2009). On the other hand, the extracts of *Angelica archangelica* and *Imperatoria ostruthium* exhibited LC$_{50}$ values lower than 70 mgl^{-1}, while *Seseli tortuosum* and *Ferula lancerottensis* displayed moderate larvicidal activity (LC$_{50}$ values around 430 mgl^{-1}). The only inactive Apiaceae plant tested was *Ferula assa-foetida* (LC$_{50}$ value higher of 1000 mgl^{-1}), with the EO of *Ferula galbaniflua* exhibiting the weakest activity against *Culex quinquefasciatus* and *Anopheles stephensi* (mortality level less than 14% of dead larvae after 48 hours, Amer & Mehlhorn, 2006a). The same authors also reported that *Anopheles stephensi* was the most resistant to dill (*Anethum graveolens*), while the *Culex quinquefasciatus* the more sensitive. Dill was also evaluated for persistency to larvicidal effects under different conditions for 1 month after the preparation of its solutions. In all cases (open, closed, in light or in dark) the EO was active only when was used immediately after preparation (Amer & Mehlhorn, 2006b).

Finally, an interesting result was obtained during the study of several EOs using coupled gas chromatography-electroantennographic detection (GC-EAD), on the hypothesis that compounds can be detected by the antennae of the yellow fever mosquito, *Aedes aegypti*. Thus, cumin aldehyde and cumin alcohol the *Cuminum cynimum* EO components were identified as such molecules. It must be noted that for both components, their EO (cumin oil) was also EAD-active (Campbell et al., 2011)

2.3 Greek Umbelliferae (Apiaceae) plants extract activities against Culex pipiens mosquitoes

The larvicidal activity of the EO obtained from the stem of Greek *Foeniculum vulgare* was determined against *Culex pipiens* larvae, while methyl chavicol was determined as its main component (more than 32%). Although the LC_{50} value of methyl chavicol was more than 80 mgl⁻¹, the respective EO was determined as 2.1-fold more toxic (Manolakou et al., 2009). *Culex pipiens* larvae were also used to test the mosquito control properties of EO from various naturally growing plants throughout Greece, belonging to the following six different Apiaceae family taxa: *Heracleum sphondylium, Seseli montanum, Conopodium capillifolium, Bupleurum fruticosum, Oenanthe pimpinelloides, Eleoselinum asclepium*. All EOs tested displayed good larvicidal activities with LC_{50} values ranging from 40.26-96.96 mgl⁻¹ (Evergetis et al., 2009)

As a continuation of our ongoing efforts to exploit the use of natural products for the development of environmentally friendly means for the mosquito population control, our interest was stimulated on the investigation of Umbeliferae (Apiaceae) plants EOs. In this context, we report herein the chemical composition and larvicidal activity results for 14 EOs originated from different taxon obtained during Greek Umbelliferae biodiversity studies (**Table 1**).

3. Materials and methods

3.1 Plant material

Fourteen different taxa of the Umbeliferae (Apiaceae) family, Apioideae subfamily belonging to seven tribes and twelve different genera have been collected during the present study. Representatives of the Apieae Tribe are *Pimpinella peregrina* L., and 5 Greek endemics, namely *Athamanta densa* Boiss. & Orph., *Pimpinella tragium* ssp *tragium* Vill., *Pimpinella rigidula* (Boiss. & Orph.) H. Wolf, *Seseli parnassicum* Boiss. & Heldr. and *Thamnosciadium junceum* (Sibth. & Sm.) Hartvig.; of Smyrnieae tribe *Scaligeria cretica* (Miller) Boiss. and *Smyrnium rotundifolium* Miller; of Angeliceae tribe *Angelica sylvestris* L.; of Scandiceae tribe the Greek endemic *Chaerophyllum heldreichii* Orph. Ex Boiss.; of Peucedaneae tribe *Ferulago nodosa* (L.) Boiss., *Peucedanum neumayeri* (Vis.) Reichenb, *Peucedanum officinale* L., and of Laserpitieae tribe the Greek endemic *Laserpitium pseudomeum* Orph., Heldr. & Sart. Ex Boiss (Pimenov & Leonov, 1993; Tutin et al., 1968).

Full collection details are provided in **Table 2**. A voucher specimen of each plant is deposited in the herbarium of the Agricultural University of Athens, Athens, Greece.

3.2 Essential oils isolation

The freshly collected plant materials (steams, leaves and flowers) were washed thoroughly, chopped off finely and subjected to steam distillation in a Clevenger-type apparatus, using the Microwave Accelerated Reaction System (MARS 5) at 1400 W for 40 min with 3 L of H_2O in order to obtain their EOs. The resulting oils were dried over anhydrous sodium sulphate and stored at 4 ºC. The EO yield of each plant is included in **Table 3**.

Species	Abbreviation	Vegetative Stage	Date	Location
Angelica sylvestris L.	AS	Flowering	05.09.2004	Mt. Parnon, Peloponnisos, forest streams
Athamanta densa Boiss. & Orph. *	AD	Flowering	15.06.2005	Mt. Parnassos, Sterea Hellas, vertical cliffs
Chaerophyllum heldreichii Orph. Ex Boiss. *	CH	Flowering	25.07.2004	Mt. Parnon, Peloponnisos, forest clearings
Ferulago nodosa (L.) Boiss.	FN	Flowering	02.05.2005	Antikyra, Sterea Hellas, olive groves
Laserpitium pseudomeum Orph., Heldr. & Sart. Ex Boiss.*	LP	Flowering	15.07.2004	Mt. Oiti, Sterea Hellas, rocky slopes
Peucedanum neumayeri (Vis.) Reichenb	PN	Flowering	28.08.2004	Mt. Smolikas, Hepiros, forest clearings
Peucedanum officinale L.	PO	Flowering	15.07.2004	Mt. Oiti, Sterea Hellas, rocky slopes
Pimpinella tragium ssp tragium Vill. *	PT	Flowering	15.07.2004	Mt. Oiti, Sterea Hellas, rocky slopes
Pimpinella peregrina L.	PP	Flowering	14.05.2005	Iraklio, Is. Crete, olive groves
Pimpinella rigidula (Boiss. & Orph.) H. Wolf *	PR	Flowering	17.08.2004	Molai, Peloponnisos, roadside
Scaligeria cretica (Miller) Boiss.	SC	Flowering	22.05.2005	Vouliagmeni, Sterea Hellas, seaside
Seseli parnassicum Boiss. & Heldr. *	SP	Flowering	15.07.2004	Mt. Oiti, Sterea Hellas, forest clearings
Smyrnium rotundifolium Miller	SR	Flowering	02.05.2005	Distomo, Sterea Hellas, roadside
Thamnosciadium junceum (Sibth. & Sm.) Hartvig *	TJ	Flowering	25.07.2004	Mt. Parnnassos, Sterea Hellas, alpic ravine

*=Greek Endemic.

Table 2. Collection data.

Species	Part distilled	Weight of aerial parts (g)	Volume of oil (mL)
Angelica sylvestris L.	Aerial	920	0,5
Athamanta densa Boiss. & Orph. *	Aerial	450	0,5
Chaerophyllum heldreichii Orph. Ex Boiss. *	Aerial	530	0,7
Ferulago nodosa (L.) Boiss.	Aerial	400	0,7
Laserpitium pseudomeum Orph., Heldr. & Sart. Ex Boiss. *	Aerial	270	0,9
Peucedanum neumayeri (Vis.) Reichenb	Aerial	600	0,5
Peucedanum officinale L.	Aerial	*180*	*0,8*
Pimpinella tragium ssp tragium Vill. *	Aerial	650	1,5
Pimpinella peregrina L.	Aerial	500	0,4
Pimpinella rigidula (Boiss. & Orph.) H. Wolf *	Aerial	235	0,7
Scaligeria cretica (Miller) Boiss.	Aerial	200	0,5
Seseli parnassicum Boiss. & Heldr. *	Aerial	*200*	*0,5*
Smyrnium rotundifolium Miller	Aerial	530	0,9
Thamnosciadium junceum (Sibth. & Sm.) Hartvig *	Aerial	600	2,0

*=Greek Endemic.

Table 3. Essential oils yields.

3.3 Gas Chromatography-Mass Spectrometry (GC-MS) analyses

Gas Chromatography (GC). All GC analyses were carried out on a Agilent Technologies 7890A gas chromatograph, fitted with a HP 5MS 30m x 0.25mm x 0.25µm film thickness capillary column and FID. The column temperature was programmed from 60 to 280 °C at a initial rate of 3 °C/min. The injector and detector temperatures were programmed at 230 and 300 °C, respectively. Helium was used as the carrier gas at a flow rate 1 ml/min.

Gas Chromatography-Mass Spectrometry (GC-MS). The GCMS analyses were performed on the same instrument using the Agilent 5957C, VL MS Detector with Triple-Axis Detector system operating in EI mode (equipped with a HP 5MS 30m x 0.25mm x 0.25µm film thickness capillary column), using He (1 ml/min) as the carrier gas. The initial temperature of the column was 60 °C. The column was heated gradually to 280 °C with a 3 °C/min rate. The identification of the compounds was based on comparison of their retention indices (RI) (Van den Dool & Kratz, 1963), obtained using various n-alkanes (C9-C24). Also, their EI-mass spectra were compared with the NIST/NBS and Wiley library spectra and the literature (Adams, 1995; Massada, 1976). Additionally, the identity of the indicated phytochemicals was confirmed by comparison with available authentic samples.

3.4 Mosquito rearing

A colony of the species *Culex pipiens* biotype *molestus* is maintained for more than 25 years in the laboratory of Entomology of the Benaki Phytopathological Institute, Kifissia, Greece. Adult mosquitoes are kept in wooden framed cages (33x33x33 cm) with a 32x32 mesh at 25±2 °C, 80±2% relative humidity and photoperiod of 14:10 (L:D) h. Cotton wicks saturated with 10% sucrose solution are used as food source. Females lay eggs in round, plastic containers (10 cm diameter x 5 cm depth) filled with 150 ml of tap water. Egg rafts are removed daily and placed in cylindrical enamel pans (with diameter of 35 cm and 10 cm deep), in order to hatch. Larvae are reared under the same conditions of temperature and light and are fed daily with baby fish food (TetraMin, Baby Fish Food) at a concentration of 0.25 gl^{-1} of water until pupation. Pupae are then collected and introduced into the adult rearing cages.

3.5 Larvicidal bioassays

Stock solutions of EOs tested were prepared in ethanol and maintained in a freezer as 1% mgl^{-1}solutions. They were dissolved in double distilled water to produce solutions of the tested materials in concentrations ranging from 5 to 150 mgl^{-1}. Prior to biological determinations the toxicity of each EO was evaluated (data not shown).

The larval mortality bioassays were carried out according to the test method for larval susceptibility, proposed by the World Health Organization (WHO, 1981). Twenty 3rd to 4th instar larvae of the species *Culex pipiens* biotype *molestus* were collected from the colony, placed in a glass beaker with 250 ml of aqueous suspension of the tested material at various concentrations and an emulsifier was added in the final test solution (less than 0.05%). Four replicates were made per each concentration and a control treatment with tap water and emulsifier was also included. Beakers with larvae were placed at 25±2 ºC, 80±2% relative humidity and photoperiod of 14:10 h (L:D).

3.6 Data analysis

Larvicidal effect was recorded 48 h after treatment. Data obtained from each dose–larvicidal bioassay (total mortality, mgl^{-1} concentration in water) were subjected to probit analysis in which probit-transformed mortality was regressed against log$_{10}$-transformed dose; LC$_{50}$, LC$_{90}$ values, and slopes were calculated (SPSS 11.0).

4. Results and discussion

4.1 Phytochemical analysis

Fourteen distinct Umbeliferae taxa (twelve genera) are studied herein, one of which is endemic to Greece (*Thamnosciadium* Hartvig). It must be noted that there are no literature reports and studies on the EOs and their chemical compostitions for the material obtained from the plants *Athamanta densa* Boiss. & Orph. (AD), *Chaerophyllum heldreichii* Orph. Ex Boiss. (CH), *Laserpitium pseudomeum* Orph., Heldr. & Sart. Ex Boiss. (LP), *Peucedanum neumayeri* (Vis.) Reichenb (PN), *Pimpinella tragium* ssp tragium Vill. (PT), *Pimpinella rigidula* (Boiss. & Orph.) H. Wolf (PR), *Scaligeria cretica* (Miller) Boiss. (SC), *Seseli parnassicum* Boiss. & Heldr. (SP) and *Smyrnium rotundifolium* Miller (SR). In addition, the discussion section on the related *taxa* EOs compositions includes ten (out of twelve) genera studied herein, since there are also no previous reports on the composition of EOs obtained from *Conium* L. and *Thamnosciadium* Hartvig genera.

In total seventy phytochemicals, representing 76.64 to 99.83 % of the respective EOs samples have been identified as their constituents using combined GC and GC/MS analyses and in certain occasions verified by NMR studies. The detailed qualitative and quantitative analytical data of the main constituents of steam volatiles (and their respective retention indices) are presented in **Table 4**.

Components	RI	PN	AS	TJ	SP	PO	CH	LP	PT	SC	PP	PR	FN	SR	AD	Identification
trans-2-hexanal	803												0.96		0.27	a, b
a-pinene	939	21.27	24.65	2.80		2.14	1.67	49.58	1.21	8.76			30.85	0.93	0.46	a, b, c
camphene	954	2.99	3.32			1.72							4.36			a, b, c
sabinene	975	2.76				1.15	71.76	24.73	4.30	13.74			1.96	0.93		a, b, c
β-pinene	979	2.66	1.33						8.51				1.79		8.86	a, b, c
myrcene	991	3.93	4.75	1.52		0.81	1.51	1.82					6.68	11.25	0.92	a, b, c
a-phellandrene	1003	2.53	2.35	3.83									1.33			a, b, c
a-terpinene	1017		3.58										0.32			a, b, c
p-cymene	1025	4.71						0.74								a, b, c
o-cymene	1026			0.91												a, b, c
limonene	1029	4.71		40.75		2.78			1.42	1.43					0.66	a, b, c
β-phellandrene	1030		12.76	42.96	1.42		10.86	6.73					10.20			a, b, c
cis-ocimene	1037	4.78		18.59									2.76		1.66	a, b, c
trans-ocimene	1050		0.82										1.01		5.16	a, b, c
γ-terpinene	1060	32.25					2.54	1.42		1.40			0.41			a, b, c
cis-sabinene hydrate	1070							2.53								a, b, c
terpinolene	1089			12.97												a, b, c
linalool	1097														0.50	a, b, c
trans-limonene oxide	1137		0.46													a, b
geijerene	1143								10.23							a, b
a-terpineol	1189						3.35	2.43		0.78						a, b, c
pregeijerene	1287								5.13							a, b
1-bornyl acetate	1289		3.84			81.13							0.52			a, b
2,3,4-trimethyl benzaldehyde	1359		2.16			4.68										a, b
2,3,6-trimethyl benzaldehyde	1371		0.62													a, b
isoledene	1376										0.69					a, b
a-copaene	1377									0.67			0.42		0.26	a, b
β-cubebene	1388													0.86		a, b
β-elemene	1391				10.85					0.50			0.51	2.09	0.33	a, b
aristolene	1407										19.92					a, b
calarene	1411										3.40				0.40	a, b
β-caryophyllene	1419		1.73		2.76					0.92	3.07		2.10			a, b, c
a-bergamontene	1435											62.15				a, b, c
γ-elemene	1437			3.22												a, b, c
β-humulene	1439			3.30						0.40						a, b, c
β-farnesene	1457									29.27		2.47			1.17	a, b, c
C14H30O (m/z: 189, 147, 105, 91, 204)	1483														19.80	b
a-amorphene	1484											0.32				a, b
C14H22O (m/z: 119, 91, 105, 145, 131)	1485								20.39							b
germacrene D	1487	2.55	4.42	0.89	13.02		2.53	0.71	0.87	28.37	0.85		6.42		1.21	a, b, c
β-selinene	1490				5.14						3.78	23.80			3.95	a, b, c
a-selinene	1498					0.95						3.53		5.28		a, b, c
a-zingiberene	1499											7.75				a, b
bicyclogermacrene	1500				6.25					3.51	1.21		4.04			a, b
a-farnesene	1506		1.84													a, b, c
β-bisabolene	1506				2.85					1.18	4.16				12.72	a, b, c
myristicin	1519											6.72			4.32	a, b, c
β-sesquiphellandrene	1523				30.39						1.04	1.80	0.92			a, b, c
δ-cadinene	1524	1.30	3.02							0.38						a, b, c
germacrene B	1561				10.64				19.28					2.13	0.98	a, b, c
spathulenol	1578				1.52						0.46					a, b, c
caryophyllene oxide	1583				1.02						0.61					a, b, c
β-elemenone	1601								1.72							a, b
isofuranogermacrene	1648													1.28		a, b
furanodiene	1649													11.81		a, b
a-bisabolol	1686		2.04													a, b
germacrone	1694								23.33					5.62		a, b
trans-isomyristicin	1721		10.14									7.74				a, b
trans-pseudoiso-eugenyl 2-methylbutyrate	1774											7.74				a, b
trans-epoxypseudoisoeugenyl 2-methylbutyrate	1783											26.72				a, b
furanoeremophil-1-one	1880													6.42	0.91	a, b
1β-acetoxyfuranoeudesm-4(15)-ene	1889													8.87		a, b
1β-acetoxyfuranoeudesm-3 ene	1911													20.72		a, b
C12H29O2N (m/z: 91, 55, 115, 129, 77)	1923														2.94	b
C12H29O2N (m/z: 91, 115, 55, 129, 77)	1943														8.58	b
C13H27O2N (m/z: 91, 115, 55, 129, 159)	2030														12.17	b
n-heneicosane	2100														0.59	a, b, c
tricosane	2300														0.32	a, b, c
pentacosane	2500														0.25	a, b, c
Total		99.20	99.83	96.46	92.38	95.36	94.22	99.20	88.80	93.35	94.22	92.75	76.64	79.83	89.39	

[a]Comparison of mass spectra with MS libraries and retention times
[b]Comparison of experimental RI with reported RI
[c]Comparison with authentic compounds
RI: Retention indices calculated against C_8 to C_{24} n-alkanes on the HP 5MS column.

Table 4. Chemical constituents of the essential oils tested.

The determined chemical composition of the EO from the aerial part of *Angelica sylvestris* L. (AS) is consistent with the literature reports for EOs obtained from its seeds (Bernard, 2001) and roots (Bernard & Clair, 1997), with α-pinene and β-phellandrene being the major components. Same compounds were reported as the prevailing phytochemicals in the EOs of *A. archangelica* L. *sensu lato* (Bernard, 2001; Nykanen et al., 1991; Bernard & Clair, 1997; Chalcat & Garry, 1997; Nivinskiene et al., 2005), while the EO of *A. glauca* is reported to contain β-phellandrene as major component and only small portions of α-pinene (Aghinotri et al., 2004; Kaul et al., 1996). Other *Angelica* L. taxa, such as *A. sinensis* (Dung et al., 1996; Kim et al., 2006), *A. gigas* (Kim et al., 2006), *A. acutiloba* (Kim et al., 2006), *A. heterocarpa* (Bernard, 2001; Bernard & Clair, 1997) and *A. tenuissima* (Ka et al., 2005) display a completely different, both qualitative and quantitative, EO composition profile.

In addition to *a*-pinene, which is the main constituent as previously reported by Demetzos et al. (2000), the studied EO of *Ferulago nodosa* (L) Boiss. (FN) was found to contain thirteen new components for the *taxon's* EO. More specifically, the molecules of *trans*-2-hexenal, myrcene, α-phellandrene, α-terpinene, β-phellandrene, *cis*-ocimene, *trans*-ocimene, γ-terpinene, bornyl acetate, β-elemene, β-caryophylene, germacrene D and bicyclogermacrene were also determined as constituents of this EO. With the exception of *trans*-2-hexenal all the abovementioned compounds have been assayed in the EOs of the following *Ferulago* W.D.J. Koch taxa; *F. asparagifolia* (Baser et al., 2001), *F. phialocarpa* (Masoudi et al., 2004b), *F. macrocolea* (Rustaiyan et al., 2005), *F. galbaniflua* (Rustaiyan et al., 2002a), and *F. thirkeana* (Baser et al., 2002).

The EO of *Peucedanum officinale* L. (PO) is dominated by bornyl acetate, which was previously found only in *P. scoparium* (Masoudi et al., 2004a). It is also characterized by the presence of 2,3,4-trimethyl benzaldehyde, which has not been previously reported as constitutent of *Peucedanum* L. EOs. In addition, the EO tested was found to contain five molecules, namely *a*-pinene, sabinene, myrcene, limonene and β-selinene, never reported in a EO of *P. officinale* (Jaimand et al., 2006). These five phytochemichals are abundant in the general profile of *Peucedanum* L. EOs, as reported for *P. scoparium* (Masoudi et al., 2004a), *P. zenkeri* (Menut et al., 1995), *P.vertcillare* (Fraternale et al., 2000), *P. petiolare* (Rustaiyan et al., 2001) and *P. cervariifolium* (Bazgir et al., 2005).

The EOs of *Pimpinella* L. have been thoroughly studied, mainly because the application of their several taxa as culinary herbs and/or spices. Though the EOs of fourteen (14) *taxa* were studied, only one (Tabanca et al., 2005) refers to PP (*Pimpinella peregrina* L.) and none to PT (*Pimpinella tragium* ssp tragium VIII) and PR (*Pimpinella rigidulla* Boiss. & Orph. H. Wolf). The main constituent of EO of PO is *a*-bergamontenene, reported so far only for *P. anagodendron* (Velasco-Negueruela et al. 2005) and *P. anisum* (Santos et al., 1998). Two additional components determined herein, β-bisabolene and β-sesquiphellandrene, have not been reported in previous studies for PP but are well documented for *P. anagodendron* (Velasco-Negueruela et al. 2005), *P. junoniae* (Velasco-Negueruela et al. 2003), *P. anisum* (Santos et al., 1998), *P. anisetum* (Baser et al., 1999; Tepe et al., 2006) and *P. tragioides* (Askari & Sefidcon, 2007). New entries, for this genera EO components list, are isoledene, aristolene, calarene and β-selinene which were also assayed in the EO of PP. On the contrary, the EO of PR is characterized by the complete absence of monoterpenes, advocating previous record of β-selinene and introducing *a*-amorphene, *a*-selinene and *trans*-isomyristicin as components of the *Pimpinella* L. EOs. Finally, the EO of PT has only two differences as compared to the genus EO components, an unidentified component and β-elemenone. In general, its composition is in accordance with the phytochemical profiles reported for the EOs of *P. aromatica* (Baser et al., 1996), *P. serbica* (Ivanic et al., 1983), *P. flabellifolia* (Tepe et al., 2006), *P. aurea* (Tabanca et al., 2005; Assadian et al., 2005), *P. acuminata* (Melkani et al., 2006), *P. barbata* (Fakhari & Sonboli, 2006), *P. rupicola* (Velasco-Negueruela et al. 2005), *P. corymbosa* and *P. puberula* (Tabanca et al., 2005).

Major components of the EO of *Scaligeia cretica* (Miller) Boiss (SC) are a-pinene, β-farnesene and germacrene D, which have also been detected in previous studies on the EOs of *Scaligeria* DC. In this respect, the EO of *S. lazica* contains β-farnesene as major and a-pinene, germacrene D as minor components (Baser et al., 1993). On the contrary, the EO of *S. tripartite* contains β-farnesene and germacrene D as minor compounds, while a-pinene is absent (Tabanca et al., 2007). Compounds assayed herein and never reported before in the EOs of *Scaligeria* DC are a-terpineol, β-elemene and β-humulene.

The EOs of *Laserpitium pseudomeum* Orph. Heldr. & Sart Ex Poiss. (LP) contains a-pinene, β-pinene, sabinene and β-phellandrene as major components, all well known constitutents of the EOs of *Laserpitium* L. Previous literature reports indicated that the EOs of *L. latifolium* contains a-pinene and β-pinene as major components (Borg-Karlson et al., 1994), the *L. petrophilum* a-pinene and sabinene (Baser et al., 1997), while the molecule of β-phellandrene is present in traces in both EOs. On the contrary, the phytochemical profile of the EO of *L. siler* is completely different containing mainly limonene and perillaldehyde (Chizzola et al., 1999)

The EO composition of *Smyrnium* L. has also been scarcely investigated, since only three *taxa's* EOs, namely *S. perfoliatum* (Molleken et al., 1998a; Tirillini et al., 1996; Tirillini & Tosi, 1992), *S. cordifolium* (Amiri et al., 2006) and *S. olusatrum* (Molleken et al., 1998b), have been studied todate. The studied of EO of *Smyrnium rotundifolium* Miller (SR) contains 7 major components, with the molecule of a-selinene reported for the first time as EO component of *Smyrnium* L.. Other compounds present in large quantitites are furanodiene (reported as major constitutent in *S. olusatrum*), myrcene, furanoeremophil-1-one, 1β-acetoxyfuranoeudesm-4(15)-ene, 1β-acetoxyfurano eudesm-3-ene (detected in *S. olusatrum* and *S. perfoliatum*, Molleken et al., 1998) and germacrone (present in *S. cordifolium*).

The phytochemical profile of *Chaerophyllum* L. EOs was studied previously for *C. macropodum* (Baser et al., 2006), *C. crinitum* (Baser et al., 2006; Nematollahi et al., 2005), *C. macrospermum* (Sefidcon & Abdoli, 2005; Rustaiyan et al., 2002b, Mamedova, 1994), *C. bulbosum sensu lato* (Mamedova & Akhmedova, 1991; Kokkalou & Stefanou, 1989), *C. aksekiense* (Baser et al., 2000b), *C. coloratum* (Vajs et al., 1995), *C. azoricum* (Pedro et al., 1999) and *C. prescotii* (Letchamo et al., 2005). The more significant differentiation among the literature results and the assayed herein EO of *Chaerophyllum heldreichi* Orph. Ex Boiss (CH) comprises the identification for first time of a-terpineol as main component of EO of *Chaerophyllum* L..

The EO of *Seseli parnassicum* Boiss. & Heldr. (SP) was found to contain three new compound entries, β-humulene, β-selinene and β-sesquiphellandrene, as compared with the EO of the Seseli L. *taxa* (also including the synonymous *Lomatopodium* Fisch. et C.A. Mey *taxa*). The remaining components are in accordance with the EO content of same *taxa* plants, such as *S. montanum* (Evergetis et al., 2009), *S. campestre* and *S. peucedanoides* (Baser et al. 2000a;

Bulatovic et al. 2006) and in *S. buchtormence*. These compounds were also present in the EOs of *S. resinosum* and *S. tortuosum*, obtained from the fruits and not the herbal part of the plants (Dogan et al. 2006). The *L. khorassanicum* and *L. staurophyllum* EOs were assayed to contain mostly aliphatic terpenes, while the corresponding cyclic terpenes were present in smaller amounts compared to EOs of Seseli L. (Sedghat et al. 2003; Sefidkon et al. 1997).

Finally, the investigated EO of *Athamanta densa* Boiss. & Orph., contains as major constituents myristicin and various unidentified alkaloids, which account for almost 24 % of its weight. The literature reports of EOs of *Athamanta* L. indicate that they mainly contain either myristicin, such as the EOs of *A. sicula* (Camarda & Di Stefano, 2003), *A. turbith sensu lato* (Tomic et al., 2009), *A. macedonica* (Verykokidou et al., 1995) and *A. haynaldi* (Zivanovic et al., 1994), or apiole as in *A. sicula* (Camarda & Di Stefano, 2008).

4.2 Larvicidal assays

The investigated EOs were evaluated —for the first time— in respect to their larvicidal activities against 3rd- 4th instar larvae of *Culex pipiens*. The relative results expressed as the respective LC_{50} and LC_{90} values are included in **Table 5**. Among the EOs tested only two were rather inactive (AS and PP, displaying LC_{50} values above 150 mgl^{-1}), while the EOs of SC and SP were moderately active displaying LC_{50} values above 100 mgl^{-1} (111.99 and 122.54 mgl^{-1} respectively).

Essential Oils tested	LC_{50} (95% CL)[a]	LC_{90} (95% CL)[a]	Slope (±SE)
Athamanta densa	10.15 (9.49-10.73)	15.75 (14.52-17.76)	6.72±0.80
Pimpinella tragium ssp tragium	40.13 (32.43-45.95)	71.10 (61.51-91.00)	5.15±0.52[b]
Pimpinella rigidula	40.31 (34.75-43.64)	60.41 (55.66-70.57)	7.29±1.44[b]
Thamnosciadium junceum	44.17 (41.52-46.62)	64.42 (59.94-71.28)	7.82±0.86[b]
Peucedanum neumayeri	47.40 (40.25-54.15)	81.47 (68.63-113.57)	5.44±0.53[b]
Chaerophyllum heldreichii	53.61 (50.29-56.55)	75.96 (71.53-82.15)	8.46±0.87
Laserpitium pseudomeum	56.73 (53.50-59.60)	79.59 (75.18-85.71)	8.46±0.86
Ferulago nodosa	67.39 (64.17-70.41)	95.59 (89.90-103.94)	8.43±0.84
Smyrnium rotundifolium	80.32 (76.88-84.16)	105.30 (98.33-116.61)	10.89±1.29
Peucedanum officinale	86.46 (82.27-90.30)	125.05 (117.23-136.95)	7.99±0.84
Scaligeria cretica	111.99 (107.86-115.47)	133.83 (128.35-143.21)	6.58±0.73[b]
Seseli parnassicum	122.54 (115.54-141.06)	167.15 (143.83-268.76)	6.30±0.68
Angelica sylvestris	>150		
Pimpinella peregrina	>150		

[a] LC values are expressed in mgl^{-1} and they are considered significantly different when 95% CL fail to overlap.
[b] Since goodness–of–fit test is significant (P<0.05), a heterogeneity factor is used in the calculation of confidence limits (CL)

Table 5. LC_{50} and LC_{90} values for the tested essential oils against larvae of *Culex pipiens* biotype *molestus*.

The EO derived from the endemic in Greece plant *Athamanta densa* was determined as the most active since displayed the highest toxicity against mosquito larvae, with LC_{50} value 10.15 mgl^{-1}. The EO tested contains a series of compounds which were not found in the other EOs tested, such as bisabolene and the unidentified compounds $C_{14}H_{30}O$, $C_{12}H_{25}O_2N$ and $C_{13}H_{27}O_2N$, which have to study more thoroughly in order to determine their activities. The remaining EOs (PR, TJ, PT, PN, CH, LP, FN, SR and PO) displayed LC_{50} values ranging from 40.31 to 86.46 mgl^{-1}. No significant relationship between toxicity and phytochemical content was detected.

5. References

Adams, R. (1995). Identification of essential oil components by Gas Chromatography / Mass Spectroscopy. Carol Stream, Allured Publishing, Illinois, USA. ISBN 0931710421.

Aghinotri, V.K., Thappa R.K., Meena, B., Kapaphi, B.K. Saxena, R.K., Qazi, G.N. & Agarwal S.G. (2004). Essential oil composition of aerial parts of *Angelica glauca* growing wild in North-West Himalaya (India). *Phytochemistry*, 65, 2411-2413.

Alves, S.N., Serro, J.E. & Melo, A.L. (2010). Alterations in the fat body and midgut of *Culex quinquefasciatus* larvae following exposure to different insecticides. *Micron*, 41, 592-597.

Amer, A. & Mehlhorn, H. (2006a) Larvicidal effects of various essential oils against *Aedes*, *Anopheles*, and *Culex* larvae (Diptera: Culicidae). *Parasitology Research*, 99, 466-472.

Amer, A. & Mehlhorn, H. (2006b) Persistency of larvicidal effects of plant oil extracts under different storage conditions. *Parasitology Research*, 99, 473-477.

Amiri, H., Khavari-Nejad R.A., Masoud, S., Chalabian, F. & Rustaiyan, A. (2006). Composition and antimicrobial activity of the essential oil from stems, leaves, fruits and roots of *Smyrnium cordifolium* Boiss. from Iran. *Journal of Essential Oil Research*, 18, 574-577.

Askari, F. & Sefidkon, F. (2007). Essential oil composition of *Pimpinella tragioides* (Boiss.) Benth et Hook from Iran. *Journal of Essential Oil Research*, 19, 54-56.

Assadian, F., Masoudi, S., Nematollahi, F., Rustaiyan, A., Larijani, K. & Mazloomifar, H. (2005). Volatile constituents of *Xanthogalum purpurascens* Ave-Lall., *Eryngium caeruleum* M.B. and *Pimpinella aurea* DC. three Umbelliferae herbs growing in Iran. *Journal of Essential Oil Research*, 17, 243-245.

Ateyyat, M.A., Al-Mazra'awi, M., Abu-Rjai, T. & Shatnawi, M.A. (2009). Aqueous extracts of some medicinal plants are as toxic as Imidacloprid to the sweet potato whitefly, *Bemisia tabaci*. *Journal of Insect Science*, 9:15, 6pp.

Baser, K.H.C. Demirci, B., Demirci, F., Hashimoto, T., Asakawa, Y. & Noma, Y. (2002). Ferulagone: a new monoterpene ester from *Ferulago thirkeana* essential oil. *Planta Medica*, 68, 564-567.

Baser, K.H.C., Demirci, B. & Duman, H. (2001). Composition of the essential oil of *Ferulago asparagifolia* Boiss. from Turkey. *Journal of Essential Oil Research*, 13, 134-135.

Baser, K.H.C. & Duman, H. (1997). Composition of the essential oil of *Laserpitium petrophilum* Boiss. et Heldr. *Journal of Essential Oil Research*, 9, 707-708.

Baser, K.H.C., Ozek, G., Ozek, T. & Duran, A. (2006). Composition of the essential oil of *Chaerophyllum macropodum* Boiss. fruits obtained by microdistillation. *Journal of Essential Oil Research*, 18, 515-517.

Baser, K.H.C., Ozek, T., Kurkcuoglu, M.K. & Guner. A. (1993). The essential oil of *Scaligeria lazica* Boiss. *Journal of Essential Oil Research*, 5, 463-464.

Baser, K.H.C., Ozek, T., Duman, H. & Guner, A. (1996). Essential oil of *Pimpinella aromatica* Bieb. from Turkey. *Journal of Essential Oil Research*, 8, 463-464.

Baser, K.H.C., Ozek, T. & Tabanca, N. (1999). Essential Oil of *Pimpinella anisetum* Boiss. et Bal. *Journal of Essential Oil Research*, 11, 445-446.

Baser, K.H.C., Ozek, T., Kurkcuoglu, M. & Aytac, Z. (2000a) Essential oil of *Seseli campestre* Besser. *Journal of Essential Oil Research*, 12, 105-107.

Baser, K.H.C., Tabanka, N., Ozek, T., Demicri, B., Duran, A. & Duman, H. (2000b). Composition of the essential oil of *Chaerophyllum aksekiense* A. Duran et Duman, a recently described endemic from Turkey. *Flavour and Fragrance Journal*, 15, 43-44.

Bazgir, A., Shaabani, A. &, Sefidkon, F. (2005). Composition of the essential oil of *Peucedanum cervariifolium* C.A. Mey. from Iran. *Journal of Essential Oil Research*, 17, 380-381.

Becker, N., Petric, D., Zgomba, M., Boase, C., Dahl, C., Lane, J., Kaiser, A. Mosquitoes and their control, Kluwer Academic/Plenum Publishers: New York, 2003. ISBN 0306473607.

Bermard, C. (2001). Essential oils of three *Angelica* L. species growing in France. Part II: fruit oils. *Journal of Essential Oil Research*, 13, 260-263.

Bernard, C. & Clair, G. (1997). Essential oils of three *Angelica* L. species growing in France. I. Root oils. *Journal of Essential Oil Research*. 9, 289-294.

Borg-Karlson, A-K., Valterova, I. & Anders Nilson, L. (1994). Volatile compounds from flowers of six species in the family Apiaceae: Bouquets for different pollinators? *Phytochemistry*, 35, 111-119.

Boyer, S., Serandour, J., Lemperiere, G., Raveton, M. &, Ravanel, P. (2006). Do herbicide treatments reduce the sensitivity of mosquito larvae to insecticides? *Chemosphere*, 65, 721-724.

Bulatovic, V.M., Savikin-Fodulovic, K.P., Zdunic, G.M. & Popovic, M.P. (2006). Essential oil of *Seseli peucedanoides* (MB) Kos.-Pol. *Journal of Essential Oil Research*, 18, 286-287.

Camarda, L. & Di Stefano, V. (2003). Essential oil of leaves and fruits of *Athamanta sicula* L. (Apiaceae). *Journal of Essential Oil Research*, 15, 133-134.

Camarda, L., Di Stefano, V. & Pitonzo, R. (2008). Chemical composition of essential oils from *Athmanta sícula*. *Chemistry of Natural Compounds*, 44, 532-533.

Campbell, C., Gries, R. & Gries, G. (2011) Forty-two compounds in eleven essential oils elicit antennal responses from *Aedes aegypti*. *Entomologia Experimentalis et Applicata*, 138, 21-32.

Caquet, T., Roucaute, M., Le Goff, P. & Lagadic, L. (2011). Effects of repeated field applications of two formulations of *Bacillus thuringiensis* var. *israelensis* on non-target saltmarsh invertebrates in Atlantic coastal wetlands. *Ecotoxicology and Environmental Safety*, 74, 1122-1130.

Chakraborty, S., Singha, S. & Chandra, G. (2010). Mosquito larvicidal effect of orthophosporic acid and lactic acid individually or their combined form on *Aedes aegypti*. *Asian Pacific Journal of Tropical Medicine*, 3, 954-956.

Chalchat, J.C., Garry, R.P., (1997). Essential oil of angelica roots (*Angelica archangelica* L.): optimization of distillation, location in plant and chemical composition. *Journal of Essential Oil Research*, 9, 311-319.

Chizzola, R., Novak, J. & Franz, C. (1999). Fruit oil of *Laserpitium siler* L. grown in France. *Journal of Essential Oil Research*, 11, 197-198.

Choi, W-S., Park, B-S., Lee, Y-H., Jang, D.Y., Yoon, H.Y. & Lee, S-E. (2006). Fumigant toxicities of essential oils and monoterpenes against *Lycoriella mali* adults. *Crop Protection*, 25, 398-401.

Choochote, W., Tuetun, B., Kanjanapothi, D., Rattanachanpichai, E., Chaithong, U., Chaiwong, P., Jitpakdi, A., Tippawangkosol, P., Riyong, D. & Pitasawat, B. (2004) Potential of crude seed extract of celery, *Apium graveolens* L., against the mosquito *Aedes aegypti* (L.) (Diptera: Culicidae). *Journal of Vector Ecology*, 29, 340-346.

Chowdhury, N., Laskar, S. & Chandra, G. (2008). Mosquito larvicidal and antimicrobial activity of protein of *Solanum villosum* leaves. *BMC Complementary and Alternative Medicine*, 8, 62, doi: 10.1186/1472-6882-8-62. Available online http://www.biomedcentral.com/1472-6882/8/62.

Conti, B., Canale, A., Bertoli, A., Gozzini, F. & Pistelli, L. (2010). Essential oil composition and larvicidal activity of six Mediterranean aromatic plants against the mosquito *Aedes albopictus* (Diptera: Culicidae). *Parasitology Research*, 107, 1455-1461.

Daaboub, J., Ben Cheikh, R., Lamari, A., Ben Jha, I., Feriani, M., Boubaker, C. & Ben Cheikh, H. (2008). Resistance to pyrethroid insecticides in *Culex pipiens pipiens* (Diptera: Culicidae) from Tunisia. *Acta Tropica*, 107, 30-36.

Del Pilar Corena, M., Van Den Hurk, P., Zhong, H., Brock, C., Mowery, R., Johnson, J.V. & Linser, P. J. (2006). Degradation and effects of the potential mosquito larvicides methazolamide and acetazolamide in sheepshead minnow (*Cyprinodon variegatus*). *Ecotoxicology and Environmental Safety*, 64, 369-376.

Delaunay, P., Jeannin, C., Schaffner, F. & Marty, P. (2009). Actualites 2008 sur la presence du moustique tigre *Aedes albopictus* en France metropolitaine. *Archives de Pediatrie*, 16, Supplement 2, 66-71.

Demetzos, C., Perdetzoglou, D., Gazouli, M., Tan, K. & Economakis, C. (2000). Chemical analysis and antimicrobial studies on three species of *Ferulago* from Greece. *Planta Medica*, 66, 560-563.

Dogan, E., Duman, H., Tosun, A., Kurkcuoglu, M. & Baser K.H.C. (2006). Essential oil composition of the fruits of *Seseli resinosum* Freyn et Sint. and *Seseli tortuosum* L. growing in Turkey. *Journal of Essential Oil Research*, 18, 57-59.

Dung, N.X., Cu, L.D., Moi, L.D. & Leclercq, P.A. (1996). Composition of the leaf and flower oils from *Angelica sinensis* (Oliv.) diels cultivated in Vietnam. *Journal of Essential Oil Research*, 8, 503-506.

Eckenbach, U., Lampman, R.L., Seigler, D.S., Ebinger, J. & Novak, R.J. (1999). Mosquitocidal activity of acetylenic compounds from *Cryptotaenia Canadensis*. Journal of Chemical Ecology, 25, 1885-1893.

Evergetis, E., Michaelakis, A., Kioulos, E., Koliopoulos, G. & Haroutounian, S. A. (2009). Chemical composition and larvicidal activity of essential oils from six Apiaceae family taxa against the West Nile virus vector *Culex pipiens*. *Parasitology Research*, 105, 117–124.

Fakhari, A.R. & Sonboli, A. (2006). Essential oil composition of *Pimpinella barbata* (DC.) Boiss. from Iran. *Journal of Essential Oil Research*, 18, 679-681.

Fraternale, D., Giamperi, L., Rici, D. & Manunta, A. (2000). Composition of the essential oil of *Peucedanum verticillare*. *Biochemical Systematics and Ecology*, 28, 143-147.

Gu, J., Liu, M., Deng, Y., Peng, H. & Chen, X. (2011). Development of an Efficient Recombinant Mosquito Densovirus-Mediated RNA Interference System and Its Preliminary Application in Mosquito Control. *PLoS One*, 6, doi: 10.1371/journal.pone.0021329.

Hancock, P.A., Sinkins, S.P. & Godfray, H.C.J. (2011). Strategies for Introducing Wolbachia to Reduce Transmission of Mosquito-Borne Diseases. *PLoS Neglected Tropical Diseases*, 5, doi: 10.1371/journal.pntd.0001024.

Hayes, S.R., Hudon, M. & Park, H.W. (2011). Isolation of novel *Bacillus* species showing high mosquitocidal activity against several mosquito species. *Journal of Invertebrate Pathology*, 107, 79-81.

Imbahale, S.S., Fillinger, U., Githeko, A., Mukabana, W.R. & Takken, W. (2010). An exploratory survey of malaria prevalence and people's knowledge, attitudes and practices of mosquito larval source management for malaria control in western Kenya. *Acta Tropica*, 115, 248-256.

Isman, M.B. (2000). Plant essential oils for pest and disease management. *Crop Protection*, 19, 603-608.

Ivanic, R., Savin, K. & Robinson, F.V. (1983). Essential oil from *Pimpinella serbica* fruits. *Planta medica*, 48, 60-61.

Jaimand, K., Ashorabadi, E.S. & Dini, M. (2006). Chemical constituents of the leaf and seed oils of *Peucedanum officinale* L. cultivated in Iran. *Journal of Essential Oil Research*, 18, 670-671.

Ka, M.H., Chol, E.H., Chun, H.S. & Lee, K.G. (2005). Antioxidative activity of volatile extracts from *Angelica tenuissimae* roots, peppermint leaves, pine needles, and sweet flags leaves. *Journal of Agricultural and Food Chemistry*, 53, 4124-4129.

Kanzok, S.M. & Jacobs-Lorena M. (2006). Entomopathogenic fungi as biological insecticides to control malaria. *Trends in Parasitology*, 22, 49-51.

Kaul, P.N., Mallavarapu, G.R. & Chamoli, R.P. (1996). The essential oil composition of *Angelica glauca* roots. *Planta medica*, 62, 80-81.

Khater, H.F. & Shalaby, A.A.S. (2008). Potential of biologically active plant oils to control mosquito larvae (*Culex pipiens*, Diptera: Culicidae) from an Egyptian locality. *Revista do Instituto de Medicina Tropical de São Paulo*, 50, 107-112.

Kilama, W.L. (2009). Health research ethics in public health: Trials and implementation of malaria mosquito control strategies. *Acta Tropica*, 112, Supplement 1, 37-47.

Killeen, G.F. & Smith, T.A. (2007). Exploring the contributions of bed nets, cattle, insecticides and excitorepellency to malaria control: a deterministic model of mosquito host-seeking behaviour and mortality. *Transactions of the Royal Society of Tropical Medicine and Hygiene*, 101, 867-880.

Kim, D-H., Kim, S-I., Chang, K-S. & Ahn, Y-J. (2002). Repellent Activity of Constituents Identified in *Foeniculum vulgare* Fruit against *Aedes aegypti* (Diptera: Culicidae). *Journal of Agricultural and Food Chemistry*, 50, 6993-6996.

Kim, M.R., El-Aty A.M.A., Kim, S-I. & Shim J.H. (2006). Determination of volatile flavor components in danggui cultivars by solvent free injection and hydrodistillation followed by gas chromatographic-mass spectrometric analysis. *Journal of Chromatography A*, 1116, 259-264.

Kim, S-I., Park, C., Ohh, M-H., Cho, H-C. & Ahn, Y-J. (2003a). Contact and fumigant activities of aromatic plant extracts and essential oils against *Lasioderma serricorne* (Coleoptera: Anobiidae). *Journal of Stored Products Research*, 39, 11-19.

Kim, S-I., Roh, J-Y., Kim, D-H., Lee, H-S. & Ahn, Y-J. (2003b). Insecticidal activities of aromatic plant extracts and essential oils against *Sitophilus oryzae* and *Callosobruchus chinensis*. *Journal of Stored Products*, 39, 293-303.

Knio, K.M., Usta, J., Dagher, S., Zournajian, H. & Kreydiyyeh, S. (2008). Larvicidal activity of essential oils extracted from commonly used herbs in Lebanon against the seaside mosquito, *Ochlerotatus caspius*. *Bioresource Technology*. 99, 763-768.

Kokkalou, E. & Stefanou, E. (1989). The volatiles of *Chaerophyllum bulbosum* wild in Greece. *Pharmaceutica Acta Helvetiae*, 64, 133-134.

Kwon, J-H. & Ahn, Y-J. (2002). Acaricidal Activity of Butylidenephthalide Identified in *Cnidium officinale* Rhizome against *Dermatophagoides farinae* and *Dermatophagoides pteronyssinus* (Acari: Pyroglyphidae). *Journal of Agricultural and Food Chemistry*, 50, 4479-4483.

Lamiri, A., Lhaloui, S., Benjilali, B., Berrada, M. (2001). Insecticidal effects of essential oils against Hessian fly, *Mayetiola desstructor* (Say). *Field Crops Research*, 71, 9-15.

Lavialle-Defaix, C., Apaire-Marchais, V., Legros, C., Pennetier, C., Mohamed, A., Licznar, P., Corbel, V. & Lapied, B. (2011). *Anopheles gambiae* mosquito isolated neurons: A new biological model for optimizing insecticide/repellent efficacy. *Journal of Neuroscience Methods*, doi:10.1016/j.jneumeth.2011.06.003.

Lee, H-S. (2004). Acaricidal Activity of Constituents Identified in *Foeniculum vulgare* Fruit Oil against *Dermatophagoides* spp. (Acari: Pyroglyphidae). *Journal of Agricultural and Food Chemistry*, 52, 2887-2889.

Lee, H-S. (2006). Mosquito larvicidal activity of aromatic medicinal plant oils against *Aedes aegypti* and *Culex pipiens pallens*. *Journal of the American Mosquito Control Association*, 22, 292-295.

Letchamo, W., Korolyk, E.A. & Tkachev, A.V. (2005). Chemical screening of essential oil bearing flora of Siberia. V. Composition of the essential oil of *Chaerophyllum prescottii* DC tops from Altai region. *Journal of Essential Oil Research*, 17, 560-562.

Lima, E.P., Paiva, M.H.S., de Araújo, A.P., da Silva, E.V.G., da Silva, U.M., de Oliveira, L.N., Santana, A.E.G., Barbosa, C.N., de Paiva Neto, C.C., Goulart, M.O.F., Wilding, G.S., Ayres, C.F.J. & de Melo Santos, M.A.V. (2011). Insecticide resistance in *Aedes aegypti* populations from Ceará, Brazil. *Parasites & Vectors*, 4, doi: 10.1186/1756-3305-4-5.

Mamedova, S.A. & Akhmedova, E.R. (1991). Essential oil of turnip-root chervil. *Chemistry of Natural Compounds*, 27, 248-249.

Mamedova. S.A. (1994). Essential oil of *Chaerophyllum macrospermum*. *Chemistry of Natural Compounds*, 30, 267-277.

Manolakou, S., Pitarokili, D., Koliopoulos, G., Michaelakis, A. & Tzakou, O. (2009). Essential Oil Composition of Different Parts of Greek *Foeniculum vulgare* and Larvicidal Activity of the Stem Oil. In: "Essential Oils and Aromas: Green Extraction and Application" edit by Prof. Farid Chemat, "Har Krishan Bhalla and Sons" (Publisher Journal of Essential Oil Bearing Plants), France.

Masoudi, S., Akhgar, M.R. & Rustaiyan, A. (2004a). Essential oils of *Peucedanum scoparium* (Boiss.) Boiss. and *Serotinocarpum insignis* Mozaffarian. from Iran. *Journal of Essential Oil Research*, 16, 117-119.

Masoudi, S., Rustaiyan, A. & Ameri, N. (2004b). Volatile oils of *Ferulago phialocarpa* Rech. f. et H. Reidl. and *Leutea elbursensis* Mozaffarian from Iran. *Journal of Essential Oil Research*, 16, 143-144.

Massada, Y. (1976). Analysis of essential oil by Gas Chromatography and Spectrometry. New York: Wiley.

Melkani, A.B., Javed, M.S., Melkani, K.B., Dev. V. & Beauchamp, P.S. (2006). Terpenoid composition of the essential oil from *Pimpinella acuminata* (Edgew.) CB Clarke. *Journal of Essential Oil Research*, 18, 312-314.

Menut, C., Mve-Mba, C.E., Lamaty, G., Amvan Zollo, P.H., Tchoumbougnang, F. & Bessiere, J.M. (1995). Aromatic plants of tropical central Africa. XVIII: Essential oils of leaf and rhizome of *Peucedanum zenkeri* Engl. from Cameroon. *Journal of Essential Oil Research*, 7, 77-79.

Miyazawa, M., Tsukamoto, T., Anzai, J. & Ishikawa, Y. (2004). Insecticidal Effect of Phthalides and Furanocoumarins from *Angelica acutiloba* against *Drosophila melanogaster*. *Journal of Agricultural and Food Chemistry*, 52, 4401-4405.

Moleken, U., Sinnwell, V. & Kubeczka, K.H. (1998a). The essential oil composition of fruits from *Smurnium perfoliatum*. *Phytochemistry*, 47, 1079-1083.

Moleken, U., Sinnwell, V. & Kubeczka, K.H. (1998b). Essential oil composition of *Smurnium olusatrum*. *Phytochemistry*, 49, 1709-1714.

Momin, R.A. & Nair, M.G. (2001). Mosquitocidal, nematicidal, and antifungal compounds from *Apium graveolens* L. seeds. *Journal of Agricultural and Food Chemistry*, 49, 142-145.

Nematollahi, F., Akhgar, M.R., Larijani, K., Rustaiyan, A. & Masoudi, S. (2005). Essential oils of *Chaerophyllum macropodum* Boiss. and *Chaerophyllum crinitum* Boiss. from Iran. *Journal of Essential Oil Research*, 17, 71-72.

Nivinskiene, O., Butkiene, R. & Mockute, D. (2005). The chemical composition of the essential oil of *Angelica archangelica* L. roots growing wild in Lithuania. *Journal of Essential Oil Research*, 17, 373-377.

Nykanen, I. & Nykanen. L. (1991). Composition of angelica root oils obtained by Supercritical CO_2 extraction and steam distillation. *Journal of Essential Oil Research*, 3, 229-236.

Orozco, O.L., Lentz, D.L. (2005). Poisonous plants and their uses as insecticides in Cajamarca, Peru. *Economic Botany*, 59, 166-173.

Pandey, S.K., Upadhyay, S. & Tripathi A.K. (2009). Insecticidal and repellent activities of thymol from the essential oil of *Trachyspermum ammi* (Linn) Sprague seeds against *Anopheles stephensi*. *Parasitology Research*, 105, 507-512.

Papachristos, D.P. & Stamopoulos D.C. (2002). Repellent, toxic and reproduction inhibitory effects of essential oil vapours on *Acanthoscelides obtectus* (Say) (Coleoptera: Bruchidae). *Journal of Stored Products Research*, 38, 117-128.

Patersson, R.S., Lofgren, C.S. & Boston, M.D. (1968). The sterile-male technique for control of mosquitoes: a field cage study with *Anophles quadrimaculatus*. *The Florida Entomologist*, 51, 77-82.

Pavela, R. (2008). Larvicidal effects of various Euro-Asiatic plants against *Culex quinquefasciatus* Say larvae (Diptera: Culicidae), *Parasitology Research*, 102, 555-559.

Pavela, R. (2009). Larvicidal effects of some Euro-Asiatic plants against *Culex quinquefasciatus* Say larvae (Diptera: Culicidae), *Parasitology Research*, 105, 887-892.

Pedro, L.G., Da Silva, J.A., Barroso, J.G., Figueiredo, A., Cristina, D., Stanley, G., Looman, A. & Scheffer, J.J.C. (1999). Composition of the essential oil of *Chaerophyllum azoricum* Trel., an endemic species of the Azores archipelago. *Flavour And Fragrance Journal*, 14, 287-289.

Pennetier, C., Chabi, J., Martin, T., Chandre, F., Rogier, C., Hougard, J-M. & Pages, F. (2010). New protective battle-dress impregnated against mosquito vector bites. *Parasites & Vectors*, 3, 81.

Peterson, R.K.D., Macedo, P.A. & Davis R.S. (2006). A Human-Health Risk Assessment for West Nile Virus and Insecticides Used in Mosquito Management. *Environmental Health Perspectives*, 114, 366–372.

Pimenov, M.G. & Leonov, M.V. (1993). The genera of the Umbelliferae, Royal Botanic Gardens Kew & Botanical Garden of Moscow State University, Kent U.K., p. 156.

Pitasawat, B., Champakaew, D., Chochote, W., Jitpakdi, A., Chaithong, U., Kanjanapothi, D., Rattanachanpichai, E., Tippawangkosol, P., Riyong, D., Tuetun, B. & Chaiyasit, D. (2007). Aromatic plant-derived essential oil: an alternative larvicide for mosquito control. *Fitoterapia*, 78, 205-210.

Prajapati, V., Tripathi, A.K., Aggarwal, K.K. & Khanuja, S.P.S. (2005). Insecticidal, repellent and oviposition-deterrent activity of selected essential oils against *Anopheles stephensi*, Aedes aegypti and *Culex quinquefasciatus*. *Bioresource technology*, 96, 1749-1757.

Promsiri, S., Naksathit, A., Kruatrachue, M. & Thavara, U. (2006). Evaluations of larvicidal activity of medicinal plant extracts to *Aedes aegypti* (Diptera: Culicidae) and other effects on a non target fish. *Insect Science*, 13, 179-188.

Raghavendra, K., Barik, T.K., Bhatt, R.M., Srivastava, H.C., Sreehari, U. & Dash, A.P. (2011). Evaluation of the pyrrole insecticide chlorfenapyr for the control of *Culex quinquefasciatus* Say. *Acta Tropica*, 118, 50-55.

Rajanarendar E., Reddy, M.N., Murthy, K.R., Reddy, K.G., Raju, S., Srinivas, M., Praveen, B. & Rao, M.S. (2010). Synthesis, antimicrobial, and mosquito larvicidal activity of 1-aryl-4-methyl-3,6-bis-(5-methylisoxazol-3-yl)-2-thioxo-2,3,6,10b-tetrahydro-1H-pyrimido[5,4-c]quinolin-5-ones. *Bioorganic & Medicinal Chemistry Letters*, 20, 6052-6055.

Ray, D. (2010). Organochlorine and Pyrethroid Insecticides. *Comprehensive Toxicology*, 13, 445-447.

Regnault-Roger, C., Hamraul, A. (1994). Inhibition of reproduction of *Acanthoselides obtectus* Say (Coleoptera), a kidney bean (*Phaseolus vulgaris*) bruchid, by aromatic essential oils. *Crop Protection*, 13, 624-628.

Riaz, M.A., Poupardin, R., Reynaud, S., Strode, C., Ranson, H. & David, J-P. (2009). Impact of glyphosate and benzo[a]pyrene on the tolerance of mosquito larvae to chemical insecticides. Role of detoxification genes in response to xenobiotics. *Aquatic Toxicology*, 93, 61-69.

Rustaiyan, A., Komeilizadeh, H., Mojab, F., Khazaie, A., Masoudi, S. & Yari, M. (2001). Essential oil composition of *Peucedanum petiolare* (DC) Boiss. from Iran. *Journal of Essential Oil Research*, 13, 49-50.

Rustaiyan, A., Monfared, A., Masoudi, S. & Ameri, N. (2002a). Essential oils of the stem and root of *Ferula galbaniflua* Boiss. et Bushe. from Iran. *Journal of Essential Oil Research*, 14, 286-287.

Rustaiyan, A., Neekpoor, N., Rabani, M., Komeilizadeh, H., Masoudi, S. & Monfared, A. (2002b). Composition of the essential oil of *Chaerophyllum macrospermum* (Spreng.) Fisch. and C.A. Mey. from Iran. *Journal of Essential Oil Research*, 14, 216-217.

Rustaiyan, A., Nadimi, M., Mazloomifar, H, & Massudi, S. (2005). Composition of the essential oil of *Ferula macrocolea* (Boiss.) Boiss. from Iran. *Journal of Essential Oil Research*, 17, 55-56.

Santos, P.A.G., Figueiredo, A.C., Oliveira, M.M., Barroso, J.G., Pedro, L.G., Deans, S.G., Younous, A.K.M. & Scheffer, J.J.C. (1998). Essential oils from hairy root cultures and from fruits and roots of *Pimpinella anisum*. *Phytochemistry*, 48, 455-460.

Sedghat, S., Rustaiyan, A., Khosravi, M., & Masoudi, S. (2003). Chemical constituents of the essential oil of *Lomatopodium khorassanicum* Mozaffarian-a species endemic to Iran. *Journal of Essential Oil Research*, 15, 416-417.

Sefidkon, F., Khajavi, M.S. & Mirza, M. (1997). Essential oil of *Lomatopodium staurophyllum* (Rech. f.) Rech. f. *Journal of Essential Oil Research*, 9, 471-472.

Shaalan, E., Canyon, D., Younes, M-W., Abdel-Wahad, H. & Mansour, A-H. (2005). A review of botanical phytochemicals with mosquitocidal potential. *Environment International*, 31, 1149-1166.

Stevens, M.M., Burdett, A.S., Mudford, E.M., Helliwell, S. & Doran, G. (2011). The acute toxicity of fipronil to two non-target invertebrates associated with mosquito breeding sites in Australia. *Acta Tropica*, 117, 2 125-130.

Sukumar, K., Perich, M-J. & Boobar, L-R. (1991). Botanical derivatives in mosquito control: a review. *Journal of the American Mosquito Control Association*, 7, 210-237.

Tabanca, N., Demirci, B., Baser, K.H.C., Mincsovics, E., Khan, I.A., Jacob, D.E. & Wedge, D.E. (2007). Characterization of volatile constituents of *Scaligeria tripartita* and studies on the antifungal activity against phytopathogenic fungi. *Journal of Chromatography B*, 850, 221-229.

Tabanca, N., Demirci, B., Kirimer, N., Baser, K.H.C., Bedir, E., Khan, I.A. & Wedge, D.E. (2005). Gas chromatographic-mass spectrometric analysis of essential oils from *Pimpinella aurea*, *Pimpinella corymbosa*, *Pimpinella peregrina* and *Pimpinella puberula* gathered from Eastern and Southern Turkey. *Journal of chromatography A*, 1097, 192-198.

Talontsi, F.M., Matasyoh, J.C., Ngoumfo, R.M. & Chepkorir, R. (2011). Mosquito larvicidal activity of alkaloids from *Zanthoxylum lemairei* against the malaria vector *Anopheles gambiae*. *Pesticide Biochemistry and Physiology*, 99, 82-85.

Tedesco, C., Ruiz, M. & McLafferty, S. (2010). Mosquito politics: Local vector control policies and the spread of West Nile Virus in the Chicago region. *Health & Place*, 16, 1188-1195.

Tepe, B., Akpulat, H.A., Sokmen, M., Daferera. D., Yumrutas Aydin, E., Polissiou, M. & Sokmen, A. (2006). Screening of the antioxidative properties of the essential oils of *Pimpinella anisetum* and *Pimpinella flabellifolia* from Turkey. *Food Chemistry*, 97, 719-734.

Tilquin, M., Paris, M., Reynaud, S., Despres, L., Ravanel, P., Geremia, R.A., & Gury, J. (2008). Long Lasting Persistence of *Bacillus thuringiensis* subsp. *israelensis* (Bti) in Mosquito Natural Habitats. *PLoS ONE*, 3, e3432.

Tirillini, B.B., Stoppini, M.A.M. & Pellegrino, R.R. (1996). Essential oil components in the epigeous and hypogeous parts of *Smyrnium perfoliatum* L. *Journal of Essential Oil Research*, 8, 611-614.

Tirillini, B. & Tosi, B. (1992). Presence of *a*-pinene in plant callus cultures of *Smyrnium perfoliatum* L. *Journal of Essential Oil Research*, 4, 431-432.

Tomic, A., Petrovic, S., Pavlovi, M., Tzakou, O., Couladis, M., Milenkovic, M., Vučicevic, D. & Lakušic, B. (2009). Composition and antimicrobial activity of the rhizome essential oils of two *Athamanta turbith* subspecies. *Journal of Essential Oil Research*, 21, 276-279.

Trabousli, I, El-Haj, A.F., Tueni, S., Taoubi, M., Nader, K. & Mrad, N.A. (2005). Repellency and toxicity of aromatic plant extracts against the mosquito *Culex pipiens molestus* (Diptera: Culicidae). *Pest Management Science*, 6, 597-604.

Tsukamoto, T., Ishikawa, Y. & Miyazawa, M. (2005). Larvicidal and Adulticidal Activity of Alkylphthalide Derivatives from Rhizome of *Cnidium officinale* against *Drosophila melanogaster. Journal of Agricultural and Food Chemistry*, 53, 5549-5553.

Tuetun, B., Choochote, W., Kanjanapothi, D., Rattanachanpichai, E., Chaithong, U., Chaiwong, P., Jitpakdi, A., Tippawangkosol, P., Riyong, D. & Pitasawat, B. (2005). Repellent properties of celery, *Apium graveolens* L., compared with commercial repellents, against mosquitoes under laboratory and field conditions. *Tropical Medicine & International Health*, 10, 1190-1198.

Tuetun, B., Choochote, W., Pongpaibul, Y., Junkum, A., Kanjanapothi, D., Chaithong, U., Jitpakdi, A., Riyong, D., Wannasan, A. & Pitasawat, B. (2009). Field evaluation of G10, a celery (*Apium graveolens*)-based topical repellent, against mosquitoes (Diptera: Culicidae) in Chiang Mai province, northern Thailand. *Parasitology Research*, 104, 3, pp 515-521

Tunc, I., Berger, B.M., Erler, F. & Dagh, F. (2000). Onicidal activity of essential oil from five plants against two stored-product insects. *Journal of Stored Product Research*, 36, 161-168.

Tutin, T.G., Heywood, V.H., Burges, N.A., Moore, D.M., Valentine, D.H., Walters, S.M., Webb, D.A., Ball, P.W, Chater A.O., & Ferguson I.K. (1968). *Flora Europaea* vol. 2, Cambridge University Press, pp 315-375.

Vajs, V., Milosavljevic, S., Tesevic, V., Zivanovic, P., Jancic, R., Todorovic, B. & Slavkovska, V. (1995). *Chaerophyllum coloratum* L.: essential oils of ripe fruits and umbels. *Journal of Essential Oil Research*, 7, 529-531.

Van Breukelen, F.R., Haemers S., Wijffels R.H. & Rinzema, A. (2011). Bioreactor and substrate selection for solid-state cultivation of the malaria mosquito control agent *Metarhizium anisopliae. Process Biochemistry*, 46, 751-757.

Van den Dool, H. & Kratz, P.D.A. (1963). Generalization of the retension index system including linear temperature programmed gasliquid partition chromatography. *Journal of Chromatography A*. 11, 463-471.

Velasco-Negueruela, A., Perez-Alonso, M.J., Perez de Paz, P.L., Pala-Paul, J. & Sanz, J. (2003). Analysis by gas chromatography-mass spectrometry of the essential oil from the aerial parts of *Pimpinella junoniae* Ceb. & Ort., gathered in La Comera, Canary Islands, Spain. *Journal of Chromatography A*, 1011, 241-244.

Velasco-Negueruela, A., Pérez-Alonso, M.J., Pérez de Paz, P.L., Palá-Paúl, J. & Sanz, J. (2005). Analysis by gas chromatography-mass spectrometry of the essential oils from the aerial parts of *Pimpinella anagodendron* Bolle and *Pimpinella rupicola* Svent., two endemic species to the Canary islands, Spain. *Journal of Chromatography A*, 1095, 180-184.

Verykokidou, E., Tzakou, O., Loukis, A. & Roussis, V. (1995). Chemical composition of the essential oil of *Athamanta macedonica* (L.) Sprengel subsp. *macedonica*, from Greece. *Journal of Essential Oil Research*, 7, 335-336.

WHO. (1981). Instructions for determining the susceptibility or resistance of mosquito larvae to insecticides. Vol. WHO/VBC/81.807. 1981, Geneva: World Health Organization. p. 6.

Yadouleton, A., Martin, T., Padonou, G., Chandre, F., Asidi, A., Djogbenou, L., Dabiré, R., Aïkpon, R., Boko, M., Glitho, I. & Akogbeto, M. (2011). Cotton pest management practices and the selection of pyrethroid resistance in *Anopheles gambiae* population in Northern Benin. *Parasites & Vectors*, 4, 60, doi: 10.1186/1756-3305-4-60.

Zivanovic, P., Djokovic, D., Vajs, V., Slavkovska, V., Todorovic, B., & Milosavljevic, S. (1994). Essential oils flowers and fruits of *Athamanta haynaldii* Bort. Et Uchtr. (Apiaceae). *Pharmazie*, 49, 463-464.

14

Advances in Aerial Application Technologies and Decision Support for Integrated Pest Management

Ian M. McLeod[1], Christopher J. Lucarotti[2,3,*], Chris R. Hennigar[3],
David A. MacLean[3], A. Gordon L. Holloway[4],
Gerald A. Cormier[1] and David C. Davies[1]

[1]*Forest Protection Limited, Fredericton International Airport, Lincoln,*
[2]*Natural Resources Canada, Canadian Forest Service -*
Atlantic Forestry Centre, Fredericton,
[3]*Faculty of Forestry and Environmental Management and*
[4]*Department of Mechanical Engineering, The University of New Brunswick, Fredericton*
Canada

1. Introduction

The first aerial applications of a pesticide against forest insect pests in Canada used calcium arsenate to protect forest stands from defoliation by the spruce budworm (*Choristoneura fumiferana*). The program was first conducted in Nova Scotia in 1927, and then in Ontario in 1928 and 1929 (Randall, 1975). There were few developments during the 1930s, but World War II and the need to protect military personnel from mosquitoes helped establish aerial spray techniques using a variety of aircraft (Randall, 1975). Extensive outbreaks of spruce budworm populations in Canada were countered by aerial application of DDT (dichloro-diphenyl-trichloro-ethane) in 1944 in Ontario (Howse & Sippell, 1975), then in 1952 in Quebec (Blais et al., 1975) and New Brunswick (Miller & Kettela, 1975).

Insecticidal powders were first released from hoppers mounted directly on the aircraft, but liquid formulations of pesticides led to the development of gravity-flow, open-pipe systems, boom and nozzle systems and rotary atomizers variously distributed on the aircraft fuselage and wings (Randall, 1975). In the 1950s and 1960s, hydraulic boom and nozzle systems were used predominantly in forestry, but with research suggesting that 100-μm and smaller droplets are more effective against forest pest insects, rotary atomizers are now favoured (Weisner, 1995).

Aircraft guidance and direction to spray blocks was initially accomplished using topographic maps and features on the landscape. In the case of small blocks, helium-filled balloons, deployed from the ground, were sometimes used to mark the corners. Forest Protection Limited (FPL) (Lincoln, New Brunswick) was the first to use "spotter" aircraft that communicated information on block location and boundaries to spray pilots in flight (Flieger, 1964). Electronic guidance systems for the positioning of spray aircraft were first used in 1959 (Randall, 1975), and the availability of satellite-based, global positioning system (GPS) technologies onboard aircraft immensely improved aircraft guidance and navigation.

Since its incorporation in 1952, FPL has been a leader in the development of technologies to increase the efficacy, accuracy and safety of aerial application and fire control operations. In January 2007, FPL - in partnership with AG-NAV Inc., BioAtlantech, the Canadian Forest Service (CFS), Sylvar Technologies Inc., and the University of New Brunswick (UNB), received a 5-year grant from the Atlantic Innovation Fund (AIF) of the Atlantic Canada Opportunities Agency (ACOA) to further develop aerial application technologies and baculoviruses for use in controlling forest insect pests. In this review, we describe progress on the research that has led to the Accuair™ suite of aerial application technologies and services. For more detailed information on aerial application practices and technologies generally, readers are directed to articles referenced in this review (e.g., Payne, 1995; Weisner, 1995; Kilroy et al., 2003, Mierzejewski et al., 2007).

2. The development of Accuair™ technologies

2.1 Overview

Accuair™ is the brand for an integrated system that is capable of increasing aerial spray program efficiency. One component is the Accuair Aerial Management System (AMS) – an onboard guidance, navigation and control system that optimizes spray lines on a treatment block to compensate for changes in wind conditions and aircraft altitude at the time of spraying. The AMS uses spray droplet size information gathered from a second component, the Accuair Wind Tunnel. Droplet size is influenced by a host of factors whose effects cannot be easily predicted or generalized. A wind tunnel provides a means of measuring droplet sizes under a set of conditions that replicate actual operational situations. With these data from the wind tunnel, the AMS drift simulation software can develop more realistic predictions. Accuair™ Forest Protection Optimization System (ForPRO), the third component, is a planning software that uses forest stand information and risk factor overlays to identify areas within a region that are most in need of protection from forest insect pests.

2.2 Aerial management systems

The AMS that led to Accuair™ had its beginning in the early 1990s with the availability of inexpensive GPS and other sensing equipment that could be easily installed and used on spray aircraft. As early as 1991, the usefulness of newly available radar altimeters for determining of aircraft altitude above the ground and forest canopy was recognized (Mickle & Robinson, 1991). Knowledge of aircraft altitude above ground is a critical parameter in predicting spray drift, and such data are not provided by standard aircraft altimeters that use air pressure to determine height above sea level. A summary by Davies (1994) described trials of a system that made it possible to log flight data—including aircraft altitude and attitude, boom on/off functions, atomizer rotational speed, and application rate—that allowed managers to visualize these parameters on a computer interface, assess spray quality and make improvements. At the same time, Riley (1994) reported on work that had examined factors affecting the deposition and drift of pesticide sprays, including droplet size, atomizer configuration, influence of evaporation and product volatility, release height and atmospheric turbulence. Using the insights from these reports, Mickle (1998, 1999) used geographic information systems (GIS) and GPS-based guidance systems to improve the accuracy of aircraft positioning relative to the spray block. Wind data were also recorded on spotter aircraft using an Aircraft-Integrated Meteorological Measurement System (AIMMS-10; Aventech Research Inc., Barrie, ON) to obtain wind speed and direction at the altitude of the aircraft and at the time of spraying. In addition, a spray droplet dispersion model was used to determine optimal droplet size distributions to deliver a maximum number of droplets of pesticide to the target area for a variety of spray strategies. It was concluded that this practice delivered a more uniform application because it compensated for low upwind deposition as a result of wind-driven drift (Mickle, 1998, 1999). A second conclusion was that better efficacy was achieved if smaller droplets were used, as drifting sprays inherently give more uniform coverage of the target area. This second conclusion agreed with the report by Picot (1994) describing simulations of spray drift using a dispersion model where spray efficacy was evaluated based on delivering the maximum number of droplets to a block. It was found that increasing the small droplet content (<20 μm diameter) delivered the most uniform coverage over the target area.

In 2000, spray aircraft were equipped with an auto-flow system that automatically adjusted the flow rate of the control product to maintain a constant application rate as the aircraft speed changed during flight (Mickle, 2000). An auto-flow system takes information from the GPS system to determine aircraft speed relative to ground. This parameter is critical in determining application rate but is not available from standard aircraft airspeed indicators. A comparison was made on blocks sprayed with, and without, auto-flow and it was found that auto-flow technology yielded a significant reduction in application rate variability (Mickle, 2000).

The aircraft-mounted AIMMS-20 probe, an evolution of the AIMMS-10 system, was flown at different altitudes, and results were compared with SODAR (sonic detection and ranging) (Mickle, 2005). SODAR systems use sound waves to estimate the wind speed profile above ground. This study is of particular significance because it validated the ability of the AIMMS-20 probe to accurately measure wind speed and demonstrated that a single-point measurement of wind speed from a single location on the ground is insufficient to characterize the wind speed at altitude. In a project led by FPL, an AMS that incorporated real-time meteorology measurements and flight line offsetting was used on spray aircraft (Cormier, 2005; 2006; Mickle et al., 2007). The system was used to spray 8-ha triangular blocks (a difficult shape to treat), and a second set of 64-ha blocks was also sprayed with, and without, flight-line offsets. This early version of the AMS was the first operational system of its kind in agricultural/forest spraying, and results of this research demonstrated that offset spraying gave better deposit on the target area than spraying with no offset of flight lines in crosswind scenarios (Cormier, 2005, 2006; Mickle et al., 2007).

2.3 Wind tunnel

In parallel with AMS development, a wind tunnel facility was established to provide accurate droplet size information to the AMS. In 1975, the UNB Department of Chemical Engineering was invited by FPL to become involved in studies related to the prevention of conifer defoliation by the spruce budworm in the forests of New Brunswick. Professors Jules Picot and David Kristmanson constructed a wind tunnel specifically designed to measure spray droplet spectra incorporating then state-of-the-art particle size measurement systems. The 1-m diameter portable tunnel was completed in 1984. Newly emerging particle measuring systems were assessed, and in 1991, a Malvern Instruments Inc. (Malvern, UK) Fraunhofer laser diffraction spectrometer was purchased to determine droplet spectra from spray atomizers. Throughout this period, the work was coordinated with the New Brunswick Spray Efficacy Research Group (later SERG-International), which provided timely focus on the spray atomizer characterization requirements. In 2004, a partnership agreement was signed between UNB, FPL and CFS, whereby FPL became the caretaker of the wind tunnel in its new permanent location at the CFS-owned Acadia Research Forest in Noonan, NB. Extensive upgrades to the original tunnel and its new home were made at this time. In 2007, a new experimental and computational research program on aerial sprays was initiated by the UNB Department of Mechanical Engineering and FPL. The program was funded by the Natural Sciences and Engineering Research Council (NSERC) together with AIF funding to further improve the wind tunnel facility and to purchase a Sympatec (Clausthal-Zellerfeld, Germany) particle size analysis system. Since 2008, FPL and its research partners at UNB have been involved in the establishment of standards for the characterization of droplet spectra from atomizers for aerial spray applications.

2.4 Forest protection planning systems

Spruce budworm is the most widespread and economically important forest insect pest in eastern North America, affecting over 40 million hectares of forest at the peak of the last (1970s–1980s) outbreak. Repeated defoliation during budworm outbreaks, which typically last about 10 years, results in up to 90% growth reduction and 40-85% mortality in forests containing high quantities of spruce and balsam fir. As a result, mitigation measures such as insecticide spraying and salvage of impacted stands have been widely used during past infestations. Nevertheless, growth and mortality losses wreak havoc with forest management plans, creating considerable uncertainty about future forest conditions. Both theory and past experience imply that another spruce budworm outbreak is due across the northern forest regions of North America. The spruce budworm outbreaks of the 1970s and 1980s stimulated major insecticide spray programs as well as extensive salvage of vulnerable, dead and dying stands in some regions. It is vital that land managers responsible for forests understand potential consequences of the next spruce budworm outbreak on their wood supplies, land values and management plans. This provided the rationale for development of a forest protection planning system or PROPS. PROPS is the software component of the Spruce Budworm Decision Support System (SBWDSS) used to assist with spruce budworm population management. The concept of the SBWDSS was developed by Erdle (1989), and the software application was developed by the CFS between 1992 and 1996. From 1996 to 1999, it was operationally implemented, on a cost-shared basis with industry and the provincial government, on all 6 million hectares of forest in New Brunswick (MacLean et al., 2001, 2002). The SBWDSS provides the conceptual basis for calculating marginal timber supply benefits (m³/ha) of alternative foliage protection scenarios for each stand in a forest. PROPS software is used to implement these methods and has GIS tools to visually display SBW projected volume impacts on inventory at set time periods in the future or at the time of planned harvest; it facilitates manual spatial blocking of aerial bio-insecticide operations. PROPS allows users to determine effects of different foliage protection strategies on forest development and timber harvests (Erdle, 1989;

MacLean et al., 2000, 2002). PROPS was implemented for all forests in New Brunswick and for test areas in four other Canadian provinces.

3. Accuair™ – Components

3.1 Accuair™ aerial management system

Over the years, a variety of fixed-wing aircraft have been developed or adapted for use in aerial pesticide application programs (see Randall, 1975; Kilroy et al., 2003; Estey, 2004; Mierzejewski et al., 2007). Air Tractor Inc. (Olney, Texas) manufactures a number of single-engine aircraft that are used for pesticide application and in fire control operations. Currently, FPL owns and operates six AT-802F (Figs. 1A and B) that can be used both for aerial application and fire suppression and an AT-802F Fire Boss that is equipped with amphibious floats, allowing it to scoop surface water from lakes and rivers near a wildfire whereas other tankers must return to the airstrip for refilling with fire-retardants.

The current version of the AMS integrates several components that make real-time spray optimization possible. The AIMMS-20 probe measures and records air velocity and direction on the wing of the aircraft (Fig. 1C). Using data from the probe and information on aircraft position and orientation, as determined by GPS, the AIMMS system calculates wind speed at the altitude of the aircraft and the direction of the wind at this level relative to the ground. In addition to wind speed and direction, the altitude of the aircraft above the target area is another important parameter in predicting the distance that spray droplets will drift once they leave the atomizers. The altitude of the aircraft is measured using radar or laser altimeters that use radio waves or laser beams, respectively, to determine the altitude of the aircraft above ground level instead of using air pressure, which determines altitude above sea level.

A light bar (AG-NAV Inc., Newmarket, ON) (Fig. 1D), mounted on the exterior and in front of the cockpit, provides the pilot with information related to spray parameters, e.g., total number of swaths, current swath, total area to be sprayed, area sprayed, application and flow rates, ground speed, course deviation indicator and obstacle warning messages.

Two types of atomizers are routinely used for aerial application: hydraulic nozzles and rotary atomizers. With hydraulic nozzles, liquid is atomized as it is forced through a small orifice. The design of the orifice determines the shape of the spray, the size of drops and ultimate usefulness of the nozzle to a given application scenario. Rotary atomizers may be wind driven or electrically powered. During flight, wind flowing over the wing drives small propeller blades of wind-driven rotary nozzles that are attached to a cylindrical wire cage (Fig. 1E). Liquid enters the cage and is broken up into droplets by centrifugal force as it hits the rotating cylinder. The droplet spectrum, generated by rotary atomizers, is determined by the physical characteristics of the spray liquid, the liquid flow rate and the speed of the rotors. The revolutions per minute (rpm) of the rotary atomizer can be adjusted by changing the pitch of the propeller blades. Boom on/off functions and adjustment of flow rates from atomizers (e.g., to compensate for aircraft speed) are made automatically using the AG-FLOW system (AG-NAV Inc., Newmarket, ON) (Fig. 2). The required spray offset is calculated by a central, onboard processor based on droplet spectrum information (from wind tunnel data) and inputs from aircraft instrumentation including aircraft altitude, attitude and speed, relative to the ground and wind speed and direction at altitude (Fig. 2).

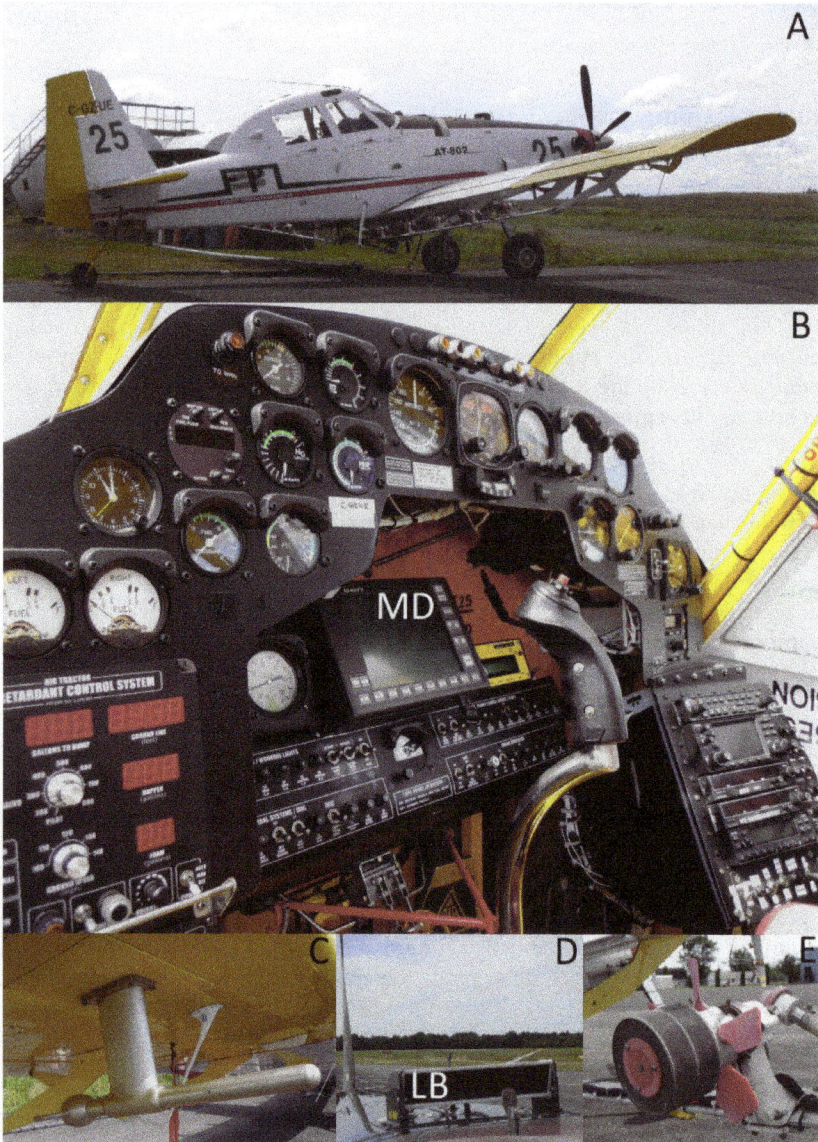

Fig. 1. Air Tractor Inc. AT-802F and components of the Accuair™ Aerial Management System. **A.** Forest Protection Limited (FPL) AT-802F aircraft. **B.** Cockpit instrument display of the FPL AT-802F shown in A. The AG-NAV differential GPS navigation and moving map display system (MD) is centrally located within the instrument display. **C.** The air-data probe is located on the underside of the wing near the wing-tip. **D.** The light bar (LB) is exterior to the cockpit on the nose cowling of the aircraft. **E.** A Micronair AU 4000 rotary atomizer.

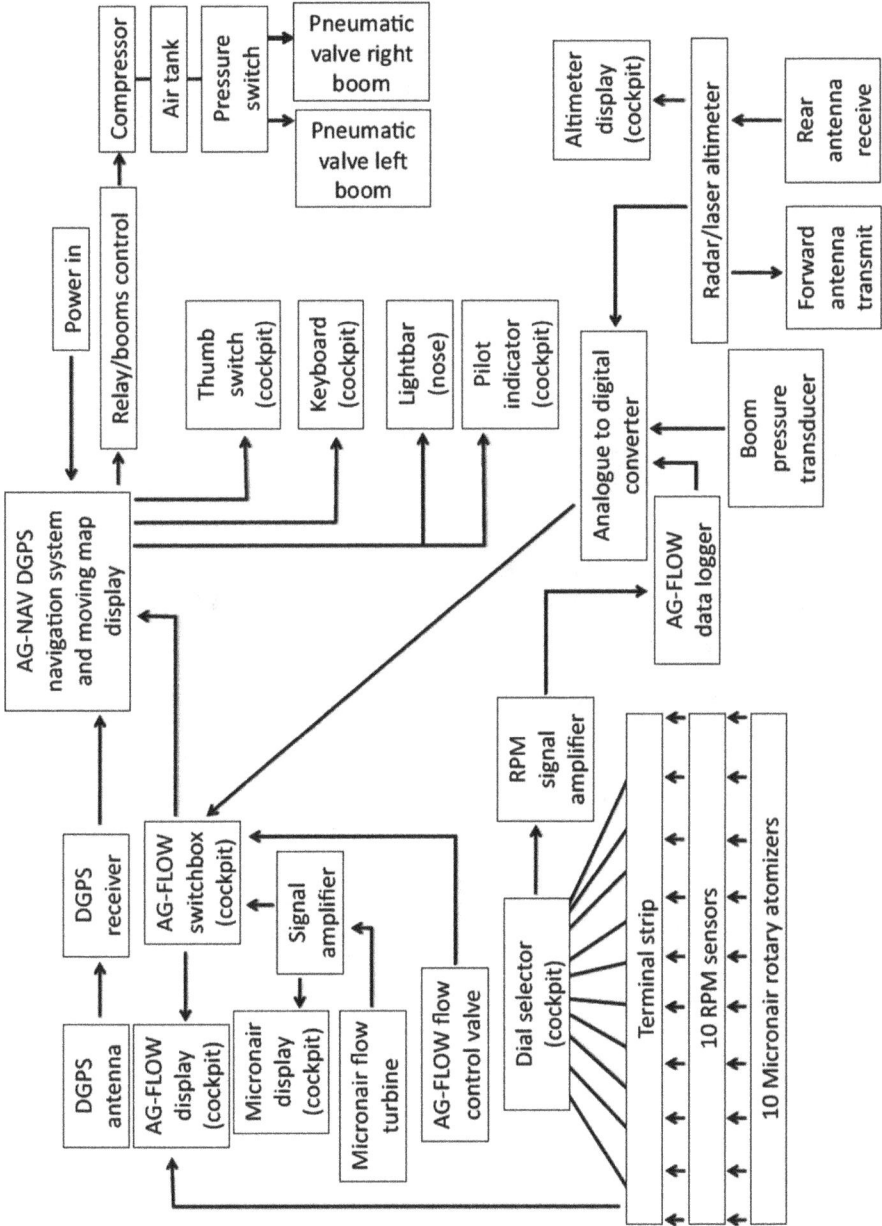

Fig. 2. Schematic diagram of the Accuair™ Aerial Management System (AMS).

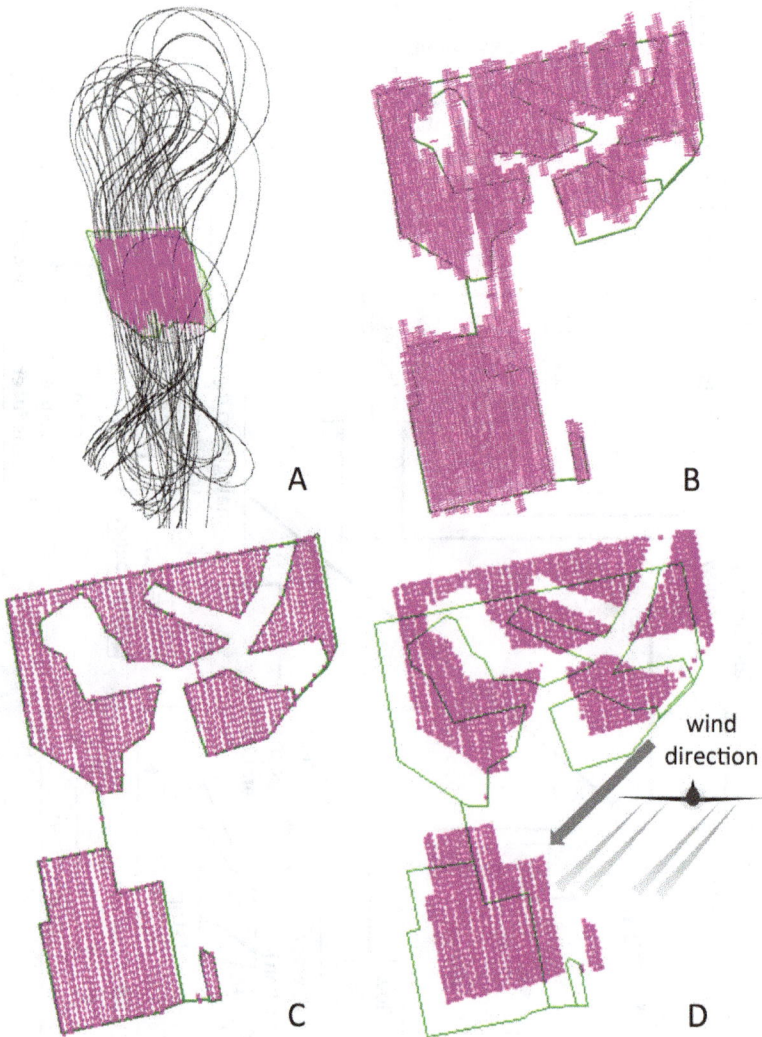

Fig. 3. Printouts of spray operation records. **A**. The boundaries of the block are input into the on-board computer from GPS data. Spray booms are automatically turned on as the aircraft enters the airspace over the block and turned off as the airplane exits. The black tracings are records of the aircraft flight path and the pink shows where the spray booms were turned on. **B** and **C**. Logs showing the accuracy of aerial applications when the booms were turned on (pink) and off manually by the pilot (**B**) and automatically by the GPS-directed autoboom controller (**C**) over an irregularly shaped block. **D**. Simulation of an optimized spray scenario over the same block as Figs. B and C, where spray lines have been off-set to allow ambient wind conditions to drift the atomized spray into the block.

As it flies through the air, an aircraft's propulsion system and wings generate a strong aerodynamic wake. This wake extends hundreds of meters behind the aircraft and entrains much of the fine component of a spray in long horizontal vortices. The size and shape of the aircraft wake depend on aircraft wingspan, velocity and weight. The wake downwashes (moves downward) toward the ground at a typical rate of 0.3–1.0 m/s (Payne, 1995), which is much faster than the settling rate of a fine spray (40-80 µm) in still air. The downwashing wake is one of the primary influences on transport of fine sprays downward into the forest canopy (Mickle & Rousseau, 1999). With the dissipation of the aircraft wake, spray droplet trajectories are influenced by wind speed, direction and turbulence. In turn, these are influenced by air temperature and humidity, solar radiation, time of day and characteristics of the ground cover (Mierzejewski et al., 2007). Using empirical data from numerous pesticide deposition and off-target drift studies, researchers with the United States Department of Agriculture Forest Service (USDA-FS) developed the AGDISP model for use by pesticide applicators to simulate and assess spray deposition and drift. The AGDISP model has been validated and is widely used within the aerial pesticide application industry worldwide (Mierzejewski et al., 2007).

In most aerial application scenarios, treatment blocks, buffer zones and exclusion zones are established by program managers and are then entered into the AMS computer as GPS coordinates (Fig. 3A). Aircraft spray along flight lines within the block boundaries, and the boundaries act as triggers to the auto-boom system. This type of system achieves much greater accuracy than manual pilot control (Figs. 3B and 3C). Spray drift can result in significant amounts of the control product being transported and deposited outside of the target area. Procedures to optimize spray deposit were developed where the control product is released outside or at the edge of the target block so it can drift into the block (Fig. 3D) resulting in greater amounts of product actually landing within the treatment area (Mickle & Rousseau, 1999; Cormier, 2005, 2006; Mickle et al., 2007).

3.2 Accuair™ wind tunnel

Spray testing is done in wind tunnels to obtain reliable control over wind speed and experimental conditions and to facilitate the use of bulky, vibration-sensitive laser drop-sizing equipment that cannot easily be used to measure sprays in an operational setting. A wind tunnel enables researchers to understand how fluids are atomized under real conditions and to measure the size spectra of spray droplets produced by various atomizers. This information is essential for accurate predictions using spray drift models and for optimizing the effectiveness of sprays using modern AMS. Spray drift models are also used to guide regulatory decisions and to determine the content of pesticide labels (i.e., instructions and restrictions for use), which are of interest to operators and manufacturers alike. Knowledge of spray droplet-size spectra and the ability to calibrate atomizers to produce desired droplet-size spectra enable more accurate estimates of spray drift, increase the accuracy of flight line offset predictions, increase spray efficacy, reduce costs and reduce the environmental impact of spray programs.

The Accuair™ Wind Tunnel (Fig. 4) is owned and operated by FPL and is one of only a few wind tunnels in the world capable of testing pesticide spray products. In order to simulate aerial application conditions, the Accuair™ Wind Tunnel was designed to produce a highly uniform, low-turbulence airstream within its 1-m diameter, 5-m long test section at speeds of up to 300 km/h. In the wind tunnel, sprays are measured to determine the characteristics of the droplet size spectrum. These data are vital to understanding the behaviour of droplets released into the environment and in predicting the amount of spray that will be deposited on a target. Fine sprays composed primarily of small droplets are more prone to drift but provide higher efficacy in many cases, whereas coarser sprays provide better drift control but sacrifice some efficacy (Weisner, 1995). Depending on the application and the type of pesticide product used, either efficacy or drift or both may be of primary concern. It is important, therefore, to know what sort of droplet spectrum a particular nozzle or atomizer will produce under actual spray conditions. Once the performance characteristics of a nozzle or atomizer are known, the information can then be used in computer drift simulations to determine the magnitude of flight line offsets and to classify the nozzle or atomizer according to its drift potential. Drift potential is used to determine the size of buffer zones that restrict how close operators may spray near sensitive areas. Manufacturers of atomizers are motivated by regulatory authorities to incorporate drift-reducing technologies in their products that spray program operators can then use to obtain reduced buffer zone requirements. Qualification as a drift-reducing technology requires wind tunnel testing and subsequent favourable comparison of data to those obtained for standardized or reference atomizers.

Fig. 4. The Accuair wind tunnel facility.

The measurement of droplet size spectrum is easily distorted by aerodynamic obstructions so measurements are typically made using laser drop-sizing instruments that pose no obstruction to the airflow or droplet trajectory in the wind tunnel. The most common form of expressing the droplet size distribution (DSD) for forestry and agricultural sprays is the volumetric DSD. This is represented in histograms where each histogram bar is assigned a droplet-size range, and height of the bar indicates the percentage of the emitted spray volume composed of droplets in that size range (volume fraction). The droplet size corresponding to the point where the cumulative sum of the volume fractions reaches 50% is the volumetric median drop diameter (VMD or D_{V50}). The D_{V50} provides a basic measure of spray droplet size but it is used along with other parameters such as D_{V10} and D_{V90}, which are the points at which the cumulative volume fraction sum reaches 10% and 90% of the total, respectively, indicating how widely the spray droplet sizes are distributed about the median.

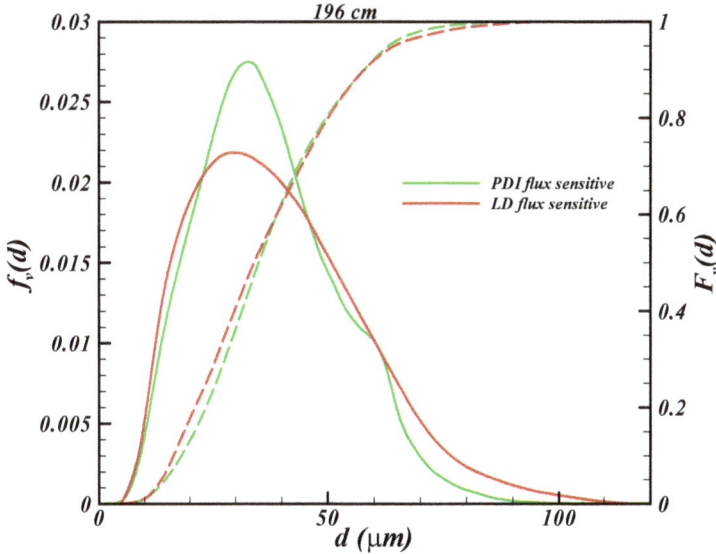

Fig. 5. Sample droplet size distribution (DSD) from the laser diffraction (LD) system and phase Doppler instrument (PDI).

The Accuair™ Wind Tunnel employs two instruments to characterize emitted droplet spectra. The laser diffraction (LD) system provides rapid measurements of a DSD across the entire plume that is based on droplet concentration and can be applied to opaque droplets, making it a good choice for characterization of pesticide spray products (Fig. 5). The phase Doppler (PD) system provides measurements of a DSD that are based on droplet flux and measurements of droplet velocity, which is vital for data interpretation. For the DSD results to be meaningful, sampling must take into account the flux (flow rate) of droplets through the measurement location because droplet flux rather than concentration is correlated to spray coverage and deposition The requirement to transform LD measurements to flux, based DSD poses a significant problem for spray characterization in wind tunnels where droplet velocity measurements are not available. The Accuair™ Wind Tunnel facility, with its complete set of instrumentation, has played a central role in developing guidelines to address issues related to LD spray measurement. These guidelines were compiled in a standard testing methodology submitted to the American Society for Testing and Materials (ASTM) in 2011.

Research into full physics computational fluid dynamic (CFD) modelling (Figs. 6 and 7) can provide a means of examining spray development beyond the wind tunnel environment. Typically, CFD models require validation and calibration using a set of measurements from limited numbers of test cases. For example, CFD models validated against measurements of turbulent droplet transport in the wind tunnel may be extended to simulations of the near wake of the aircraft (10–250 m downwind) and ultimately to predict transport far beyond the wake of the aircraft. These are long-term goals of the present computational research program of UNB and FPL.

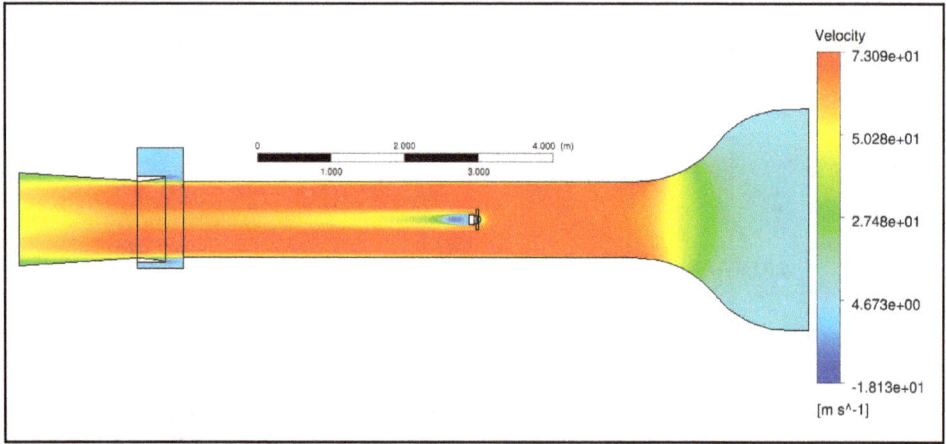

Fig. 6. Computational fluid dynamic (CFD) simulation of the axial velocity field in the wind tunnel.

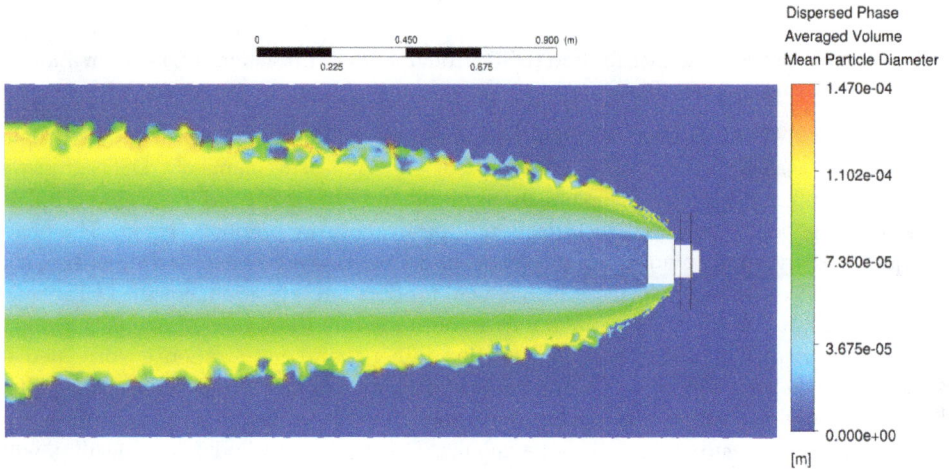

Fig. 7. Computational fluid dynamic (CFD) simulation of the dispersed droplet phase in the wind tunnel.

3.3 Accuair™ ForPRO

The Accuair™ Forest Protection Optimization System (ForPRO) is a new software package that allows integration between forest management planning and optimization models and underlying tree impact information derived from pest management decision support tools. ForPRO includes defoliation impact modelling data for a number of insect species, including spruce budworm, jack pine budworm (*Choristoneura pinus*), eastern hemlock looper (*Lambdina fiscellaria fiscellaria*) and balsam fir sawfly (*Neodiprion abietis*). Potential users of ForPRO include government agencies responsible for forest protection and maintenance of sustainable timber supplies, forest companies that conduct insecticide spray programs, and private landowners interested in quantifying vulnerability of their forest to insects and prioritizing ways to reduce future losses. Proactive assessment of vulnerability to, and implications of, current or possible future insect outbreaks is essential to develop and justify effective policies, response strategies and appropriate infrastructure before these events unfold. ForPRO can assist land managers in quantifying marginal benefits of protecting forest stands against insect defoliation (e.g., in terms of timber volume in m^3/ha or value as \$/ha). Protection cost:benefit analyses can be conducted using existing forest inventory and insect monitoring data in combination with forest management planning models to project the effects of foliage protection strategies on forest development and forest values.

ForPRO is composed of a number of specialized tools that allow users to simulate insect impacts on trees, stands, and forests. These tools leverage stand growth modelling capabilities from the FORUS Simulation Framework (FORUS Research, Fredericton, NB) and allow forest impact analyses to be conducted with existing strategic forest management optimization models (Remsoft Spatial Planning System; Remsoft, Inc. Fredericton, NB). These capabilities permit efficient exploration of cost-effective foliage protection or salvage scenarios.

ForPRO tools can be divided into those used in estimating stand impacts for strategic forest impact analysis (non-spatial tools) and those used for optimal spatial design of operational spray blocks (spatial tools). Non-spatial tools can be used to calibrate and implement SBWDSS. In 2009, ForPRO was used by the New Brunswick Department of Natural Resources (NBDNR) to estimate timber supply impacts on all Crown lands in New Brunswick for a variety of spruce budworm outbreak and foliage protection scenarios. These results were also used in the analyses of the effects of spruce budworm outbreaks on direct and indirect economic benefits from forests (Chang et al., 2012). The generalized framework of the SBWDSS has also been applied to insects other than spruce budworm. ForPRO was used to calibrate and predict timber supply impacts of balsam fir sawfly and hemlock looper defoliation scenarios in Newfoundland, Canada (Iqbal et al., 2011c), and NBDNR subsequently used ForPRO to help prioritize protection treatments for a localized balsam fir sawfly outbreak in New Brunswick.

Spatial tools include the 'Blocking Assistant' and ArcGIS extensions to allow forest protection priority maps (expected volume loss/area) to be spatially blocked for aerial application of biological insecticides. The Blocking Assistant uses meta-heuristic optimization algorithms and information about aircraft cost and flight constraints to help users search for blocking arrangements that increase the cost:benefit ratio of an aerial protection program.

The ForPRO insect decision support framework includes several software tools and steps to translate measured or predicted annual defoliation levels into estimated tree, stand and forest volume impacts. Some of these tools can be used to calculate tree growth reduction and survival from annual defoliation estimates for specific insect pest species or to classifying stands based on host tree species susceptibility to defoliation by certain insect species. Other tools do not include input of insect-related data and are used for such functions as averaging projected stand volume impacts according to stand-type classes to build the Stand Impact Matrix (MacLean et al., 2001).

ForPRO provides three software tools to assist with stand-level impact modelling and compilation for use in forest-level impact analyses.

3.3.1 Stand model multiplier builder

Modelling insect effects on stand dynamics within ForPRO is typically conducted using the stand development model, STAMAN (New Brunswick Growth and Yield Unit [NBGYU], 2004) to forecast defoliation impacts on merchantable volume (Erdle & MacLean, 1999), estimate salvageable volume (Hennigar et al., 2007) and determine changes to stand harvest timeframes. For spruce budworm and jack pine budworm, ForPRO facilitates modelling via the Stand Model Multiplier Builder. This tool converts user-defined estimates of annual defoliation by species into periodic 5-year mean estimates weighted by the proportion of total foliage mass by age class on a healthy balsam fir crown (MacLean et al., 2001), and then uses 5-year periodic defoliation and defoliation-impact relationships to calculate percentage tree growth or survival multiplier values relative to undefoliated conditions. ForPRO has pre-defined defoliation-damage multiplier files derived for spruce budworm (Erdle & MacLean, 1999) and jack pine budworm (Iqbal et al., 2011b), but users can develop their own damage multiplier files for other insects or adjust existing ones.

The FORUS Simulation Framework can be used to execute stand development models such as STAMAN with, or without, defoliation-impact multiplier commands for one or thousands of forest inventory plots. The Simulation Framework also provides a means to summarize tree-level projection output files (tree lists or stand tables) into other measures such as merchantable volume over time by species or species groups. Summarized reports can be output to a database table or text file. Although ForPRO, STAMAN and the FORUS Simulation Framework can all work together, they can also act as stand-alone programs.

3.3.2 Stand composition classifier

The Stand Composition Classifier tool provides a forest stand-type class link between stand impacts and forest inventory attributes. Classes are generally derived based on stand species composition, site, development stage (young, mature, old), canopy density, historical silviculture regime and geographic region. Similar classification logic is used to aggregate plot-level impact forecasts into stand-type classes in order to minimize projected volume impact variance within types. Typically, stand impacts vary the most by species composition and age because these attributes, in general, most significantly influence tree defoliation - damage relationships (MacLean et al, 2001; Hennigar et al., 2007, 2008; Iqbal et al., 2010, 2011a, 2011b).

The ForPRO Stand Composition Classifier tool facilitates the classification of percent tree species composition information from a GIS layer or yield table stored in a database into stand-type impact classes. The stand-type impact classes calculated for each record are written back to the same database in a single column, which can then be manually merged with other forest classification attributes common to both the inventory plots and forest estate model or GIS. The combination of criteria used to classify stands is left up to the user. The species classification algorithm and species information required also depend on the insect species of concern.

3.3.3 Stand impact matrix builder

Once the stand impact class link has been established between data sets, differences (percent reductions) between volume projections by tree species and plot with, and without, defoliation are calculated using the ForPRO Stand Impact Matrix Builder. The Stand Impact Matrix Builder tool also averages impact multipliers by grouping plots into stand-type impact classes. Model outputs or key performance indicators can be any stand measure (e.g., carbon, wildlife habitat index), but are typically timber volume over time by tree species. Using these data, the Stand Impact Matrix Builder quantifies relative differences between the base (undefoliated) and defoliated stand growth, by scenario, stand and time period:

%volume change = defoliation volume/base volume x 100

Once stand-level impacts are calculated, the Stand Impact Matrix Builder automatically averages impacts for each scenario and stand measure by stand impact class and period. Options allow for some degree of control of this averaging algorithm. If stand measures of dead volume are included with live volume estimates, additive (insect-caused) or total salvageable volume relative to base live volume yields can be calculated using the Stand Impact Matrix Builder. The resulting stand-type class, salvageable volume can be referenced against base-yield tables in the forest estate model to estimate forest-level salvageable volume available over time. The Stand Impact Matrix is essentially a large lookup table of time-dependent volume impact multipliers by defoliation scenario, tree species and stand impact class that can be readily linked to forest inventory records and yield tables stored in a relational database. Once these information sources are linked, calculations of projected volume impacts at time of planned harvest or forest inventory reduction caused by a defoliator.

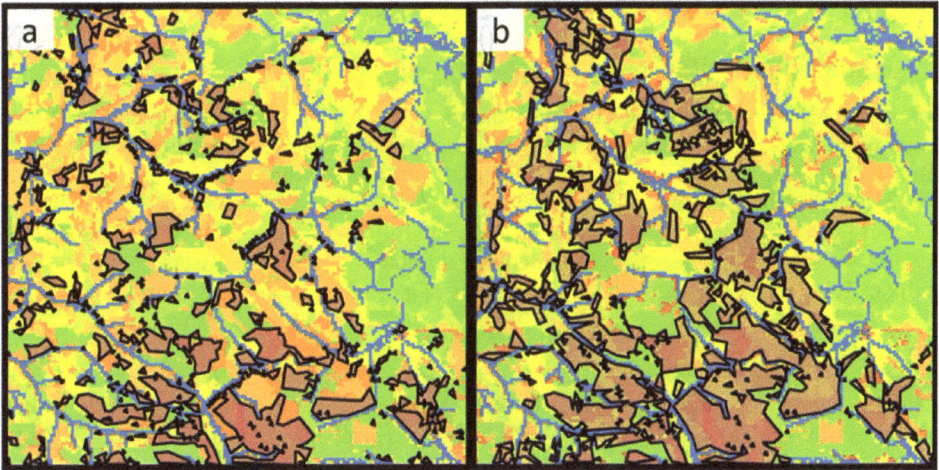

Fig. 8. Comparison of two $1 million aerial foliage protection programs scheduled for: a) highest volume loss protected first (non-spatial), and b) optimal spatial blocking of aerial operations to maximize the economic cost:benefit ratio of individual block flight plans. Each grid cell is 1 ha.

For land managers who use the Remsoft Spatial Planning System (RSPS; Remsoft, 2010), much of the database work described above can now be avoided by using the ForPRO plug-in for RSPS. The ForPRO plug-in allows yield tables in the RSPS forest estate model to be classified and linked to the Stand Impact Matrix with only minor changes to the RSPS model. This integration allows users to utilize the management optimization capabilities of RSPS with yield table impact information produced by ForPRO. The ForPRO framework allows for more efficient communication and integration between existing strategic forest planning optimization models and pest impact information. This can ultimately extend and accelerate implementation of the impact analysis methods used in the SBWDSS (Erdle, 1989; MacLean et al., 2001). ForPRO has also been expanded to allow for the automated spatial blocking of aircraft flight plans to optimize the net cost:benefit ratio of any foliage protection program (Fig. 8). The blocking algorithm blends strategic-level information on land value with industry-calibrated aircraft operational constraints and costs. A blocking cost algorithm that considers aircraft flight plan and spray constraints was refined with input from FPL and the Pest Management Branch of NBDNR.

4. Concluding remarks

The funding provided by the ACOA AIF and NSERC and the additional financial contributions of the research partners have allowed for the consolidation of a number of independent research efforts that has resulted in the Accuair™ suite of aerial application technologies and services (AMS, wind tunnel, and ForPRO). Additionally, these funds and this partnership have advanced the development of baculovirus-based biopesticides for use in forestry (not presented here). The results of this 5-year research project have established Canada as a leader in the development and practical use of aerial application technologies and decision support so that pesticides can be applied accurately, effectively and with minimal impact on the environment.

5. Acknowledgements

The financial support provided by the Atlantic Canada Opportunities Agency – Atlantic Innovation Fund, The Natural Sciences and Engineering Research Council of Canada, the Canadian Forest Service (Natural Resources Canada), the membership of SERG-International, AG-NAV Inc., BioAtlantech, Forest Protection Limited, Sylvar Technologies Inc. and the University of New Brunswick is gratefully acknowledged. A great number of people assisted in this project, and their contributions are also gratefully acknowledged. We thank C. Riley and E.G. Kettela for their reviews of an earlier draft of the manuscript.

6. References

Blais, J.R., Benoit, P., and Martineau, R. 1975. Aerial control operations against the spruce budworm in Quebec, In: *Aerial Control of Forest Insects in Canada*, M.L. Prebble (Editor), pp. 113-125. Department of the Environment, Ottawa, Canada.

Chang, W-Y., Lantz, V.A., Hennigar, C.R., and MacLean, D.A. 2011. Benefit-cost analysis of spruce budworm (*Choristoneura fumiferana* Clem.) control: incorporating market and non-market values. *Journal of Environmental Management* 93: 104-112.

Cormier, G. 2005. An aerial management system. *SERG-International / Canadian Forest Service Workshop: Forest pests and sustainable forest management in Canada 1994-2014*, p. 115. Fredericton, NB, Canada, February 7-10, 2005.

Cormier, G. 2006. Aerial application to PROPS polygons. *SERG-International / USDA Forest Service Workshop*, pp. 206-207, Banff, AB, Canada, February 13-16, 2006.

Davies, D. 1994. Some operational parameters for consideration in aerial pesticide applications and their potential for future development. *CAPCO Technical Session: Overview of Spray Drift and Deposit. Canadian Association of Pesticide Control Officials (CAPCO) Annual Meeting*, pp. 53-55, ISBN 0-662-22637-2, Ottawa, ON, Canada, October 25, 1994.

Erdle, T.A. 1989. *Concept and practice of integrated harvest and protection design in the management of eastern spruce fir forests.* Ph.D. Dissertation, University of New Brunswick, Fredericton, NB, Canada. 161 pp.

Erdle, T.A., and MacLean, D.A. 1999. Stand growth model calibration for use in forest pest impact assessment. *Forestry Chronicle* 75: 141-152.

Estey, R.H. 2004. Canadian use of aircraft for plant protection. *Phytoprotection* 85: 7-12.

Flieger, B.W. 1964. A new method for guiding spray aircraft. *Department of Forestry, Forest Entomology and Pathology Branch, Bi-monthly Progress Report* 20: 2-4.

Hennigar, C.R., MacLean, D.A., Porter, K.B., and Quiring D.T. 2007. Optimized insecticide application and harvest planning to reduce volume losses to spruce budworm. *Canadian Journal of Forest Research* 37: 1755-1769.

Hennigar, C.R., MacLean, D.A., Quiring, D.I., and Kershaw, J.A. Jr. 2008. Differences in spruce budworm defoliation among balsam fir and white, red, and black spruce. *Forest Science* 54: 158-166.

Howse, G.M. and Sippell, W.L. 1975. Aerial control operations against the spruce budworm in Ontario, In: *Aerial Control of Forest Insects in Canada*, M.L. Prebble (Editor), pp. 85-93, Department of the Environment, Ottawa, Canada.

Iqbal, J., MacLean, D.A., and Kershaw, J.A. 2010. Quantitative impacts of hemlock looper defoliation on growth and survival of balsam fir, black spruce and white birch in Newfoundland, Canada. *Forest Ecology and Management* 261: 1106-1114.

Iqbal, J., MacLean, D.A., and Kershaw, J.A. 2011a. Balsam fir sawfly defoliation effects on survival and growth quantified from permanent sample plots and dendrochronology *Forestry*. 84: 349-362

Iqbal, J., Amos-Binks L.J., and MacLean, D.A. 2011b. Meta-analysis of growth and mortality response of jack pine to jack pine budworm defoliation. *Canadian Journal of Forest Research* (under review).

Iqbal, J., Hennigar C.R., and MacLean, D.A. 2011c. Contrasting insecticide protection versus forest management approaches to reducing balsam fir sawfly and hemlock looper damage. *Agricultural and Forest Entomology* In press.

Kilroy, B., Karsky, D., and Thistle, H. 2003. Aerial application equipment guide 2003. *FHTET-2003, USDA-FS*, Morgantown, WV.

MacLean, D.A., K.B. Porter, W.E. MacKinnon, and K.P. Beaton. 2000. Spruce Budworm Decision Support System: lessons learned in development and implementation. *Computers and Electronics in Agriculture* 27: 293-314.

MacLean, D.A., Erdle, T.A., MacKinnon, W.E., Porter, K.B., Beaton, K.P., Cormier, G., Morehouse, S., and Budd, M. 2001. The spruce budworm decision support system: forest protection planning to sustain long-term wood supply. *Canadian Journal of Forest Research* 31: 1742-1757.

MacLean, D.A., Beaton, K.P., Porter, K.B., MacKinnon, W.E., and Budd, M.G. 2002. Potential wood supply losses to spruce budworm in New Brunswick estimated using the spruce budworm decision support system. *The Forestry Chronicle* 78: 739-750.

Mierzejewski, K., Reardon, R.C., Thistle, H., and Dubois, N.R. 2007. Conventional application equipment: aerial application. In: *Field Manual of Techniques in Invertebrate Pathology*, pp. 99-126, L.A. Lacey and H.K. Kaya (Editors), Springer, ISBN 978-1-4020-5931-5, Dordrecht, The Netherlands.

Mickle, R. 1998. Developing effective strategies for managing current and new epidemics of spruce budworm. *Report (unpublished) to Forest Protection Limited*, 100 pp. Lincoln, NB, Canada.

Mickle, R. 1999. Optimization trials for insecticide spraying into small blocks, Murdochville results, *Report (unpublished) to Forest Protection Limited*, 36 pp. Lincoln, NB, Canada.

Mickle, R. 2000. An evaluation of the Auto-Cal flow controller. *Report (unpublished) to Forest Protection Limited*, 15 pp. Lincoln, NB, Canada.

Mickle, R. 2005. Evaluation of the AIMMS-20 airborne meteorological package. *Report (unpublished) to Forest Protection Limited*, 19 pp. Lincoln, NB, Canada.

Mickle, R., and Robinson, A.G. 1991. An evaluation of a radar altimeter for forestry spraying. *Report (unpublished) to Forest Protection Limited*, 6 pp. Lincoln, NB, Canada.

Mickle, R.E. and Rousseau, G. 1999. Optimization trials for insecticide spraying into small blocks - Murdochville results. *Spray Efficacy Research Group (SERG) Report #1998/04*, 37 pp.

Mickle, R., Rousseau, G., and Cormier, G. 2007. Optimized aerial application to PROPS forest polygons. *SERG-International Workshop*, pp. 206-207, Quebec, QC, Canada, February 12-14, 2007.

Miller, C.A. and Kettela, E.G. 1975. Aerial control operations against the spruce budworm in New Brunswick. In: *Aerial Control of Forest Insects in Canada*, M.L. Prebble (Editor), pp. 94-112 Department of the Environment, Ottawa, Canada.

New Brunswick Growth and Yield Unit. 2004. *STAMAN 5.5 user's guide*, 15 pp, New Brunswick Department of Natural Resources, Fredericton, NB, Canada.

Payne, N.J. 1995. Spray dispersal, deposition, and assessment. In: *Forest Insect Pests in Canada*, J.A. Armstrong and W.G.H. Ives (Editors), pp. 465-478, Natural Resources Canada, Canadian Forest Service, ISBN 0-660-15945-7, Ottawa, Canada.

Picot, J. 1994. Deposit and drift in aerial spraying. *CAPCO Technical Session: Overview of Spray Drift and Deposit. Canadian Association of Pesticide Control Officials (CAPCO) Annual Meeting*, pp. 33-52, ISBN 0-662-22637-2, Ottawa, ON, Canada, October 25, 1994.

Randall, A.P. 1975. Application technology. In: *Aerial Control of Forest Insects in Canada*, M.L. Prebble (Editor), pp. 34-55, Department of the Environment, Ottawa, Canada.

Remsoft Inc. 2010. Remsoft Spatial Planning System. *Woodstock 2010.5 user guide*, Remsoft Inc., Fredericton, NB, Canada.

Riley, C. 1994. Factors affecting the deposition and drift of pesticide sprays. *CAPCO Technical Session: Overview of Spray Drift and Deposit. Canadian Association of Pesticide Control Officials (CAPCO) Annual Meeting*, pp. 13-24, ISBN 0-662-22637-2, Ottawa, ON. October 25, 1994.

Weisner, C.J. 1995. Review of the role of droplet size effects on spray efficacy. In: *Forest Insect Pests in Canada*, J.A. Armstrong and W.G.H. Ives (Editors), pp. 493-496, Natural Resources Canada, Canadian Forest Service, ISBN 0-660-15945-7, Ottawa, Canada.

15

Flourensia cernua DC: A Plant from Mexican Semiarid Regions with a Broad Spectrum of Action for Disease Control

Diana Jasso de Rodríguez[1], F. Daniel Hernández-Castillo[1],
Susana Solís-Gaona[1], Raúl Rodríguez- García[1]
and Rosa M. Rodríguez-Jasso[2]
[1]*Universidad Autónoma Agraria Antonio Narro*
(UAAAN), Buenavista, Saltillo, Coahuila
[2]*Centre of Biological Engineering,*
University of Minho,Campus Gualtar
[1]*México*
[2]*Portugal*

1. Introduction

Mexico has an extensive variety of plants, it is the world´s fourth richest country in this aspect. Some 25,000 species are registered, and it is thought that there are approximately 30, 000 not described. Particularly the regions of the north of Mexico, with their semiarid climate, have a great number and variety of wild plants grown under extreme climatic conditions. Wild species which have compounds with flavonoid structures, sesquiterpenoids, acetylenes, *p*-acetophenones, benzofurans, and benzopyrans grow in these regions. The polyphenolic compounds include tannins and flavonoids which have therapeutic uses due to their anti-inflammatory, antifungal, antibacterial, antioxidant, and healing properties.

Flourensia cernua is an endemic species which grows in semiarid zones of Mexico and contains polyphenolic, lactone, benzofuran, and benzopyran compounds which give it a potential use for disease control.

In this work, *F. cernua* is reviewed in terms of its geographical distribution in Mexico, traditional uses, bioactive compounds identified for controlling fungi, bacteria, and insects, as well as cytotoxic activity.

2. Common names

It is commonly known in different ways, as it is found in the United States of America as well as in Mexico. The names given in the United States of America are: tarbush, hojase, American-tarbush, black-brush, varnish-brush, and hojasen (Correl and Johnston, 1970; Vines, 1960). In Mexico, it is known as hojasen, tarbush, black-brush (Arredondo, 1981).

3. Geographical distribution

In Mexico, *F. cernua* is found in the Chihuahuan and Sonoran deserts, as well as in the states of Coahuila, Chihuahua, Durango, Hidalgo, Nuevo León, San Luis Potosí, Sonora, and Zacatecas (Valdés, 1988; Martínez, 1993) (Fig. 1). In the United States of America, it is found West of Texas and South of New Mexico and Arizona (Vines, 1960).

It is found in altitudes ranging from 1000 to 2000 masl. In studies carried out in Mexico, it is observed that the prevailing altitude for this species is 1900 masl and slopes from 1 to 6 per cent (Arredondo, 1981) although it is also found in altitudes from 300 to 400 Northeast of Coahuila.

4. Associated species

The main shrub species to which *Flourensia cernua* (Fig.2) is related to are: *Larrea tridentata*, *Yucca filifera*, *Atriplex canescens*, *Castela texana*, *Acacia fernesiana*, *Prosopis juliflora*, *Agave lechuguilla*, *Parthenium incanum*, *Fouquieria splendens*, and *Acacia constricta* (Comisión Técnico Consultiva para la Determinación de los Coeficientes de agostadero, 1979).

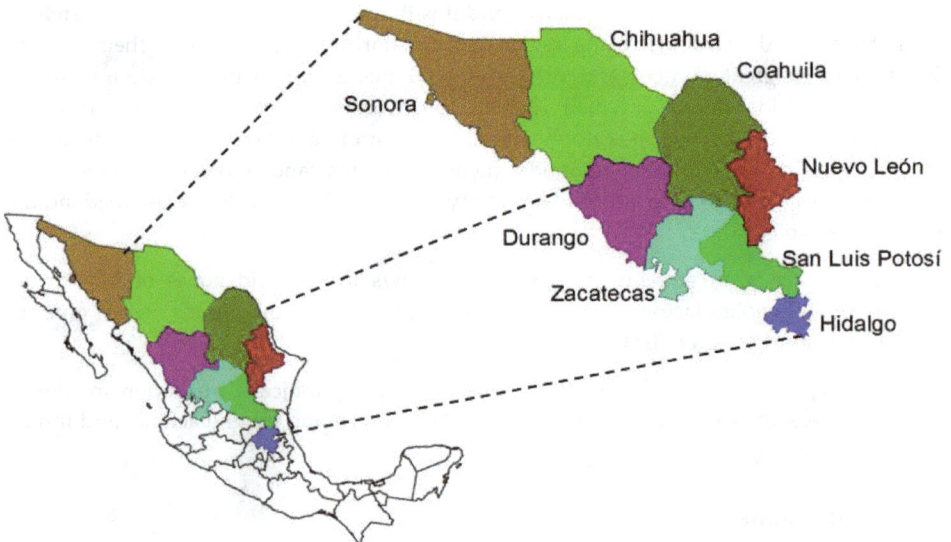

Fig. 1. Geographical distribution of *Flourensia cernua* DC in Mexico.

Fig. 2. *Flourensia cernua* plant at su in a wilderness site in Northeast Mexico.

5. Phytochemical analysis

The *Flourensia genus* is important due to the great amount of secondary metabolites it possesses; these are widely used for biological and ecological applications. Nine species of *Flourensia* have been reported, being *Flourensia cernua* the one with the highest number of chemicals (Aregullín and Rodríguez, 1983) with economical potential. The authors correlated the presence of benzofurans and benzopyrans with biological activity. The fact that these secondary metabolites are not present in other species, led to a correlation between the ecographical distribution and a possible chemical adaptation to the environment.

6. Properties and documented actions

Several medicinal properties have been reported for the tea obtained from the leaves or flowers for indigestion and gastrointestinal problems (Arredondo, 1981).

The green fruits are innocuous for cattle. However, dry fruits are toxic and when consumed at approximately 1% of the animal weight they cause death during the first 24 hours (Sperry et al., 1968).

7. Active chemicals

Fractionation of a CH_2Cl_2-MeOH (1:1) extract of the aerial parts of *Flourensia cernua* led to the isolation of three phytotoxic compounds, namely dehydroflourensic acid (Fig.3a), flourensadiol (Fig.3b), and methyl orsellinate (Fig.3c) and seven hitherto unknown γ-lactones were obtained (Fig.3d), these being tetracosane-4-olide, pentacosane-4-olide, hexacosane-4-olide, heptacosane-4-olide, octacosane-4-olide, nonacosane-4-olide and triacontane-4-olide. Besides, a previously known flavonoid, ermanin (Dominguez et al., 1973). Also there are benzopyrans (Fig.3e) and benzofurans (Fig.3f).

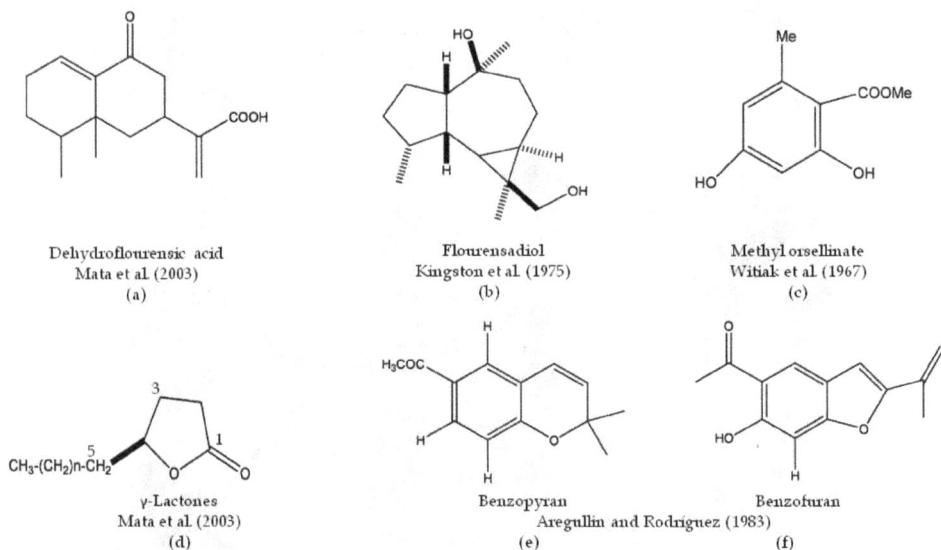

Fig. 3. Chemical of six active compounds from *Flourensia cernua*.

8. Plant fungicide, bactericide, and insecticide activity

8.1 Antifungal activity

Fungicide activity *in vitro* of leaves extracts at solution concentration of 1,000 mg L⁻¹ on *Rhizoctonia solani*, *Pythium* sp. and *Fusarium oxysporum* was reported (Saeedi-Ghomi & Maldonado, 1982). The leaf fractions of hexane, diethyl ether and ethanol were active against *Colletotrichum fragariae* Brooks, *C. gloesporioides* Penz and Sacc. The essential oils from the hexane fraction were active at 1 µg doses, whereas the diethyl ether and ethanol fractions were active at 10 µg doses. The ethanol fraction was active against *C. accutatum* Simmons only at 400 µg. (Tellez et al., 2001).

Gamboa et al. (2003) used an extraction method by soxhlet to obtain methanolic extracts which were evaluated on soil pathogene *Rhizoctonia solani* and on phytopathogene algae *Phytophthora infestans*, 20,000 µl L⁻¹ were the required dose for 86% pathogene inhibition.

Mata et al. (2003) reported that the fractionation of an extract of the aerial parts of *F. cernua* led to the isolation of three phytotoxic compounds namely: Flourensadiol, methyl orsellinate, and dehydroflourensic acid.

In a study carried out at our lab the inhibitory effect of ethanolic extracts was evaluated for three *Flourensia* species: *F. cernua*, *F. microphylla*, and *F. retinophylla* on three pathogens: *Alternia* sp., *Rhizoctonia solani* (Fig.4), and *Fusarium oxysporum*, which attack commercial cultivars (Jasso de Rodriguez et al., 2007). The variance analysis on the pathogen mycelial development showed highly significant differences (p≤0.01) on extract, dose, and extract interactions x dose. *F. microphylla* inhibited *Alternaria* sp 42.5% at a 10 µl L⁻¹, reaching 76.8% at a 100 µl L⁻¹. *F. cernua* and *F. retinophylla* showed a similar effect for high concentrations, however, inhibition was slightly lower than *F. microphylla* at low concentration. The highest inhibition level was 98.6%.

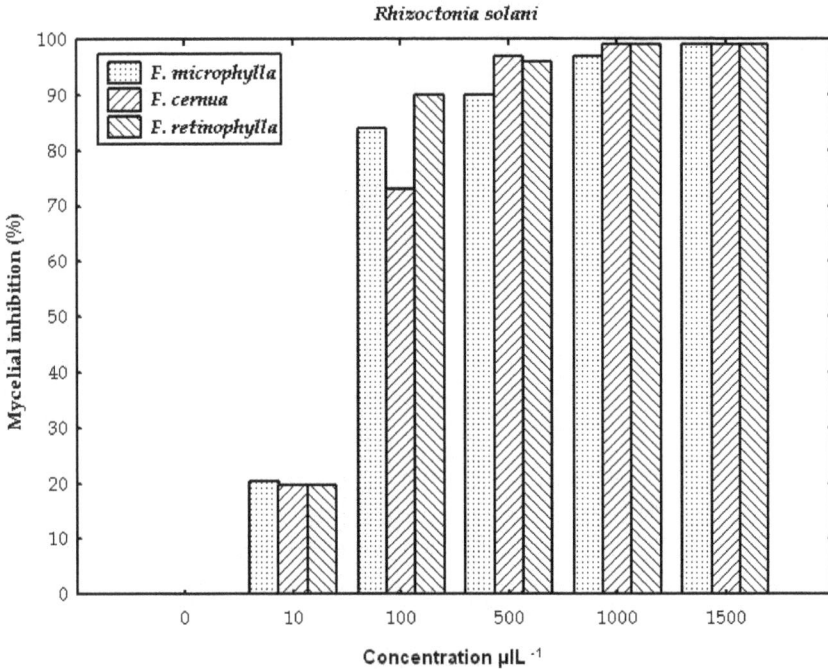

Fig. 4. Mycelial growth inhibition (percentage of *Rhrizoctonia solani*) as a function of *Flourensia* spp. extracts concentration.

In studies carried out by our research group, Guerrero-Rodriguez et al. (2007) proved the *F. cernua* effect on mycelial inhibition of *Alternaria alternata*, *Colletotrichum gloesporoides*, and *Penicillum digitatum*, where methanol:chloroform (1:1) solvents and sequential extractions with hexane, diethyl ether, and ethanol were used. *A. alternata* reported the highest mycelial inhibition, took place with hexane fractions (91.9%) and methanol: chloroform (88.4%) at 4000 mg L^{-1}. 2000 mg L^{-1} ethanol extract caused the lowest production of *C. gloesporoides* conidium. The four extracts reduced conidium production for *P. digitatum*, however they didn´t present statistical differences. In general, the ethanolic extract was the most efficient for inhibiting mycelial growth of *C. gloesporoides* and *P. digitatum*, (Fig. 5).

Studies carried out in the UAAAN by Castillo et al. (2010) on plant extract of *Larrea tridentata*, *Flourensia cernua*, *Agave lechuguilla*, *Opuntia* sp, and *Yucca* sp; obtained with alternative organic solvents (lanolin and cocoa butter) and water were tested against *Rhizoctonia solani* pathogens. The obtained results were as follows: The *L. tridentata* and *F. cernua* extracts by means of lanolin and cocoa butter at a 2000 and 1000 ppm total tannins inhibited the growth of *R. solani* 100%. Lanolin and cocoa butter solvents allowed a high recovery of polyphenolic molecules of strong antifungal activity against *R. solani* thus offering an alternative production of antimicrobial agents.

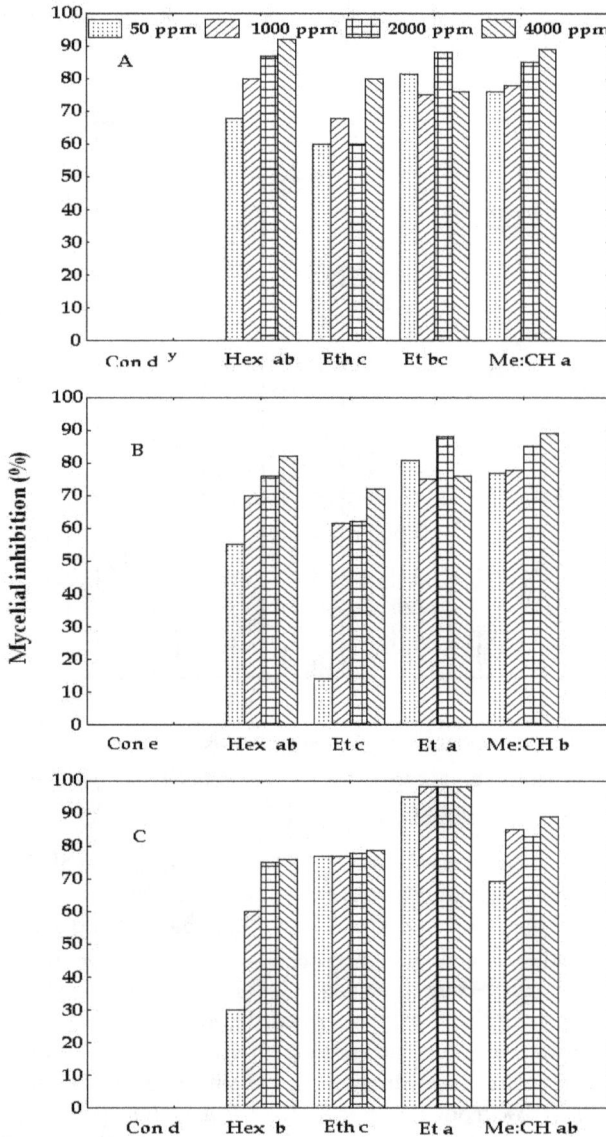

Fig. 5. Percentage of mycelial inhibition of four extracts from *Flourensia cernua* DC at four concentrations on three postharvest pathogens: A) *Alternaria alternata*, B) *Colletotrichum gloeosporioides*, and C) *Penicillium digitatum*. yP ≤0.01 (LSD). Con=Control; Hex=Hexane; Eth=Ether; Et=EtOH, and Me:CH=MeOH-CHCl₃.

8.2 Antibacterial activity

The mixtures of benzofurans and benzopyrans of *F. cernua*, were tested against Gram positive and Gram negative bacteria, fungi and *Saccharomyces* under two experimental conditions: One where the inoculated media was kept in darkness and the other one where the inoculated media was UV irradiated (280-400 nm) for 15 min previous to incubation in darkness. Bioactivity was greatly increased by the UV irradiation (Aregullin & Rodriguez, 1983; Towers et al., 1975).

Molina-Salinas et al. (2006), evaluated the crude extracts effect of methanol, acetone, and hexane of the aerial parts of *Artemisa ludoviciana* Nutt., *Chenopodium ambrosioides* L., *Murrubium vulgare* L., *Mentha spicata* L., and *Flourensia cernua* DC to inhibit the growth or death of *Mycobacterium tuberculosis* strains H37Rv and CIBIN:UMF:15:99. Results showed that from the evaluated plants, *F. cernua* was the only active plant which inhibited and killed *Mycobacterium tuberculosis* strains H37Rv and CIBIN:UMF:15:99. The hexane extract showed a minimal inhibitory concentration (MIC) 50 and 25 µg L-1 against sensitive and resistant strains, respectively; acetone extract was active against and only CIBIN:UMF:15:99 (MIC=100 µg L-1). It may be concluded that hexane extract of *F. cernua* leaves could be an important source of bactericidal compounds against multidrug-resistant *M. tuberculosis*.

In a study carried out by our research group, the antibacterial activity of *F. cernua* obtained with hexane, ether, ethanol, and metanol-chloroform mixture, at different dose, was evaluated on *Pseudomonas cichorii* (Pc), *Xanthomonas axonopodis* pv. phaseoli (Xap), and *Pectobacterium caratovora* subsp. atroseptic (Pca) (Peralta, 2006). All the extracts showed activity on Xap and Pc, however, none of them showed any inhibition effect when evaluated on Pca. *F. cernua* hexane extract at a 4000 µl ml-1 (P≤0.05) concentration showed the highest inhibition on Xap (82.51%) and Pc (83.96%). The other extracts showed a lower activity (Table 1).

Extracts and concentrations (µl L-1)		Inhibition of CFU ml-1								
		Pc	A	B	Xap	A	B	Pca	A	B
Hexane	500	1.89		a	37.23		a	0		a
	1000	26.41	B	ab	51.55	C	ab	0	A	a
	2000	44.34		b	74.77		b	0		a
	4000	83.96		c	82.51		c	21.73		B
Diethyl ether	500	0		a	2.41		a	11.01		c
	1000	12.95	A	ab	0	B	a	5.29	A	b
	2000	17.27		b	2.08		a	0		a
	4000	47.48		c	63.44		b	3.96		Ab
Ethanol	500	18.7		a	0		a	4.58		a
	1000	20.87	A	ab	0	A	a	5.6	B	ab
	2000	43.04		b	1.47		a	13.49		c
	4000	52.17		c	14.73		b	9.92		B
Meth-Chlor	500	0		a	0		a	5.93		c
	1000	6.77	A	b	0	A	a	3.56	A	a
	2000	0		a	17.46		b	3.86		a
	4000	0		a	0		a	4.91		b

Table 1. Interactions of *F. cernua* extracts x concentrations of the CFU ml-1 inhibition precentage, of three bacteria at 24 h of incubation. Meth-Chlor= Methanol-Chloroform; Pc = *Pseudomonas cichorii*; Xcp = *Xanthomonas axenopodis* pv. Phaseoli; Pca = *Pectobacterium caratovora* subsp. Atroseptica; A: Extract, ** = p≤ 0.05; B: Concentration ** = p≤ 0.05.

8.3 Insecticide activity

The insecticide activity of the benzofuran 7-methoxy-2-isopropenyl-5-acetyl-2, 3-dihydrobenzofuran-3-ol-cinnamate proved its activity as a juvenile hormone causing anatomic malformation, juvenile characteristics retention and sterility in the insects treated from their second to fourth stages of development (Towers et al., 1975). The results were similar to those reported by Bowers (1971) with precosene. Termiticidal activity of hexane, diethyl ether and ethanol fractions was found by Tellez et al. (2001).

In lab studies carried out by our research team, when evaluating the bioinsecticide activity of crude extracts from *F. cernua* leaves extracted with solvents of variable polarity on three insect plagues of agronomical importance: *Sitophilus oryzae* (Linneaus), *Phthorimaea operculella* (Zeller), and *Brevicoryne brassicae* (Linnaeus), as well as the repellent or attraction effect on *Sitophilus oryzae* (Linneaus) (Martinez, 2006). The following results were obtained: extracts didn´t provoke mortality at 24 and 48 h (P≤0.05) on *Sitophilus oryzae* (Linneaus), *Phthorimaea operculella* (Zeller); mortality for cabbage plant louse (*B. brassicae*), by effect of all extracts was observed, although some required a higher concentration to kill beings 100%. The extract that presented insecticide effect potential against *B. brassicae* was hexane which had 100% mortality from concentration at 10,000 µl L^{-1} (P≤0.05) at 24 h. Besides, hexane fraction showed insectistatic effect when inciting repellency to *S. oryzae* at 5 and 45 days, in raffia sacks as well as in jute sacks (Fig. 6). The repellency effect incited by the hexane fraction may be due to the volatile substances borneol and camphor it contains.

8.4 Antioxidant activity

Salazar et al. (2008) in a study carried out in order to evaluate the antioxidant potential of six species of the Northeast of Mexico, reports that leaves, stem, root, and flowers of *F. cernua* possess antioxidant activity, due to the content of phenolic compounds showing up in the different parts of the plant. Stems and roots of this species report the highest contents of phenolic compounds.

8.5 Cytotoxic activity

Pure benzopyrans and benzofurans of *F. cernua* (Figs. 3 e and f) have been studied for cytotoxic activity using blood red cells and measuring the hemoglobin released on cell destruction. The benzopyrans were more active than benzofurans although no clear correlation between activity and structure has been obtained. The UV irradiated compounds showed higher cytotoxic activity than the non-irradiated ones (Towers et al., 1980).

Benzopyrans and benzofurans react with L-cystein (Towers et al., 1979) and the microbicide and cytotoxic activities may be associated with the alkyl formation capacity. Significant inhibition of radicle growth of *Amaranthus hypochondriacus* was reported by Mata et al. (2003). The crude extracts and their fractions are cytotoxic against five human breast cancer cell lines (Molina-Salinas et al., 2006).

9. Relevant achievements of *F. cernua* research

The crude extracts and fractions of different polarity inhibit mycelial development of *Rhizoctonia solani*, *Pythium* sp., *Fusarium oxysporum*, *Colletotrichum fragariae*, *Colletotrichum gloeosporioides* Penz, *Colletotrichum accutatum* Simmons, *Phytophthora infestans*, *Alternaria* sp., *Alternaría alternata*, *Penicillum digitatum*.

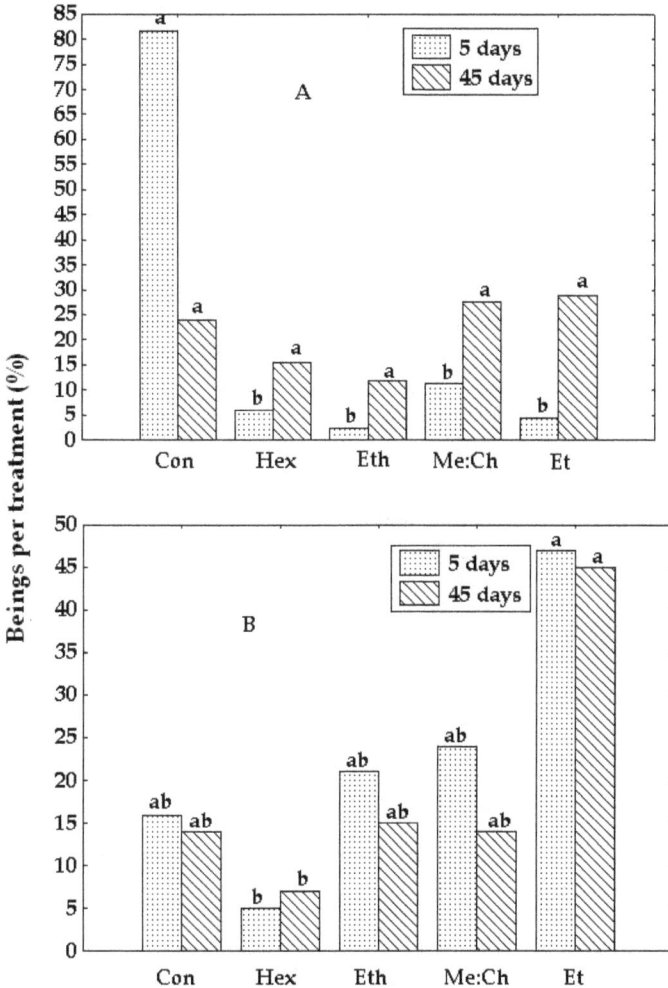

Fig. 6. Percentage of beings found in maize grains treated with four extracts of *Flourensia cernua* at 20,000 μl L⁻¹ in (A) raffia sacks and (B) jute sacks. Con=Control; Hex=Hexane; Eth=Ether; Me:Ch=Methanol-Chloroform; Et= Ethanol.

Flourensia cernua extracts have a high bioinsecticide activity for controlling *B. brassicae* and insectistatic effect when provoking repellency of *S. oryzae* and high degree of antitermite activity.

The antibacterial activity from the hexane extract of the leaves against *M. tuberculosis* suggests that *F. cernua* could be an important source of non-polar compounds with bactericidal activity.

Crude extracts and its fractions are cytotoxic against five human breast cancer cell lines.

The antioxidant activity of this plant is considered to increase the fungicide activity of the extracts.

The institutional (UAAAN) research results as to the bactericidal and insecticide activity are scientifically reported for the first time.

10. Conclusions

The secondary metabolites of semidesert plants as a result of genetic, climatic, and soil factors, vary greatly from plants which develop under less extreme environmental conditions.

Flourensia cernua is a Mexican semidesert plant which proved to have the capacity to control *Rhizoctonia solani, Pythium* sp., *Fusarium oxysporum, Colletotrichum fragariae, Colletotrichum gloeosporioides* Penz, *Colletotrichum accutatum* Simmons, *Phytophthora infestans, Fusarium oxysporum, Alternaria* sp., *Alternaría alternata, Penicillum digitatum,* diseases which cause great worldwide losses in field and postharvest production of high commercial value cultivars as tomato, potato, apple, avocado, papaya, banana. As bioinsecticide it also controls the activity of *Brevicoryne brassicae,* which attacks different species of the *Cruciferae* family. Besides, it's a repellent for *Sitophilus oryzae* which attacks stored grains.

The hexane fraction of *F. cernua* leaf mainly contains monoterpenoids, while the ethanol fraction mainly contains sesquiterpenoids, volatile compounds. This extract also contains an unknown number of molecules next to the active principle.

Crude extracts and its fractions of this plant are cytotoxic against five human breast cancer cell lines

Phenolic compounds found in extracts could be responsible of this species antioxidant activity, besides, these extracts don't show cytotoxicity.

Flourensia cernua phenolic compounds may be used as food additives with the purpose of preventing oxidation and in general, granting health benefits.

Taking into account the biological importance of this plant compounds, average and long term research must continue focusing on the following: 1. Isolation and identification of the plant active compounds; 2. *In vivo* evaluation of isolated extracts and compounds activity; 3. Product formulation for its industrial and pharmacological use.

Experimental results show a good correlation with the use of this plant in traditional Mexican medicine.

Because *F. cernua* is an endemic species, in the future it's necessary to carry out a domestication for the commercial production of this species to be able to have enough vegetal material for its industrialization and commercialization.

11. References

Aregullin, M., & Rodríguez, E. (1983). Phytochemical studies of the genus *Flourensia* from the Chihuahuan desert. Department of Developmental and Cell Biology. University of California. Irvine, CA. 92717. Oct. *Proceedings, Second Chihuahuan Desert Symposium*

Arredondo, V. D. G. (1981). *Componentes de la vegetación del Rancho demostrativo Los Angeles.* Tesis licenciatura. Universidad Autónoma Agraria Antonio Narro, Saltillo Coahuila México

Bowers, W. S. 1971. Discovery of insect antiallatotropins. In: *Naturally occurring insecticides.* Jacobson, M., & Croby, D. G. Eds. Marcel Dekker Inc. New York. p. 394-405

Castillo, F., Hernández, D., Gallegos, G., Méndez, M., Rodríguez, R., Reyes, A & Aguilar, C. N. (2010). *In vitro* antifungal activity of plant extracts obtained with alterntive organic solvents against *Rhizoctonia solani* Kuhn. *Industrial Crops and Products* 32(2010) 324-328, ISSN: 0926-6690

Comisión Técnico *Consultiva para la Determinación de los Coeficientes de agostadero* (COTECOCA). (1979). Coahuila, Secretarla de Agricultura y Recursos Hidráulico, México. 255 pp

Correll, S. D., & Johnston, C. M. (1970). *Manual of the Vascular Plants of Texas.* Texas Research Foundation. Renner, Texas. p. 1879, 1981.

Domínguez, X. A., Escarria, S., & Butruille, D. (1973). The methyl-3,4′-kaempferol de *Cordia boissieri. Phytochemistry* 12, 724-725, ISSN: 0031-9422

Gamboa-Alvarado, R., Hernández-Castillo, F.D., Guerrero-Rodríguez E., Sánchez-Arizpe, A., & Lira-Saldívar, R.H. (2003). Inhibición del crecimiento micelial de *Rhizoctonia solani* Kuhn y *Phytophthora infestans* Mont (De Bary) con extractos vegetales metanólicos de hojasen (*Flourensia cernua* D.C.), mejorana (Origanum majorana L.) y trompetilla [*Bouvardia ternifolia* (Ca.) Schlecht]. *Revista Mexicana de Fitopatología.* 21, 13-18, ISSN: 0185-33-09

Guerrero, R.E. Solís, G.S., Hernández, C.F.D., Flores, O.A., Sandoval, L.V., & Jasso de Rodríguez, D. (2007). Actividad biológica in vitro de extractos de *Flourensia cernua* D.C en patógenos de postcosecha: *Alternaria alternata* (Fr.:Fr) Keiss., *Colletotrichum gloesporoides* (Penz.) Penz y Sacc. y *Penicillum digitatum* (Pers.:Fr.). *Revista Mexicana de Fitopatología.* 25, 48-53, ISSN: 0185-33-09

Jasso de Rodríguez, D., Hernández, CD., Angulo, S.J.L., Rodríguez, G.R., Villarreal, Q.J.A., & Lira, S.R.H. (2007). Antifungal activity *in vitro* of *Flourensia* spp. extracts on *Alternaria* sp., *Rhizotocnia solani*, and *Fusarium oxysporum. Industrial Crops and Products.* 25, 111-116, ISSN: 0926-6690

Kingston, D. G. I., Rao, M. M., Spittler, T. D., Pettersen, R. C., & Cuulen, D. L. (1975). Sesquiterpenes from *Flourensia cernua. Phytochemistry* 14,2033-2037, ISSN: 0031-9422

Martínez, M. (1993). *Las plantas medicinales de México.* Ed. Botas. México, D. F. p. 656

Martinez, Z.J. (2006). *Actividad de extractos vegetales* de *Flourensia cernua D.C. contra Brevcorine brassicae L y de Flourensia cernua D.C., Agave lechuguilla* Torr., *Azadirachta indica L., Argemone Mexicana L.* y *Larrea tridentata D.C. contra Sitophilus oryzae* L. Tesis de Licenciatura. Universidad Autónoma Agraria Antonio Narro, Saltillo Coahuila México

Mata, R., Bye, R., Linares, E., Macías, M., Rivero-Cruz, I., Pérez, O., & Timmerman, B. N. (2003). Phytotoxic compounds from *Flourensia cernua. Phytochemistry* 64,285-291, ISSN: 0031-9422

Molina-Salinas, G.M., Ramos-Guerra, M.C., Vargas-Villarreal, J., Mata-Cardenas, B.D, Becerril-Montes, P. & Said-Fernández, S. (2006). Bactericidal activity of organic extracts from *Flourensia cernua* DC against strains of *Mycobacterium tuberculosis. Archives of Medical Research,* 37, 45-49, ISSN: 0188-4409

Peralta, B.J.E. (2006). *Evaluación de la actividad de extractos de hojasen (Flourensia cernua D.C), in vitro en el control de las bacterias fitopatogenas Xanthomona campestris pv. phaseoli (Smith) Dye, Erwinia carotovora pv. atroseptica (Van Hall) Dye y Pseudomona cichorri (Swingle) Stapp.* Tesis de Licenciatura. Universidad Autónoma Agraria Antonio Narro, Saltillo Coahuila México

Saeedi-Ghomi, H., & Maldonado, R. (1982). *Potencial de la flora de las zonas áridas.* En: Ciencia y Desarrollo, (Ed. CONACYT) 47, 98-110. México

Salazar, R., Pozos, M.E., Cordero, P., Pérez, J., Salinas, M.C & Waksman, N. (2008). Determination of the Antioxidant Activity of Plants from Northeast Mexico. *Pharmaceutical Biology*, Vol.46, No. 3: 166-170, ISSN: 1388-0209

Sperry, D. E., Dollahite, J. W., Hoffman, G. O., & Camp, V. J. (1968). *Texas plants poisonous to livestock.* Texas A&M University. Texas Agriculture. Ext. Serv. USA. Bull. B- 1028, 59 pp

Tellez, M., Estelle, R., Frederickson, E., Powell, J., Wedge, D., Schrader, K., & Kobaisy, M. (2001). Extracts of *Flourensia cernua* (L.), Volatile constituents and antifungal, antialgal, antitermite bioactivities. *Journal of Chemical Ecology.* 27, 2263-2273, ISSN: 0098-0331

Towers, G. H. N., Macrae, W.D., Irwin, D. A. J. & Bisalputra,T. (1980). Membrane lesions in human erytrocytes induced the naturally occurring compounds alfa-terthienyl and phynylheptatryne. *Photobiochemistry and Photobiophysics.* 1, 309-318.

Towers, G.H.N., Picman, A. R., & Rodríguez, E. (1979). Formation of adducts of parthenin and related sesquiterpene lactones with cysteine and glutathion. *Chemico-Biological Interactions.* 28, 83-89.

Towers, G. H. N., Shan, G. F. Q., & Mitchell, J. C.(1975). Ultraviolet mediated antibiotic activity of thiophene compounds of *Tagetes. Phytochemistry* 14, 2295-2296, ISSN: 0031-9422

Valdes, S. I. (1988). *Efecto de la densidad de Flourensia cernua en la producción de forraje de pastizal mediano abierto.* Tesis profesional. Universidad Autónoma Agraria Antonio Narro, Saltillo Coahuila México

Vines, R.A. (1960). *Trees, shrubs and woody vines of the southwest.* Univ. of Texas, press. USA. 1104 pp

Witiak, D., Patel, D.,& Lin, Y. (1967). Nuclear magnetic resonance. Influence of substituents on the long-range spin-spin coupling constant between benzylic and ring protons in the orcinol series. *Journal American Chemistry Society.* 89, 1908-1911, ISSN: 0002-7863

Permissions

The contributors of this book come from diverse backgrounds, making this book a truly international effort. This book will bring forth new frontiers with its revolutionizing research information and detailed analysis of the nascent developments around the world.

We would like to thank Marcelo L. Larramendy and Sonia Soloneski, for lending their expertise to make the book truly unique. They have played a crucial role in the development of this book. Without their invaluable contribution this book wouldn't have been possible. They have made vital efforts to compile up to date information on the varied aspects of this subject to make this book a valuable addition to the collection of many professionals and students.

This book was conceptualized with the vision of imparting up-to-date information and advanced data in this field. To ensure the same, a matchless editorial board was set up. Every individual on the board went through rigorous rounds of assessment to prove their worth. After which they invested a large part of their time researching and compiling the most relevant data for our readers. Conferences and sessions were held from time to time between the editorial board and the contributing authors to present the data in the most comprehensible form. The editorial team has worked tirelessly to provide valuable and valid information to help people across the globe.

Every chapter published in this book has been scrutinized by our experts. Their significance has been extensively debated. The topics covered herein carry significant findings which will fuel the growth of the discipline. They may even be implemented as practical applications or may be referred to as a beginning point for another development. Chapters in this book were first published by InTech; hereby published with permission under the Creative Commons Attribution License or equivalent.

The editorial board has been involved in producing this book since its inception. They have spent rigorous hours researching and exploring the diverse topics which have resulted in the successful publishing of this book. They have passed on their knowledge of decades through this book. To expedite this challenging task, the publisher supported the team at every step. A small team of assistant editors was also appointed to further simplify the editing procedure and attain best results for the readers.

Our editorial team has been hand-picked from every corner of the world. Their multi-ethnicity adds dynamic inputs to the discussions which result in innovative outcomes. These outcomes are then further discussed with the researchers and contributors who give their valuable feedback and opinion regarding the same. The feedback is then collaborated with the researches and they are edited in a comprehensive manner to aid the understanding of the subject.

Apart from the editorial board, the designing team has also invested a significant amount of their time in understanding the subject and creating the most relevant covers. They scrutinized every image to scout for the most suitable representation of the subject and create an appropriate cover for the book.

The publishing team has been involved in this book since its early stages. They were actively engaged in every process, be it collecting the data, connecting with the contributors or procuring relevant information. The team has been an ardent support to the editorial, designing and production team. Their endless efforts to recruit the best for this project, has resulted in the accomplishment of this book. They are a veteran in the field of academics and their pool of knowledge is as vast as their experience in printing. Their expertise and guidance has proved useful at every step. Their uncompromising quality standards have made this book an exceptional effort. Their encouragement from time to time has been an inspiration for everyone.

The publisher and the editorial board hope that this book will prove to be a valuable piece of knowledge for researchers, students, practitioners and scholars across the globe.

List of Contributors

Arturo Goldazarena and Sergio López
Neiker-Basque Institute of Agricultural Research and Development, Spain

Pedro Romón
FABI- Forestry and Agricultural Biotechnology Institute, South Africa

Dalva Luiz de Queiroz
Embrapa Florestas, Colombo-PR, Brazil

Daniel Burckhardt
Naturhistorisches Museum, Basel, Switzerland

Jonathan Majer
Department of Environment and Agriculture Biology, Curtin University of Technology, Perth, Australia

Carlos Vásquez and José Morales-Sánchez
Universidad Centroccidental Lisandro Alvarado, Decanato de Agronomía, Departamento de Ciencias Biológicas. Barquisimeto, Estado Lara, Venezuela

Fernando R. da Silva
University of Amsterdam, Institute for Biodiversity and Ecosystem Dynamics (IBED), Research Group of Population Biology, Amsterdam, The Netherlands

María Fernanda Sandoval
Instituto Nacional de Salud Agrícola Integral (INSAI), Av., Principal Las Delicias. Edif. INIA – Maracay, Estado Aragua, Venezuela

Ahmed Sallam
Plant Protection Dept., Faculty of Agriculture, Sohag University, Sohag, Egypt

Nabil El-Wakeil
Pests & Plant Protection Dept. National Research Centre, Dokki, Cairo, Egypt

Can Li
Department of Biology and Engineering of Environment, Guiyang University, Guiyang, People's Republic of China

Vanina Andréa Rodriguez, Mariano Nicolás Belaich and Pablo Daniel Ghiringhelli
LIGBCM-AVI (Laboratorio de Ingeniería Genética y Biología Celular y Molecular Area Virosis de Insectos), Universidad Nacional de Quilmes/Departamento de Ciencia y Tecnología, Argentina

Ricardo Antonio Polanczyk, Sergio Antonio De Bortoli and Caroline Placidi De Bortoli
Universidade Estadual Paulista (UNESP), Brazil

Renée Lapointe and David Thumbi
Sylvar Technologies Inc., New Brunswick, Canada

Christopher J. Lucarotti
Natural Resources Canada, Canadian Forest Service-Atlantic Forestry Centre, New Brunsick, Canada

Shad D. Nelson
Texas A&M University-Kingsville, Kingsville, USA
Texas A&M University -Kingsville Citrus Center, Weslaco, USA

Catherine R. Simpson
Texas A&M University -Kingsville Citrus Center, Weslaco, USA

Husein A. Ajwa
University of California-Davis, Salinas, USA

Clinton F. Williams
US Department of Agriculture-Agricultural Research Service, Maricopa, USA

Nabil El-Wakeil and Nawal Gaafar
Pests and Plant Protection Department, National Research Centre, Dokki, Cairo, Egypt
Institute of Agric. and Nutritional Sciences, Martin-Luther-University Halle-Wittenberg, Germany

Christa Volkmar
Institute of Agric. and Nutritional Sciences, Martin-Luther-University Halle-Wittenberg, Germany

Mostafa El-Wakeil
Kuwaiti Ministry of Interior – Information Systems Directorate, Dajij, Farwania, Kuwait

Marcelo Ramalho-Ortigão and Iliano Vieira Coutinho-Abreu
Kansas State University, Department of Entomology, USA

Fernando Zúñiga-Navarrete, Alejandra Bravo, Mario Soberón and Isabel Gómez
Instituto de Biotecnología-UNAM. Depto. Microbiología Molecular.Cuernavaca, Morelos, Mexico

Epameinondas Evergetis and Serkos A. Haroutounian
Chemistry Laboratory, Agricultural University of Athens, Athens, Greece

Antonios Michaelakis
Department of Entomology, Benaki Phytopathological Institute, Athens, Greece

Ian M. McLeod, Gerald A. Cormier and David C. Davies
Forest Protection Limited, Fredericton International Airport, Lincoln, Canada

Christopher J. Lucarotti
Natural Resources Canada, Canadian Forest Service - Atlantic Forestry Centre, Fredericton, Canada
Faculty of Forestry and Environmental Management, Canada

Chris R. Hennigar and David A. MacLean
Faculty of Forestry and Environmental Management, Canada

A. Gordon L. Holloway
Department of Mechanical Engineering, The University of New Brunswick, Fredericton, Canada

Diana Jasso de Rodríguez, F. Daniel Hernández-Castillo, Susana Solís-Gaona and Raúl Rodríguez- García
Universidad Autónoma Agraria Antonio Narro (UAAAN), Buenavista, Saltillo, Coahuila, México

Rosa M. Rodríguez-Jasso
Centre of Biological Engineering, University of Minho, Campus Gualtar, Portugal